THE HANDBOOK OF LABOUR UNIONS

THE HANDBOOK OF LABOUR UNIONS

Edited by

GREGOR GALL

To John Kelly – thanks for your friendship, comradeship and help and advice over the course of 30 years. It is very much appreciated.

First published in 2024 by Agenda Publishing

Agenda Publishing Limited
PO Box 185
Newcastle upon Tyne
NE20 2DH
www.agendapub.com

ISBN 978-1-78821-551-0

British Library Cataloguing-in-Publication Data
A catalogue record for this book is available from the British Library

Typeset by Newgen Publishing UK
Printed and bound in the UK by TJ Books

CONTENTS

ACKNOWLEDGEMENTS

My thanks are to all the contributors for their patience and perseverance. Increasing pressures of work in the academy were heightened by the pandemic, where so much teaching moving online meant a universally steep and arduous learning curve. This delayed the writing of the chapters in many cases as well as the production of the handbook itself. My thanks are also due to Jane Holgate, Alison Howson and John Kelly for giving advice along the way.

Gregor Gall

CONTRIBUTORS

Chiara Benassi is Reader in Comparative Employment Relations at the King's Business School, King's College London. Her research is in the area of comparative political economy and industrial relations, looking in particular at the role of industrial relations institutions for skill development, work organization and pay in the workplace. She currently holds a grant from the Leverhulme Trust to explore the effect of industrial relations on the export strategies of selected European countries. Her writing has been published in, among others, the *British Journal of Industrial Relations*, *ILR Review* and *Socio-Economic Review*.

Stefan Berger is Professor of Social History and Director of the Institute for Social Movements at Ruhr-Universität Bochum. He is also executive chair of the Foundation History of Ruhr in Bochum and an Honorary Professor at Cardiff University. Before taking up his current position in Bochum in 2011 he held professorial positions at the University of Manchester (2005–11) and the University of South Wales (then the University of Glamorgan, 2000–05). He has published six monographs and more than 30 edited collections on labour history, the history of nationalism, the history of historiography and historical theory as well as on British-German relations, border studies, deindustrialization studies, industrial heritage and social movement studies. His most recent monograph is entitled *History and Identity: How Historical Theory Shapes Historical Practice* (2022).

Christine Bischoff is a Lecturer in Sociology at the University of the Witwatersrand, Johannesburg. She holds a PhD in Sociology from the University of Pretoria. She has published on labour in publications such as the *International Labour Review* and *Work and Occupations*. Her areas of research interest are unions and employment relations.

Mark Bray is Emeritus Professor at the University of Newcastle and Honorary Professor at RMIT University, both in Australia. With Johanna Macneil, he has published several case studies of successful transitions from adversarial management–union relations to cooperation, especially when change has been facilitated by industrial tribunals.

His other research interests include industry patterns of labour regulation, trends in collective bargaining (both union and non-union) and the creation and operation of awards. He is also lead author of five editions of Australia's most successful university textbook, *Employment Relations: Theory and Practice.*

Marissa Brookes is Associate Professor of Political Science and Faculty Co-Director of the Inland Empire Labor and Community Center at the University of California, Riverside. Her research focuses on labour movements, transnational activism, theories of power and the politics of work in the global economy, and has appeared in *Labor Studies Journal, Global Labour Journal* and *Comparative Political Studies.* She is also Principal Investigator on the Transnational Labor Alliances Database Project, which documents over 150 transnational labour campaigns from the 1990s to 2020s. She earned her PhD in Political Science from Northwestern University.

David Camfield is Associate Professor in the Labour Studies Program (of which he is currently the coordinator) and the Department of Sociology and Criminology at the University of Manitoba. He is the author of *Canadian Labour in Crisis: Reinventing the Workers' Movement* (2011), *We Can Do Better: Ideas for Changing Society* (2017) and *Future on Fire: Capitalism and the Politics of Climate Change* (2022), and a member of the Editorial Board and the Editorial Advisory Committee of *Labour/Le Travail.*

Heather Connolly is Associate Professor of Employment Relations at Grenoble Ecole de Management. Her research explores the possibilities for union renewal, and how unions across Europe shape, and are constrained by, their institutional contexts. Other areas of research include relations between unions and social movements, worker representation and social dialogue in France, union renewal in UK local government, and the politics of equality in the UK and Europe. Her publications include *The Politics of Social Inclusion and Labor Representation Immigrants and Trade Unions in the European Context* (2019).

Pauline de Becdelièvre is an Assistant Professor at ENS Paris-Saclay (IDHES). She specializes in two main areas: unions and professional transition. In the field of unions, she has contributed to various publications in *Relations Industrielles/Industrial Relations.* Her research focuses on topics such as the employability of union activists, the effects of the Covid-19 crisis on collective bargaining, and the utilization of digital tools in negotiation processes. Additionally, she has published articles in French journals such as *Revue de Gestion des Ressources Humaines*, exploring the strategies employed by independent workers in platforms to overcome challenges.

Roland Erne obtained his PhD in Political and Social Sciences at the European University Institute in Florence, Italy. He is Professor of European Integration and Employment Relations, Principal Investigator of the European Research Council (ERC) project "Labour Politics & the EU's New Economic Governance Regime" at University College Dublin and Adjunct Professor of International & Comparative Labor at the ILR School, Cornell University.

Jack Fiorito is J. Frank Dame Professor of Management at Florida State University. He received his PhD from the University of Illinois and previously held faculty appointments at Illinois, Oklahoma State University, University of Iowa, University of Stirling and University of Hertfordshire. In 2021 he was named a Labour and Employment Relations Association Fellow "for exceptional contributions to research in Labor and Employment Relations". His research interests include worker and public attitudes towards unions, unions as organizations, and human resource (HR) management policies and their effects on worker attitudes.

Lorenzo Frangi is Professor of Employment Relations at the School of Management, University of Quebec in Montreal. His research examines how new employment relations settings and social dynamics affect both union actions and working conditions, mainly through an international comparative perspective. He analyses different data sources with the goal of highlighting resources that foster union actions, opportunities to improve labour conditions, and industrial relations strategies to promote effective change. His contributions have appeared in the *British Journal of Industrial Relations, Work Employment and Society, European Journal of Industrial Relations* and *Economic and Industrial Democracy* among others.

Gregor Gall is a Visiting Professor of Industrial Relations at the University of Leeds and an Affiliate Research Associate at the University of Glasgow. He is also a writer and commentator, being the author and editor of 30 books, mainly about labour unions and about politics in Scotland. He writes regularly for *The National* and the journal of the ASLEF train drivers' union. His latest book is a biographical and sociological study of a rail union leader, *Mick Lynch: The Making of a Working-Class Hero* (2024). Before this he published a biography of another rail union leader, *Bob Crow: Socialist, Leader, Fighter* (2017).

Chris Howell is the James Monroe Professor of Politics at Oberlin College, and the Visiting JY Pillay Professor at Yale-NUS College. He is the author of three books, *Regulating Labor: The State and Industrial Relations Reform in Post-War France* (1992), *Trade Unions and the State: The Construction of Industrial Relations Institutions in Britain, 1890–2000* (2005) and (with Lucio Baccaro) *Trajectories of Neoliberal Transformation: European Industrial Relations Since the 1970s* (2017), as well as numerous articles on industrial relations, comparative political economy and left parties.

Christian L. Ibsen is Associate Professor at FAOS, University of Copenhagen. His research focuses on collective bargaining, vocational education and training, and the future of work and employment relations. He has previously received research grants for projects on employer associations and trade unions, and is currently working on a project exploring green skills in vocational education and one on wage formation and unionization. His work has been published in, among others, *World Politics, Socio-Economic Review, British Journal of Industrial Relations, ILR Review* and *European Sociological Review.*

Andrew Keyes is an Organizational Behavior and Human Resources doctoral student at Florida State University. His research explores attitudes towards labour unions and HR management policies and practices. His research is published in academic journals, including *Economic and Industrial Democracy* and *Personnel Psychology*. Prior to attending Florida State University, he worked in HR and labour relations in the telecommunications industry.

Johanna Macneil is Professor of People, Organisation and Work at RMIT University, Australia. Her academic career has focused on pluralist policies and practices (of unions, employers and governments) to promote workplace cooperation. With Mark Bray and Andrew Stewart, she published *Cooperation at Work: How Tribunals Can Help Transform Workplaces* (2017) and with Bray and John Budd, "The many meanings of co-operation in the employment relationship and their implications" in the *British Journal of Industrial Relations* in 2019. Her current research interest is in informed and cooperative approaches to addressing psychosocial hazards at work.

Miguel Martínez Lucio has worked on comparative industrial relations and the development of HR management in political terms since the 1980s. He has focused upon the way representation and regulation have been changing and how the move to a more marketized approach has generated new tensions and conflict across a broader range of workers and agents. He is engaged with the way the political dimensions of industrial relations are becoming broader and involve a range of actors and spaces. He has worked at various universities within the UK and is currently a member of the Work & Equalities Institute of the University of Manchester. He is involved in the CRIMT network of the University of Montreal and others such as the Critical Labour Studies network.

Jeremy Morris is Professor of Global Studies at Aarhus University, Denmark. His most recent book is *Everyday Postsocialism* (2016). He has also published widely on Russian labour politics, the informal economy in the post-Soviet space, and many other anthropological and sociological topics relevant to the region. He is currently completing a book on capitalism realism and micropolitics in Russia.

Ronaldo Munck is Professor of Sociology and Director of the Centre for Engaged Research at Dublin City University (DCU) and a member of the Council of Europe Task Force on The Local Democratic Mission of Higher Education. He was the first Head of Civic Engagement at DCU and drove the "third mission" alongside teaching and research. As a political sociologist, he has written widely on the impact of globalization on development, changing work patterns and migration. Recent works include *Rethinking Global Labour: After Neoliberalism* (2018) and the co-edited *Migration, Precarity and Global Governance* (2015).

Jörg Nowak obtained his PhD in Political Sciences at Kassel University, Germany. He is a Visiting Professor at University of Brasilia, Brazil and an External Research Associate within the ERC project "Labour Politics & the EU's New Economic Governance Regime" at University College Dublin.

Greg Patmore is Emeritus Professor of Business and Labour History at the University of Sydney Business School. He is the author of *The Innovative Consumer Co-operative: The Rise and the Fall of Berkeley* (2020). He is currently writing a history of Australian cooperatives with Nikola Balnave and Olivera Marjanovic.

Jenny K. Rodriguez is a Senior Lecturer in Employment Studies at Alliance Manchester Business School and lead of the Equality, Diversity and Inclusion research theme at the University of Manchester's Work & Equalities Institute. Her research explores intersectional inequality in work and organizations, and the regulation of work and employment in the Global South. Her published work has reported on these issues in Latin America, the Hispanic Caribbean and the Middle East.

Zachary A. Russell is an Associate Professor and Chair of the Department of Management and Entrepreneurship in the Williams College of Business at Xavier University. He earned a PhD in Business Administration, with a focus in organizational behaviour and HR, from Florida State University. His research focuses on reputation, attitudes towards labour unions, and organizational politics.

Stefan Schmalz is head of the Heisenberg research group Sociology of Globalization at the University of Erfurt, Germany. His research areas include organized labour, strikes and global political economy. He is currently working on a project on Chinese direct investment in Germany and its impact on labour relations and economic governance.

Ed Snape is Emeritus Professor at the Hong Kong Baptist University (HKBU). He was previously Dean and Chair Professor in Management at HKBU, and before that Professor of Management at Durham University. He is a chartered member of the Chartered Institute of Personnel and Development. His research interests include HR strategy, employee commitment, leadership, and union participation. He has published in management, organizational behaviour and industrial relations journals, including *Academy of Management Journal, British Journal of Industrial Relations, Human Relations, Industrial and Labor Relations Review, Industrial Relations, Journal of Applied Psychology* and *Organization Science.*

Maite Tapia is Associate Professor at the School of Human Resources and Labor Relations, Michigan State University. Her research focuses on worker voice within the workplace, as well as worker organizing and movement building within the broader society, paying specific attention to workers' social identities and structural racism. Her latest work, with Tamara Lee, is on intersectional organizing in *Industrial Relations* and on the need of Critical Race Theory and Intersectionality within the field of industrial relations in *ILR Review.* Currently, she examines the employment conditions of Amazon warehouse workers in the US.

Alan Tuckman is currently an honorary fellow at the University of Keele after holding posts at numerous universities in the UK. His longest-term research interest has been into workers' control and industrial disputes, particularly factory occupations and sit-ins in the UK. He has contributed widely in journals and book chapters, and is also the author of *Kettling the Unions* (2018). He is a founder of workerscontrol.net and is currently researching the history of Institute for Workers' Control. He remains an active trade unionist and is currently Chair of the Nottingham Unite community branch.

Kurt Vandaele is a political scientist and senior researcher at the European Trade Union Institute (ETUI) in Brussels. His research interests include the history and sociology of the labour union movement in Europe; the collective action repertoire of workers; the platform economy and labour; and the political economy of Belgium and the Netherlands. He has recently co-edited *A Modern Guide to Labour and the Platform Economy* (2021) and *Trade Unions in the European Union: Picking Up the Pieces of the Neoliberal Challenge* (2023). He is currently co-editing books on app-based food delivery in Europe and on strikes and labour conflicts in the twenty-first century.

Horen Voskeritsian is Lecturer in Management at Birkbeck University of London, having earned his PhD from the London School of Economics and Political Science. His research interests fall within the broad field of industrial relations, ranging from the philosophical and historical foundations of the field to collective bargaining, the digitalization of labour and the future of work, workers' participation and industrial conflict. He has extensively researched the transformation of employment relations in Greece as a result of the 2010–16 bailout programmes. He is currently researching the role of new automation technologies in the transformation of work and employment. His work has appeared in numerous international conferences and international publications, such as *Economic and Industrial Democracy, International Journal of HRM, Relations Industrielles/Industrial Relations*, and *Labour History*.

Edward Webster is Distinguished Research Professor at the Southern Centre for Inequality Studies and the past director of the Global Labour University at the University of Witwatersrand, Johannesburg. His most recent publication is *Recasting Workers' Power: Work and Inequality in the Shadow of the Digital Age* (2023).

Geoffrey Wood is DANCap Private Equity Chair, Professor and Department Chair at DAN Management, Western University, Canada. He has published widely on employment relations, HR management, international business and corporate finance. He is editor-in-chief of *Human Resource Management Journal* and the *Academy of Management Perspectives*.

Jamie Woodcock is a researcher based in London and a Senior Lecturer at the University of Essex. He is the author of books including *Troublemaking* (2023), *Employment* (2023), *The Fight Against Platform Capitalism* (2021), *The Gig Economy* (2019), *Marx at the Arcade* (2019) and *Working the Phones* (2017). His research is inspired by workers' inquiry and focuses on labour, work, the gig economy, platforms, resistance, organizing and videogames. He is on the editorial boards of *Notes from Below* and *Historical Materialism*.

Tingting Zhang is Assistant Professor in the School of Labor and Employment Relations at the University of Illinois, Urbana-Champaign. She received her PhD in Industrial Relations from University of Toronto. Her research interest focuses on social media communication in labour movements and union renewal. Another stream probes how workers use non-traditional channels (e.g. social media), individually and collectively, to voice workplace issues and lead organizational changes. A recent project examines how labour unions respond to workplace changes in the context of the future of work. Her work has appeared in *British Journal of Industrial Relations, International Migration Reviews* and *Journal of Labor Research* among others.

INTRODUCTION

Gregor Gall

Publishing this handbook on labour unions towards the middle of the second decade of the twenty-first century comes at an opportune time. This is because there is something of a glint in the eyes of many members, activists and officers of labour unions and their supporters. This glint is one of cautious and tempered hope. It arises because, after the tremendous disruption of the Covid pandemic to economies and societies (and, thus, including employment matters), there is a sense that many unions and their members have found their "mojo" again. In the battle to at this stage protect – rather than advance – members' interests in the cost-of-living crisis that has afflicted most economies around the world, unions have had an opportunity to "stand up and be counted" in order to reassert their historic purpose of pursuing members' self-interest combined with social justice. This has most obviously taken the form of strike and street mobilizations. In the UK, for example, the number of days not worked as a result of strikes – officially designated as "days lost" – in 2022 was nearly ten times higher than in the previous years for which there is full data, namely 2017 and 2018 (ONS 2023a). And from June 2022 to May 2023, data showed some 3.93 million days were not worked due to strikes (ONS 2023a), this being the largest uptick in strike activity in the UK since the late 1980s.

In the case of the UK – but which has implications for unions elsewhere in other countries – although halting and hesitant as well as far from encompassing all unions and all workers (mostly obviously including non-union workers), the hope is that by doing so unions can begin the process of renewing and revitalizing themselves internally and externally. The former refers to levels and types of membership participation and the latter refers to augmenting recruitment (based upon high levels of retention). Both should make unions "bigger" and "better" in their ability to advance – rather than just protect – their members' economic and political interests. The stimulus of striking to this internal and external generation is held to be true by both practitioners and scholars. Kate Bell, the Assistant General Secretary at the Trades Union Congress, told the Bakers, Food and Allied Workers' Union annual conference in 2023: "We know that industrial action usually drives membership" (*Morning Star* 13 June 2023). Meanwhile, Hodder *et al.* (2017) summarized the four strands of the salient literature as finding a broadly positive relationship between strikes and recruitment (even if based upon relatively few robust studies and for different reasons) before, from the analysis of their own data for a specific union, also finding a broadly positive relationship. So, there seems to be some support for showing that strikes and struggles do build unions and certainly membership numbers.

And yet the evidence for this hope seems rather to be somewhat a case of "hoping against hope", at least if continuing with the case of the UK is anything to go by (which is a big Anglophone as well as Eurocentric "if" – see Chapter 2). Recovery – with the jury still out on the issues of recomposition and reorientation – remains somewhat elusive. Union membership did not rise in 2022 and neither did participation rates in 2023. Membership fell in relative and absolute terms: to 22.3 per cent down from 23.1 per cent in 2021 (which itself was down 0.6 per cent on 2020), and by 200,000 members to 6.25 million members in 2022, being only 6,000 members more than in 2017 (ONS 2023b).

Participation remained doggedly dreadful. In the election for a new National Education Union (NEU) General Secretary, held between February and March 2023, the turnout was just 9 per cent, even though the union recruited a mass of new members in pretty much the same period: "Since announcing its strike ballot results in January, the NEU [with 457,143 members on 1 January 2022] … gained around 50,000 mostly teaching members, with 7,000 of those joining over the five weeks since the first national strike day at the beginning of last month" (*TES* 4 March 2023). It cannot be credibly argued that the time constraints of organizing strike action got in the way of filling out a ballot paper and putting it in the post. Elections for union general secretaries in the previous years of the 2020s routinely saw between just 5 per cent to 10 per cent of members voting.

Turnouts for elections to unions' national executives showed no discernible uptick in 2023 either. For the NEU, Public and Commercial Services, Unison and Unite unions, turnouts remained very low at between just 5 and 7 per cent. These voting behaviours stand in stark contrast to the experience of most unions in easily winning lawful mandates for action from most of the ballots for strike and industrial action that they organized. Since 2017, the Trade Union Act 2016 has required that voting turnouts must be at least 50 per cent and in several sectors deemed "essential" those voting for action must also equate to 40 per cent of all those entitled to vote (hence, non-voters are essentially held to be no-voters). It is then reasonable to speculatively suggest that the high turnouts and high levels of "yes" votes in industrial action ballots indicates that members see a more tangible and direct connection between voting in such ballots and their own actions in attempting to defend their own interests, which they do not with ballots for (internal) elected positions within their own unions. At present there is no data available on how many new shop stewards (or workplace representatives) unions saw created as a result of the strikes and whether this represented a net gain. The same is true for data on whether members' consciousnesses have been in any way transformed by the experience of organizing to strike and them striking itself (see also Kelly 1988).

In the US, union density fell to 10.1 per cent in 2022 from 10.3 per cent in 2021, the lowest level on record since comparable recording began in 1983 (BLS 2023a). This was despite a 1.9 per cent increase in absolute membership since 2021 and some notable breakthroughs in recent years at the likes of anti-union employers in the private service sector such as Amazon, Apple and Starbucks. In terms of striking, although 2022 saw an increase on 2021, which saw an increase itself on 2020 in terms of the number of strikes, days not worked and workers involved, this only took the situation back to where it was in 2018 and 2019 (BLS 2023b).

A similar situation exists in Australia, with absolute union membership increasing between 2017 and 2021 but density falling from 15.6 per cent in 2016 to 12.5 per cent in 2022 (ABS 2022). There has been a noticeable uptick in strike activity in Australia in 2022 on all three measures of number of strikes, days not worked and workers involved (ABS 2023). However, this was essentially a spurt of activity during the summer (Northern Hemisphere) period with all three measures returning in March 2023 to the levels experienced in March 2022. More importantly, since 2007 strike activity by these three measures in Australia has continued, since 2007, to be around a quarter of that which existed in the mid-1980s to early 1990s.

And in France, which has witnessed continual general strike waves, based around general strikes and mass street mobilizations in, for example, 1986–87, 1995, 2003, 2005, 2010, 2018, 2019–20 and 2023, there seems to be no positive relationship between striking and union membership. From 1995 to 2019, according to DARES (2021), union density has hovered around 10 per cent and never been more than 11 per cent, with some employer organizations putting it as low as 8 per cent in 2022 (L&E Global 2022).

If it may be suggested that general strikes against government policies as a form of political exchange are not the right tool by which to test such a strike hypothesis, then other DARES (2023) data on enterprise-level strikes shows no positive relationship either. In 2013, density was 11.0 per cent while 1.2 per cent of enterprises of ten or more workers experienced a strike. In 2019, density was 10.1 per cent when 2.5 per cent of enterprises of ten or more workers experienced a strike. Neither for the UK, the US, Australia nor France is this to consider the impact of the strikes themselves upon their propensity to be "recruiting sergeants" for unions. Nonetheless, it can be ventured that the outcomes in the form of possible positive and negative demonstration effects will be far from unimportant but it is beyond this brief discussion in an introductory chapter to consider this issue. With the possible exception of France as result of the mass strikes against pension reform in particular,[1] what can be ventured is that while the contemporary uptick in strike action has often been the largest in a generation, if longwave theory still has palpable purchase, it should be noted that the contemporary strike wave is weak – in terms of scale and length – by comparison with previous wages cycles. However, this weakness brings into doubt the purchase of longwave theory (see Gall & Holgate 2018: 571–2).

Prior to the cost-of-living crisis, the Covid pandemic was believed by some writers to witness the welcome death knell of neoliberalism by virtue of the huge and unprecedented levels of economic and social state intervention. This was a forlorn hope for neoliberalized states, as with their response to the global financial crisis of 2008–09, they intervened precisely to stabilize the existing system and not to presage its

1. Greece is no longer an exception, having experienced a significant downturn in general strikes and other strike activity since the first half of the 2010s.

transmogrification into social democracy or some form of post-neoliberalism. For that reason, the terrain of neoliberal capitalism has remained an equally challenging one for unions and their members. This has been most starkly illustrated in the continually growing levels of wage and income inequality while membership levels and collective bargaining coverage have fallen in an "effect and cause" relationship. Further challenges for unions and their members are already in train and set to strengthen, whether that be the precariatization of many sections of the labour force, the spread of artificial intelligence leading to proletarianization of many in the highly unionized professional middle-classes or how to expand the remit of collective bargaining and political exchange to respectively "green" their members' workplaces and gain a "Just Transition".

All this means – unfortunately for those concerned with the worth and potential of labour unions – that chapters written for this handbook are as salient now as when they were first conceived and written prior to, during and after the pandemic. With these elements of context explained, the remainder of this introductory chapter lays out the rationale and structure of this handbook.

But first a word on terminology and its significance. If this handbook had been written 20 or more years go, the nomenclature – although still inaccurate – would have, undoubtedly, been "trade unions" and not "labour unions". For the vast majority of writers and researchers the former is still the preferred parlance, whether used consciously or unconsciously. This may seem like a semantic point to make. And yet it speaks to the key issues contemplated in this handbook, such as identity, appeal, form, constituencies of interest and resources. To put it bluntly, this seeming semanticism is about correctly consigning the organizing principle of each union or type of unionism. For many decades the vast majority of unions have long since ceased to be *trade* unions or even unions of *trades* following mergers and amalgamations. There are still some specialist unions for occupations and professions but they are not *trades* unions. Most unions, through amalgamations and mergers as a result of declining membership and the impact of changing technologies, are now hybrids, general, sectoral or industrial unions of labourers. In many of the countries of the Anglophone world, *trades* unions may only still have salience as the terminology still used in legislative and legal texts.

LABORIOUS LABOURING OVER LABOUR UNIONS

Increasingly in the new millennium, much mainstream progressive public opinion as well as many public intellectuals, thinktanks and political parties see labour unions as a key force for civilizing both an increasingly uncivil economy and society. Some, echoing past proselytizing, go further in this present period and see labour unions as the primary force for helping choose socialism (sometimes known as social democracy) over barbarism (the unregulated, crisis-prone neoliberal variant of capitalism). In this sense, there is some faith in – and for – labour unions and their civilizing and equalizing role. And yet notwithstanding the aforementioned contemporary context, labour unions are still at their lowest ebb in most corners of the globe after nearly five decades of the neoliberal variant of capitalism and despite heavy investment in the turn to "union organizing" in the last 30 years (see, e.g. Gall & Fiorito 2011). Consequently, dreams of civilizing or of socialism are, in fact, more distant than at any time since the postwar period. So, the paradox is that this potential historic role of – and for – unions is ever more highlighted by their weakness (and that of associated ideologies like social democracy and socialism) and the strength of counter-forces such as capital and the state (and associated ideologies like neoliberalism and right-wing populism and nationalism).

As an edited collection, *The Handbook of Labour Unions* is both a primer-cum-overview and active intervention for two engaged audiences. First, advanced undergraduate and postgraduate students concerned not just with the political economy of work and employment but also those student milieux concerned with political and social sciences examining the potential for particular subordinate social actors to act as forces for democratization and egalitarianism. Second, academics, public intellectuals and policy analysts in similar areas of study and interest who would wish for an intellectual thesis to argue for the continued economic, social and political relevance of labour unions to progressive societal change and the conditions under which this project can be realized. The task for *The Handbook of Labour Unions* is then to fuse understanding of the past and present to provide some kind of general guide to shaping the future – of course, rather than any "route map" as such.

This handbook is both primer and overview because it brings together an array of experts to summarize the salient literature in terms of concepts and theories allied to research findings. Having done that, the chapters critically but sympathetically engage with debates and discussions about the role and purpose of unions and the array of means by which they seek to attain these, thereby constituting the active intervention. Overall, and in these two senses, as a handbook *The Handbook of Labour Unions* provides an understanding of the workings of labour unionism as well as insights into how challenges and problems may be surmounted. This is a comprehensive treatment but, inevitably given the limit on pagination and the availability of authors, there are some aspects of topics that would have warranted further consideration.

So, these two tasks of providing a primer and a critical intervention are carried out by analysing the past, present and future purposes, roles and activities of labour unions in three ways, with the ever-present analytical foundation of the troika of components of power, material interests and ideology.

First, thematically with regard to ideal types as well as manifest variations in practice. This is a heuristic undertaking for broadening and deepening the understanding and analysis of what are complex organizations. Inevitably, it means that the constitution of these individual chapters creates some artificial division between what are interlocking and related issues, much like the circles in a Venn diagram scenario that are overlapping each other. These thematic chapters should be especially read together. They comprise first-order-level subject matter, namely the consideration of the appeal and identity of unionism, the interests and ideologies of unionism, the forms of unionism, the resources of unionism, the governance of unionism, the (external) relations of unionism and the terrain of unionism. Put together, these provide a comprehensive way to view the totality of the actuality and potentiality of labour unionism as well as the obstacles that prevent the realization of labour unionism's historic purpose.

The appeal and identity of unionism concern which constituencies unions organize and seek to organize (wider/narrower, horizontally/vertically across time and space). Closely related to this, but not synonymous with it, is the ideology of unionism, be it, for example, social democratic, socialist or liberal democratic. The form of unionism concerns how the appeal and identity of unionism marry with the ideology of unionism into the differing shapes unions take. The (power) resources of unionism concern the influence, both actual and potential, in terms of campaigning and bargaining leverage with agencies external to themselves and that are derived from their memberships (subscriptions, attachment, participation) and union employees' expertise. The governance of unionism concerns the democratic aspect of the participation of members in the processes that shape and influence (if not determine) the actions of unionism, and out of which organizational legitimacy and membership commitment can be derived. The relations of unionism concern interactions with other workers as well as workers (or producers) as citizens and consumers and relations with capital, state and political parties. The terrain for unionism concerns the temporal, spatial and existential dimensions of the ground upon which unions have to operate. The first of these dimensions concerns the predominant periodization of conditions when labour unionism has been stronger or weaker, in terms economic growth, state intervention, social democratic and socialist ideologies, wider social turmoil and working-class struggle. The second dimension concerns the varying geographical aspects of the development of labour unionism, which do not necessarily accord with global trends, from corporatist incorporation (in the former Soviet Union, China, many Latin American countries, and fascist Germany and Italy, for example) to militant social movementism (in, for example, Brazil, South Africa, South Korea prior to democratization) and subjugation and pacification (such as countries of the former Soviet Union). The third dimension concerns unions as secondary organizations whose rationale and behaviour are often reactive to that of the primary organizations of the employment relationship, namely employers, be they state or capital. From this categorization, the potential and obstacles of and for labour unionism realizing the more or less limited aforementioned aspirations across economy, polity and society can be assessed.

Second, and following the aforementioned seven-fold schema, by beginning to lay out how these components have dynamically interacted with each other to produce an overall balance of power, ideology and material interests between contending social forces in society and economy. This is operationalized in the examination of the liberal capitalist low point (1880–1945), the social democratic high point (1945–75), the "socialist" high point (1945–90) and the neoliberal low point (1975–1990 onwards).

Then, third, and upon this basis, by assessing numerous key aspects of labour unionism and labour unions, such as their influence on the frontiers of control over pay, work organization, working time and workplace democracy; how unions relate to, and organize on the basis of sectionalities such as class, race, gender, migration and generation; how unions as organizations are subject to processes of birth, ossification and renewal; how unions may be schools for (self-)teaching citizenship; why unions are, ordinarily in practice, economistic despite many radical avowed political aims; whether the reflowering of unions is dependent upon other social movements; and how unions respond to the reconfiguration of work and employment under capitalism.

These three ways lead this handbook to have three parts. The first concerns the components, characteristics and context of labour unions and labour unionism. The second concerns space, power and periodization and the third the practice of building presence and power.

This transnational – across national boundaries rather than international between international boundaries – thematic approach to assessing labour unions and labour unionism in terms of their past, present and future is believed to be more productive than: (1) the usual alternatives of single-country studies (including whether comparative analysis is conducted or whether standardized areas of inquiry are deployed); (2) critical case studies (even within the parameters of the recent rise of global labour studies such as the *Global Labour Journal* (established in 2010), *Global Labour Studies* (Taylor & Rioux 2017) and *Rethinking Global Labour* (Munck 2018)); (3) analysing specific aspects such as strikes, struggles and social movements (see, e.g. Azzellini & Kraft 2018; Bieler *et al.* 2015; Grote & Wagemann 2018; Nowak, Dutta & Birke 2018); and (4) studying through case studies of unions (or any of their subsets) the connected processes and outcomes of labour unionism in a "cause and effect" manner. This is because this handbook attempts to go beneath and beyond the relative superficialities of the surface expressions and articulations of labour unionism in order to provide a comprehensive and high-level analysis of labour unions across time and space that synthesizes extant literature (rather than drawing upon empirical case studies, which are likely to have significant idiosyncrasies that make generalization fraught). This means that, for example, and following from the seven initial foundational chapters, the issues of leadership and followership are dealt with in several chapters (such as Chapters 15, 16 and 17 especially).

Throughout the handbook, each chapter not only seeks to build upon the preceding chapters but also to refer to and engage with others. Sometimes this is explicitly stated, sometimes it is not. Nonetheless, this overlapping nature of chapters means that the most productive way to benefit from them is to read them as a whole and in the sequential order they are set out in.

REFERENCES

ABS 2022. "Trade union membership". Australian Bureau of Statistics, 14 December. www.abs.gov.au/statistics/labour/earnings-and-working-conditions/trade-union-membership/latest-release

ABS 2023. "Industrial Disputes, Australia". March 2023. www.abs.gov.au/statistics/labour/earnings-and-working-conditions/industrial-disputes-australia/latest-release

Azzellini, D. & M. Kraft 2018. *The Class Strikes Back: Self-Organized Workers' Struggles in the Twenty-First Century*. Leiden: Brill.

Bieler, A. *et al.* (eds) 2015. *Labour and Transnational Action in Times of Crisis*. London: Rowman & Littlefield.

BLS 2023a. "Union Members Summary – 2022". Washington, DC: Bureau of Labor Statistics. www.bls.gov/news.release/archives/union2_01192023.htm

BLS 2023b. "Annual work stoppages involving 1,000 or more workers, 1947 – Present". Washington, DC: Bureau of Labor Statistics. www.bls.gov/web/wkstp/annual-listing.htm

DARES 2021. "La syndicalisation". Paris: Direction de l'Animation de la Recherche, des Etudes et des Statistiques. https://dares.travail-emploi.gouv.fr/donnees/la-syndicalisation

DARES 2023. "Les grèves en 2021". Paris: Direction de l'Animation de la Recherche, des Etudes et des Statistiques. https://dares.travail-emploi.gouv.fr/publication/les-greves-en-2021

Gall, G. & J. Fiorito 2011. "The forward march of labour halted? Or what is to be done with 'union organizing'? The cases of Britain and the US". *Capital & Class* 35(2): 231–50.

Gall, G. & J. Holgate 2018. "Rethinking industrial relations: appraisal, application and augmentation". *Economic and Industrial Democracy* 39(4): 561–76.

Grote, J. & C. Wagemann 2018. *Social Movements and Organized Labour: Passions and Interests.* London: Routledge.

Hodder, A. *et al.* 2017. "Does strike action stimulate trade union membership growth?". *British Journal of Industrial Relations* 55(1): 165–86.

Kelly, J. 1988. *Trade Unions and Socialist Politics.* London: Verso.

L&E Global 2022. *France: Trade Unions and Employers Associations – Brief Description of Employees' and Employers' Associations.* Brussels: L&E Global.

Munck, R. 2018. *Rethinking Global Labour: After Neoliberalism.* Newcastle upon Tyne: Agenda Publishing.

Nowak, J., M. Dutta & P. Birke 2018. *Workers' Movements and Strikes in the Twenty-First Century: A Global Perspective.* London: Rowman & Littlefield.

ONS 2023a. "Labour disputes UK; total working days lost; all inds. & services (000's)". London: Office for National Statistics. www.ons.gov.uk/employmentandlabourmarket/peopleinwork/employmentandemployeetypes/timeseries/bbfw/lms

ONS 2023b. *Trade Union Membership, UK 1995–2022: Statistical Bulletin.* London: Office for National Statistics.

Taylor, M. & S. Rioux 2017. *Global Labour Studies.* Cambridge: Polity.

PART I

COMPONENTS, CHARACTERISTICS AND CONTEXT

CHAPTER 1

UNION IDENTITY AND APPEAL

Lorenzo Frangi and Tingting Zhang

ABSTRACT

Union identity is a highly contested and multiplex terrain. Union leaders aim at shaping a union's official, projected identity, but union identity is also shaped by actual union actions in the everyday efforts of numerous actors besides union officials. Taking stock of extant literature, we tease out relevant analytical dimensions along which union identity can be studied and compared. We highlight how certain macro-institutional, meso-organizational, and micro-quotidian forces affect the characteristics of these dimensions. We then analyse union identity and its appeal to different populations. Finally, we propose a comprehensive and original analytical scheme for union identity that incorporates the analysed elements. We indicate some challenges for the debate on union identity and appeal and trace strategies to overcome them.

Keywords: Union identity; union appeal; analytical framework; challenges for unions

INTRODUCTION

"Who are we?" and "What do we stand for?" These fundamental questions define the identity of a union (Albert & Whetten 1985). Answers reflect a union's core values and goals; that is, its "essence" (Hodder & Edwards 2015) or "very nature" (Hyman 2001: 1), and distinguish a union from other organizations and other unions (see, e.g. Ravasi & Schultz 2006). Self-referential meanings that constitute union identity are used to project union organizational patterns of actions (see Gioia *et al.* 2013) and to define union appeal among members to motivate their commitment, participation and activism to achieve union goals (Gall & Fiorito 2012). Moreover, union identity can appeal to a wider range of potential supporters, such as non-member workers and even non-workers, thus giving unions access to a wider constituency whose resources can contribute to a greater impact (e.g. Marino 2012; Scott & Lane 2000).

A better understanding of union identity and appeal may yield deeper insights into specific unions, as well as the union movement as a whole today (Smale 2020). Unions have traditionally been understood as wielding a "sword of justice" in the promotion of labour rights (Flanders 1975). However, there is more heterogeneity in union identity than this common understanding suggests. Union identity has included a wide variety of values, ideologies and goals expressed across and within countries over time by different unions, based on different understandings of the structure of the economic, social and political sphere within which a particular union must operate (Gumbrell-McCormick 2013; Hodder & Edwards 2015; Hyman 2001). Moreover, the crumbling of the Fordist production regime and its homogeneous, male-dominated working class and the emergence of flexible production arrangements brought into being an increasingly diverse workforce, more complex labour market structures and new employment conditions (Tapia, Lee & Filipovitch 2017). All these changes have challenged unions to adapt and revise their answers to questions of "Who are we?" and "What do we stand for?"

Union identity is a highly contested and multiplex terrain (e.g. Gioia *et al.* 2013; Scott & Lane 2000). Union leaders aim at shaping a union's official, projected identity through official statements, documents and communications that pattern the content, direction and scale of union strategies (Smale 2020). Beyond this, union identity is also shaped by actual union actions in everyday efforts by a larger number of actors besides union officials (Smale 2020). A union's official and actual identities are fundamentally shaped by three forces: first, external, macro-institutional forces set the space within which each union identity can develop; second, within this space, internal, meso-organizational strategy forces steer the definition of a specific union identity; third, daily, micro-forces in the workplace and beyond contribute to union identity.

Taking stock of extant literature, we elaborate upon a more comprehensive analytical framework to understand union identity and appeal. Thus, we highlight gaps in the literature and suggest related developments. More specifically, section two teases out relevant analytical dimensions along which union identity can be studied and compared. Section three highlights some relevant forces that affect these dimensions. Section four explains union identity appeal and mechanisms among different populations. Section five proposes an analytical scheme that can take stock of the analysed elements. Finally, section six indicates some challenges for the debate on union identity and appeal and outlines ways forward in studying them.

ANALYTICAL DIMENSIONS OF UNION IDENTITY

The diverse and multiplex identities that unions assume can be studied along a set of analytical core dimensions: interests, scale, functioning and power.

Interests

Defining elements in determining "who we are as a union" are the occupational characteristics of union members and the related employment relations interests that unions must include in their agendas (Hyman 1975). Turner (1962) describes differences in unions' identity based on the degree of their closeness to representing various interests. Using the same line of reasoning, Hyman (1975) differentiates three types of unions: those representing only employees in specific trades/professions (i.e. craft unions); those whose membership is from a specific sector (i.e. industrial unions); and those who organize all workers indiscriminately from trades or sectors (i.e. general unions). Craft unions tend to promote and preserve the interests of a specific group of members who share common skills. In this case, unions stand for a tight control of the barriers to access the occupation to preserve their status. Industrial unions attempt to create homogeneous working conditions in a sector, so that salaries are upwardly standardized throughout similar firms. Finally, general unions fight for fundamental labour rights, the enhancement of working conditions for less well-off workers and fairer conditions in the labour market. This categorization can be further refined, distinguishing unions by the type of worker (blue- or white-collar) and the comprehensiveness of the industrial sector represented (narrow sub-sector, sub-sector, general sector) or unions whose identity goes above and beyond all these differences; that is, general unions (Visser 2012).

The social movement unionism approach has added to this debate (Johnston 1994; Sullivan 2010; Tapia & Alberti 2018). Unions' self-reflection about "who we are" may go beyond a membership-only focus to include non-members and non-workers' interests (e.g. Johnston 1994; Sullivan 2010), creating a broad-based constituency with different demographical, social and work-interest characteristics. A plethora of new social actors beyond members can be networked to constitute part of union identity. Such actors include civil society organizations (Croucher & Wood 2017; Martínez & Perret 2009), community groups (Holgate, Keles & Kumarappan 2012), militant individuals (Holgate, Simms & Tapia 2018), and a large number of marginalized and unorganized people who are not involved with unions in traditional terms (McAlevey 2016).

Scale

Union identity is also defined by a specific scale (Smale 2020). This is first and foremost the geographical dimension of union representation, such as a specific workplace, a region, a sector, a country or, as in the case of global union federations, the entire world. However, in addition to this main facet of scale, two others aspects need to be considered. The first is the extent to which a union has structural relationships with other unions that expand its geographical scope, such as the typical relationship between a local union and a sectoral federation, or union networks across different workplaces, regions and nations (Lévesque & Murray 2010). The second is the extent to which a union, in defining what it is and stands for, extends its geographical scope at a higher level. For instance, many local unions define their identity in terms of national policy or even global goals (Frangi & Zhang 2022).

Functioning

Union identity is fundamentally shaped by the features of its internal functioning (Hyman 1994); that is, the degree of democracy. This entails focusing on the balance between a union constituency's power versus union leaders' power to set the agenda. Child, Loveridge & Warner (1973) analytically distinguish union identities based upon the relevance of goal formation through representation versus goal implementation through administration. Heery and Kelly (1994) propose a tripartite classification on a continuum linking the vivacity of membership participation at one end to the organizational ossification due to Michels' (1915/2001) "iron law of oligarchy" at the other end: in "managerial" unions, members are conceived as reactive consumers; "professional" unions are mainly official-led organizations; in "participative" unions, bottom-up forces prevail. Social movement analysis furthers the debate on the participative forces of union functioning, highlighting how unions democratically function when they are based on the engagement of a broad-based constituency led by grassroots activists rather than by bureaucratic union officials (de Turberville 2004; Heery 2005; Holgate, Simms & Tapia 2018).

Power

Power is a fourth dimension of union identity. This refers to union ability to achieve its objectives and influence others' perceptions of it (Hodder & Edwards 2015; Hyman 1994). Union power has been measured along three dimensions: size, activism and impact (e.g. Lévesque & Murray 2010; Peetz & Pocock 2009). The size essentially entails headcounts of union members or, from a social movement unionism perspective, the coverage of a broad-based constituency. Unions with a large number of members project an identity as "big", resourceful and powerful. More than simple counting, unions have a powerful identity when their rank-and-file members are more active (Fiorito, Padavic & DeOrtentiis 2015). The level of membership activism is considered a critical element for union effectiveness (Clark 2009), the "soul" (Budd 2004) and "fabric" (Gordon *et al.* 1980) of unions. Activists reach out to and involve more members, represent members' interests and join union actions (Fiorito, Padavic & DeOrtentiis 2015). The proportion of committed and active members influences union identity and its ability to achieve more ambitious bargaining goals (de Turberville 2004; Kelly & Kelly 1994). Finally, the perception of a union as a powerful institution depends on its ability to influence employment relations and, thus, its impact. Bargaining power in terms of number of workers affected and the improvement of their working conditions is constitutive of union power identity. Besides union-to-employer bargaining power at the workplace or higher levels, unions can assert power vis-à-vis the government in their ability to leverage parties' support, typically among labour-inclined parties (see Blackburn 1967).

MACRO-, MESO- AND MICRO-FORCES

Several forces can shape the specific characteristics of the analytical dimensions of union identity. Arguably, the most relevant are macro-institutional, meso-organizational, and micro-individual forces. As macro-forces have been extensively discussed, in what follows we briefly discuss macro-forces, emphasize meso-organizational ones and trace some possible micro-individual ones.

Macro-, external institutional forces

To shed light on the variety of union identities, Hyman (2001) emphasizes the effect of the macro-, external institutional environment. His analytical approach allows a specific union identity to be positioned along the three edges of a "triangle": business (unions as bargaining agents of a workplace), class (unions as institutions for all workers) and society (unions as agents of social justice). He shows how specific union identity equilibria are centrally linked to path-dependent institutional, legal industrial relations settings in different countries. For instance, unions' identities in the UK are located between market and class, German unions' identities between society and market, and Italian unions' identities between class and society. Macro-institutional forces, especially at the national level, undoubtedly contribute to shape union identities. Institutional differences between North American, Nordic, central and southern European models, or the variety of institutional settings in developing countries, define sets of constraints, as well as opportunities, within which different sets of union identities can develop. A rich debate, mostly developed around the variety of capitalism framework, has explored several aspects of how institutions shape the analytical dimensions of union identity (Frege & Kelly 2013; Frege, Kelly & Kelly 2004).

Meso-, internal organizational forces

While union identities tend to be similar in some instances in a given country, much diversity also characterizes union identity in each of these contexts. Union identity is not just the reflection of national environments (Lillie & Martínez Lucio 2004). Meso-, internal organizational strategies shed light on these differences. Indeed, even in the same institutional context, unions have distinguishable organizational core values. Over time unions can remain path-dependently constrained to their original values (Ross & Martin 1999) or, alternatively, unions can modify their core values and assume distinctive, new identity traits (Gahan & Pekarek 2013; Gall 2003). Unions demonstrating different abilities over time could strategize around, plan and innovatively act upon these value legacies (Hyman 2007; Smale 2020). Many actors can shape union values, but leaders' narratives about "what values we stand for" are central due to their organizational legitimacy and ability to deliberately influence others' sense-making process about a union (Gioia *et al.* 2010; Hatch & Shultz 2002; Hodder & Edwards 2015). Change in leadership is often characterized by introducing new or revised organizational values, making some more latent, or dropping others from the organizational claims (see Albert & Whetten 1985; Cheney & Christensen 2001).

Union identity tends to rest upon a mix of "just for us" and "for others" (Fiorito, Padavic & Russell 2018). Values held and communicated by union leaders can affect the mix of these two values. When union identity moves from a strict "just for us" to more encompassing core values, it becomes shaped in terms of different interests relating to a set of members, potential members and individuals in the society who share part of union ideology (Frangi & Zhang 2022; Tapia, Lee & Filipovitch 2017). For instance, framing more encompassing values, starting from a simple extension of the content of "us" in "just for us", might allow unions to set goals that are not exclusively defined by their current members' specific needs (Gall & Fiorito 2016). Further steps in the framing of "us" can include other groups of workers similar to the present union members, an occupationally more diverse workforce, with workers from different trades and sectors, thus moving narrow union interest representation progressively towards a *general* union interest (Smale 2020). A typical example of this enlargement of identity by leadership framing happens during a union merging process (Behrens & Pekarek 2012; Hoffman, Kahmann & Waddington 2007). Organizational framing forces can take a definitive step towards embracing core values of "for others", connecting with a broad-based consistency beyond members and workers (Tapia & Alberti 2018) and including the representation of interests of more marginalized workers, such as young, unrepresented, precarious and oppressed groups (Holgate, Simms & Tapia 2018; Tapia, Lee & Filipovitch 2017).

Within the same institutional environment, unions can have different established internal functioning routines and take strategic actions to change these routines and shape identity by moving between two extreme models: a top-down servicing or a bottom-up organizing model (Carter 2000; de Turbeville 2007; Heery *et al.* 2000). For instance, a specific union's identity can be shaped by the extent to which the union deploys structures to improve engagement and participation at the rank-and-file level (Hyman 2007). This is reflective of a bottom-up union identity (Johnson & Jarley 2005). Many unions distinguish their identity vis-à-vis other unions by promoting internal democracy and larger participation, leveraging the new technologies as deliberative tools (Hyman 2007). Others take a more conservative, mostly top-down and informative approach to the use of new technology, missing the opportunity to deeply reshape the internal functioning of their identity (Frangi, Zhang & Hebdon 2020; Martínez Lucio, Walker & Trevorrow 2009).

Finally, a union's organizational strategies at the meso-level can affect the level of power that characterizes a specific union identity. Unions with organizing strategies centred around the "traditional" member profile are more impacted by a stagnating or declining membership (Hyman 2007). Unions that embrace internal differences and proactively stimulate the desire for representation of new worker groups have opportunities to maintain or increase their membership (Freeman 2007). Unions that deliberately forge coalitions with other unions and nurture alliances with civil society organizations and activist individuals extend their constituency, weaving relationships within and across workplaces and beyond (Murray 2017; Tattersall 2005).

Union strategies to identify members who are central "nodes" in their network, who have more traction over colleagues, and then provide them with support to involve more members in union activities can shape union commitment and militantism towards higher levels (Johnson & Jarley 2005). Targeted union strategies to leverage as well as capacitate more militant members in the use of new communication technologies can have important impacts on members' commitment, and willingness to defend, implement and enhance union goals (i.e. militantism) (Houghton & Hodder 2021). The intake of non-member militant individuals into the union fold can increase union militantism and mobilization power (Holgate, Simms & Tapia 2018).

Finally, meso-level union strategies can shape the power facet of union identity. Familiar activities, such as typical strike actions, have become more difficult to deploy and have limited impact on advancing working conditions (Johnston 1994). Union power identity is much more defined by innovative repertoires of contention (Tilly 2006). Union international alliances allow unions facing declining membership to become powerful organizations able to stand against multinational corporations and globalization forces (Bronfenbrenner 2007; Brookes 2019). Some unions have gained the ability to influence public policies by studying the global production network to find possible allies or points of vulnerability, or shifting the locus of action to increase labour rights from the workplace to the streets, embracing contentious politics (Martínez Lucio & Stuart 2009; Simms 2012).

Micro-, individual

Micro-actions by union stewards and members contribute to define the answers to "who we are as a union and what we stand for" in terms of projected and actual identity. Values demonstrated by union members, especially stewards in their everyday interactions, project a specific union identity. This is the lively part of the union identity as perceived by less active union members, non-members and, to some extent, the community around a specific workplace and the network of people around union stewards and active members. For instance, the kind of reception a non-member receives from a union steward helps define the perception of a union's identity as "just for us" or "for others" (Gall & Fiorito 2016). Daily exchanges, formal and informal, about the union, employment and social issues between union stewards and more active members, on the one hand, and non-members and even non-workers, on the other, help to confirm, modify or challenge the union's official identity in terms of represented interests, functioning, scale and power. Indeed, these micro-interactions are de facto the base of union power. They can solicit more union members to become activists, encourage more workers to join unions and create a larger base of activists beyond the workplace. For instance, the micro-, informal and supportive interactions between union stewards in a workplace and outsourced workers in the same workplace about their working conditions and issues extends the interests, scale and power of a union's identity (Benassi & Dorigatti 2020; Drahokoupil 2015).

Summary

The industrial relations literature includes numerous studies about the macro-effect of the institutional environment on union identity, including the effect of unions' meso-organizational strategies in shaping identity, especially through the rich debate about union renewal. The study of micro-forces, in other words of the union identity as experienced by the everyday worker, has hitherto received scant attention in the industrial relations literature.

UNION IDENTITY APPEAL

"Identity is what identity does" (Whetten & Mackey 2002: 393). Social identity theory suggests unions appeal to individuals whose social identity is congruent with union identity (Cregan, Bartram & Stanton 2009; Iverson & Buttigieg 1997). When a union appeals to those individuals, the union can benefit from their interests and support. Given these resources, union appeal forms the foundation of union power and influence. Although it is fundamental for union vitality and efficacy, union appeal per se is an underdeveloped concept and has not been directly empirically analysed. Yet some related constructs in the debate about unionism shed some light on union appeal; one of these is union image (Craft & Abboushi 1983; Youngblood *et al.* 1984). Union appeal is composed of three main analytical dimensions related to the different populations to whom union can appeal: union members, non-union fellow employees and non-workers.

Among union members, some are less attracted to union identity while others are more compelled by it. The latter are more likely to support union activities and commit to unions (Redman & Snape 2014). A union identity appeals to those members who perceive that their union effectively represents, responds to and enhances their interests at the bargaining table. Moreover, members commit to union actions when the appeal of union is centred around servicing them properly in daily employment relations dynamics in their workplace (Redman & Snape 2014). Even in a context of generalized union membership decline, the instrumental appeal of unions attracts more participation and commitment (Bamberger, Kluger & Suchard 1999; Redman & Snape 2014; Waddington & Whitston 1997).

However, union instrumentality is not the only appealing aspect of a union identity to members. Union identity promoting mutual support among members has been reported to encourage union participation (Waddington & Whitston 1997). Moreover, unions that have recently been able to act upon, expand or reshape their identity to embrace diversity and inclusion values appeal to more marginalized groups of members, such as females, minorities and migrant workers (Kirton 2005; Tapia, Lee & Filipovitch 2017). More specifically, those groups of members feel more inclined to commit to a union identity that not only promotes diversity and inclusion values in the union's projected identity but also takes actual, effective actions to create opportunities for them to self-organize and have a voice, become more involved in key union roles and pursue leadership positions (e.g. Colgan & Ledwith 2002; Kirton 2005). Union identities that fulfil instrumental, supportive and diversity values can appeal to a wider share of members, solicit a larger commitment and, in turn, gain more power.

Union identity appeals to some non-member fellow employees. These employees are more inclined to join or support a union. While specific industrial relations institutional settings, sector and firm characteristics, and individual employment conditions explain variations in joining unions in different countries (Kranendonk & De Beer 2016), union identity appeal might add a further explanation of this phenomenon. When controlling for institutional and organizational characteristics, studies have found unions appeal the most to workers who have more pro-social values, who feel their identity is more aligned with unions and who feel more compelled to join a union (Kirmanoğlu & Başlevent 2012; Scheuer 2011). This appeal has become more important since the start of union decline, not only in terms of membership count but also in terms of the ability to achieve definitive instrumental gains. When union appeal among employees is measured beyond the barriers of union membership, and pro-union attitudes are assessed, interesting evidence emerges. In most countries, unions appeal to a much wider share of employees than just members (Hadziabdic & Frangi 2022). In places where union membership is marginal, such as in the US, union appeal is at the base of an important "frustrated" demand to become a union member (i.e. employees who would like to join unions but for whom there are too many roadblocks to membership) (Bryson & Gomez 2005; Givan & Hipp 2012; Gomez, Lipset & Meltz 2001).

Moreover, union appeal through time, while not always at the same level, seems to have a fairly stable trend (Brenan 2021) or even a growing one (Frangi, Koos & Hadziabdic 2017). In addition to groups of employees attracted to possible union instrumentality effects (Jarley & Kuruvilla 1994), studies consistently report a positive relationship between left-wing ideology and union support (Colgan & Ledwith 2002; Riley 1997; Waddington & Kerr 2002). Individuals with social values congruent to union values (e.g. collectivist orientation, parents' perceived attitudes to unions, association with union members in the workplace, or previous collectivism) are more likely to be attracted to trade unions, even if they cannot become union members because of contextual constraints (Cregan & Johnston 1990; Kelly & Kelly 1994; Kirton 2005). More recently, marginalized employees, low-skilled employees with a more precarious employment status (e.g. Kirton 2005; Silverblatt & Amann 1991), and racialized employees are increasingly attracted to union identities (Healy, Bradley & Mukherjee 2004; Riley 1997).

Unions are very much present in public debate (Baccaro, Hamann & Turner 2003; Streeck & Hassel 2003) and their identity seems to appeal to some individuals in the society at large more than others. Many of these individuals are *non-workers*. Union identities appeal to students and younger people (Cregan & Johnston 1990; Waddington & Kerr 2002), recent immigrants, retirees and the unemployed (Frangi, Koos & Habsiabdic 2017; Oliver & Morelock 2020). While different in their nature, those groups are more likely to find union identities appealing as organizations able to project values and take actual action to improve fairness. Unions might appeal to students, younger individuals and recent immigrants hoping to enter into a fairer labour market. Retirees and the unemployed might appreciate union values of promoting welfare measures to protect economically more vulnerable people. Some of these individuals might share certain identity values with unions, such as the promotion of multi-culturalism, the fight against global warming or a set of community-specific social issues that unions include in their identity when collaborating with civil society organizations (Healy, Bradley & Mukherjee 2004).

Union identity is increasingly projected towards a wide audience – composed of a few members, many employees, but also non-workers – by the increased use of social media. Societal issues, coalitions with civil society organizations, encompassing visions of how to improve society, might appeal to more than those with narrow interests in instrumental representation (Frangi, Masi & Poirier 2022; Heckscher & McCarthy 2014). These new horizons may push unions to incorporate new elements into their organizational identity to appeal to untapped segments of society. This new framework of union identity and appeal can become a cornerstone in the development of social movement unionism (Turner & Hurd 2001). Although the creation of a supportive constituency beyond union members can place high demands on union resources to develop a dense, resourceful network (Engeman 2015), unions can strategically leverage the different types of societal power embedded in the relationships they develop with individual and collective civil society actors. Importantly, unions' ability to harness these relationships in a dense network may increase the circulation of resources, expand union power and help to empower non-union workers trapped in unfair conditions (Frangi, Masi & Poirier 2022).

A COMPREHENSIVE ANALYTICAL SCHEME

Figure 1.1 graphically summarizes the elements discussed above. Macro-institutional forces define the space in which a specific union identity can develop. Union official, projected identity (on the left) and actual identity (on the right) are both shaped by meso-organizational forces (mainly union leadership and official narratives) and micro-dynamics (union stewards' and active members' action). These forces shape the different union identity characteristics along four dimensions: interest, power, functioning and scale. A union has to develop strategies (arrow in the middle) to cope with a possible gap between the two identities and the consequences of such a gap. Finally, other stakeholders (employers, civil society organizations, other unions and global organizations, etc.) can have an impact on how union identity is shaped along the four core dimensions and how it appeals to members, workers and a larger set of societal actors.

CHALLENGES AND WAYS FORWARD

Revisiting the debate around union identity and appeal leads us to highlight three main challenges and related ways forward. These are conceptual, analytical and empirical.

Conceptual

The concept of union identity in employment relations has been discussed in an extensive debate and numerous core analytical dimensions are highlighted. Union appeal has not been studied to the same extent, thus representing a significant gap in studies of unionism. While it is a key aspect in attracting support and resources, and it clearly affects union impact, union appeal remains highly under-conceptualized. In section

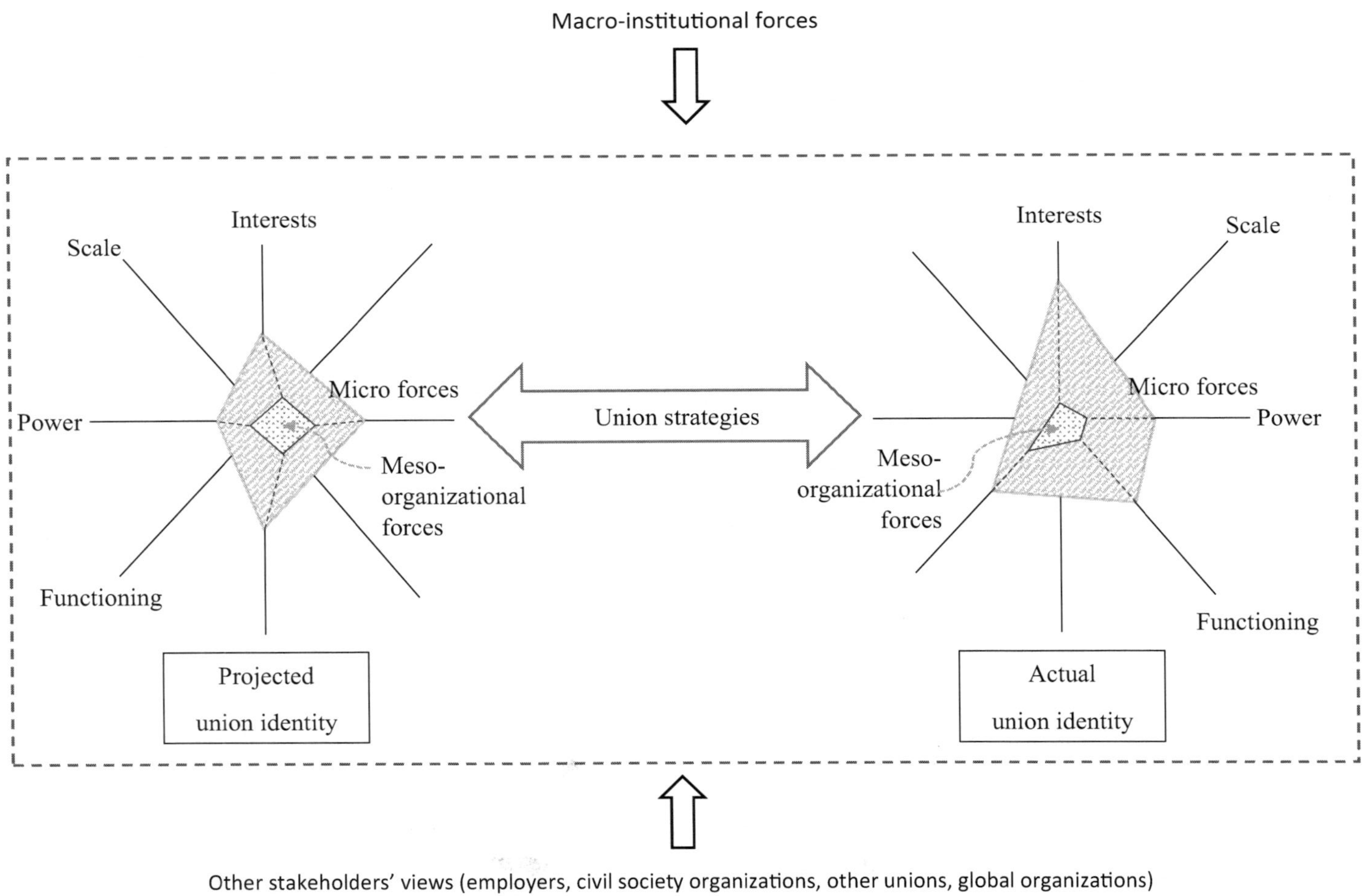

Figure 1.1 Representation of environmental, institutional and organizational forces

four, we highlighted some possible dimensions of union appeal, drawing on specific facets of unionism. This requires further development. The conceptualization of union appeal can benefit from organizational studies' frameworks developed to explain organizational appeal (e.g. Gioia *et al.* 2013; Kjærgaard, Morsing & Ravasi 2011; Ravasi & Schultz 2006) and adapt them to the specific reality of unions. Such conceptualization could include specific sub-concepts from the employment relations debate; we mentioned some of these in section three. This conceptualization is a fundamental step if we are to empirically investigate this construct and understand its antecedents and consequences.

Analytical

Figure 1.1 proposes a more comprehensive analytical framework to study union identity. We are aware that this is just a small step, and many more need to be taken to consolidate and refine it. More importantly, we suggest that the study of union identity should explore new analytical perspectives to better understand the phenomenon. Studies about union identity rely upon internal points of view, such as official documents, leaders' speeches, and actions. Moving from the simple analysis of "who we are as a union", the debate on union identity can draw further insights when the points of view of other relevant stakeholders are analysed. This shift in perspective has two advantages. First, it will enrich the understanding of the positioning of a specific union along several identity dimensions, and, second, the effect of stakeholders' narratives in actively shaping union identity and appeal can be more easily assessed (e.g. Scott & Lane 2000; Zellweger *et al.* 2013). Narratives about specific traits of the identity of a union that are made by other similar, competing unions in a specific geographical scale are also essential to understand union identity. A second relevant stakeholder point of view is that of the employers and managers with whom a union is in direct relations. Through the "for others" perspective, in addition to more details about a union's perceived official characteristics, much more can be added about a union's actual modus operandi in daily employment relations dynamics. Several civil society organizations, especially those promoting labour rights or who have developed relationships with unions, can add to the understanding of union identity and appeal. Last but not least, the points of view of many social media users, especially influencers, about a union can help in understanding "who they are as a union" and pinpoint the appeal of a specific union.

Empirical

While many analytical dimensions along which union identity can be studied have been highlighted, and possible ideal types of union identity have been developed, the debate about union identity lacks thick empirical evidence on which analyses of union identity characteristics and its determinants can be built (Smale 2020).

We foresee three ways forward. First, it is a matter of defining the object of empirical investigation. The debate about union identity and appeal can be enhanced if the empirical investigation moves beyond surveying "unions" as generally defined institutions. The discussion should not simply privilege a "specific union" investigation; it should include studies about the appeal of different types of union identity (e.g. top-down, servicing, participative, or social movement unionism). Importantly, empirical investigation should discern which specific characteristics of each analytical dimension of union identity appeal to different members and societal groups. Second, it is a matter of data collection. Several sources can provide data to empirically study union-projected, official identity; these can complement each other to achieve fine-grained insights into the nature and distinguishing elements of a specific union identity. A union's official constitution, website, newsletters and social media accounts are fundamental sources of observable characteristics of union projected identity (Frangi & Zhang 2022; Smale 2020). A union's internal documents, minutes and union leader narratives can provide important data on the positioning of union identity along the analytical dimensions previously identified (e.g. Gioia *et al.* 2010; Gioia, Corley & Hamilton 2013; Scott & Lane 2000). These sources can be complemented by surveying the internal point of view of officials and members on "what their union is and what it stands for" and the distinguishing elements of their union identity. However, data on actual identity are limited. Narratives about daily, informal micro-interactions about union, employment and societal issues between union stewards and active members, on the one hand, and less active members, non-members and non-workers, on the other hand, can provide important empirical evidence about the actual union identity, the daily "doing union" and the relevance of micro-forces in shaping union identity (Smale 2020). Finally, it is a matter of collecting longitudinal and comparative empirical evidence. The collection of data on union official and actual identity over time can support longitudinal analyses about changes in union identity, gaps and possible tensions between "official, projected" and the "actual, de facto" identity deployed in everyday actions and interactions, and provide insights into the relevance of macro-, meso- and micro-forces in shaping union identity.

CONCLUSION

Ever-changing societal, political and economic conditions will constantly challenge labour unions to adjust their identities and tactics to attract support from members, non-member employees and other social actors. The traditional vision of union identity as composed of uniform interests is increasingly an over-simplification that does not acknowledge the growing diversity of union constituencies and the increasing complexity of union organizational identities. Our multidimensional, multilevel, and multi-actor conceptual framework to evaluate union identity and appeal is a step forward towards developing a systematic analytical approach. We leave it to future industrial relations researchers to validate, expand and deploy it empirically.

REFERENCES

Albert, S. & D. Whetten 1985. "Organizational identity". *Research in Organizational Behavior* 7: 263–95.

Baccaro, L., K. Hamann & L. Turner 2003. "The politics of labour movement revitalization: the need for a revitalized perspective". *European Journal of Industrial Relations* 9(1): 119–33.

Bamberger, P., A. Kluger & R. Suchard 1999. "The antecedents and consequences of union commitment: a meta-analysis". *Academy of Management Journal* 42(3): 304–18.

Behrens, M. & A. Pekarek 2012. "To merge or not to merge? The impact of union merger decisions on workers' representation in Germany". *Industrial Relations Journal* 43(6): 527–47.

Benassi, C. & L. Dorigatti 2020. "Out of sight, out of mind: the challenge of external work arrangements for industrial manufacturing unions in Germany and Italy". *Work, Employment and Society* 34(6): 1027–44.

Blackburn, R. 1967. *Union Character and Social Class: A Study of White-collar Unionism*. London: Batsford.

Brenan, M. 2021. "At 65%, approval of labor unions in U.S. remains high". GALLUP poll report. https://news.gallup.com/poll/318980/approval-labor-unions-remains-high.aspx

Bronfenbrenner, K. (ed.) 2007. *Global Unions: Challenging Transnational Capital Through Cross-Border Campaigns* (No. 13). Ithaca, NY: Cornell University Press.

Brookes, M. 2019. *The New Politics of Transnational Labor: Why Some Alliances Succeed*. Ithaca, NY: Cornell University Press.

Bryson, A. & R. Gomez 2005. "Why have workers stopped joining unions? The rise in never-membership in Britain". *British Journal of Industrial Relations* 43(1): 67–92.

Budd, J. 2004. *Employment with a Human Face*. Ithaca, NY: Cornell University Press.

Carter, B. 2000. "Adoption of the organising model in British trade unions: some evidence from manufacturing, science and finance (MSF)". *Work, Employment and Society* 14(1): 117–36.

Cheney, G. & L. Christensen 2001. "Organizational identity: linkages between internal and external communication". In *The New Handbook of Organizational Communication: Advances in Theory, Research, and Methods*, edited by F. Jablin & L. Putnam, 231–69. London: Sage.

Child, J., R. Loveridge & M. Warner 1973. "Towards an organizational study of trade unions". *Sociology* 7(1): 71–91.

Clark, I. 2009. "Private equity in the UK: job regulation and trade unions". *Journal of Industrial Relations* 51(4): 489–500.

Colgan, F. & S. Ledwith 2002. "Gender and diversity: reshaping union democracy". *Employee Relations* 24(2): 167–89.

Craft, J. & S. Abboushi 1983. "The union image: concept, programs and analysis". *Journal of Labor Research* 4(4): 299–314.

Cregan, C., T. Bartram & P. Stanton 2009. "Union organizing as a mobilizing strategy: the impact of social identity and transformational leadership on the collectivism of union members". *British Journal of Industrial Relations* 47(4): 701–22.

Cregan, C. & S. Johnston 1990. "An industrial relations approach to the free rider problem: young people and trade union membership in the UK". *British Journal of Industrial Relations* 28(1): 84–104.

Croucher, R. & G. Wood 2017. "Union renewal in historical perspective". *Work, Employment and Society* 31(6): 1010–20.

de Turberville, S. 2004. "Does the 'organizing model' represent a credible union renewal strategy?" *Work, Employment and Society* 18(4): 775–94.

de Turberville, S. 2007. "Union organizing: a response to Carter". *Work, Employment and Society* 21(3): 565–76.

Drahokoupil, J. (ed.) 2015. *The Outsourcing Challenge: Organizing Workers Across Fragmented Production Networks*. Brussels: European Trade Union Institute.

Engeman, C. 2015. "Social movement unionism in practice: organizational dimensions of union mobilization in the Los Angeles immigrant rights marches". *Work, Employment and Society* 29(3): 444–61.

Fiorito, J., I. Padavic & P. DeOrtentiis 2015. "Reconsidering union activism and its meaning". *British Journal of Industrial Relations* 53(3): 556–79.

Fiorito, J., I. Padavic & S. Russell 2018. "Pro-social and self-interest motivations for unionism and implications for unions as institutions". In *Advances in Industrial and Labor Relations: Shifts in Workplace Voice, Justice, Negotiation and Conflict Resolution in Contemporary Workplaces*, edited by D. Lewin & P. Gollan, 185–211. Bradford: Emerald.

Flanders, A. 1975. *Management and Unions: Theory and Reform of Industrial Relations*. London: Faber & Faber.

Frangi, L., S. Koos & S. Hadziabdic 2017. "In unions we trust! Analysing confidence in unions across Europe". *British Journal of Industrial Relations* 55(4): 831–58.

Frangi, L., C. Masi & B. Poirier 2022. "From unwoven societal relationships to a broad-based movement? Union power in societal networks in Quebec (Canada)". *Work, Employment and Society* 37(5). https://doi.org/10.1177/09500170221092546

Frangi, L. & T. Zhang 2022. "Global union federations on affiliates' websites: forces shaping unions' global organisational identity". *British Journal of Industrial Relations* 60(2): 444–66.

Frangi, L., T. Zhang & R. Hebdon 2020. "Tweeting and retweeting for fight for $15: unions as dinosaur opinion leaders?" *British Journal of Industrial Relations* 58(2): 301–35.

Freeman, R. 2007. *Do Workers Still Want Unions? More than Ever*. Washington, DC: Economic Policy Institute.

Frege, C. & J. Kelly (ed.) 2013. *Comparative Employment Relations in the Global Economy*. Abingdon: Routledge.

Frege, C., J. Kelly & J. E. Kelly (eds) 2004. *Varieties of Unionism: Strategies for Union Revitalization in a Globalizing Economy*. Oxford: Oxford University Press.

Gahan, P. & A. Pekarek 2013. "Social movement theory, collective action frames and union theory: a critique and extension". *British Journal of Industrial Relations* 51(4): 754–76.

Gall, G. 2003. *Union Organizing: Campaigning for Trade Union Recognition*. Abingdon: Routledge.

Gall, G. & J. Fiorito 2012. "Union commitment and activism in Britain and the United States: searching for synthesis and synergy for renewal". *British Journal of Industrial Relations* 50(2): 189–213.

Gall, G. & J. Fiorito 2016. "Union effectiveness: in search of the Holy Grail". *Economic and Industrial Democracy* 37(1): 189–211.

Gioia, D., K. Corley & A. Hamilton 2013. "Seeking qualitative rigor in inductive research: notes on the Gioia methodology". *Organizational Research Methods* 16(1): 15–31.

Gioia, D. *et al.* 2010. "Forging an identity: an insider-outsider study of processes involved in the formation of organizational identity". *Administrative Science Quarterly* 55(1): 1–46.

Gioia, D., *et al.* 2013. "Organizational identity formation and change". *Academy of Management Annals* 7(1): 123–93.

Givan, R. & A. Hipp 2012. "Public perceptions of union efficacy: a twenty-four country study". *Labor Studies Journal* 37(1): 7–32.

Gomez, R., S. Lipset & N. Meltz 2001. "Frustrated demand for unionisation: the case of the United States and Canada revisited". Centre for Economic Performance Working Paper 492, London School of Economics and Political Science.

Gordon, M. *et al.* 1980. "Commitment to the union: development of a measure and an examination of its correlates". *Journal of Applied Psychology* 65(4): 479–99.

Gumbrell-McCormick, R. 2013. "The international trade union confederation: from two (or more?) identities to one". *British Journal of Industrial Relations* 51(2): 240–63.

Hadziabdic, S. & L. Frangi 2022. "Rationalizing the irrational: making sense of (in)consistency among union members and non-members". *European Journal of Industrial Relations* 28(2): 147–74.

Hatch, M. & M. Schultz 2002. "The dynamics of organizational identity". *Human Relations* 55(8): 989–1018.

Healy, G., H. Bradley & N. Mukherjee 2004. "Individualism and collectivism revisited: a study of black and minority ethnic women". *Industrial Relations Journal* 35(5): 451–66.

Heckscher, C. & J. McCarthy 2014. "Transient solidarities: commitment and collective action in post-industrial societies". *British Journal of Industrial Relations* 52(4): 627–57.

Heery, E. 2005. "Sources of change in trade unions". *Work, Employment and Society* 19(1): 91–106.

Heery, E. & J. Kelly 1994. "Professional, participative and managerial unionism: an interpretation of change in trade unions". *Work, Employment and Society* 8(1): 1–22.

Heery, E. *et al.* 2000. "Organizing unionism comes to the UK". *Employee Relations* 22(1): 38–57.

Hodder, A. & P. Edwards 2015. "The essence of trade unions: understanding identity, ideology and purpose". *Work, Employment and Society* 29(5): 843–54.

Hoffman, J., M. Kahmann & J. Waddington 2007. *A Comparison of the Trade Union Merger Process in Britain and Germany: Joining Forces?* Abingdon: Routledge.

Holgate, J., J. Keles & L. Kumarappan 2012. "Visualizing 'community': an experiment in participatory photography among Kurdish diasporic workers in London". *Sociological Review* 60(2): 312–32.

Holgate, J., M. Simms & M. Tapia 2018. "The limitations of the theory and practice of mobilization in trade union organizing". *Economic and Industrial Democracy* 39(4): 599–616.

Houghton, D. & A. Hodder 2021. "Understanding trade union usage of social media: a case study of the Public and Commercial Services Union on Facebook and Twitter". *New Technology, Work and Employment* 36(2): 219–39.

Hyman, R. 1975. *Industrial Relations: A Marxist Introduction*. London: Macmillan.

Hyman, R. 1994. "Changing trade union identities and strategies". In *New Frontiers in European Industrial Relations*, edited by R. Hyman & A. Ferner, 108–39. Oxford: Blackwell.

Hyman, R. 2001. *Understanding European Trade Unionism: Between Market, Class and Society*. London: Sage.

Hyman, R. 2007. "How can trade unions act strategically?" *Transfer: European Review of Labour and Research* 13(2): 193–210.

Iverson, R. & D. Buttigieg 1997. "Antecedents of union commitment: the impact of union membership differences in vertical dyads and work group relationships". *Human Relations* 50(12): 1485–510.

Jarley, P. & S. Kuruvilla 1994. "American trade unions and public approval: can unions please all of the people all of the time?" *Journal of Labor Research* 15(2): 97–116.

Johnson, N. & P. Jarley 2005. "Unions as social capital: the impact of trade union youth programmes on young workers' political and community engagement". *Transfer: European Review of Labour and Research* 11(4): 605–16.

Johnston, P. 1994. *Success While Others Fail: Social Movement Unionism and the Public Workplace*. Ithaca, NY: ILR Press.

Kelly, C. & J. Kelly 1994. "Who gets involved in collective action? Social psychological determinants of individual participation in trade unions". *Human Relations* 47(1): 63–88.

Kirmanoğlu, H. & C. Başlevent 2012. "Using basic personal values to test theories of union membership". *Socio-Economic Review* 10(4): 683–703.

Kirton, G. 2005. "The influences on women joining and participating in unions". *Industrial Relations Journal* 36(5): 386–401.

Kjærgaard, A., M. Morsing & D. Ravasi 2011. "Mediating identity: a study of media influence on organizational identity construction in a celebrity firm". *Journal of Management Studies* 48(3): 514–43.

Kranendonk, M. & P. De Beer 2016. "What explains the union membership gap between migrants and natives?" *British Journal of Industrial Relations* 54(4): 846–69.

Lévesque, C. & G. Murray 2010. "Understanding union power: resources and capabilities for renewing union capacity". *Transfer: European Review of Labour and Research* 16(3): 333–50.

Lillie, N. & M. Martínez Lucio 2004. "The role of national union approaches". In *Varieties of Unionism: Strategies for Union Revitalization in a Globalizing Economy*, edited by C. Frege J. Kelly, 159–80. Abingdon: Routledge.

Marino, S. 2012. "Trade union inclusion of migrant and ethnic minority workers: comparing Italy and the Netherlands". *European Journal of Industrial Relations* 18(1): 5–20.

Martínez Lucio, M. & R. Perrett 2009. "Meanings and dilemmas in community unionism: trade union community initiatives and black and minority ethnic groups in the UK". *Work, Employment and Society* 23(4): 693–710.

Martínez Lucio, M. & M. Stuart 2009. "Organising and union modernisation: narratives of renewal in Britain". In *Union Revitalisation in Advanced Economies*, edited by G. Gall, 17–37. Basingstoke: Palgrave Macmillan.

Martínez Lucio, M., S. Walker & P. Trevorrow 2009. "Making networks and (re)making trade union bureaucracy: a European-wide case study of trade union engagement with the Internet and networking". *New Technology, Work and Employment* 24(2): 115–30.

McAlevey, J. 2016. *No Shortcuts: Organizing for Power in the New Gilded Age*. Oxford: Oxford University Press.

Michels, R. 1915/2001. *Political Parties: A Sociological Study of the Oligarchical Tendencies of Modern Democracy*. Original 1911 in German: *Zur Soziologie des Parteiwesens in der modernen Demokratie; Untersuchungen über die oligarchischen Tendensen des Gruppenlebens*. Trans. Eden Paul & Cedar Paul 1915. Kitchener, ON: Batoche Books.

Murray, G. 2017. "Union renewal: what can we learn from three decades of research?" *Transfer: European Review of Labour and Research* 23(1): 9–29.

Oliver, R. & A. Morelock 2020. "Insider and outsider support for unions across advanced industrial democracies: Paradoxes of solidarity". *European Journal of Industrial Relations* 27(2): 167–83.

Peetz, D. & B. Pocock 2009. "An analysis of workplace representatives, union power and democracy in Australia". *British Journal of Industrial Relations* 47(4): 623–52.

Ravasi, D. & M. Schultz 2006. "Responding to organizational identity threats: exploring the role of organizational culture". *Academy of Management Journal* 49(3): 433–58.

Redman, T. & E. Snape 2014. "The antecedents of union commitment and participation: evaluating moderation effects across unions". *Industrial Relations Journal* 45(6): 486–506.

Riley, N. 1997. "Determinants of union membership: a review". *Labour* 11(2): 265–301.

Ross, G. & A. Martin 1999. "European unions face the millennium". In *The Brave New World of European Labor*, edited by A. Martin & G. Ross, 1–25. New York: Berghahn.

Scheuer, S. 2011. "Union membership variation in Europe: a ten-country comparative analysis". *European Journal of Industrial Relations* 17(1): 57–73.

Scott, S. & V. Lane 2000. "A stakeholder approach to organizational identity". *Academy of Management Review* 25(1): 43–62.

Silverblatt, R. & R. Amann 1991. "Race, ethnicity, union attitudes, and voting predilections". *Industrial Relations: A Journal of Economy and Society* 30(2): 271–85.

Simms, M. 2012. "Imagined solidarities: where is class in union organising?" *Capital and Class* 36(1): 97–115.

Smale, B. 2020. *Exploring Trade Union Identities: Union Identity, Niche Identity and the Problem of Organising the Unorganised*. Bristol: Bristol University Press.

Streeck, W. & A. Hassel 2003. "Trade unions as political actors". In *International Handbook of Trade Unions*, edited by J. Addison & C. Schnabel, 335–65. Cheltenham: Edward Elgar.

Sullivan, R. 2010. "Labour market or labour movement? The union density bias as barrier to labour renewal". *Work, Employment and Society* 24(1): 145–56.

Tapia, M. & G. Alberti 2018. "Social movement unionism: a toolkit of tactics or a strategic orientation? A critical assessment in the field of migrant workers campaigns". In *Social Movements and Organized Labour*, edited by J. Grote & C. Wagemann, 109–27. Abingdon: Routledge.

Tapia, M., T. Lee & M. Filipovitch 2017. "Supra-union and intersectional organizing: an examination of two prominent cases in the low-wage US restaurant industry". *Journal of Industrial Relations* 59(4): 487–509.

Tattersall, A. 2005. "There is power in coalition: a framework for assessing how and when union-community coalitions are effective and enhance union power". *Labour and Industry: A Journal of the Social and Economic Relations of Work* 16(2): 97–112.

Tilly, C. 2006. *Regimes and Repertoires*. Chicago, IL: University of Chicago Press.

Turner, H. 1962. *Trade Union Growth Structure and Policy: A Comparative Study of Cotton Unions in England*. Toronto: University of Toronto Press.

Turner, L. & R. Hurd 2001. "Building social movement unionism". In *Rekindling the Movement: Labor's Quest for Relevance in the 21st Century*, edited by L. Turner, H. Katz & R. Hurd, 9–26. Ithaca, NY: ILR Press.

Visser, J. 2012. "The rise and fall of industrial unionism". *Transfer: European Review of Labour and Research* 18(2): 129–41.

Waddington, J. & A. Kerr 2002. "Unions fit for young workers?" *Industrial Relations Journal* 33(4): 298–315.

Waddington, J. & C. Whitston 1997. "Why do people join unions in a period of membership decline?" *British Journal of Industrial Relations* 35(4): 515–46.

Whetten, D. & A. Mackey 2002. "A social actor conception of organizational identity and its implications for the study of organizational reputation". *Business and Society* 41(4): 393–414.

Youngblood, S. *et al.* 1984. "The impact of work environment, instrumentality beliefs, perceived labor union image, and subjective norms on union voting intentions". *Academy of Management Journal* 27(3): 576–90.

Zellweger, T. *et al.* 2013. "Why do family firms strive for nonfinancial goals? An organizational identity perspective". *Entrepreneurship Theory and Practice* 37(2): 229–48.

CHAPTER 2

UNION INTERESTS AND IDEOLOGIES

Ronaldo Munck

ABSTRACT

The changing composition of labour at a global level is posing both challenges and opportunities for labour unions. The doubling in numbers of workers worldwide since the advent of globalization creates the conditions for a revitalization of the labour movement despite the challenges posed. This chapter asks the following: How has this global working class organized or not organized when we think of both ends of the spectrum, the informal and the high-tech sectors? How have labour unions – and other non-traditional forms of labour representation – risen to the challenges posed by neoliberal globalization? Based upon an understanding of the identity and appeal of labour unionism, this chapter examines what labour unions try to achieve, and what frameworks govern their strategic choices. It is then the ideologies of labour unionism that determine how the various interests of labour – economic, political and social – are articulated and pursued on behalf of their members.

Keywords: Union interests; union ideology; analytical framework; challenges for unions

INTRODUCTION

Now that we understand better the identity and appeal of labour unionism (see Chapter 1), we can seek to uncover what it is that labour unions then try to achieve, and what frameworks govern their strategic choices. It is then the ideologies – as set of comprehensive of values and beliefs – of labour unionism that determine how the various interests – as set of concerns and desires – of labour – economic, political and social – are articulated and pursued on behalf of their members. That is the purpose of this chapter, building upon the findings of Chapter 1, which developed a multidimensional, multilevel, and multi-actor conceptual framework to evaluate union identity.

We need to bear in mind when pursuing this analysis that the composition of the labour force is complex and involves horizontal and vertical divisions and pre-existing solidarities.

There are many key questions that need to be asked. Prime among them is: How does labour unionism create a unified set of interests and an ideology that can carry it forward to achieve its aims and objectives? Historical precedents matter here, and the formative phase of the industrial working class is always relevant, but we must be prepared to rethink labour, its organizations and ideologies if labour unionism is to have a future.

To answer the question posed requires starting from the point of the changing composition of labour at a global level, the massive quantitative expansion of labour since the year 2000 and the emergence of a global working class. At present, there are only approximately 10 per cent of those workers organized through labour unions. Many observers have, thus, concluded that labour unions as we have known them, in terms of articulating the common collective interests of labour, are defunct or no longer fit for purpose in the twenty-first century. What is proposed in this chapter is to explore the prospects for a revitalization of labour unions and a reigniting of the labour movement through a broad comparative survey that is not reduced to the very specific North Atlantic region and its very particular history of unions, parties and politics that should not be taken as a universal model. The existence of a global working class calls for a global approach and not a Eurocentric or otherwise limited view.

The accelerated development of capitalism on a global scale since around 1990 has generalized the internationalization of the capital/wage labour relations through the real and not just the formal subsumption (or subordination) of labour in the state socialist and national developmental state regimes. That has created a massive new global working class. So, it needs to be now asked: How has it organized or not organized when we think of both ends of the spectrum, the informal and the high-tech sectors? And how have labour unions – and other non-traditional forms of labour representation – risen to the challenges posed by neoliberal globalization?

To develop the analysis, a basic framework of economic (i.e. market-oriented), political (state-oriented) and social (society or class-oriented) unionism is proposed, which needs to be operationalized through a comparative perspective (see a similar model in Hyman 2001). This is not presented as yet another model or tired ideal-type typology, but as an active frame to establish the parameters of both the challenges and possibilities for labour unionism in the twenty-first century. The analysis will proceed through a series of interlocked steps from an examination of the essence of labour unions, followed by the context within which labour unionism is set, through to the crisis of labour unionism towards the end of the twentieth century. It then moves on to the recent debates concerning the revitalization of labour unionism, through the organizing approach, for example, to then focus in some detail upon the various strategies of labour unionism prevailing in different parts of the world, and the vital spatial dimension in a section called "spaces of unionism" that adds a much needed degree of complexity to the analysis.

ESSENCE OF LABOUR UNIONS

The classic definition of trade (or labour) unions by the Webbs in 1894 was that of "a continuous association of wage earners for the purpose of maintaining or improving the conditions of their employment" (Webb & Webb 1920: 3). These wage earners are represented by labour unions in the format that they take in the capitalist economy. They are, in that sense, a passive reflection of the existing order and not, intrinsically, a challenge to it. As Anderson (1967: 265) once put it: "Trade unions are essentially a *de facto* representation of the working class at its workplace. Formally, they are voluntary organisations, but in actual practice they are much more like institutional reflections of their environments." The labour union is, thus, in essence, a mirror of the existing social order, and its maximum weapon an organized absence of work, namely the strike. Within the Marxist tradition, this perspective still exists as some kind of common sense, seen as obvious, even if a "class" or socialist role for unionism is advocated and called for, with unions seen as "schools for socialism".

The Italian autonomist tradition also presented labour unions in a negative light, as it were, being seen as a central element in the capitalist regulation and organization of workers. For Tronti (2019: 109), "the very organisations of workers acquire a decisive importance for the real interests of capital. There is a time in which modern capital cannot do without a modern union, in the factory, in society, and directly in the state." The union brings coherence to the underlying anarchy of class under capitalism. Far from being autonomous from capital, for Tronti (2019: 117) unions actually represent "the most perfect form of integration of the working class within capitalism". This vision more or less encapsulates the pessimistic view of labour unions, even though it does leave space for the autonomy of workers' imagination, organization and action against capital, a contradiction that will be returned to below.

A different perspective could be gained by rejecting the commodity character of labour as Polanyi does. Against some Marxist readings, Polanyi (2001: 185) seeks to "safeguard the human character of the alleged commodity, labour". For Polanyi:

> To argue that ... trade unions have not interfered with the mobility of labour and the flexibility of wages, as is sometimes done, is to imply that [they] have entirely failed in their purpose, which was exactly that of interfering with the law of supply and demand in respect of human labour, and removing it from the orbit of the market. (Polanyi 2001: 185)

That role is, indeed, part of the essence of labour unionism today: to remove labour from the laws of the market and capitalist competition. This logic of social protection rejects the employers' demands for labour mobility and wage flexibility in a system where labour is but a commodity. Labour unionism can, thus, follow a logic of "decommodification" and not just passively reflect the contours of the capitalist order. This approach would bring labour struggles closer to the many varied forms of struggle against dispossession typical of neoliberal globalization.

This non-structuralist understanding of the essence of labour unionism needs to be concretized by setting its interests and ideologies in their complex socio-spatial context. As Hyman notes, unions "can help shape workers' own definitions of their individual and collective interests" (Hyman 1994: 122). Union members are not passive recipients of the labour organization's understanding of their interests. Rather, they influence them and their identities and ideologies for social transformation. Unions are shaped by their history, for sure, but mainly by the complex interaction with the market, society and the state, and the feedback loops these create. Finally, unions have an internal life as well as an external orientation. Thus, the degree of internal democracy, engagement and debate plays a vital role in establishing the interests of unions and forging their strategies (see Chapter 1). It is also a key area where a conscious strategy of revitalization can play an effective role in changing the way unionism is practised.

Taking a broad theoretical perspective, Hodder and Edwards (2015) have posited a framework of union essence that provides some key concepts to be unravelled in the pages that follow, albeit it in a slightly different language. For Hodder and Edwards (2015: 848), the essence of unionism evolves as follows:

> Identity embraces interests and causal powers at a fundamental level ... Identity and the degree of market or class orientation then affect ideology ... The outcome of the interactions of society with market, class and ideology generates the empirical basis of a union. These items establish its purposes and overall objectives ... Finally, strategies generate outcomes.

However, we cannot assume class as a category in this way; that is, as a given rather than something that must be constructed, but the logic overall does provide the dynamic parameters for the study of labour unions that sets in motion identities, interests and ideologies that interact with the market and society in a complex way to generate the objectives and strategies of a union.

CONTEXT OF LABOUR UNIONISM

As Chapter 1 argues, the identity of unions is shaped, in the first instance, by the macro-institutional forces that sets their context. Interests are also shaped within this context and cannot be taken as a given or read off from occupational categories. What this section does is set the parameters that labour unions then respond to through diverse meso-level strategies (also mentioned in Chapter 1).

To paraphrase, Marx once said that workers make their own history but not under conditions of their own choosing. At the end of the twentieth century, international labour unions were faced with a tragic paradox. There was a bigger workforce than ever before – embracing three billion wage workers worldwide – but rarely had labour unionism been that weak – organizing less than 10 per cent of those workers. Neoliberal globalization implied the simultaneous shift of the locus of global production to the Global South, the undermining of traditional job security and union rights, and the decline or disappearance of labour's traditional communist and social democratic political allies. Many analysts saw this as the death knell of the labour movement, but from 2000 onwards there was growing recognition in union circles that it was time to change its defensive attitudes. Danish unions, for example, declared that "the time has come for ... unions to use the positive side of globalisation [such as the communications revolution] to the advantage of workers and poor people all over the world" (SID 1997: 23).

In the first wave of globalization in the 1990s, it was, however, capital that held the upper hand. The number of workers doubled in that decade as the workers of China, India and the ex-Soviet bloc were brought under the sway of the capital/wage-labour relation directly. This world historic shift in terms of the size and composition of the global labour force had a massive impact upon the labour movement. This new model capitalism, as it moved into the 2000s and expanded worldwide, was based largely on the expansion of the informal sector, and not on what the Northern-centric United Nations' body, the International Labour Organisation (ILO), called "decent jobs". The new transnational corporate elite was told that "just as managers speak of world markets for products, technology and capital, they must now think in terms of a world market for labour" (Johnston 1991: 115).

As a result of this massive expansion of the labour market, the ratio of capital to labour was reduced decisively, compared, for example, to the situation that prevailed in the North Atlantic region from the 1950s to the mid-1970s. The expansion of labour that followed what Freeman (2001: 1) called the "great doubling" had an obvious impact on labour unions. In the Global North, the greatly reduced ratio of capital to labour shifted the global balance of power to capital, and in the Global South it accelerated economic growth and reduced absolute poverty in some countries (e.g. China) but most of the new jobs were in the informal or precarious sector.

It would, however, be wrong to interpret this dynamic/destructive period of capitalist development as purely unilinear in terms of its impact upon labour. The notion of a "race to the bottom" was for a time popular in union circles, with its image of an inexorable downward spiral of terms and conditions based on intensified competition between workers worldwide. As Silver (2003: 48) points out, this "is a one-sided focus that does not recognize the continuous unmaking but also remaking of the working classes on a world-scale". So, while in the Global North labour unions faced a dramatic "unmaking" of the working classes they have been based upon, in the Global South we have witnessed a dramatic "remaking" of a new working class that closely mirrors Marx's analysis in *Das Kapital*. Capitalism always had a creative/destructive aspect. Thus, Schumpeter (1942: 2) describes creative destruction as the "process of industrial mutation that incessantly revolutionizes the economic structure from within, incessantly destroying the old one, incessantly creating a new one". And, as seen in the case of China as a leading example, capitalist development always creates new working classes and generates new capital/wage labour conflicts. That process is, indeed, inexorable.

There is, however, a general consensus that this new global context accelerated the secular decline of labour organizing practically everywhere. Thus, in analysing "the crisis of world labour", Van der Linden (2015: 1) declares bluntly that "on a global scale union density is almost insignificant" bar a few exceptions such as Canada and Norway. The cause, for Van der Linden (2015: 1) is simple: "Old-style trade-unionism … can no longer cope with the challenges offered by the contemporary world." If by this we mean the model of labour unions and labour/social democratic parties, we can agree, and no one doubts the challenges for global labour are manifestly real. Where a more nuanced picture needs to be painted is in terms of the general validity of the 10 per cent union density figure frequently mentioned, and the writing off of the Chinese labour movement because China's union federation is controlled by the state. We must also note that union power cannot be reduced to union density as though this was the only determinant, devoid of any context.

According to ILO data, of the three billion people currently in employment, 516 million, or 17 per cent are union members (Visser 2019). However, if the self-employed and family workers are excluded, there is a global union density of 27 per cent. Union membership in the advanced industrial societies has, indeed, declined but in the rest of the world it has risen since 2000 – and especially since 2008 in the aftermath of the great financial crisis. Indeed, since 2008 there has been a "spectacular" union growth with "significant" membership gains in South America and South East Asia in particular. Conversely, in Western Europe the steady decline continued and in post-communist Eastern Europe this trend accelerated. In North America, the unions in the US lost members, while those in Canada gained them.

This picture is very uneven and cannot just be reduced to one of inevitable decline, nor can figures be taken in isolation without exploring the growing organization (and not always through labour unions) of the informal sector workers, for example. While the analysis that labour unionism is at a crossroads should be dismissed, it is important to go beyond headline statistics to get at the texture of union organizing worldwide. Likewise, one should not read off one country or region's experience and assume it is universal. The uneven and combined development of capitalism finds its reflection in a similar complex pattern for labour unions and there is always the need to foreground the role played by human agency and recognize that labour unions are not necessarily passive subjects.

THE CRISIS OF LABOUR UNIONISM

The crisis of labour unionism impacts upon both the "official and actual" (Chapter 1) or "formal versus real" forms of identity. This multifaceted crisis redefines the "interests" of labour and questions the effectiveness of pre-existing ideologies of labour, be they syndicalist, socialist, communist or nationalist in form. Identity does not simply reflect interests but is shaped discursively in the crisis period and in its resolution. Crisis can be read as a positive normal, as it were, and not the sign of irreversible decline. Ideologies are not set in stone for they develop, and agency is always to the fore in these periods.

The recomposition of capitalism after the crisis of the 1970s, driven by the neo-liberal offensive of the 1980s and 1990s, led to a new mode of capital accumulation and regulation, commonly known as globalization. One of the early analysts of this new informational digitalized global capitalism, Castells (1996: 475) noted its impact on labour and labour unions: "Labor is disaggregated in its performance, fragmented in its organisation, diversified in its existence divided in its collective action." Labour, thus, loses its collective identity and labour unions lose their capacity to organize this more diverse workforce. Capital and labour now live in different worlds, as it were. One is global and the other painfully local, so they are playing by different rules. Labour unions are, thus, seen as prisoners of an earlier mode of capitalism that no longer exists – they are tilting at windmills. For Castells, and for many others, labour unions, and the labour movement more broadly, were seen as a fading force in the mid-1990s – they might retain corporate power but they could never again offer an alternative to the established order.

Labour unions have always been a specific, institutionalized form of representation of workers interests and the forging of collective identities and bargaining. The changing nature of capitalism in the 1990s led to a virtual collapse of collective identity and, thus, created a crisis of representation for the unions. As work became more disaggregated and individualized so, in response, many unions shifted from collective goals to those of an individual service provider. Certainly, solidarity does not spring automatically from cooperative organizational practices, but must be continually reconstructed. Labour unions should no longer be seen as semi-automatic or simple reflective representations of a homogeneous social group. More broadly, *contra* Castells, Catalano (1999: 34) argues that "unions in spite of their integration within system regulations, are a bearer of multiple contradictions and ambiguities, even in a context of deep crisis and the reconstruction of labour relations". Although their representativity has been seriously compromised, or de-centred, they continue to reflect the basic contradiction between capital and labour, with its increasing power asymmetries.

The challenge of representativity posed for the labour unions was most acute in the vast and growing informal sector of the Global South. The mythical "standard employment contract" had, in fact, never prevailed here, and now between two-thirds to three-quarters of the total labour force falls outside state regulation. Nevertheless, there have been many efforts, since 2000 in particular, to organize informal workers, street traders, the newly unemployed and, sometimes, migrant workers. The impact of informalization/precarity has been such that unions have needed to adapt or perish. Many unions have worked successfully to "organize the unorganized", sometimes in partnership with non-governmental organizations and these workers' own organizations, to influence government policy and improve social security. The smartest unions today in many areas of the Global South are making common cause with migrant and precarious workers in general (Munck 2019). This has helped combat racist and divisive political forces and has reinforced the democratic credentials of the union movement as it counters of the divide-and-rule strategy of managements.

Unions were also placed on the defensive in the Global North when confronted with the rise of the "gig economy" and the challenge of organizing platform workers (see Chapters 7 and 21). The obstacles in the way of unionizing workers who did not have a set place of work seemed daunting. It seemed for a while that the shift towards a northern version of precarity would finally finish off labour unions. Yet in the 2000s several successful efforts by unions to organize that sector were undertaken. Significant victories have been won through impacting public opinion through social media and through the courts; for example, the ruling that Uber workers were not "self-employed". Sometimes deals were done by unions with employers in this sector; for example, with delivery companies in the UK. As one report put it: "The threat of the gig economy has given [labour union] new reason to invest in their old strengths: organising workers on the ground and sitting across the table to thrash out a deal" (*FT* 26 February 2019).

What is now clear from even a cursory examination of the crisis of representation facing labour unions after 2000, is that "solidarity" is not a long-lost original condition that can simply be regained. When it has emerged, it has been, as Richards (2001) notes, a constructed and contingent phenomenon rather than a structurally predetermined given. Labour unionism has, in fact, always excluded sections of workers as much as it has included its "core" constituency. The unionized versus non-unionized division among workers has become acute and thus the collective role of unions is brought into question. As Richards (2001: 56) puts it: "Contemporary trade unionism is beset by an acute crisis of representation, rooted in a division between those whom the unions have, historically, sought and claimed to represent, and those whom they now actually represent." To address this crisis of representation many analysts and activists have called for a "revitalization" of the union movement, a topic that is now turned to.

REVITALIZING LABOUR UNIONISM

The debate around strategies for union revitalization (see Frege & Kelly 2003) has provided a potential alternative to the structural pessimism of those who read off from the changes in capitalism in the 1990s an inevitable period of decline; indeed, probably terminal for labour unions. These strategies are not a panacea to stop the decline of labour unions, but they do open up a new period through political action, coalition building, union reform, greater internationalism and, above all, a concerted organizing drive. There are, overall, two ideological strands in the revitalization debate: those who couch it in terms of a renewed "partnership" strategy and those for whom building union power through organizing is part of a radical reform strategy. In one review of the revitalization debate, Ibsen and Tapia (2017: 58) argue that "union revival only has a chance if the state accepts that union involvement in employment relations provides positive "externalities" for society and is willing to intervene in favour of unions". With that proviso in mind, the implications of the revitalization debate can now be reviewed for the purpose of this chapter – which is to critically deconstruct the notions of the interests and ideologies of labour unionism.

Examination is needed of the ways in which labour unions can change, be revitalized and even reinvent themselves. Labour is still a social force to be reckoned with, and labour unions can transform their repertoire of strategies and tactics. While a return to "business unionism" and "partnership" is the logical outcome of a servicing model of unionism, the organizing perspective, while not offering a panacea (and unions are often resistant to change) has opened up a possible way out of terminal decline. It is distinct from the partnership model in that it seeks a transformation of unions themselves. It is different in its means but also in its ends. While recognizing its advance in terms of revitalizing unions, albeit to different degrees, Carter (2006: 416) correctly adds a notion of caution that "the organizing approach leaves unanswered a series of questions about the strategic direction of the labour movement, including the constituencies of unions and how issues of globalization and international solidarity are addressed". It is curiously apolitical in a way, insofar as it seems to bracket some of the bigger political questions around labour unions and their identities.

What has become clear since at least 2000 is that "business as usual" is not a viable strategy for labour unions or the renewal of them. To survive, let alone prosper, labour unions would need to adopt a social movement orientation (see Turner & Hurd 2001).

If we take a long-term view of the labour movement, we can see that its revitalization has in the past always occurred during periods of popular mobilization and protest and in coalition with other social movements. Institutional decline and decay usually follows in periods of "normal" industrial relations. Friedman (2008: 35) has called this the "supreme irony of collective-bargaining unionism" in that the way unions grow and broaden their influence "undermines future union growth by deflating militancy and separating workers from their own organizations". In other words, the business or "bread-and-butter" unionism will never be able to reinstall the "fire in the belly" of labour unions that would allow them to play the role of "sword of justice" once again.

Revitalizing labour unionism may well involve a redefinition of what its interests are and what ideologies may serve it best. While labour unions exist to represent their members' interests, it is not clear how these are interpreted. Is it possible to create new interests? Can the notion of "solidarity" take on new life as a discursive space shared by the labour and "new" social movements? Clawson (2003: 3) argues for the US case that to create a new upsurge, labour "must fuse with social movements concerned with race, gender and global justice". The "democratic equivalent" (or common denominator) might be that of a radical democracy that could act at once as a framing paradigm for the labour movement and the new social movements, despite their different constituencies and methods. Its tools for social transformation would include the strike, the sit-down and the riot but also the new online and transnational means to build democracy and solidarity on a global scale.

Anyone with a passing familiarity with labour unions will have heard of the slogan "An injury to one is an injury to all" and the call of "Solidarity forever". These can be empty and quite stale slogans without a practice of solidarity that can help create a new collective identity. As Richards (2001: 58) notes: "Traditional appeals to class solidarity alone will not resolve the crisis [of labour unionism]." Thus, the starting point must be recognition that unions have divided as much as they have united the various fragments or fractions of the working class. There have been persistent and deep divides along the lines of gender and ethnic identity, Global North/South divisions and even the old slogan of "last in first out" is clearly divisive. A revitalized unionism for the twenty-first century will, of necessity, need to be more inclusive, explicitly addressing the intersectionality of social divisions and forging a new form of syndicalism as an effective labour union ideology that empowers and inspires.

One gain from the revitalization debate is that we can no longer limit ourselves to the simple index of union density as a measure of labour union vitality. While it may appear as a simple, quite technical term, the notion of "union density" across time and space must be deconstructed. Most often this apparently simple indicator is used to show that unions are in terminal decline, at least in the advanced industrial societies since the onset of neoliberal policies in the 1980s. Yet union density is not the only factor affecting the bargaining power of a union. Employer–state negotiations are equally important, not to mention the political power of the unions. Kelly (2003) goes further in decoupling union density from the prospects for union revitalization, arguing on the basis of complex data that "despite the problems they have faced in recent years and their denigration as a merely sectional interest group, unions remain a powerful force both for egalitarianism and for democracy". In shaping their role in society, unions may reshape their identity and goals, fundamentally altering their economic, political and social roles and influencing society as a consequence. In brief, the revitalization debate has reintroduced the element of human agency into the analysis of labour unions in the current era.

STRATEGIES OF LABOUR UNIONISM

To situate the debate around the possible strategies for labour unionism today, we need to start from an understanding of how social life is constructed through both social and spatial relations well captured in the concept of a "socio-spatial dialectic" in which both aspects are mutually constitutive of each other. Labour studies, labour history and industrial relations theory have all operated in a more or less aspatial world in the past. Globalization accentuated this tendency with its triumphalist proclamation of the "death of distance" due to time-space compression. In this new "borderless world", it seemed that geography was now irrelevant, and capital and labour operated and existed in quite different spheres: the first was hyper-mobile and fluid, the second was fixed firmly (at least on the whole) in places, and was also "sticky"; that is, embedded in traditional social relations. In practice, the mobility of one (footloose and fancy-free capital) and the immobility of the other (this was the "era of migration", after all) were exaggerated, but the conceptual problem was a deeper one.

The concept of a socio-spatial dialectic (see Soja 1980) that sets the terms of labour union possibilities is then required. One force field that we can detect is that of market/society/state triad that sets the social parameters for labour strategies, but it is overlaid by a spatial force field that can be simplified here to the local, national and global domains. The labour unions sit in the middle and are pulled various ways at different times, sometimes in contradictory ways. This may help us envisage how labour strategies emerge and have relevance.

The rough double triangle diagram shown in Figure 2.1 can only assist us if it is considered dynamically, with all its elements interacting. Thus, in terms of the politics of scale, workers and their organizations think and operate at local, national, regional and global scales concurrently. Unions may also make alliances – tactical to strategic – with community organizations and even with political organizations when their interests align. Global unions may have a strong alliance with official organizations such as the ILO and the European Commission, while also at the same time be forging alliances with grassroots coalitions in the Global South. Clearly, unions – like any other body – do not simply choose between a market and a society orientation, rather they interact with a range of hybrid forms that are pulled one way or the other at different times and in different places. This is a non-linear diamond with organizations constantly adapting to changing situations. Just by way of example, Evans (2014: 1) argues that "assessing the connections among national labor movements and the new global organizational infrastructure that has emerged under neoliberalism is a necessary foundation for building theories of labor's evolving contestation with global capital". But the local and regional dimensions are equally important to this emerging complex scenario.

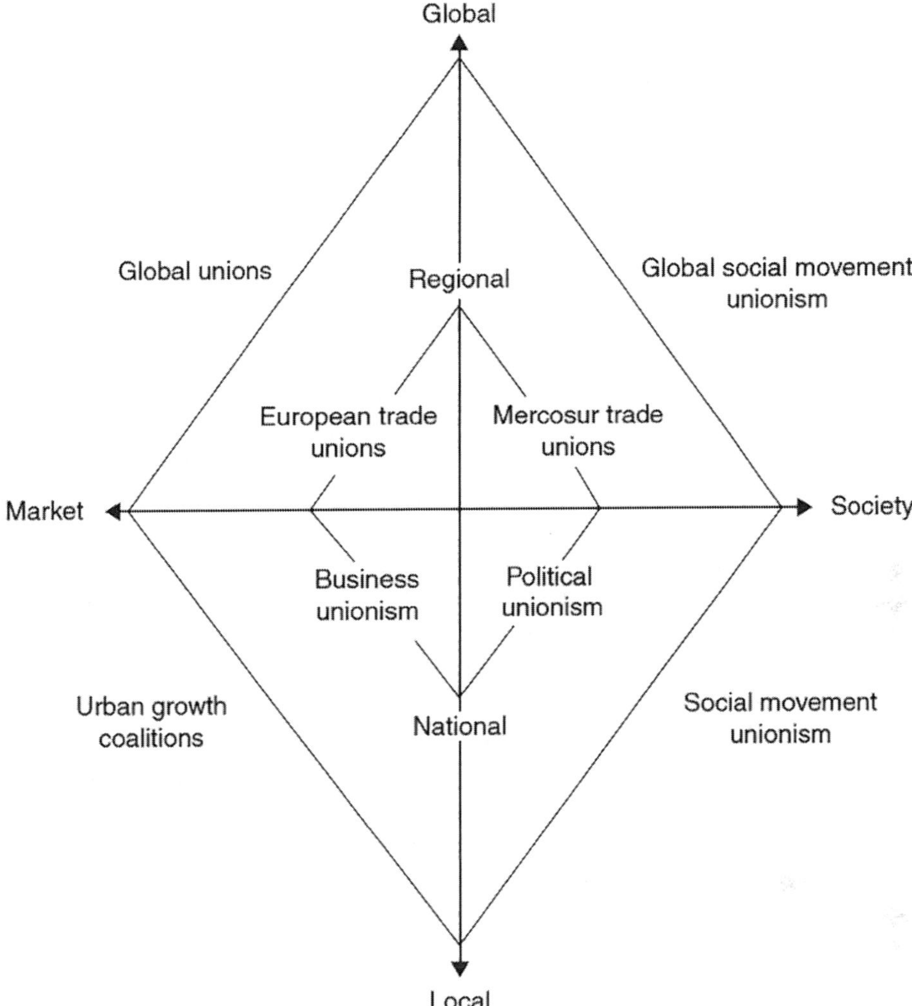

Figure 2.1 The socio-spatial force fields impacting on labour unions
Source: Munck (2018: 228).

The market/society/state force field also needs to be considered in a dynamic and interactive way and not just a static model. Thus, labour movements can be empowered, notes Lawrence (2014), when they are able to "translate" their gains from one arena of struggle to another. Thus, expanded membership for labour unions can "translate" from workplace power into economic power. Translation can also operate through a scale shift from local to global or vice-versa. This addition to the common state/society/market or the alternative state/market/class schema is most useful and adds a much needed complexity to our analysis. The spatial dimension is dealt with below but for now the main strategies for labour unionism that spring from the market/state/society model are explained.

There are three main labour union strategies arising from their orientation primarily to the market, the state or society. Economic unionism faces towards the market and takes labour as a commodity that needs to seek its best economic value; political unionism orients towards the state and bargains as a political actor; and social (or social movement) unionism sees labour as primarily a social actor and looks to the broader society for support.

The "economic unionism" strand, strategy or ideology is best exemplified by "business unionism" in the US, developed in the 1890s as the American Federation of Labour moved from a class struggle perspective to an accommodation with capital. Union leader Samuel Gompers, in particular, began to preach the benefits for workers of an accommodation with capital. Workers could aspire to an "American" standard of living that included a decent home, food and clothing and education for their children. Based on his UK and US experience Gompers prioritized "collective bargaining" with the employers in pursuit of higher wages, better conditions and shorter hours. This model worked for a whole period in the North Atlantic region. As Aglietta (1998: 58) notes: "Labour standards linked to full-employment equilibria, determined themselves by multi-year collective bargaining, were the economic lynchpins of wage society during the Fordist era." By contrast, in the post-Fordist era economic integration has forced companies to compete on labour costs to meet their profitability objectives.

The trajectory of US "business unionism" under Gompers and beyond illustrates the findings of Western (1997), who, in a wide- ranging study of labour unions "between class and market", finds that labour unions thrive best when they are insulated through political support from market forces. This was also why Polanyi advised unions in the 1930s to seek their removal from the influence of market forces. However, the situation in which collective bargaining could take wages out of competition were circumscribed both historically and geographically. As a universally applicable model, this does not translate well to the Global South nor can it serve as a benchmark through the notion of a "standard employment contract". Nevertheless, most unions still practise an element of "economic unionism" when they negotiate with employers around wages and conditions in the workplace. As such, it is still part of the ideology and practice of most unions.

Then, the "political unionism" referred to above finds a typical expression in Latin American labour unionism in the 1950s and onwards (see Payne 1965). In a situation of early industrialization, workers may be in a weak position vis-à-vis employers but these in turn may be in a weak position in relation to the state. This allows for a process of "political bargaining" where workers can make gains through direct action that threatens political stability. This settled down into a form of political unionism in which nationalist/populist parties established a strongly corporatist model of industrial relations. In recent decades, the hold of corporatism has weakened – for example in Brazil – with the development of more independent unions based on considerable rank-and-file activation. This created, for a while, the possibility of a class struggle unionism, although that path was eventually closed off in most countries.

Today, political unionism can take many forms with varying degrees of corporatism or state-ness and backing of nationalist, social democratic or other political parties. Overall, as Flew (1989: 79) puts it, based on the experience of Australia, "It is seen as crucial that the labour movement shifts the arena of class conflict from the industrial sphere, engaging on democratic class struggle." The economic dominance of capital can, thus, be counterbalanced by translating the political strength of the labour movement into concrete gains. In some formulations, there is an expectation that democracy can be expanded from the political sphere to the economic sphere and the workplace. Now, while most unions engage in some degree of political bargaining – even among those most committed to economic unionism – there is still little evidence that this strategy has fundamentally assisted the revitalization of the labour movement, only being able to manage decline, as it were.

As to "social unionism", one case that has become paradigmatic is that of South Africa during the struggle against apartheid (see Chapter 5). The development of independent unions in the 1980s soon led to these taking up issues beyond the workplace and also forging alliances with township organizations struggling against the regime and its impacts upon society. In Brazil during the same period of time, the struggle against the dictatorship that came to power in 1964 led to a flourishing of independent unions that broke with corporatism and became part of a broader movement for democracy. The community, and at times the city, became a space of solidarity with the wider struggles of workers often fusing the demands of production and reproduction. Day-to-day issues around running water and health care, for example, became union issues. And broad circles of the community, often including religious organizations and communities, became firm supporters of the workers' cause. Another such example can be found in South Korea based upon resistance to the authoritarian regime (see Minns 2001).

Social movement unionism then took off internationally with some advocating it as a new form of unionism that could be developed globally and displace the failed economic and political models of unionism (Waterman 1993). This led to the concept becoming both "stretched" to cover too many situations and codified as a set of principles that had to be satisfied in order to qualify. Von Holdt (2002), from a South African perspective, stressed its very particular roots and also how, in practice, the solidarity it had generated fractured along old and new lines. Social unionism as advanced here means simply a shift away from the market to civil society. It has, albeit translated into the term "community unionism", travelled to the Northern Hemisphere in the late 1990s. In this form it has generated a very rich seam of strategic thinking that recovers some of the formative experiences of the labour movement in the community and how this strategy has effectively been deployed by unions and their allies; for example, in organizing migrant workers (see Munck 2011).

Of course, these three broad strategic paths for labour unions are not mutually exclusive, given the complexity of the tasks facing labour unions today. Lévesque and Murray (2010: 335) highlight how "the multiplication of social identities and employment status, particularly as regards the explosion of both precarious and migrant workers, poses a huge challenge of melding collective identities and deliberative mechanisms to yield internal solidarity". Emphasis must be, therefore, on the contingent nature of union members' interests because they are always fluid and constructed. The capacity of the labour unionism to act as a progressive intermediary and seek to forge a new hegemonic set of interest is, of course, open to question. Union organizers often have as a default position the conservative nostrum that their job is "to represent the members I have and not the members I would like to have".

Nevertheless, taking a broad view of the way labour unions are mobilizing their power resources to confront the challenges of globalized capital, we can see "new windows of opportunity" emerging for labour (see Fichter *et al.* 2018). Thus, while the tendency towards outsourcing and offshoring may, of course, weaken the bargaining power of unions, it also creates opportunities to attack the "choke points" of "just-in-time" supply lines. Transnational labour union linkages may also leverage in powerful allies and create a "whipsawing" effect that benefits workers rather than employers (see Greer & Hauptmeier 2015). The disaggregation of labour and the generalizing of precarious employment conditions can make a traditional labour union ideology more or less defunct, but it can also prompt revitalization and a democratic realization that international solidarity "begins at home", as it were, by incorporating migrant workers.

Thus, migrant labour is both a challenge and an opportunity for organized labour and unions. Migrants may often be introduced into a settled labour force with the intention of undercutting wages and conditions, which unions have naturally opposed. Recent examples, however, point to the positive role that unions can have in terms of migrant integration and to the possible impact of migrant workers upon union revitalization, giving them a strong democratic and organizational boost. Economic migration poses social and political challenges, but it also has the power to transform the future of societies in the Global North beset by declining birth rates and an aging population. The current "migration crisis" in Europe shows it is an issue that can dramatically divide, not least across the working classes, and give rise to massive political crises directly or indirectly.

SPACES OF LABOUR UNIONISM

Apart from the market/society/state triad, the other major framing concept to bear in mind is the spatiality of labour in all its aspects. The "spatial turn" in the social sciences belatedly had an impact upon labour studies. Now, as McGrath, Herod and Rainnie (2010) argue, the vision of Ohmae (1995) and others of a "flat world" emerging under neoliberal globalization was simply wrong and we must recognize how labour markets are regulated in place-specific ways, social relations are constructed over space, and workers are subject to spatial constraints. The triumphalist visions of Ohmae *et al.* that globalization meant the end of distance/space, and the pessimism of others who saw labour mired in a static local world were now being replaced by a more complex understanding of the various interlinked scales on which labour was formed, organized and struggled, from the local to the global, through national sub-regional and supra-regional, and not excluding the household level. Workers could also, in this regard, "skip scales" and take local labour struggles to the transnational domain as happened with the Liverpool dockers' strike of 1995–98.

Our social place in society is also a geographical space so we need to accept that labour unionism does not follow a universal model. Labour relations and labour movements, thus, need to be situated geographically, which is part of their path dependency trajectory. In the same way that Indian labour scholars, among others, have sought to "provincialize Europe" (Chakrabarty 2008), we need to situate the dominant industrial relations frame in its specific North Atlantic postwar home ground. Nor should it be seen as the norm against which other situations or spaces are judged and found wanting. Just one example would be that the move towards organizing in the informal sector should not assume that "formalization" is the only or the best strategy. Collective bargaining is not always the best way for labour unions to advance and represent the interests of their members and certainly the partnership model of industrial relations promoted by the ILO and most unions at an international level is not the best way forward.

A thumbnail sketch of labour unions and their constraints today starts with the BRICS (Brazil, Russia, India, China and South Africa) grouping, and especially the "emerging economies" of China and India. With 30 per cent of the world's labour force China will play a pivotal role in the future of the global labour movement. In the last three decades an unprecedented process of industrialization has moved from the interior and the north of the country to the coast and south where, as King-Chi Chan (2012: 5) puts it, "young temporary migrant workers who work in conditions of low pay, long working hours, despotic management and appalling environment, have gradually replaced veteran urban permanent SOE [state-owned enterprise] workers with their iron rice bowl". We have, thus, seen defensive labour struggles waged by the "old" working class in the SOEs being laid off and the start of a new wave of strikes since 2010 among the new and increasingly militant working class formed through internal migration.

It is often assumed in Western union circles that the party "transmission belt" function of the huge ACFTU (All-China Federation of Trade Unions) deprives it of any legitimate role in representing workers' interests. While the co-option intent of union recognition is clear it would be wrong to assume that the functions delegated to them by management are meaningless. It is certainly the case that, by and large, the labour unions do not proactively promote workers' grievances and that they see themselves as "mediators", much like the famous definition of unions by Wright Mills (1960) as "managers of discontent". Nevertheless, the Chinese unions are, arguably, no longer simple transmission belts for Communist Party directives and, while they have limited room for manoeuvre, they have gained "much greater authority and significance" (Clarke 2005: 3). Their role is clearly to prevent, or at least delay, the emergence of autonomous workers' organizations but the outcome will depend upon circumstances moving forward.

In India, the other major "emerging economy", the starting point is a very low level of union density overall, usually estimated at around 13 per cent for the casual and regular labour market. Labour unions in India have faced an unprecedented level of challenge due to the impact of neoliberal globalization and a privatization drive in particular. Downsizing, restructuring and draconian labour laws have driven many workers out of the formal economy into the informal sector where probably 90 per cent of workers are now located. What has emerged is a combination of "old" unions with a long history of organizing, often under distinctive (and divisive) political banners, and a plethora of new forms of organizing in the informal sector, including cooperatives and more ephemeral but very militant mobilizations. Divided by caste, gender and regions,

India's vast working classes have continued to reorganize and mount resistance to the untrammelled rule of capital. Labour unions will be part of that future but will not be the exclusive agents of social change, at least in their present form.

Elsewhere in the Global South, labour unions continue to thrive in both old and new modalities. In Argentina, where there are labour unions that hark back to the national-populist era, there are also exemplars of the "new" forms of mobilization, especially after the collapse of the economy in 2001. International attention focused on the latter, especially the organizations of the unemployed (the *piqueteros* or pickets), yet in the last decades the labour unions have come back to the fore again (see Atseni & Grigera 2019). Under neo-populist governments, they have reclaimed much ground within the state as representatives of workers but also as political power brokers. The post-2001 grassroots organizations, the *piqueteros* and the occupied factory movement, have faded by comparison. The unemployed workers' organizations were largely co-opted by the government and its benefits system while the occupied factories came to an inevitable end in a capitalist private property regime.

One thing we learn from Argentina is about the resilience of the labour union form. In periods of democracy and a degree of economic stability, the organized labour unions are able to reclaim political space. They retain deep roots within the working classes and are able to intelligently represent their members' interests. But the period of semi-insurrectional upheaval (2001–02) has also had an impact, renewing militant grassroots traditions going back to the Cordobazo civil uprising of 1969. The economic crisis and the politics of the post-crisis regimes has significantly altered the composition and modus operandi of the labour unions. For one, they are much more likely to accept a more social definition of their role captured in the popular radical union slogan that "the factory is the neighbourhood" (*la fábrica es el barrio*). This echoes the Italian autonomist slogans of the 1960s, albeit it in a very different context. Likewise, the bureaucratic union leaderships have accepted the need to formally integrate the organizations representing the informal sector workers into a broad negotiating front in a tacit recognition of the value of a social movement unionism orientation, given the changing composition of the working class.

In Asia, the case of Indonesia teaches us that path dependency can only go so far in explaining the current role of a labour union or labour movement. The movement in Indonesia was, until recently, in a structurally weak position and the unions had shown little inclination towards engagement with politics. The effects of the 1965–66 massacre of Indonesia's communists had cast a long shadow and the legacy of authoritarianism was a heavy one. Union density was low and the internal divisions within the labour unions were not conducive to unified action. Yet in the period since the fall of Suharto in 1998, as Caraway and Ford (2020: 1) recount, "Indonesia's labour movement achieved many of its goals first through the disruptive power of contentious politics and later by combining street and electoral politics." This should not be read as a simple vindication of political unionism, but rather as a warning against structural determinism and not understanding the nature of labour unions as evolving and non-static social organizations.

Turning now to the Global North, the original industrializing countries, we find a mixed picture in terms of labour union activities. The challenges to a renewed labour unions solidarity are still marked after the long period of neoliberal hegemony. The results of either business unionism or political unionism are hard to find. Most labour movements – for example, those in the UK and the US – have at best managed to ameliorate decline. And few hold their breath at the prospects of a Democratic or Labour government helping to return the labour unions to their old positions of influence. However, there have been significant changes in the labour movements in these two countries that are of international significance. The victory of the "New Voice" slate in the American Federation of Labor and Congress of Industrial Organization elections of 1995 marked a significant turn in the US labour movement. This was a movement in decline, with its membership being almost halved and its strategic vision almost non-existent. A process of change began with a massive recruitment drive and a long overdue emphasis on questions of gender and "race". The language of class was also rediscovered and there was a significant turn towards "community unionism", defined by Fine (1998: 120) as "union organizing that takes place across territorial and industrial communities much larger than a single workplace [that] recognizes that a worker's identities are much broader than just who they work for or what they do". A major impetus to this shift was made, significantly, by the SEIU (Service Employees International Union) that had led the landmark struggles of the Justice for Janitors in the 1990s that successfully organized precarious migrant workers.

In the UK too there have been moves since the turn of century towards a form of "community unionism", particularly oriented towards migrant workers. This was part of a broader emphasis upon securing new members through a concerted organizing drive. Reaching beyond the workplace and the traditional collective bargaining mechanisms, some unions, particularly in London, began to reach out to migrant unorganized workers. The London minimum wage campaign of the 2000s saw intense union engagement for a time, with strong grassroots support, and won considerable concessions. Unions found common ground with migrant communities and faith-based organizations and began to change their mindsets and the way they operated. However, an overall assessment of the organizing drive makes for sobering reading: "It's hard to avoid the conclusion that thirteen years of organising activity has made comparatively little impact on formal, aggregate measures of union power" (Heery 2015: 5). That should not detract from the innovations that revitalization brought about and the dynamics for change it unleashed within unions. It may, of course, caution us against any facile uplifting narratives.

Set within an international context, and not elevated to a normative position, the case of these two countries is of great interest. No labour movement, however conservative or set in its ways, is incapable of change. Social unionism is shown to be relevant beyond the confines of authoritarian developing countries. Ultimately, they also show that progress for a labour movement is subject to setbacks and there is no simple linear path to union revitalization.

COMPLEX FUTURES

We have now set the parameters for the main current ideologies of labour unionism that, in turn, determine how the various interests of labour – economic, political and social – are articulated and pursued on behalf of their members. The articulation of the complexity lens and the socio-spatial dialectic helps steer clear of universal truths; for example, regarding labour's decline as a social movement. It is common to hear that union density worldwide today is below 10 per cent; therefore, organized labour is unlikely to have a future as an effective counter to the new capitalism. We can easily build a whole theoretical edifice of labour co-option through consumerism, labour division through precaritization and union incorporation through bureaucratization to convince ourselves that labour unions have no future. But in a world characterized by uneven development, we cannot take a so-called global trend on union density as one that determines the future of a very complex movement across all regions. Nor can we ever confuse trends with an overall situation: there are always counter-trends, new factors enter into the equation, and uneven development, as always, means that complexity prevails. We might also add that academic disciplines are slow to change and tend to maintain paradigms for quite some time after the world around them has moved on.

To be future-oriented is also to be constantly examining the making and remaking of the working classes. If we accept that class is fluid, then we cannot work productively with a static analysis. Nor can we live with images of a labour past that no longer exists. Labour history is replete with myths about this or that group of workers and this or that militant union but the fact is that the world of workers is constantly shifting, being remade and reimagined. It is worth noting how Negri and Hardt (2009) carry out an interesting analysis of the Bolivian labour movement and the replacement of the once hegemonic miners by other fractions of the working classes. In 1952 there was a national revolution in Bolivia led by the miners and the peasants that created a myth (in the Sorelian sense) of working-class power. In 2001 Bolivia once again witnessed a formidable insurgency, the so-called "Water Wars", when the Indigenous communities of Cochabamba successfully resisted the privatization of water. This time, though, there was no phalanx of marching miners in tin hats carrying sticks of dynamite as was the case in 1952. Any mythical connection between 1952 and 2001 that was made by commentators was denied by the dramatic deconstruction of the Bolivian working class under neoliberalism in the 1980s and 1990s.

Bolivia, like the rest of the world, is a country where temporalities are mixed – the past is the present and the present is the future – and society takes on a motley or variegated character (in local terms it is a *sociedad abigarrada*). In the 1980s and early 1990s, many of the mines – in what was an economy based on mining – closed and the Bolivian miners lost their central role in national politics. Miners became coca cultivators or small-scale producers or were simply cast into a world of migration and joblessness. As with other working classes, the Bolivian working class became more flexible and mobile as it lost its centre. Miners could no longer – as autoworkers or steelworkers elsewhere – represent hegemonically the broader working class through vertical structures. But, crucially, as Negri and Hardt stress (2009: 110): "This shift … signals no farewell to the working class or even a decline in worker struggles, but rather an increasing multiplicity of the proletariat and a new physiognomy of struggle." In this new, more multitudinous array of working classes, no single sector can provide hegemony, and horizontal, as against vertical, forms of organization emerge, which are more fitted to the networked society of the global era. Bolivia is not, of course, the UK but the implications of this analysis are clear: our analysis of the interests and ideologies of labour unions needs to start with the changing composition of labour under capitalism.

REFERENCES

Aglietta, M. 1998. "Capitalism at the turn of the century: regulation theory and the challenge of social change". *New Left Review* I/232.

Anderson, P. 1967. "The limits and possibilities of trade union action". In *The Incompatibles: Trade Union Militancy and the Consensus* edited by R. Blackburn & A. Cockburn, 265. Harmondsworth: Penguin.

Atseni, M. & J. Grigera 2019. "The revival of Labour Movement Studies in Argentina: old and lost agendas". *Work, Employment and Society* 33(5): 865–76.

Caraway, T. & M. Ford 2020. *Labor and Politics in Indonesia*. Cambridge: Cambridge University Press.

Carter, B. 2006. "Trade union organizing and renewal: a response to de Turberville". *Work, Employment and Society* 20(2): 415–26.

Castells, M. 1996. *The Rise of Network Society*. Oxford: Blackwell.

Catalano, A. 1999. "The crisis of trade union representation: new forms of social integration and autonomy-construction". In *Labour Worldwide in the Era of Globalization*, edited by R. Munck & P. Waterman. Basingstoke: Palgrave Macmillan.

Chakrabarty, D. 2008. *Provincializing Europe: Postcolonial Thought and Historical Difference*. Princeton, NJ: Princeton University Press.

Clarke, S. 2005. "Post-socialist trade unions: China and Russia". *Industrial Relations Journal* 36(1): 2–18.

Clawson, D. 2003. *The Next Upsurge: Labour and the New Social Movements*. Ithaca: ILR Press.

Evans, P. 2014. "National labor movements and transnational connections: global labor's evolving architecture under neoliberalism". *Global Labour Journal* 5(3):1–25.

Fichter, M. *et al.* 2018. *The Transformation of Organised Labour Mobilising Power Resources to Confront 21st Century Capitalism*. Friedrich Ebert Foundation.

Fine, J. 1998. "Moving innovation from the margins to the center for a new American labor movement". In *A New Labor Movement for the New Century*, edited by G. Mantsios. New York: Monthly Review Press.

Flew, T. 1989. "The limits to political unionism". *Journal of Australian Political Economy* 24(Mar): 77–99.

Freeman, R. 2001. "The new global labour market". *Focus* 2(6): 1–6.

Frege, C. & J. Kelly 2003. "Union revitalization strategies in comparative perspective". *European Journal of Industrial Relations* 9(1): 7–24.

Friedman, G. 2008. *Reigniting the Labor Movement: Restoring Means to Ends in a Democratic Labor Movement*. London: Routledge.

Greer, I. & M. Hauptmeier 2015. "Management whipsawing: the staging of labor competition under globalization". *Industrial and Labor Relations Review* 69(1): 29–52.

Heery, E. 2015. "Unions and the organising turn: reflections after 20 years of organising". *Economic and Labour Relations Review* 26(4): 545–60.

Hodder, A. & P. Edwards 2015. "The essence of trade unions: understanding identity, ideology and purpose". *Work, Employment & Society* 29(5): 848–9.

Hyman, R. 1994. "Changing trade union identities and strategies". In *New Frontiers in European Industrial Relations*, edited by R. Hyman & A. Ferner, 122. Oxford: Blackwell.

Hyman, R. 2001. *Understanding European Trade Unionism: Between Market, Class and Society*. London: Sage.

Ibsen, C. & M. Tapia 2017. "Trade union revitalisation: where are we now? Where to next?" *Journal of Industrial Relations* 59(2): 170–91.

Johnston, W. 1991. "Global Work Force 2000: the new world labor market". *Harvard Business Review* 69(2): 115–27.

Kelly, J. 2003. "Labour Movement Revitalization? A Comparative Perspective". Dublin: Countess Markievics Memorial Lecture.

King-Chi Chan, C. 2012. *The Challenge of Labour in China: Strikes and the Changing Labour Regime in Global Factories*. London: Routledge.

Lawrence, A. 2014. *Employer and Worker Collective Action: A Comparative Study of Germany, South Africa, and the United States*. Cambridge: Cambridge University Press.

Lévesque, C. & G. Murray 2010. "Understanding union power: resources and capabilities for renewing union capacity". *Transfer* 16(3): 333–50.

McGrath-Champ, S., A. Herod & A. Rainnie (eds) 2010. *Handbook of Employment and Society: Working Space.* Cheltenham: Edward Elgar.

Minns, J. 2001. "The labour movement in South Korea". *Labour History* 81(3): 175–95.

Munck, R. 2011. "Beyond North and South: migration, informalization, and trade union revitalization". *Working USA* 14(1): 1–4.

Munck, R. 2018. *Rethinking Global Labour: After Neoliberalism.* Newcastle upon Tyne: Agenda Publishing.

Munck, R. 2019. "Organising precarious workers: a view from the South". Open Democracy. www.opendemocracy.net/en/beyond-trafficking-and-slavery/organising-precarious-workers-view-south/

Negri, A. & M. Hardt 2009. *Commonwealth.* Cambridge, MA: Harvard University Press.

Ohmae, K. 1995. *The Borderless World: Power and Strategy in an Interdependent Economy.* New York: Harper Business.

Payne, J. 1965. *Labor and Politics in Peru: The System of Political Bargaining.* New Haven, CT: Yale University Press.

Polanyi, M. 2001. *The Great Transformation.* Boston, MA: Beacon Press.

Richards, A. 2001. "The crisis of union representation". In *Can Class Still Unite?* Edited by G. Van Guyes, H. De Witte & P. Pasture. Aldershot: Ashgate.

Schumpeter, J. 2014 [1942]. *Capitalism, Socialism and Democracy.* 2nd edn. Floyd, VI: Impact Books.

SID 1997. "A new global agenda: visions and strategies for the 21st century". SID Global Labour Summit, Copenhagen, 31 May–1 June.

Silver, B. 2003. *Forces of Labor: Workers' Movements and Globalization since 1870.* Cambridge: Cambridge University Press.

Soja, E. 1980. "The socio-spatial dialectic". *Annals of the Association of American Geographers* 70(2): 207–25.

Tronti, M. 2019. *Workers and Capital.* London: Verso.

Turner, L. & R. Hurd 2001. "Building social movement unionism: the transformation of the American labor movement". In *Rekindling the Movement: Labor's Quest for Relevance in the Twenty-first Century* edited by L. Turner, H. Katz & R. Hurd, 9–26. Ithaca, NY: Cornell University Press.

Van der Linden, M. 2015. "The crisis of global labour". *Against the Current* No. 176.

Von Holdt, K. 2002. "Social movement unionism: the case of South Africa". *Work, Employment and Society* 16(2): 283–304.

Visser, J. 2019. "Trade unions in the balance". ILO ACTRAV Working Paper.

Waterman, P. 1993. "Social movement unionism: a new model for a new world order". *Review* 16(3): 245–78.

Webb, S. & B. Webb 1920. *History of Trade Unionism.* London: Longman.

Western, B. 1997. *Between Class and Market.* Princeton, NJ: Princeton University Press.

Wright Mills, C. 1960. "Letter to the New Left". *New Left Review* I/5 (Sep/Oct).

CHAPTER 3

UNION RESOURCES: THE POWER RESOURCES APPROACH

Stefan Schmalz and Edward Webster

ABSTRACT

Several labour scholars have argued that workers are rediscovering their power by collectively mobilizing to influence the asymmetric relationship between capital and labour. As a result of these developments, the power resources approach (PRA) emerged as a research heuristic. In this chapter we will outline the discussion on the PRA and present a scheme to differentiate workers' power into structural, associational, institutional and societal power. We discuss two drivers of structural change – globalization and digitalization – and their implications for workers' power. By referring to Silver's work and the PRA, we analyse emerging forms of transnational organizing. A prominent example of transnational organizing is that of global union federations. We then focus on labour organizing in the platform economy. We observe the emergence of hybrid forms of union-like associations in the transportation sector. We conclude with some questions for a research agenda on workers' power and global capitalism in the twenty-first century.

Keywords: Union resources; analytical frameworks; challenges for unions

INTRODUCTION

Over the past two decades, many scholars have challenged the "end of work" paradigm in labour studies and research on unions by arguing that unions are strategic actors and, therefore, able to revitalize. Thus, even under the conditions of an increasingly globalized economy, organized labour can develop and mobilize power to protect workers' interests. As a result of this discussion, the power resources approach (PRA) emerged as a research heuristic to identify different sources of workers' power. The PRA argues that organized labour can build its power through the collective mobilization of power resources to advance their interests in the structurally asymmetric and antagonistic relationship between capital and labour.

Today, several conceptions and varieties of the PRA have emerged, which identify a variety of power resources (Brookes 2013; Gumbrell-McCormick & Hyman 2013; Refslund & Arnholtz 2022; Schmalz, Ludwig & Webster 2018; Von Holdt & Webster 2008), strategic capabilities (Lévesque & Murray 2010, 2013) and spaces of action (Ford & Gillan 2021) of workers' power. The PRA has, therefore, become an important research tool in global labour studies and inspired rich empirical studies on successful organizing strategies (Rhomberg & Lopez 2021), campaigning (Pannini 2023) and on labour relations (Rego 2022). However, there have also been critical accounts pointing out some of the weaknesses of the PRA, such as its neglect of structural constraints on workers' power (Bieler 2018), the role of the capital–labour relationship (Gallas 2018; Nowak 2022) and a one-sided focus on unions as actors (Atzeni 2021).

There are three main reasons for this surge of interest in PRA-inspired studies. First, the PRA, with its focus on agency, has been useful in challenging existing theoretical approaches that tend to focus upon the institutional setting of labour relations or structural trends such as the impact of globalization on labour. In contrast, the PRA has emphasized the ability of organized labour to act, which has changed the way that scholars deal with the issues of union action, labour conflict and the strategic innovations of the labour movement. Second, many of the studies inspired by the PRA were the result of close collaboration between unions and scholars and can be seen as a form of "organic public sociology in which the sociologist works in close connection with a visible, thick, active, local and often counter-public" (Burawoy 2005: 7). Third, the PRA terminology of power resources has served as a common cross-border language for both scholars and union activists to help facilitate knowledge transfers on union strategies, which has consequently contributed to a transnational learning process between unions.[1]

In this chapter we provide a brief overview of the recent developments of, and challenges to, the PRA. We first outline the PRA by focusing on our conception of the PRA before presenting the more recent debates on the PRA by discussing several adoptions, critiques and innovations of the approach. These debates illustrate various innovative adoptions of the PRA, and we demonstrate how the approach can be contextualized and applied. We discuss two major drivers of ongoing structural change – globalization and digitalization – and their implications for labour unionism, and argue that although both processes tend to weaken workers' power globally, they have also spurred new forms of worker organization. A prominent example of transnational organizing in the globalized economy are the global union federations (GUFs), which have notable success stories such as GUF Education International's campaign in Kenya and Uganda. In turn, digitalization has led to the emergence of self-organized, hybrid forms of union-like associations in the transport sector in different regions across the world. We conclude that it remains uncertain whether it is possible to sustain

1. An example of transnational learning between union activists across the world is the global research project run by the Friedrich Ebert Stiftung called Trade Unions in Transformation, where 25 successful accounts of union revitalization are identified and analysed through the power resources approach (Herberg 2018). The follow-up project, Trade Unions in Transformation 4.0, comprised 12 studies on how organized labour has successfully responded to the challenges of digital capitalism.

these new associational forms and re-embed markets in society in a way that would ensure the future of labour.

POWER RESOURCES APPROACH: DEVELOPMENT AND CONCEPTS[2]

The origins of the PRA date back to concepts that were created by Wright (2000) and Silver (2003). The notions of structural power that arose from labour's position in the economic system, as well as the associational power that resulted from the collective actions of workers' associations, were the starting point for a global debate on workers' power that would lay the foundation of the PRA. Scholars from the US (Brookes 2013; Chun 2005), the Global South (Webster, Lambert & Bezuidenhout 2008), and Europe (Dörre, Holst & Nachtwey 2009) added conceptual innovations to the approach and adjusted it to the diverging realities of their home countries. For instance, researchers from South Africa argued that workers in the informal sector can mobilize logistical power through roadblocks and other forms of joint action by unions and social movements (Webster, Lambert & Bezuidenhout 2008: 12ff.).

Our conceptualization of workers' power draws from this debate to which it adds two further power resources, institutional and societal power, to the original sources of workers' power (structural and associational power). We perceive the relationship between the four power resources as complex and sometimes conflicting, as it is hardly possible for organized labour to advance all power resources at the same time. Therefore, it is not so much the extent of power resources, but rather their development and specific combinations that are crucial for successful organizing, campaigns and collective interest representation. Workers' power is discerned first and foremost as the power to do something (power to) rather than the power to determine the rules of the game (power over) (Lévesque & Murray 2010), but its use is always embedded in social relationships and power relations. The PRA can, therefore, be understood as a relational concept, as it is not primarily aimed at analysing structural power relations. Hence, workers' power must be analysed in the context of global capitalism as power resources need to be located within the strategic environment in which workers find themselves.

Structural power comes from the position of workers within the economic system (Silver 2003: 13ff.; Wright 2000: 962). It is a primary power resource, given its availability to all workers and employees, even those without collective-interest representation (Jürgens 1984: 61), and rests upon the power to disrupt capital accumulation (disruptive power) (Piven 2008: Ch. 2). Following Silver's (2003: 13) argument, one can distinguish between two forms of structural power: workplace bargaining power and marketplace bargaining power. Workplace bargaining power is mobilized by labour unrest in production, which results in rising costs for companies and can allow workers to enforce concessions. Workplace bargaining power is sometimes exerted spontaneously or in a decentralized way and includes covert forms of industrial conflict such as sabotage (Brinkmann *et al.* 2008: 27).

2. This section includes an abridged and updated version of some arguments we made in Schmalz, Ludwig & Webster (2018).

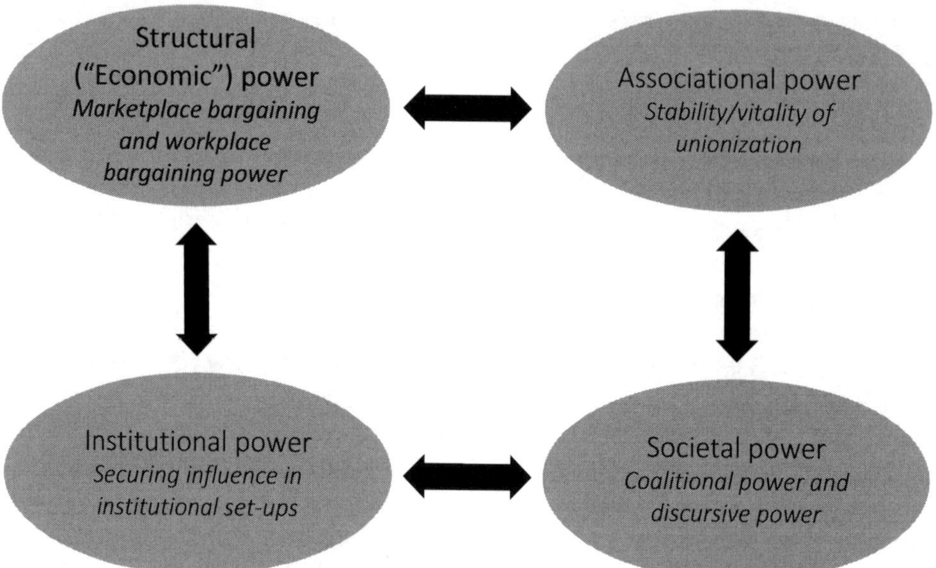

Figure 3.1 Workers' power resources
Source: Schmalz, Ludwig & Webster (2018: 116).

Workers in sectors with highly integrated production processes have a particularly high degree of workplace bargaining power, as the impact of local work stoppages causes high costs for capital and extends far beyond their workplace (Silver 2003: 13). Structural power is shaped by the development of global capitalism and new technologies. For instance, the introduction of the assembly line in the early twentieth century facilitated coordinated strike actions as workers learned to stop the assembly line and, thus, disrupt the production process (Silver 2003: 47–66). A twenty-first century version of work stoppages is a case such as that of informal food courier riders collectively disconnecting from the relevant apps to bring the transport of food to a standstill (Webster *et al.* 2021). Structural power can also be exercised outside the production process, as in the case of informal self-employed workers mobilizing through street blockades or occupations (Webster 2015: 119). This could be called logistical power, a subtype of workplace bargaining power. Thus, "Unlike marketplace bargaining power and workplace bargaining power … logistical power takes structural power outside the workplace and into the public domain" (Webster, Lambert & Bezuidenhout 2008: 13).

Marketplace bargaining power is the second form of structural power. It results from a tight labour market and, thus, from "having rare qualifications which are demanded by employers, little unemployment" and from the "capability to completely withdraw from the labour market and to live on other sources of income" (Silver 2003: 13). Marketplace bargaining power depends on the structure of the labour market; that is, its fragmentation into core workforces, precarious and informal workers and the unemployed, as well as gendered and racialized division lines. Marketplace bargaining power can be exercised both individually and collectively. For instance, when marketplace bargaining power is high, employees can simply change their job without fearing unemployment, while at the same time unions tend to reach better collective bargaining results.

Associational power is captured through the interactions between collective political or union-based workers' organizations. It is capable of partly compensating for a lack of structural power without completely replacing it. In contrast to structural power, this requires organization processes to take place and for collective actors to emerge. These actors include unions, grassroots groups at the workplace level, hybrid forms of organized labour, and political parties – with unions historically playing a pivotal role in representing labour. Wright (2000: 963f.) distinguishes between three levels at which organized labour comes into play: at the workplace – in connection with workplace bargaining power – there are works groups, shop steward councils and works councils; at the sectoral level – closely connected to marketplace bargaining power – where unions are the major players; and in the political system – in connection with societal power – where workers' parties represent collective interests. Above the levels described there are GUFs that operate transnationally at different scales.

Membership numbers are often cited as important indicators for determining associational power. Marx (1974: 91) knew that the "power of the workmen" lay in the "force of numbers". However, there are also other important factors such as infrastructural resources, organizational efficiency, internal cohesion and, of course, membership participation (Schmalz, Ludwig & Webster 2018: 118–24). For instance, without active member participation, worker organizations tend to turn into bureaucratic organizations, whereas material and human resources remain essential for carrying out their daily work. However, power resources alone are not sufficient to exert power, for organized labour must also develop strategic capabilities to be successful. As Brookes (2018: 256) argues, one can also perceive "associational power as workers' capacity to behave as a collective actor", which is embodied "in workers themselves and how they relate to one another".

Capabilities and capacities such as articulation (constructing multilevel interaction to link the local and the global across space), organizational flexibility (adopting the organization to new challenges), framing (providing credible interpretation patterns) and intermediation (consensus building within organizations) are crucial for developing associational power. The skills of pooling scarce power resources can lead seemingly weak organizations to successful outcomes, while ill-designed strategies can cause strong unions to fail (Ganz 2000).

Institutional power is usually the result of struggles and negotiation processes based upon structural power and associational power. As "a secondary form of power", such institutions are embedded in "cross-company political-economic relations" (see Jürgens 1984: 66) and tend to be only partly dependent on forms of primary power. They are often the result of a concession or an attempt at cooperation (or co-optation) on the part of capital towards labour. Institutional power is a "double-edged sword" as it has a twofold nature. Although it may grant unions rights, it can also restrict unions' power to act. For instance, the extensive rights of workers on the shop-floor level in Germany are "balanced" by a ban on political strike actions.

The relationship between strengthening and weakening labour rights is always the product of a unique power balance between capital and labour that has been "solidified" in co-determination institutions (Poulantzas 1978: 123ff.). The dual nature of institutional power brings with it the challenge of reconciling the "two faces of unionism"

(Webster 1988) – the focus on the movement, on the one hand, and institutional representation of interests on the other, or mediating between the "logic of membership" and the "logic of influence" (Schmitter & Streeck 1981). Occasionally, conflicts arise between different unions and groups of workers over the nature of institutions (Runciman 2019).

The unique feature of institutional power is its steadfastness over time. It is rooted in the fact that institutions lay down basic social compromises that transcend economic cycles and short-term political changes. Unions can continue to use institutional power resources even when their associational and structural power shrinks. One key question, therefore, is how stable institutionalized resources are. There are different time horizons that apply here. Sometimes, they are extremely far-reaching, as they, like freedom of association, are considered untouchable privileges that have constitutional or legal standing or have been enshrined by supranational regimes. Other institutional resources are, however, more fragile. Many corporatist alliances are based on (tripartite) institutionalized dialogue procedures and can be rescinded rather easily (Haipeter 2012: 117ff.).

However, although institutional power has remained formally intact in many countries, such as in Germany, the underlying economic conditions and behaviour of capital have changed. As a result, dwindling workplace bargaining and associational power can also contribute to the erosion of institutions and make negotiation processes between capital and labour increasingly asymmetrical (Dörre 2018: 894ff.); for instance, through the pressure of plant relocations or the focus on shareholder value that has been prevalent in Western Europe since the 1990s.

We understand societal power as the spaces of action that result from sustainable cooperation with other social groups and organizations, as well as from the public support for organized labour. The exercise of societal power is essentially a question of the ability to assert hegemony in the Gramscian sense; that is, to generalize the political project of organized labour within the prevailing power constellation so that society as a whole adopts it as its own. Similar to structural power, we identify two subtypes of societal power – coalitional and discursive power. Their use is interconnected and mutually reinforcing. Coalitional power results from being able to mobilize networks with other social actors for campaigning or participating in workers' actions (Brinkmann *et al.* 2008: 98ff; Frege, Heery & Turner 2004: 137ff; Turner 2006). Thus, coalitional power is based upon boosting one's associational power by harnessing the resources of other players or upon unions' ability to mobilize support of these actors. Relevant literature cites social movements, social associations, non-governmental organizations, students and churches as typical allies (Frege, Heery & Turner 2004: 151; Joynt & Webster 2016: 58–67; Milkman, Bloom & Narro 2010).

Effective discursive power, on the other hand, is the "ability to intervene successfully in areas of the public sphere that are pre-structured on a hegemonic basis" (Urban 2012: 222; see also McGuire & Scherrer 2015) to become opinion leaders for labour-related topics. However, mobilizing discursive power requires people to consider unions and workers' demands to be justified. If unions are seen as defenders of just causes, their social influence will increase. Unions can exert public pressure, particularly when the "moral economy" (Thompson 1971: 76) and norms of justice are undermined by the state or private companies. Above all, this occurs through the scandalization of injustices and unions waging classification battles over working conditions that are considered unfair (Chun 2005).

In summary, our presentation of workers' power implies that organized labour can create its strategy by choosing which power resources to develop and mobilize. In reality, organized labour does not develop its power in isolation from other important wielders of power, namely capital and the state. Consequently, the concept of power resources cannot be understood as a universal and static formula but needs to be located within the strategic environment in which workers find themselves (Gallas 2016). As the historical, political and social context differs, the use of power also takes different forms, as it is both dependent on specific institutional arrangements and historical developments, as well as continuously being restructured by the logic of global capitalism. For instance, African unions have been profoundly shaped by the politics of national liberation struggles. The political nature of African unions is made clear in the following comment: "The emergence of … unions was more than merely a response to conditions of economic exploitation. It was simultaneously a response to the conditions of political oppression created by colonialism, particularly the denial of political rights and the violation of the dignity of workers and the general population of the colonised" (Buhlungu 2010: 198)

The PRA, thus, must be understood as an intermediate concept that focuses on labour relations and actions and, therefore, the level between structural developments and agency. It tends to highlight the possibility of worker agency at a time when the dominant narrative continues to be that of union decline in the face of structural trends of globalization, digitalization, and precaritization.

POWER RESOURCES: THE DEBATE

The mobilization of power resources by organized labour is, therefore, dependent on context, history and political traditions. It has a different meaning in diverging local and regional contexts, something acknowledged by many scholars who work with the PRA to analyse labour relations, organizing processes and new forms of power in different world regions. Different varieties of the PRA have been used to understand organized labour in Europe (Dias 2021; Dörflingler, Pulignano & Valla 2021; Lehndorff, Dribbusch & Schulten 2017), the Americas (O'Brady 2021; Osorio Lavín & Velásqués 2022), Africa (Anwar & Graham 2020; Mmadi 2022), and Asia (Doutch 2021; Hui 2022). Many of these studies tend to adopt the PRA to divergent local realities by either modifying and further developing the approach (Birelma 2018; Hui 2022; Rhomberg & Lopez 2021), before combining it with other theories of labour relations and global capitalism (Ellem. Goods & Todd 2020; Marslev, Staritz & Raj-Reichert 2022) or concretizing the nature and mobilization of specific power resources (Fox-Hodess & Santibáñes Rebolledo 2020; O´Brady 2021).

Thus, these adaptations of the approach tend to highlight specific aspects of the mobilization of workers' power. There are different ways to work with the approach.

Most of the work with the PRA is focused on organizing, campaigning or innovative union strategies at the plant or sectoral level and, thereby, adopt the approach to the diverging realities of labour (Hui 2022; Rhomberg & Lopez 2021; Schmalz & Thiel 2017). For instance, Hui (2021) has used the PRA to analyse plant-level organizing approaches in the Chinese context, showing how the bottom-up organizing of migrant workers has been challenged by the party-state and employers, thereby analysing several contradictions present in the establishment of workplace bargaining power. In many cases, capital and the state tend to prefer to respond to immediate economic interests to pacify workers instead of engaging with the newly developed institutional structures of plant-level collective interest representation (Hui 2022: 106ff.).

Second, there have been several PRA-inspired studies on the institutional setting of labour relations (also on the plant-level and sectoral level) that focus on the role of organized labour in the institutional system (Dörflingler & Pulignano 2021; O'Brady 2021; Rego 2022). These studies are often concerned with institutional change that occurs through structural trends such as digitalization (Rego 2022) or the role of institutions in representing vulnerable labour market groups. For example, O'Brady (2021) has analysed inclusive union strategies in supermarket chains in Canada, Germany, Sweden and the US, discussing the impact of cost pressures in both the production and product market spheres. One of the main results of his research is the finding that unions have often been able to tap into institutional power to include and mobilize precarious workers, which effectively makes institutional power a "precursor to effective resistance to precarious working conditions through collective bargaining" (O'Brady 2021: 1105).

A third debate considers the PRA as a tool to analyse the development of labour relations and unions on the national level (Krein & Dias 2018; Lehndorff, Dribbusch & Schulten 2017). In their study on European unions in the Eurozone crisis, Lehndorff, Dribbusch & Schulten (2017: 15) integrated the PRA in a comparative political economy approach in order to analyse national "configurations of power resources" instead of "varieties of capitalism". From this perspective, national indicators and trajectories of workers' power can be described to analyse the impact of European Union (EU) labour market reforms, different patterns of corporatism and activism, and the structural changes to labour relations after the crisis. Likewise, other scholars have analysed the development of unions and labour relations in countries and regions such as Germany, Brazil and Eastern Europe (Becker 2019; Dörre 2018; Krein & Dias 2018).

Fourth, many scholars have referred to the PRA in their analysis of transnational networks and actors and have also developed new concepts for this purpose (Brookes 2013; Ellem, Goods & Todd 2020; Ford & Gillan 2021; Marslev, Staritz & Raj-Reichert 2022). A major innovation has been the work of Ford and Gillan (2021) on the GUFs. They argue that the GUFs are able to mobilize a specific form of workers' power by connecting different spatial levels of union action. Their power is multi-scalar "both in the sense that conflict can play out on multiple scales simultaneously and in the sense that actors can act strategically to shift their engagement with other actors to different scales" (Ford & Gillan 2021: 4; see also Brookes 2013). More specifically, in their global campaigns the GUFs "combine different kinds of power resources simultaneously at different scales" (Ford & Gillan 2021: 4) through supranational institutional engagement (supranational institutional power), international campaigns (societal power), local worker mobilization (structural and associational power), and local and national institutional engagement (institutional power).

In summary, most of the recent discussions on and receptions of the PRA tend to develop the approach further to analyse specific local and structural challenges to organized labour. A common feature of these studies is a creative adoption of the approach to different levels of analysis without dropping key concepts and maintaining the focus on the ability of organized labour to act.

However, there have also been several critiques of the PRA that dismiss different aspects of the approach (Atzeni 2021; Gallas 2018; Nowak 2022). Many of the critical accounts refer to the contradictions of working with an actor-centred approach that is primarily concerned with the strategic actions of organized labour. A first critical comment on the PRA is about its notion of power. Gallas has raised the concern that the PRA is turning a "relational notion of power into a compartmentalising one" (2018: 348) as it does not specifically analyse the antagonistic relation between capital and labour and tends to describe workers' power as "quantifiable capacities that are possessed by the side of labour" (Gallas 2018: 349). As a consequence, these critiques argue that the approach tends to neglect class relations and the impact of union action on a broader class relation of forces; for example, concerted state action to break important strike actions (Nowak 2022).

A similar argument is about the PRA's focus on the union form as the dominant unit of analysis for workers action (Atzeni 2021). Most union-centred perspectives, including those referring to the PRA, would comprise one-sided "union frameworks" as they have led to serious shortcomings such as a focus on traditional industries, the formal wage-labour relationship and a lack of attention to "multiple forms of working-class action and organization" (Atzeni 2021: 10), particularly in the Global South.

A third objection to the research done with the PRA is the potential neglect of workers' subjectivities while studying workers' mobilization (Menz 2017). Menz argues that specific subjective attitudes and concerns of workers are a precondition for organizing and mobilization processes and are usually not included in a PRA analysis. Likewise, several authors have argued that the spaces of action of organized labour are shaped by the structural conditions of capitalism (Bieler 2018; Nowak 2018). Consequently, a mere analysis of possibilities of workers' agency and power would be insufficient to understand processes of workers' organization as it is often "combined with a complete disregard of the structural setting within which agency takes place" (Bieler 2018: 244). Rather, the structuring conditions of global capitalism need to be analysed to better understand workers' power.

One could argue that some of these critiques can be dismissed by looking more closely at the existing literature on power resources. As already mentioned, the PRA is usually used as an intermediate concept that focuses on the level between structural developments and agency. There are several studies linking the PRA to theories of global capitalism (Ellem, Goods & Todd 2020; Schmalz & Weinmann 2016;), class (Dörre 2019), subjectivities (Becker, Kutlu & Schmalz 2017; Birelma 2018) and hybrid forms of organization (Kumar & Singh 2018; Webster *et al.* 2021). However, there is a certain element of truth in the argument that worker agency needs to be located, as

Silver (2003) does in a long-term and global framework on the future of labour. Here, we consider her contribution on this matter.

Silver (2003: 167) asks whether we should "expect this general crisis of contemporary labour movements also to be temporary", as was the case with earlier crises, or whether the mobility and restructuring of capital globally has condemned the labour movement to permanent decline. Using the global framework that she developed resulted in Silver (2003: 171) expecting to see "new working-class formation and emerging labour movements in the leading industry and industries of the 21st century … the late 20th century crisis of labour movements is temporary and will probably be overcome with the consolidation of new working classes in formation". She goes on to argue that:

> … the twentieth-century trend toward increased workplace bargaining power … is at least partially being reversed in the twenty-first century. The bargaining power of many of today's low-wage workers in producer and personal services is closer to that of workers in the nineteenth-century textile industry than that of workers in the twentieth-century automobile industry. Lacking strong structural power, textile workers successes were far more dependent on a strong (compensatory) associational bargaining power (unions, political parties, and cross-class alliances with nationalist movements). (Silver 2003: 172)

Silver (2003: 172) concludes that "we might expect the weight of associational power in the overall power strategies of labour movements to be on the increase".

Silver framed her argument in terms of what she called the Marx–Polanyi dialectic where capital overcomes impediments to accumulation through various fixes (spatial, production, financial and technological). For Silver, these fixes result in cyclical processes of capital formation that make, unmake and remake the working class. Struggles of the old sections of the working class that resist the "recommodification" or "decommodification" of their labour power through the attack on both wages and the social wage (or even its employment altogether) are framed as being of the Polanyi-type. This is generally seen as the section of the working class that had made material gains during the heyday of social democracy – essentially to decommodify their labour power by regulating its sale and ultimately taking it out of the free market. Often the attack on this section of the class results in the "unmaking" of its traditional location through privatization or offshoring. Marx-type struggles, on the other hand, are presented as emerging from new sections of the working class that are made or remade through the various capital fixes and new capital formations. Silver traced how this process unfolds on a global level as impediments to accumulation in the auto industry force it to relocate production to cheaper parts of the world, thus generating new workers and new struggles against it.

We turn now in Section 4 to an examination of the making, unmaking and remaking of different sections of the labour force, and the question of how workers respond by drawing on different sources of power.

GLOBALIZATION: CHALLENGING THE "RACE TO THE BOTTOM"

The removal of many protections in trade, investment and finance under neoliberalism have created perfect conditions for global corporations to exploit the uneven geography of labour, leading to a "race to the bottom" as corporations search for the "China price" (Lambert & Webster 2017).[3] Harvey (2002: 31) called these strategies of capital a spatial fix as these successive geographical relocations of capital only succeeded in rescheduling crises in time and space.

The collapse of communism, which began in 1989 with the fall of the Berlin Wall, the further deepening of global free trade agreements within the World Trade Organization architecture and the creation of new forms of information and communication (most recently digitalization) have accelerated the capacity of capital to globalize. The effect of liberalization has been to virtually double the size of the global labour market, creating unprecedented opportunities for the exploitation of unprotected labour. This has undermined the traditional national approach to protecting labour and thrust labour movements in industrialized countries into crisis (Webster, Lambert & Bezuidenhout 2008). The unmaking of the industrial working classes in the Western world, therefore, tended to weaken workers' power in traditional union strongholds such as in the metal and auto industries. Plant relocations to East Asia, Mexico and Eastern Europe put the large industrial unions under pressure, with unions such as Germany's IG Metall, the US's UAW, and France's FTM losing members for many years. This resulted in a decrease of both structural and associational power, and also affected existing collective bargaining and co-determination institutions (institutional power). As a consequence, industrial unions needed to reach out to new groups of the working class such as employees and precarious workers – in the case of IG Metall, see Schmalz and Thiel (2017) – but also rethink their focus on national organizing strategies.

In the case of IG Metall, this changing environment led to new attempts to organize transnationally, which thereby entailed trying to support mainly defensive struggles over rationalization and offshoring and, hence, over the unmaking of the industrial working class through the tapping of transnational and supranational power resources. For this purpose, IG Metall created a transnational department, which aims at setting up firm networks, developing transnational organizing approaches along global value chains, and new partnerships with other unions. The challenge of this work is the competitive architecture of many free trade and investment agreements, which tend to tighten locational competition in the EU and across the world (Hürtgen 2019).

IG Metall's transnational department proved to be a space for organizing experimentation: for instance, IG Metall successfully established global firm networks in companies. At the automobile supplier, Lear Corporation, IG Metall coordinated a global struggle over outsourcing, which resulted in the establishment of guest seats for worker representatives of all non-EU countries at the European Works Council. This provided a unique platform for information exchange and joint action and developing new transnational institutional power resources (Schäfers & Schroth 2020: 12ff.; Ludwig, Simon

3. Employers use the term to describe the competitive advantage China has because of cheap labour and a lack of respect for the environment.

& Wagner 2021). Similarly, IG Metall and the UAW started to build a new partnership for organizing the German transplants of the large automobile companies in the south of the US (Fichter 2018). As a new joint institution, they opened the Transnational Labour Institute in 2015 in Tennessee, which serves as "IG Metall's bridge to the UAW and its organising drives in the US automobile industry" (Fichter 2018: 190). Since then, both unions have collaborated on joint projects, such as the notable organizing drive at the VW plant in Chattanooga, Alabama that was narrowly unsuccessful.

Both examples, whether successful or not, show that the new spatial fixes of global capitalism also open up opportunities for labour to engage in transnational organizing. Besides the transnationalization of established unions, an important new development is the re-emergence of GUFs, which have undergone a paradigm shift from their position as lobbying organizations into actors of transnational organizing since the 1990s (Fichter & McCallum 2015: 69; Schmalz *et al.* 2021). Today, the GUFs are useful tools for labour when they are combined with public campaigns and the exercise of other power resources at multiple scales. This is so, Ford and Gillan (2021) argue, because multinational enterprises operate across national boundaries, which creates an incentive to engage power resources at a supranational level, as well as within the countries where they, or their suppliers, are present. They show how GUFs can play an intermediary coordinating role in supporting the exercise of power by unions at the supranational level.

GUFs are playing a similar role in our examination of the global teacher unions contesting the abuse of digital technology in Africa (Webster & Ludwig 2021). "Although teaching has been historically impervious to technological transformations", Silver (2003: 119) anticipated the "unmaking" of teaching jobs through "the internet and other advanced technologies ... [that] might be used to bring effective competitive pressures to bear on teachers, analogous to those that automation brought to bear on manufacturing". Through transnational activism in Kenya and Uganda, the GUF Education International (EI) has effectively resisted the deprofessionalization of teachers through a global campaign against the privatization of education by for-profit operator Bridge International Academies (BIA). BIA, a private for-profit corporation, is a "flagship of creative capitalism", seeking to increase access to education for the poor, according to its founder (Bakan 2020: 122). At the centre of its business model is the standardization of education through digital technology. Concerned with the impact of this "McDonaldization" of the learning process, EI conducted studies in Kenya and Uganda, places where BIA had expanded rapidly. EI found, *inter alia*, that digital technology is used to deprofessionalize teachers and drastically cut costs. BIA, rather than engaging qualified teachers, employs unqualified, low-paid personnel who transmit scripted instructions to pupils through "teacher-computers". With this form of low-fee for-profit provision being openly advanced as an alternative to public schools, BIA represents a direct challenge to public education.

Particularly relevant for the campaign's success was EI's ability to draw on societal power. It did this by successfully influencing the public discourse (e.g. through research and protest actions) and by building alliances with civil society organizations. In 2015, and every year since, civil society organizations have written an open letter to investors, donor agencies and the World Bank, urging them to stop funding BIA and commercial education providers. As a result of this joint advocacy, in 2020 the World Bank's private sector lending arm, International Financial Corporation (IFC), decided to freeze any investments into private for-profit K-12 schools. This was a major achievement as IFC had previously promoted the privatization of education as an important area for its investment policy.

So, global unions can play an important role in facilitating local unions' resistance to global corporations. Although the power of global capital makes it a deeply unequal contest, GUFs are facilitating the development of unions' counter-power, both at the local and global levels. At a time when workers across the world are engaged in the struggle to meet the multiple challenges posed by tech giants such as Uber and Amazon, these research findings stress the important role of global union activity, which has long been neglected in labour research. Through its intermediary coordinating role at the supranational level, EI was a source of power to challenge the global tech giants. Consequently, global unions can reinforce, but not replace, activities of the local unions on the ground and have an important role to play. First, transnational learning, based on research and the exchange of experiences, has been highlighted as particularly relevant for local unions and is a process facilitated by GUFs. Second, global unions can strengthen the power resources of their members by connecting different levels of union action.

DIGITALIZATION: NEW LABOUR STRUGGLES

The struggle of EI against BIA has also shown the crucial role of technology in the process of the unmaking and remaking of global working classes. Capital tends to implement major process innovations to strengthen competitiveness against other companies and also to respond to organized labour (Basualdo *et al.* 2021: 4ff.). These "technological fixes" have been contested and have led to contradictory developments. For instance, the introduction of the assembly line in the early twentieth century in the US went hand in hand with rigid Taylorist forms of labour control, but also facilitated coordinated strike actions (Marx-type struggles) in the automobile industry. Consequently, emerging industries have usually led to the "making" or "remaking" of new working classes (as in the US automobile industry in the 1920s and 1930s), while at the same time technological transformation also tends to push forwards industrial rationalization and deindustrialization and, hence, the "unmaking" of yet-existing working classes (e.g. in the US steel industry since the 1980s). As a result of this restructuring, technological change tends to transform power resources (Basualdo *et al.* 2021: 6). With the process of unmaking of working classes, workers tend to lose structural power, whereas new skilled groups of the working class often tend to command high structural power.

Likewise, digitalization plays a contradictory role today. Algorithmic management in the platform economy can be identified as a new technological fix, bypassing existing labour law and institutional employment standards (institutional power), and also leading to intense struggles between capital and labour in the platform economy (Vandaele 2018, 2020). In a study on global labour unrest in the platform economy,

Trappmann *et al.* (2020) observed a surge of struggles since 2015 with conflicts taking place in all regions across the world. These labour conflicts are often about pay, working conditions, employment status and regulatory issues because many of the "freelancer jobs" in the platform economy are disguised employment relationships.

A sector with particularly widespread collective action is that of food delivery, which sees several large companies such as Deliveroo, Uber Eats, Foodora, Glovo and Meituan dominating the global business (Trappmann *et al.* 2020: 3). For the most part, labour conflicts in this sector have taken place in Europe but have also taken place in other regions such as Latin America, East Asia and Africa. Collective action here has usually involved relatively few participants and been short-lived. A common development was that protesters needed to build associational power, and different forms of organization and networks started to emerge. This "variety of platform unionism" (Basualdo *et al.* 2021: 13) usually includes informal groups that, in some cases (in particularly the Global North), were supported by established unions in their struggle. The repertoire of contention utilized by the riders includes digital technology and they draw on new forms of structural power such as non-traditional strikes and digital protest actions.

Courier workers in South Africa are a good example of precarious platform work and new labour struggles in the sector. They labour as "gig" workers in the digital economy that are subject to algorithmic management. Instead of clocking in, they log into an app and take instructions from an external authority. In essence, it is an old form of casual, informal work (Webster & Masikane 2021). In spite of the individualization, dispersal and pervasive monitoring that characterizes work in the "gig economy", the digital technology is also generating forms of counter-mobilization. In large part, these forms of collective action are self-organized into informal digital networks with little or no support from established unions. On 18 December 2020, in a response to a fare increase and a range of other demands related to their working conditions, about 1,000 food couriers collectively logged off the platform and forced Uber Eats to halt operations across Johannesburg. Their demands had been formulated and the strike was organized through the use of mobile messaging groups. The workers' demands were not met, although the action did lead to the establishment of a forum, which has not yet met with management. A central challenge to these attempts at organization is the tech companies' denial that they are in fact in an employment relationship. Instead, the riders are misclassified as self-employed; that is, independent contractors.

Comparable organizing processes have also happened in Europe. This is quite surprising because many of these attempts of informal self-organization have taken place in highly institutionalized systems of labour relations. Similar to the South African case, collective action in European food delivery, such as the Deliveroo workers' protests in London and other European cities, or the German Gorilla delivery riders' protests, usually started as bottom-up protests. In many cases they led to some sort of union engagement. An interesting example of these actions is the Riders' Union in Netherlands, which was founded in 2017 and became a formal part of the Dutch Trade Union Federation until mid-2018 (FNV, Federatie Nederlandse Vakbeweging) (Vandaele 2020). In the case of the Riders' Union, the couriers chose to work together with the youth organization of FNV. The union not only supported the organizing process with financial resources and legal knowledge, but union organizers also "quite successfully ... reproduced the approach of grassroots unions in mobilising and organising the couriers" (Vandaele 2020: 123). The strategy turned out to be effective and a major result of the conflict was the introduction of regular employment contracts in 2019. Unlike in the Netherlands, the couriers in neighbouring Belgium opted for a different relation to the union. The Couriers' Collective, which was set up in 2015, remained autonomous but worked together with two of the most important union federations, which created their own union structures for the riders (Vandaele 2020). In summary, in Europe there is an increasing diversity of different forms of coalition building between traditional unions and grassroots-associations in the food delivery sector.

The courier protests can be perceived as Marx-type struggles resulting from the making of a new digital precariat. By technologically linking platform workers, the gig economy tends to link their working bargaining power and, thus, contributes to the emergence of self-organized, hybrid forms of union-like associations (associational power) and new partnerships with traditional unions and non-governmental organizations (coalitional power). These collaborations can take different forms and reach from largely independent grassroot networks (South Africa) to a more formal integration of the protesters into unions (Netherlands). In the Global North, established unions play a larger role in these labour struggles than in the Global South. In most cases the couriers can also draw upon discursive power to successfully articulate their demands in the public sphere by meeting in public spaces. New digital technology is a double-edged sword. On the one hand it is leading to an extension of authoritarian managerial control over workers, increasing their insecurity and deepening levels of inequality. On the other hand, by providing technological links between workers, they have increased their workplace bargaining power and provided them with the ability to develop collective solidarity and even strike.

CONCLUSION

We have suggested in this chapter that global restructuring should be seen as a process that simultaneously undermines and provides the opportunities to potentially strengthen the labour movement. However, this process of unmaking and remaking of the working class does not occur spontaneously for it requires the identification of new strategies and sources of power. We showed how, through the application of the PRA, a growing number of labour scholars have identified various innovative ways in which labour is rediscovering a variety of sources of power. Our conception of the PRA, which adds two further power resources, institutional and societal power, to the already discussed sources of power, structural and associational power, has proven useful to analyse organized labour's response to the structural trends of globalization and digitalization. Our contribution in this chapter, thus, lies in two areas. First, we showed how globalization has deepened transnational organizing through the examples of the GUF EI's campaign in Kenya and Uganda and IG Metall's transnational department. Second, we showed how the traditional notion of a union is being challenged by the emergence of self-organized, hybrid forms of union-like associations among platform-based food delivery workers in South Africa and in the revitalization of the transport union in Uganda against the background of digitalization. Often these studies have involved close collaboration between labour scholars and labour organizers, which has had a revitalizing effect on both the study of labour and the labour movement. In this chapter we have also identified a new phase of cross-border knowledge transfer among scholars and union activists. There is, as Chun (2012: 40) argues, a "growing interest in a new political subject of labour ... women, immigrants, people of color, lowpaid service workers, precarious workers, groups that have been historically excluded from the moral and material boundaries of union membership".

There is, however, a danger that, instead of opening up to experimentation and the protection of informal workers, traditional unions will simply defend the interests of permanent workers and abandon precarious ones. As a result of their dependency, they could become defenders of the existing order, becoming allies of employers in guaranteeing industrial peace and stability.[4] However, established unions also stand to benefit from lending solidarity to precarious workers. The example of the Ugandan ATGWU demonstrates how a union under pressure, through experimentation and the incorporation of informal workers, was able to deepen associational power and dramatically expand the union (Webster *et al.* 2021). Importantly, where unions have taken up the issues of informal workers, they have also undergone fundamental changes. They often become "hybrid" organizations, which include different forms of organizations and blur the distinction between traditional unionism, informal workers' associations and cooperatives.

A crucial question raised by our account of the rediscovery of workers' power is whether it is possible to sustain these new associational forms and re-embed markets in society in a way that would ensure the future of labour. While structural power may well have been at the centre historically of the rise of labour, we suggest that to rebuild the labour movement and strengthen associational power, unions will need to draw more upon societal power by forming coalitions. It will also require a more systematic use of the courts to consolidate gains through institutional power. But answering this question on the future of labour will require a more systematic investigation into how labour is responding to what is arguably the greatest challenge facing labour in the twenty-first century, climate change.

In the past, environmental activists have not always given employment issues the attention they warrant. Similarly, labour has neglected environmental issues and could be more involved in the emerging climate justice movement. In many coal mining-affected communities, the "unmaking" of coal miners through a "just transition" is forcing the industry to find ways of "remaking" its workforce (Cock 2021). The slogan "No jobs in a dead planet" is gaining popularity in overcoming the conventional dualistic way of thinking about jobs or nature. New alliances and coalitions are emerging, as revealed in the recently published *Palgrave Handbook on Environmental Labour Studies* (Räthzel, Stevis & Ussell 2021). These emerging coalitions also raise new questions about the joint power resources of environmental and labour movements (on the concept of "metabolic power", see Dörre 2021).

4. For example, in the case of Heineken breweries in South Africa, the established union in the sector defended permanent workers at the expense of many precarious workers without representation (Webster & Englert 2021).

Another neglected issue is the forms of violent labour protests that do not directly relate to the sources of labour power we have described in this chapter. Forms of collective action are emerging and new movements being created outside the scope of institutionalized conflict. "These protests are class-specific, bread-and-butter conflicts in which protesters feel powerless in the face of international financial institutions and vent their anger in the destruction of property and militant forms of action." (Schmalz, Ludwig & Webster 2018: 128). Violent labour protests seem to mobilize other disruptive forms of power to more institutionalized struggles and therefore require the development of new research strategies and concepts.

Does this growing identification of new sources of power signal the end of traditional labour unionism and the emergence of new forms of organization that are more suited to work in the digital age? These are the questions raised by the mobilization of workers' power in the twenty-first century. We have entered a period of experimentation and the PRA has emerged as a valuable tool for identifying possible directions of the labour movement. It is also, in the process, revitalizing increasingly globalized labour studies.[5]

REFERENCES

Anwar, M. & M. Graham 2020. "Hidden transcripts of the gig economy: labour agency and the new art of resistance among African gig workers". *Environment and Planning A: Economy and Space* 52(7): 1269–91.

Atzeni, M. 2021. "'Workers' organizations and the fetishism of the trade union form: toward new pathways for research on the labour movement?". *Globalizations* 18(8): 1349–62.

Bakan, J. 2020. *The New Corporation. How "Good" Corporations are Bad for Democracy*. New York: Vintage.

Basualdo, V. *et al.* 2021. *Building Workers' Power in Digital Capitalism: Old and New Labour Struggles*. Bonn: Friedrich Ebert Stiftung.

Becker, K., Y. Kutlu & S. Schmalz 2017. "Kollektive Machtressourcen im Care-Bereich: Die mobilisierende Rolle des Berufsethos". In *Sorge-Kämpfe: Auseinandersetzungen um Arbeit in sozialen Dienstleistungen*, edited by I. Artus, P. Birke, S. Kerber-Clasen & W. Menz, 255–77. Hamburg: VSA-Verl.

Becker, J. 2019. "Labour protests in Eastern Europe". In *Confronting Crisis and Precariousness: Organised Labour and Social Unrest in the European Union*, edited by S. Schmalz & B. Sommer, 189–208. Lanham, MD: Rowman & Littlefield.

Bieler, A. 2018. "Agency and the power resources approach: asserting the importance of the structuring conditions of the capitalist social relations of production". *Global Labour Journal* 9(2): 243–8.

Birelma, A. 2018. "When local class unionism meets international solidarity: a case of union revitalisation in Turkey". *Global Labour Journal* 9(2): 215–30.

Brinkmann, U. *et al.* 2008. *Strategic Unionism: Aus der Krise zur Erneuerung? Umrisse eines Forschungsprogramms*. Wiesbaden: Springer.

Brookes, M. 2013. "Varieties of power in transnational labour alliances: an analysis of workers' structural, institutional, and coalitional power in the global economy". *Labour Studies Journal* 38(3): 181–200.

Brookes, M. 2018. "Power resources in theory and practice: where to go from here". *Global Labour Journal* 9(2): 254–57.

Buhlungu, S. 2010. *A Paradox of Victory: COSATU and the Democratic Transformation in South Africa*, Scottsville: University of KwaSulu-Natal Press.

Burawoy, M. 2005. "For public sociology". *American Sociological Review* 70(1): 4–28.

Chun, J. 2005. "Public dramas and the politics of justice: comparison of janitors' union struggles in South Korea and the United States". *Work and Occupations* 32(4): 486–503.

Chun, J. 2012. "The power of the powerless: new schemas and resources for organizing workers in neo-liberal times". In *Cross-National Perspectives on Social Movement Unionism: Diversities of Labour Movement Revitalization in Japan, Korea, and the United States*, edited by A. Suzuki. Oxford: Peter Lang.

5. The authors express their gratitude to Karlotta Hein for her research assistance and editing work.

Cock, J. 2021. "Beware of the crocodile's smile: labour environmentalism in the struggle to achieve a just transition in South Africa". In *The Palgrave Handbook of Environmental Labour Studies*, edited by N. Räthzel, D. Stevis & D. Ussell, 177–97. Basingstoke: Palgrave Macmillan.

Dias, H. 2021. "The evolution of Portuguese trade unionism: political economies and power resources". *Industrial Relations Journal* 52(3): 237–54.

Dörflinger, N., V. Pulignano & S. Vallas 2021. "Production regimes and class compromise among European warehouse workers". *Work and Occupations* 48(2): 111–45.

Dörre, K. 2018. "Überbetriebliche Regulierung von Arbeitsbeziehungen". In *Handbuch Arbeitssoziologie*, edited by F. Böhle, G. Voß & G. Wachtler, 619–81. Wiesbaden: Springer.

Dörre, K. 2019. "Umkämpfte Globalisierung und soziale Klassen. 20 Thesen für eine demokratische Klassenpolitik". In *Demobilisierte Klassengesellschaft und Potenziale verbindender Klassenpolitik. Beiträge zur Klassenanalyse (2)*, edited by M. Candeias, K. Dörre & T. Goes, 11–56. Berlin: Rosa-Luxemburg-Stiftung.

Dörre, K. 2021. "Der Machtressourcenansatz – Zwischenbilanz, Reformulierung, Ausblick". In *Mosaiklinke Zukunftspfade. Gewerkschaft, Politik, Wissenschaft*, edited by B. Aulenbacher *et al.*, 85–94. Münster: Westfälisches Dampfboot.

Dörre, K., H. Holst & O. Nachtwey 2009. "Organizing: a strategic option for trade union renewal?". *International Journal of Action Research* 5(1): 33–67.

Doutch, M. 2021. "A gendered labour geography perspective on the Cambodian garment workers' general strike of 2013 and 2014". *Globalizations* 18(8): 1406–19.

Ellem, B., C. Goods & P. Todd 2020. "Rethinking power, strategy and renewal: members and unions in crisis". *British Journal of Industrial Relations* 58(2): 424–46.

Fichter, M. 2018. "Building union power across borders: the Transnational Partnership Initiative of IG Metall and the UAW". *Global Labour Journal* 9(2): 182–98.

Fichter, M. & J. McCallum 2015. "Implementing global framework agreements: the limits of social partnership". *Global Networks* 15(1): 65–85.

Ford, M. & M. Gillan 2021. "Power resources and supra-national mechanisms: the global unions and the OECD guidelines". *European Journal of Industrial Relations* 27(3): 307–25.

Fox-Hodess, K. & C. Santibáñes Rebolledo 2020. "The social foundations of structural power: strategic position, worker unity and external alliances in the making of the Chilean dockworker movement". *Global Labour Journal* 11(3): 222–38.

Frege, C., E. Heery & L. Turner 2004. "The new solidarity? Trade union coalition-building in five countries". In *Varieties of Unionism: Strategies for Union Revitalization in a Globalizing Economy*, edited by C. Frege and J. Kelly, 137–58. Oxford: Oxford University Press.

Gallas, A. 2016. "There is power in a union: a strategic-relational perspective on power resources". In *Monetary Macroeconomics, Labour Markets and Development*, edited by A. Truger *et al.*, 195–210. Berlin: Metropolis.

Gallas, A. 2018. "Class power and union capacities: a research note on the power resources approach". *Global Labour Journal* 9(3): 348–52.

Ganz, M. 2000. "Resources and resourcefulness: strategic capacity in the unionization of Californian agriculture, 1959–66". *American Journal of Sociology* 105(4): 1003–62.

Gumbrell-McCormick, R. & R. Hyman 2013. *Trade Unions in Western Europe: Hard Times, Hard Choices*. Oxford: Oxford University Press.

Haipeter, T. 2012. "Sozialpartnerschaft in und nach der Krise: Entwicklungen und Perspektiven." *Industrielle Beziehungen* 19(4): 387–411.

Harvey, D. 2002. *Spaces of Hope*. Edinburgh: Edinburgh University Press.

Herberg, M. 2018. *Trade Unions in Transformation: Success Stories from All Over the World*. Berlin: Friedrich Ebert-Stiftung.

Hui, E. 2022. "Bottom-up unionization in China: a power resources analysis". *British Journal of Industrial Relations* 60(1): 99–123.

Hürtgen, S. 2019. "The competitive architecture of European integration: European labour division, locational competition and the precarization of work and life". In *Confronting Crisis and Precariousness: Organised Labour and Social Unrest in the European Union*, edited by S. Schmalz & B. Sommer, 33–52. Lanham, MD: Rowman & Littlefield.

Joynt, K. & E. Webster 2016. "The growth and organization of a precariat: working in the clothing industry in Johannesburg's inner city". In *Neo-liberal Capitalism and Precarious Work: Ethnographies of Accommodation and Resistance*, edited by R. Lambert & A. Herod, 43–71. Cheltenham: Edward Elgar.

Jürgens, U. 1984. "Die Entwicklung von Macht, Herrschaft und Kontrolle im Betrieb als politischer Prozess – eine Problemskisse zur Arbeitspolitik". In *Arbeitspolitik. Materialien zum Zusammenhang von politischer Macht, Kontrolle und betrieblicher Organisation der Arbeit*, edited by U. Jürgens & F. Naschold, 58–91. Opladen: Westdeutscher Verlag.

Krein, J. & H. Dias 2018. "The CUT's experience during the Workers' Party governments in Brazil (2003–2016)". *Global Labour Journal* 9(2): 199–214.

Kumar, S. & A. Singh 2018. "Securing, leveraging and sustaining power for street vendors in India". *Global Labour Journal* 9(2): 135–49.

Lambert, R. & E. Webster 2017. "The China price: the all-China Federation of Trade Unions and the repressed question of international labour standards". *Globalizations* 14(2): 313–26.

Lehndorff, S., H. Dribbusch & T. Schulten 2017. *Rough Waters: European Trade Unions in a Time of Crises*. Brussels: ETUI.

Lévesque, C. & G. Murray 2010. "Understanding union power: resources and capabilities for renewing union capacity". *Transfer: European Review of Labour and Research* 16(3): 333–50.

Lévesque, C. & G. Murray 2013. "Renewing union narrative resources: how union capabilities make a difference". *British Journal of Industrial Relations* 51(4): 777–96.

Ludwig, C., H. Simon & A. Wagner 2021. *Entgrenzte Arbeit, (un-)begrenzte Solidarität? Bedingungen und Strategien gewerkschaftlichen Handelns im flexiblen Kapitalismus*. Münster: Westfälisches Dampfboot.

Marslev, K., C. Staritz & G. Raj-Reichert 2022."Rethinking social upgrading in global value chains: worker power, state-labour relations and intersectionality". *Development and Change* 53(4): 827–59.

Marx, K. 1974. *Political Writings*, Volume 3. New York: Vintage.

McGuire, D. & C. Scherrer 2015. "Providing labour with a voice in international trade negotiations". *Philippine Journal of Labour and Industrial Relations* 33: 1–23.

Menz, W. 2017."Gerechtigkeit, Rationalität und interessenpolitische Mobilisierung ... Die Perspektive einer Soziologie der Legitimation". In *Sorge-Kämpfe: Auseinandersetzungen um Arbeit in sozialen Dienstleistungen*, edited by I. Artus *et al.*, 278–305. Hamburg: VSA-Verlag.

Milkman, R., J. Bloom & V. Narro 2010. *Working for Justice: The L.A. Model of Organizing and Advocacy*. Ithaca, NY: ILR Press.

Mmadi, M. 2022. "Working-class commuters and innovative use of associational power: the case of Mamelodi train sector in South Africa". *Global Labour Journal* 13(1): 80–93.

Nowak, J. 2018. "The spectre of social democracy: a symptomatic reading of the power resources approach". *Global Labour Journal* 9(3): 353–60.

Nowak, J. 2022. "Do choke points provide workers in logistics with power? A critique of the power resources approach in light of the 2018 truckers' strike in Brazil". *Review of International Political Economy* 29(5): 1675–97.

O'Brady, S. 2021. "Fighting precarious work with institutional power: union inclusion and its limits across spheres of action". *British Journal of Industrial Relations* 59(4): 1084–107.

Osorio Lavín, S. & D. Velásques 2022. "El poder sindical en el 'Estallido social' chileno. La huelga general de noviembre de 2019". *Revista Española de Sociología* 31(1).

Pannini, E. 2023. "Winning a battle against the odds: a cleaners' campaign". *Economic and Industrial Democracy* 44(1): 68–87.

Piven, F. 2008. *Challenging Authority: How Ordinary People Change America*. Lanham, MD: Rowman & Littlefield.

Poulantzas, N. 1978. *State Theory: Political Superstructure, Ideology, Authoritarian Statism*. London: Verso.

Räthzel, N., D. Stevis & D. Ussell 2021. *The Palgrave Handbook of Environmental Labour Studies*. London: Palgrave Macmillan.

Refslund, B. & J. Arnholtz 2022. "Power resource theory revisited: the perils and promises for understanding contemporary labour politics". *Economic and Industrial Democracy* 43(4): 1958–79.

Rego, K. 2022. "Works councils and the digitalisation of manufacturing: opportunity or threat for their power position?" *Economic and Industrial Democracy* 43(4): 1911–33.

Rhomberg, C. & S. Lopez 2021. "Understanding strikes in the 21st century: perspectives from the United States". In *Power and Protest*, edited by L. Leitz, 37–62. Bingley: Emerald.

Runciman, C. 2019. "The 'double-edged sword' of institutional power: COSATU, neo-liberalisation and the right to strike". *Global Labour Journal* 10(2): 142–58.

Schäfers, K. & J. Schroth 2020. *Shaping Industry 4.0 on Workers' Terms: IG Metall's "Work+Innovation" Project*. Bonn: Friedrich Ebert Stiftung.

Schmalz, S., C. Ludwig & E. Webster 2018. "The power resources approach: developments and challenges". *Global Labour Journal* 9(2): 113–34.

Schmalz, S. & M. Thiel 2017. "IG Metall's comeback: trade union renewal in times of crisis". *Journal of Industrial Relations* 39(4): 465–86.

Schmalz, S. & N. Weinmann 2016. "Between power and powerlessness: labour unrest in Western Europe in times of crisis". *Perspectives on Global Development and Technology* 15(3): 543–66.

Schmalz, S. *et al.* 2021. "Two forms of transnational organizing: mapping the strategies of global union federations". *Tempo Social* 33(2): 143–62.

Schmitter, P. & W. Streeck 1981. "The organization of business interests: a research design to study the associative action of business in the advanced industrial societies of Western Europe". Discussion Paper, IIM and LMP 81 and 13. Berlin: IIM and LMP.

Silver, B. 2003. *Forces of Labour. Workers' Movements and Globalization since 1870.* Cambridge: Cambridge University Press.

Thompson, E. 1971. "The moral economy of the English crowd in the eighteenth century". *Past & Present* 50: 76–136.

Trappmann, V. *et al.* 2020. *Global Labour Unrest on Platforms: The Case of Food Delivery Workers.* Berlin: Friedrich Ebert Stiftung.

Turner, L. 2006. "Globalization and the logic of participation: unions and the politics of coalition building". *Journal of Industrial Relations* 48(1): 83–97.

Urban, H.-J. 2012. "Crisis corporatism and trade union revitalisation in Europe". In *A Triumph of Failed Ideas: European Models of Capitalism in the Crisis*, edited by S. Lehndorff, 219–41. Brussels: ETUI.

Vandaele, K. 2018. "Will trade unions survive in the platform economy? Emerging patterns of platform workers' collective voice and representation in Europe". ETUI Research Paper – Working Paper 2018.05.

Vandaele, K. 2020. *From Street Protest to "Improvisational Unionism": Platform-based Food Delivery Couriers in Belgium and the Netherlands.* Bonn: Friedrich Ebert Stiftung.

Von Holdt, K. & E. Webster 2008. "Organising on the periphery: new sources of power in the South African workplace". *Employee Relations* 30(4): 333–54.

Webster, E. 1988. "The rise of social movement unionism: the two faces of the Black trade union movement in South Africa". In *State, Resistance, and Change in South Africa*, edited by P. Frankel, N. Pines & M. Swilling. London: Croom Helm.

Webster, E. 2015. "Labour after globalisation: old and new sources of power". In *Labour and Transnational Action in Times of Crisis*, edited by A. Bieler *et al.* Lanham, MD: Rowman & Littlefield.

Webster, E. & T. Englert 2021. "New dawn or end of labour?: from South Africa's East Rand to Ekurhuleni., University of Witwatersrand". In *Challenging Inequality in South Africa. Transitional Compasses*, edited by M. Williams & V. Satgar London: Routledge.

Webster, E., R. Lambert & A. Bezuidenhout 2008. *Grounding Globalization: Labour in the Age of Insecurity.* Oxford: Blackwell.

Webster, E. & C. Ludwig 2021. "Contesting digital technology through new forms of transnational activism in Africa". Southern Centre for Inequality Studies: University of Witwatersrand.

Webster, E. & F. Masikane 2021. *I Just Want to Survive: A Comparative Study of Food Courier Delivery Workers in Three African Cities.* Bonn: Friedrich Ebert Stiftung.

Webster, E. *et al.* 2021. "Beyond traditional trade unionism: innovative worker responses in three African cities". *Globalizations* 18(8): 1363–76.

Wright, E. 2000. "Working-class power, capitalist-class interests and class compromise". *American Journal of Sociology* 105(4): 957–1002.

CHAPTER 4

UNION FORMS: ADAPTATION AND INERTIA

Chiara Benassi, Christian L. Ibsen and Maite Tapia

ABSTRACT

This chapter reviews the different forms unions can take and have taken across time and space as they have adapted to changes in the labour market. When examining union "forms", we focus upon their internal governance structure, the constituencies they represent, and their repertoires of representation. Specifically, we draw upon examples from Denmark, Italy, Germany and the United States, which reflect traditional union typologies. For each country, we sketch out the union forms as well as how traditional unions have changed over time. We show that some unions have changed their statutes and bargaining strategies, expanded their services or set up new structures. Finally, we compare the trajectories of labour unionism and highlight the commonalities across countries as well as the differences due to the institutional context and the strategies of the main employment relations actors.

Keywords: Union forms; union function; union identity; challenges for unions

INTRODUCTION

Unions can take and have taken on different forms across time and space as they have chosen or been compelled to adapt to a new worlds of work. In this chapter we use the term, "labour union form", as the defining organizational characteristic to distinguish labour unions analytically from each other and trace their organizational changes over time. In line with others such as Fiorito and Jarley (2008), Givan (2007) and Streeck (2005), we distinguish union forms along three dimensions: internal governance structure, constituency and repertoires of representation. Internal governance structure – also called "internal vertical structure" – refers to the way in which organizational levels are ordered and the degree of articulation or division of power between leaderships and the rank-and-file or grassroots members. Constituency – also called the "external horizontal structure" or the "who" – refers to the groups of workers the union purports to represent. It sets the scope of membership and, thus, delimits the union from other unions (Fiorito & Jarley 2008). Repertoires of representation – also called "the how" – refer to the typical representational practices that unions employ to realize the interests of their constituencies and/or the interests of the organization.

THREE DIMENSIONS OF FORM

In the following we expand upon these dimensions and their limits to spell out the "property space" (Lasarsfeld 1937) in which different unions can occupy a position. As such, we see union forms configurationally, whereby different positions on each of the three dimensions combine to constitute specific configurations.

Internal governance structures

These range from "top-down" or even oligarchic unions (Michels 1962) to "bottom-up" or member-driven unions. These structures can be highly formalized or simply reflect informal practices. Moreover, they define relationships between levels of unions; for example, through an authoritative central confederation, which may help transcend sectional divisions and integrate diverse membership interests (Hyman 1997: 519). Essentially, they distribute power to decide and implement union strategies between different actors: leaders, officials, shop stewards and members.

While some scholars speak of the degree of democracy in unions, we prefer to differentiate between different forms of democracy. Certainly there are unions that have zero member input and cannot be considered democratic; however, in many unions, it is considered a value in itself to promote – even if only on paper – democratic values. Some unions have indirect forms of democracy, only asking members to elect representatives that make the decisions and depend on union officials to implement them. In other unions, democracy is direct and members are consulted and can vote on decisions; for example, to go on strike or ratify bargaining results.

Scholars often portray internal governance structures as a trade-off between output legitimacy and input legitimacy (Scharpf 1999). Top-down unions are said to increase output legitimacy as the union can more easily choose strategies and reach agreements with counterparts. If the rank and file becomes involved, decision-making slows down and becomes more difficult, with a delay in taking necessary decisions or even not taking them at all. In contrast, bottom-up unions are said to increase input legitimacy as leadership listens to members or even lets them decide through deliberative processes, although possibly reducing organizational effectiveness in the short term (Fiorito & Jarley 2008). However, if leadership closes off members' participation, the true interests of the constituency may not enter into decision-making. Clearly, unions with voluntary membership have to balance output and input legitimacy as the "... long run suppression or neglect of democracy reduces member commitment and consequently the union's capacity for mobilization and effectiveness in the long term" (from Kochan 1980 cited in Fiorito & Jarley 2008: 194).

Constituencies

These range from small, clearly defined groups of workers; for example, occupations in a particular enterprise, to large, broadly defined body of workers, such as unskilled and skilled workers within a national territory. Specifying the constituency and the limits of the union vis-à-vis other unions continues to be one of the most defining and difficult organizational tasks for unions. Historically, unions in many European countries and North America grew out of the guilds and were based upon trades and crafts with clearly defined boundaries based on skills. Later came industrial unions representing all workers – both skilled and unskilled – in one industry. In some countries, such as Japan, enterprise unions have been the typical demarcation. Over time, general workers' unions have become more prominent, representing workers from multiple industries and skill levels. As Fiorito and Jarley (2008) state, general workers' unions are often, but not always, the result of mergers between unions in decline. Likewise, technological change can erase demarcations based on skills or industries. Some unions organize workers in particular communities, often because of economic decline in that geographical area. Finally, international unions represent workers across countries. These unions may also serve as confederations for national unions representing workers in internationalized industries, companies or supply chains.

Because the aforementioned unions typically focus on economic issues – perhaps with the exception of community unions – social identity-based unions reach constituencies that do not primarily identify as "working class" or as an "occupation" (Healy, Hansen & Ledwith 2006; Tapia, Lee & Filipovitch 2017). The constituencies of these unions unite around social identities based on gender, race, ethnicity or religion, and often arise as an alternative to traditional unions. Historically, many Christian unions grew out of a rejection of the Marxist or social democratic class-based view of workers against employers. However, more recently, identity-based unions have grown in opposition to traditional unions' bias towards improving predominately the working lives of White, male workers in standard employment. Thus, the new unions, such as the United Voices of the World union in the UK, often fill the "representational gap" for minority constituencies such as migrant workers, non-standard workers, workers of colour, or LGBTQ+ workers. Representing these constituencies also means going beyond bread-and-butter issues and taking on a more holistic approach to working-life issues, such as racism, discrimination and harassment.

Repertoires of representation

These range from institutionalized collective bargaining with employers, to political action against the government, to collective activism and mobilization of workers in the workplace or in the streets. Historically, unions grew out of activism and mobilization of workers, which often spurred major industrial unrest and long work stoppages. To curb the political and economic damage, governments and employers in most countries, albeit in different ways, conceded to institutionalizing the right to organize, take industrial action and bargain collectively over terms and conditions of employment (Crouch 1993). The result of these historical developments provided the national institutional framework for collective bargaining as the main repertoire of representation for unions during the post-Second World War era. However, as emphasized by Kelly (1998), repertoires change according to long waves of economic development, and mobilization based upon perceived injustices in an era of liberalization and union decline may overtake collective bargaining as the main repertoire (see also Turner 2003).

Recently, some scholars have argued that social movement unionism, which aims to rebuild activism among members, is an important way to revitalize a declining labour movement (e.g. Turner, Katz & Hurd 2001; Kelly 1998). Some scholars have also looked beyond unions and studied alternative types of worker representation, such as worker centres or labour-community alliances (e.g. Fine 2006; Holgate & Wills 2007; Milkman & Ott 2014; Tattersall 2010). Worker centres, social movements and union alliances with social movements often go beyond economic or workplace-specific issues and focus upon identity politics, work–family conflict, well-being and racial or social justice. Non-bargaining actors and non-legally binding collective agreements with employers characterize these new organizations (Givan 2007). Accordingly, the repertoires of these new organizations typically focus upon service, advocacy or organizing to improve the working and living conditions for the members of the community rather than winning bargaining rights with employers (Bernhardt & Osterman 2017).

WHY AND HOW DO UNION FORMS CHANGE?

Union forms have evolved over time in response to changes in the demographic, socio-economic, political and institutional context as well as shifts in employers' strategies and production structures. Shifts in the workforce composition constitute the first obvious change unions needed to respond to. Young workers, for instance, are more likely to be employed on non-standard contracts and might have different career ambitions – especially considering the expansion of higher education – than previous generations. Thus, they might have different demands and require the provision of new services (Tapia & Turner 2018). Virtuous examples of unions opening their door to young workers are the Public and Commercial Services Union in the UK, which created a young members' network supported by a full-time youth organizer in each branch (Hodder 2014) and the Solidaires union in France, which provided financial and legal support to a newly formed group of young workers in the non-profit sector called ASSO (Simms *et al.* 2018).

Increased workforce diversity and the salience of identity politics also require unions not only to expand their constituencies but also to change how workers are represented and how their demands are formulated and framed. For instance, when it comes to supporting ethnic and migrant workers, scholars have argued that unions need to acknowledge the racialized and gendered experiences of these workers in and outside the workplace (e.g. constant discrimination and lack of respect by management) (Holgate 2005; Tapia & Alberti 2018). Furthermore, scholars have pointed to the importance of identity work for creating a sense of shared community, which can no longer rest on a singular workers' identity. As full-time, male, White workers do not represent the majority of union members anymore, unions need to embrace the intersectionality of their new constituencies (Alberti, Holgate & Tapia 2013; Tapia, Lee & Filipovitch 2017). For example, the "Fight for $15" campaigns in the US – financially and strategically supported by the Service Employees International Union (SEIU) – has not just focused on economic gains, but clearly showed how issues of economic, social and racial justice are interlinked. It has mobilized together with, for example, the Black Lives Matter movement, to show how issues of police violence are likely to take place in communities that suffer from economic violence.

Government reforms deregulating the labour market, the decline of union membership and the progressive erosion of collective bargaining institutions have also prompted unions to change their form (Baccaro, Hamann & Turner 2003; Turner 2009). Union mergers represent one possible strategic response to the crisis of traditional unionism. For instance, Unison, the main public sector union in the UK, is the outcome of the merger between unions representing, respectively, health care workers, white-collar workers and blue-collar workers in the public sector, which took place in 1993 (Dempsey & McKevitt 2001). Due to the erosion of their institutional position, unions have also been found to invest increasing resources in organizing the "unorganized" – not only in Anglo-Saxon countries, where industrial relations institutions are typically weak but also in, for example, Continental Europe, where unions have traditionally benefited from strongly institutionalized bargaining rights (Greer 2008; Ibsen & Tapia 2017; Vandaele & Leschke 2010).

Changes in employer strategies and product structures have also challenged "old forms" of unionism. Workplaces have become more fragmented through global outsourcing, subcontracting and the use of agency workers and other non-standard contracts, so companies now source much of their labour from a network of third-party organizations within or beyond national borders (Anner, Fischer-Daly & Maffie 2021; Weil 2014). Thus, the traditional bargaining strategies of industrial unions and enterprise unions have become less effective (Anner, Fischer-Daly & Maffie 2021; Benassi & Dorigatti 2020).

Through workplace fissuring (Weil 2014), workers have moved from large, unionized core companies to smaller subcontractors and agencies, which are often not unionized, or they are covered by collective agreements in other sectors or companies, setting lower wages and working conditions (Doellgast & Greer 2007; Marchington *et al.* 2005). As a result, unions have been felt compelled to change. Most often, unions changed their repertoire by expanding their constituencies to workers outside the boundaries of the core firm and by including provisions in their collective agreements aimed at improving their standards (Benassi & Dorigatti 2020; Wagner & Refslund 2016). However, they have also been found to form coalitions with civil society organizations and social movements to bargain along the value chain, thus changing their "form" into a "labour network" when they enter the collective bargaining arena. For instance, in the US the Coalition of Immokalee workers forced companies along the food value chain to participate in the Fair Food Program thanks to the support of religious leaders and students (Anner, Fischer-Daly & Maffie 2021: 706), and in Bangladesh a coalition of national and international unions as well as non-governmental organizations (NGOs) forced global apparel brands to join the Accord Program (Donaghey & Reinecke 2018).

The academic debate on the aforementioned changes in union "forms" has focused both on their trajectory of change as well as on their effectiveness. On the one hand, scholars have argued that unions change following a path-dependent trajectory. In other words, unions gradually adapt their form to evolving external circumstances by building upon their traditional institutional resources and repertoires; thus, their form changes only incrementally over time and they maintain most of their distinctive features (Frege & Kelly 2003; Hyman 2001). Scholars have found that some institutional features of national labour movements might contribute to a greater chance of successful change. For instance, unions with a working-class identity have been found more likely to expand their representation domain to new constituencies than business unions (Benassi & Vlandas 2015; Papadopoulos 2014). Internal structures have also been found to matter. For example, internally democratic unions were found to be more likely to respond to changes in labour markets (Marino 2015). A good level of coordination between workers' representatives and union officials respectively at the workplace, sectoral and national level as well as between different union confederations also contributes to greater chances of success as internal divides divert union resources from organizational change and prevent the latter from being implemented across the whole union (Benassi, Dorigatti & Pannini 2018).

On the other hand, scholars have argued that unions are able to deviate from their institutionalized path and found evidence of radical change with, for example, Continental European unions adopting US-style organizing strategies. Furthermore, even traditional, typically conservative unions have formed coalitions with social movements and NGOs (Greer 2008; Tapia & Turner 2013). Pathbreaking changes were found to take place under specific circumstances: union leadership might change from a conservative leadership to an activist leadership (Voss & Sherman 2000); or the depletion of unions' traditional resources and the rise of precarious workers might threaten the organization and its ability to represent core constituencies unless unions renew their structures and repertoire (Turner 2009). Yet, even though unions' form can transform to a great extent, some elements of continuity with past union structures are likely to still be maintained (Nicklich & Helfen 2019).

Finally, while the discussion above focused on successful cases of unions changing their form, it is often the case that unions struggle to adapt to new challenges and that large segments of the workers are left without (effective) representation. As a result, workers might form new unions or join existing niche unions. Next we briefly illustrate the three dimensions of union forms (repertoire of representation, internal democratic structure and constituency) focusing specifically on Denmark, Italy, Germany and the US, and in the last section we draw upon concrete examples to highlight how and why some of these forms have changed over time as well as the cross-national similarities and differences.

DENMARK

In Denmark there are two main union confederations: the Danish Trade Union Confederation (Fagbevægelsens Hovedorganisation – FH) and the Danish Confederation of Professional Associations (Akademikernes Centralorganisation – AC). Confederations are divided along educational lines with historical traits of ideological divisions. The FH is the result of a recent merger between Landsorganisationen I Danmark (LO) and Funktionærernes og Tjenestemændenes Fællesråd (FTF) (for salaried workers and professionals with short tertiary education) in 2019. The LO was historically Social Democratic until it severed its formal ties to the Danish Social Democratic Party in the 2000s. The FTF and AC were never affiliated to any political party. The FH typically organizes unions representing blue- and white-collar workers with non-tertiary education. Recently, however, many of the white-collar professions have short-to-medium-length tertiary education as part of the general tertiarization of education in Denmark. The AC only organizes unions representing professionals with long tertiary education (at least four years at university).

Alongside the regular unions, several "alternative" unions exist. These unions do not follow educational lines and are typically not party to either collective bargaining or tripartite negotiations. Some of these unions are old, such as the Christian union (Krifa), which grew out of a Christian rejection of class-based antagonism in the early twentieth century. Others are relatively new and serve more as a service organization, such as Det Faglige Hus and ASE. Crucially for these alternative unions is that they offer unemployment insurance just as the regular unions. When the market for unemployment insurance was liberalized in 2001, the alternative unions grew dramatically as they are typically much cheaper than regular unions (Kjellberg & Ibsen 2016).

Unions in the regular two confederations are the main parties in collective bargaining with public and private employers and engage in tripartite negotiations with government. The collective bargaining coverage stands at 70–75 per cent in the private sector and there is almost full coverage in the public sector. Union density is comparatively high but it has declined since the mid-1990s. Some 68 per cent of Danish employees were union members in 2019. However, if we discount members of alternative unions, union density was 57 per cent in 2019 (Arnholtz & Navrbjerg 2021). While unions in Denmark have focused upon collective bargaining as their main repertoire of representation, membership engagement and grassroot activism have been stifled in past decades – especially in the private sector, which has not had a major strike since 1998. In the public sector, activism is more prevalent, as seen in numerous industrial actions since the 2000s, most recently with the nurses strike in 2021 by Dansk Sygeplejeråd (Andersen & Wesley Hansen 2019; Due & Madsen 2006).

The internal democratic structures are similar across unions. Unions follow occupational demarcations and most have local branches in which members elect local union officials. These branch officials constitute the voting body of the national unions and, thus, elect national leaders. The degree of decentralization varies, however, and some unions have merged local branches extensively in the recent couple of decades. National union leadership constitute the decision-making bodies of confederations. Most of the larger unions have sub-divisions based on industry and, often, leaders of these sub-divisions are the most powerful as they negotiate collective agreements with employers' associations. Members ratify collective bargaining results in a vote. For the private sector, the vote is an "all-or-nothing" decision, since all agreements across private sector industries are lumped together (Ibsen 2016). In the public sector, members typically vote on their own agreement. Members view the mixture of indirect election of leadership and direct voting on bargaining results as generally democratic and thus legitimate (Caraker *et al.* 2015).

While traditional unions have been willing and able to encompass new constituencies due to their favourable position, they have struggled with dwindling unionization and bargaining coverage in many private sector industries (Arnholtz & Navrbjerg 2021). Young people are harder to unionize, as are atypical workers. Unions have especially struggled to encompass one particular group of workers – posted workers and labour migrants from Central Eastern Europe (CEE) (Arnholtz & Andersen 2018). Minimum wages in collective agreements are usually far below actual wages due to local wage bargaining in companies. However, employers and posting companies often pay CEE migrant workers the minimum wage or below it, making unions in especially construction and transportation, such as the general workers' union, 3F, call for legislation against "social dumping". This departure from a collective bargaining repertoire of representation has caused controversy among both employers and other unions, especially Danish Metalworkers and white-collar unions that see self-regulation as a hallmark of Danish employment relations (Arnholtz & Andersen 2018). By contrast, unions in low-density industries, especially 3F, increasingly find themselves helpless against companies operating in segments of the labour with no union members, no collective agreements and no statutory minimum wage (Ibsen 2021). And these companies will often employ workers from hard-to-unionize groups; that is, young workers, minorities and migrant workers (Toubøl, Ibsen & Larsen 2015).

The difficulties of unionizing young workers, migrants and atypical workers have spurred unions in low-skill private sector industries, such as 3F and HK, to experiment with the "organizing model" (Arnholtz, Ibsen & Ibsen 2016) or non-dues membership for young people. While both have produced some positive recruitment results, the retention rate of the latter is disappointing and shows that young people leave the union when they have to begin paying dues. HK has also experimented with a union-run temp agency for freelancers, thus giving them a kind of employment relationship that qualifies freelancers for fringe benefits. The alternative unions such as Krifa, Det Faglige Hus and ASE mostly compete with the regular unions and do not typically reach out to the new constituencies.

ITALY

In Italy there are three main union confederations: the Italian General Confederation of Labour (Confederazione Generale Italiana del Lavoro – CGIL), the Italian Confederation of Workers' Trade Unions (Confederazione Italiana Sindacati Lavoratori – CISL) and the Italian Union of Labour (Unione Italiana del Lavoro – UIL). The confederations are divided along ideological lines as each of them was historically affiliated to, respectively, the Communist Party, the Christian-Democratic party and the Socialist Party. The CGIL is typically the most left-wing confederation, with a strong working-class identity, while the CISL and UIL are more moderate (Hyman 2001). While the ties with the political parties have weakened, the three confederations remained but they mostly coordinate when negotiating sectoral agreements, which cover around

80 per cent of the workforce in companies employing above ten employees. Union density is around 30 per cent of the workforce; most union members belong to sectoral affiliates of the three confederal unions even though independent union confederations, Unione Generale del Lavoro and the left-wing unions Comitati di Base and Unione Sindacale di Base, make up around 20–40 per cent of union members in the public sector (Dorigatti & Pedersini 2021).

All three confederations have a similar internal governance structure. Their union affiliates are sectoral unions. However, each confederation (and its industry affiliates) have also cross-industry horizontal organizational units at regional and local level. Both sectoral and local organizations can contribute to the position and strategies of the national unions through mechanisms of internal democracy (Dorigatti & Pedersini 2021; Marino *et al.* 2019). At the same time, however, the positions and strategies of the affiliates are coordinated by the central confederation, preventing them from pursuing particularistic interests (Oliver 2010).

In terms of constituencies, thanks to their democratic and relatively decentralized but organic structures, the three confederal unions have been able to support those disadvantaged workers' categories, which transcend sectoral boundaries such as precarious workers and migrants workers (Benassi & Vlandas 2015; Marino 2015; Marino *et al.* 2019). At the end of the 1990s, the unions set up, respectively, three unions for atypical workers aimed at increasing social protection and improving working conditions of atypical workers (Benassi & Vlandas 2015). Nuove Identitia' di Lavoro, affiliated to the working class oriented union CGIL, focused upon collective bargaining, while Federazione Lavoratori Somministrati Autonomi (CISL) and Temp@ (UIL) focused primarily upon assisting atypical workers with different contractual situations and espouse an individual servicing (Marino *et al.* 2019). At the same time, sectoral unions built on their core repertoire of representation to defend the interests of non-standard workers, as they include provisions addressing their needs in industry-level collective agreements as well as in company-level agreements (e.g. transition rules from non-standard to permanent contracts, entitlement to bonuses and other benefits) (Benassi, Dorigatti & Pannini 2018; Durazzi 2017).

While unions are still important actors in the labour market and industrial relations arena, their ability to achieve gains for both old and new constituencies as well as for investing in expanding their representation domain has declined since the 1990s. Indeed, they have been progressively marginalized from the policy-making arena. Inter-federal coordination has weakened due to political divergencies between the CGIL and the other two unions, and the economic crisis in 2008 as well as the progressive labour market flexibilization and deregulation have made the Italian labour market situation increasingly difficult (Benassi, Dorigatti & Pannini 2018; Dorigatti & Pedersini 2021; Rinaldini & Marino 2017).

GERMANY

German unions were re-established after the Second World War according to the principles of sectoral unionism and unitary unionism. The main union confederation is the Deutscher Gewerkschaftsbund (DGB), which represent 80 per cent of union members. Following several union mergers between the 1990s and early 2000s, the DGB now has eight sectoral affiliate unions. The remaining 20 per cent of union members belong to the affiliates of, respectively, the Confederation of Christian Unions (CGB) and the German Civil Service Association (DBB), organizing in the public sector and in privatized organizations (Keller & Kirsch 2020). The CGB affiliates can negotiate collective agreements with employers' associations that set different (lower) standards than DGB agreements. However, in several cases, the court found them incapable of signing the agreement due to their limited representativeness.

Collective bargaining is core to the repertoire of representation of German unions. As a result, the German unions have been more successful in improving standards of new workforce segments through the provision of services and collective bargaining rather than through recruiting or organizing new members. That said, union density has declined in Germany since the 1990s from around 35 per cent to 16 per cent in 2018. Thus, the DGB unions now represent 6 million workers (Behrens 2020). Collective bargaining, while also declining from over 70 per cent in 1995 to below 50 per cent in 2017, covers more workers than those represented by the union because coverage rates mainly depend on employers' density – rather than union density – in Germany (Behrens 2020; Keller & Kirsch 2020). Collective agreements usually cover workers in the same industry and are bargained by sectoral union affiliates. Over time, collective bargaining has become more decentralized as company-level agreements bargained by workplace union representatives and the works councils can amend sectoral standards (Baccaro & Benassi 2017).

In terms of internal democratic structure, sectoral unions have the duty to develop their collective demands following wide consultations with members at the local and company level. The collective wage committee (*Tarifkommission*), which is constituted by elected union officials, provides the platform and guidelines for the negotiations with employers, which are conducted by the bargaining committee (the *Verhandlungskommission*, elected by the *Tarifkommission*). Members, however, cannot vote on the resulting collective agreements, which need to be approved by the simple majority of the *Tarifkommission*; and in any case, the final decision on the collective agreement is left to the union board (Ver.Di online, IG Metall online). Furthermore, sectoral unions benefit of great autonomy in the development of their wage demands because the DGB union confederation does not have a bargaining mandate and has no power to coordinate wage demands across sectors. Indeed, the wage gap between workers in manufacturing and in the low-end service sector has been increasing in the last 30 years (Baccaro & Benassi 2017).

In terms of constituencies, German unions seem to struggle to recruit members beyond their traditional membership of full-time, skilled, male employees, predominantly in the manufacturing sector. Indeed, a study from 2012 showed that women, part-time workers and young workers are heavily underrepresented in German unions (Anders, Biebeler & Lesch 2015). Data from the IG Metall union, instead, show that the union has made considerable membership gains among workers with a "migration background": not only among the membership but migrant workers are also well represented among union representatives, officials and works councillors

(Benner & Ghirmasion 2017). Similar to the Danish unions, however, German unions have been struggling to reach out to cross-border workers, whom they support mainly through centres of support for migrant workers co-funded by the German government (Heinrich, Shire & Mottweiler 2020).

UNITED STATES

In the US, the American Federation of Labor and Congress of Industrial Organizations (AFL-CIO) is the largest federation of unions and consists of 57 national and international labour unions representing 12.5 million workers. The AFL-CIO was formed in 1955 after a merger between the AFL – representing predominantly craft unions – and the CIO – representing mostly industrial unions. In 2005 several unions, such as the International Brotherhood of Teamsters, the SEIU, and the United Farm Workers, broke away from the AFL-CIO to focus much more on organizing unorganized workers and formed another federation called Change to Win. These federations do not engage in collective bargaining but rather provide a political and public voice for the union movement, resolve jurisdictional dispute among members and provide a link to the international labour movement. This is the situation resulting from the continued decline in union membership over the past four decades, now standing at 10.1 per cent in 2022 (BLS 2023). There is significant regional variation with states such as Hawaii and New York having a higher density of 21.9 per cent and 20.7 per cent, respectively, whereas states in the South, such as South Carolina and North Carolina, show a continuous low-density rate of 1.7 per cent and 2.8 per cent, respectively. Union density in the public sector (33.1 per cent) is more than five times the density in the private sector (6 per cent) (BLS 2023).

Since the early 1980s collective bargaining has become more and more decentralized, resulting in a shift from multi-employer or company-wide bargaining towards plant-level bargaining. In 2022 about 11.3 per cent of workers were covered under a collective bargaining agreement (BLS 2023). Furthermore, the growth of non-union employment as well as the decentralization of collective bargaining has led to increasing variation in terms of employment and pay practices (Friedman & Godard 2020; Katz & Colvin 2021).

Even though union strikes in the US have sharply declined since the 1970s, in terms of strategies and repertoires of representation, unions have often focused upon collective bargaining with an accompanied strike threat. In fact, after three decades of decline in the number of workers involved in work stoppages, there has been a slight uptick in 2018–19. In 2019, for example, about 50,000 workers of the United Autoworkers Union (UAW) went on a 40-day strike against General Motors. In 2018–19, teachers and their unions were on strike across different states (Givan & Schrager Lang 2020). Furthermore, given the US unions' weak political and institutional support or resources (e.g. limited access to policy-making), the hostile political context they face (e.g. weak labour laws and strong employer opposition), they tend to engage in organizing unorganized workers through grassroots mobilization and engage in coalition-building to gain power vis-à-vis external actors (Baccaro, Hamann & Turner 2003; Tattersall 2010; Turner, Katz & Hurd 2001).

In terms of the union governance or the internal democratic structures, local unions carry out day-to-day activities, but the national unions are quite powerful in dictating the strategy as well as controlling the strike funds. They can also establish and disestablish local unions (Friedman & Godard 2020; Katz & Colvin 2021). Historically as well as very recently, some large unions have been involved in major corruption scandals: from the notorious corruption charges against the Teamsters and Jimmy Hoffa in the 1960s to charges of embezzlement of funds against top union officials within the UAW starting in 2017. Consequently, some workers have argued that the internal structures, and especially the way some union elections are being held, are not democratic. Within the UAW, for example, a movement of UAW activists has come together and formed the Unite All Workers for Democracy (UAWD). Their aim is to return to the militant roots of the union and build a grassroots bottom-up structure of rank-and-file workers, moreover, by changing the internal election system. In essence, the UAWD is pushing for direct elections or "one member, one vote" as a way of electing the international executive board officials of the UAW. Instead, the UAW top leadership officials are elected through a delegate system in which the power to decide who gets elected is in the hands of its long-established Administration Caucus.[1]

In terms of constituencies, the US unions were historically quite racist and exclusive and represented largely White and male workers. Demographic changes as well as the decline of unionism has put pressure upon unions to include a broader segment of the workforce. For example, Black workers continue to be more likely to join a union than White, Asian, or Hispanic workers. That said, while Black workers in the US continue to have higher membership rates than White workers,[2] this rate has dropped more drastically over the past three decades than it has for White workers (from 31.7 per cent in 1983 to 12.6 per cent in 2017 for Black workers versus from 23.3 per cent in 1983 to 10.6 per cent in 2018 for White workers) (CEPR 2016) and Black workers still face significant employment and wage gaps compared to White workers (Carruthers & Wanamaker 2017).

EXPLAINING CHANGE IN UNION FORMS

Across these four countries, we observe a decline in union density and collective bargaining coverage. The erosion, in some countries more pronounced than in others, of unions' institutional power resources, shifts in demographic composition as well as the changing production structure has to a certain extent led unions to change their forms to seek to include new workforce segments in their membership, rethink their strategies of representation and, in some cases, internally restructure themselves. As shown below, we find elements of path dependency across these changing union forms as well as more radical deviations from their institutionalized path. In three countries

1. The Administration Caucus (AC) was set up under Walther Reuther in the 1940s and it has been in place ever since. According to the UAWD website, the AC has "bribed, pressured and intimidated convention delegates and their locals to force them to vote for their candidates" and as a result, virtually all the seats have been filled by AC candidates (UAWD 2021).
2. At least since the early 1970s (CEPR 2016).

(Italy, Germany and the US), we even find distinct actors coming on to the scene to represent certain groups of workers (e.g. self-employed, Deliveroo drivers, etc.). Finally, we also illustrate how in some cases unions have to a certain extent failed to adapt their form given the changing environment.

Change through path dependency

Across the four countries, unions tend to draw upon their traditional institutional resources and repertoires to adapt to the changing context. For example, owing to the relatively stable collective bargaining system and "institutional security" (Hassel 2007) inherent in Danish corporatism, unions in Denmark have used their traditional repertoires of representation – or collective bargaining – to protect new forms of work and new constituencies. More recently, Danish unions were among the first in the world to reach a collective agreement with platform companies (Jesnes, Ilsøe & Hotvedt 2019). The first agreement of 2018 between the United Federation of Danish Workers (3F) and Hilfr, a cleaning platform, made it possible for Hilfr workers to choose whether they wanted to work as freelancers (with their own hourly rates) or as employees (with a minimum salary and other benefits) (Ilsøe 2020). In 2021, 3F and the Danish Chamber of Commerce (Dansk Erhverv) reached sector-level agreement for platform work in food delivery. The agreement ensured hourly minimum wages and other worker rights, and was immediately signed by Just Eat, the largest food delivery platform in Denmark. HK (for salaried workers) chose a different way to protect freelancers. As mentioned, HK established its own freelance agency, which charges 8 per cent on bills, and in return ensures good conditions for freelancers. The agency is not a union, but tries to protect freelancers from low-income traps and poor non-wage conditions (pensions, vacations, etc.).

In Italy the three main confederal unions have changed their structures and the content of the bargaining to adapt to changes in the labour market, as non-standard contracts became increasingly common and the workforce more diverse. For example, all three confederal unions have been similarly inclusive towards migrant workers. Even though they did not create a separate union, since the mid-90s they have promoted self-organized local structures for migrant officials and representatives (but open to everyone) so that they could coordinate and contribute to union bargaining strategies. Furthermore, unions expanded their service provisions to include those services required by migrant workers; for example, the regularization of their immigration status through their work contracts. As a result of these efforts, the unionization rate of migrant workers is higher than the rate of those with an Italian heritage, and many workplace representatives and officials have a migration background (Rinaldini & Marino 2017).

In Germany the large unions especially tend to use their traditional repertoire of representation – collective bargaining – to adapt to a changing context. For example, IG Metall contributed to reregulating temporary agency work by negotiating wage bonuses and transition rules from temporary to permanent status, and Ver.Di negotiated collective agreements for typically outsourced workers such as those workers employed for stocking goods in the retail sector (Jaehrling, Wagner & Weinkopf 2016). The same union also pressured crowdsourcing platforms based in Germany to adopt the Crowdsourcing Code of Conduct, which commits companies to "local wage standards" (Vandaele 2018) and it set up the platform Faircrowdwork for crowdworkers to network and share their experiences (Borghi 2018). This can also lead to internal structural changes. Ver.di, for example, created a dedicated unit for self-employed – Ver.Di Selbstaendige – which provides services to its members and feeds their requests and proposals to higher organizational levels in order to influence national policies and bargaining strategies (Borghi 2018).

In the US the unions have turned to "organizing" workers for the past three decades. In other words, worker organizing or actively engaging workers in an organizing campaign has been part of their repertoire – especially given the unions' weak institutional power. Recent large organizing drives, such as the 2021 Retail Workers and Department Store Union organizing drive of Amazon workers in Bessemer, Alabama, or the 2019 UAW organizing drive of Volkswagen workers in Chattanooga, Tennessee, while quite impressive in terms of worker activism and worker solidarity, ended in defeat for these unions. This has been at least partially the result, however, of the unchecked power of capital in the US and the resources employers put behind anti-union campaigns and union-busting consultancy firms. In other words, the employer strategies in the US have made organizing more and more difficult and costly for unions.

Change through deviations from the institutionalized path

Unions also had to deviate from their institutional path and traditional repertoire of representation, taking on new strategies. For example, here, perhaps ironically, some union officials of Danish unions actively imported elements of the Anglo-Saxon organizing "model" as a way to deal with membership decline and their traditional passive model of recruitment through the Ghent system of union-run unemployment benefit funds. The use of these organizing tools remained varied, however, across the different union branches, and especially opened up power struggles between different actors and levels within the union organizations, thus affecting the internal democratic structures. For example, promoters of the organizing model challenged established leaderships that were more wary of changing their traditional repertoire (Arnholtz, Ibsen & Ibsen 2016). Furthermore, many unions in low-skilled private sector industries are reaching critically low union densities and have experimented with the "organizing model" to unionize non-members – with mixed results (Arnholtz, Ibsen & Ibsen 2016).

In Italy, in the logistic sector, which employs many vulnerable and migrant workers, the rank-and-file unions, SI-Cobas and ADL-Cobas, have been successfully recruiting members. As these unions are less bureaucratic and hierarchical than the confederal unions, they left considerable room for action to the workers themselves, who organized, with the organizational support of the union, several mobilization initiatives from boycotts to warehouse blockades, which attracted the attention of the national media (Benvegnú, Haidinger & Sacchetto 2018).

In Germany, DGB unions have increasingly engaged in several initiatives and campaigns to recruit and mobilize new workforce segments, including migrant workers, young workers and precarious workers. To do so, some German unions have expanded their repertoire of contention to include American-style organizing strategies, adapting them to their strongly institutionalized industrial relations context (Nicklich & Helfen 2019). For instance, IG Metall has run a campaign for organizing temporary agency workers and bargaining on their behalf, which was not only oriented to the workers but explicitly aimed at gaining support from the media and the wider public (Benassi & Dorigatti 2015). Furthermore, IG Metall, the service union Ver.Di and the food and catering union, NGG, have also been trying to recruit workers in the gig economy through local and national organizing campaigns (Vandaele 2018).

In the US unions have taken on new approaches. Specifically, some unions have engaged in trying to build greater coalitional power by working together with a variety of community organizations to "Bargain for the Common Good". In essence, this translates into expanding the scope of bargaining beyond members' "bread-and-butter" issues, working with the community organizations from the start, have them at the bargaining table and fight around issues of racial justice. The teacher unions, for example, have extensively engaged in "Bargaining for the Common Good", demanding increased green spaces around schools, stopping racist random searches of their mostly Black and brown students, and an increase in immigrant defence support for their students (Jaffe 2019).

Change through new actors emerging

Across most of the countries – with the exception of Denmark – we also see the prominence of new actors coming into the employment relations arena. These new actors often take on distinct forms in comparison to the traditional unions. In Italy, for example, self-employed workers, who did not feel represented by the atypical workers' unions, created from the bottom up their own professional associations with the aim of influencing policy-making for improving working conditions, social protection and the tax regime, as well as providing services to their members (Semensa & Mori 2018). Workers in the gig economy have also started organizing autonomously. In particular, platform workers (e.g. riders) have constituted informal workers' collectives and organized strikes and protests without the organizational support, let alone the representation, of traditional unions (Tassinari & Maccarrone 2020). Those groups have also entered a coalition with traditional unions (RiderXiDiritti) to negotiate a collective agreement with the employers' association, Assodelivery. The latter, however, signed an agreement in September 2020 with the right-wing "yellow union", UGL, which has been trying to expand to new representation domains and to be organizing as a legitimate collective bargaining partner with employers (Mara & Pulignano 2020). However, the coalition RiderXIDiritti has openly contested the agreement, and the labour court in Bologna declared the contract unlawful because UGL unions are not "representative" of the workforce. The controversy is still ongoing at the time (2023) of writing.

In Germany, for example, independent unions have gained increased importance in the collective bargaining arena as representatives of new workforce segments. Among others, delivery riders have established their own guild called Deliverunion, which is affiliated to the grassroots anarcho-syndicalist union, FAU, in Berlin. FAU supports Deliverunion in the collective negotiations with employers as well as in court, fighting the contractual misclassification of platform workers as independent contractors (Vandaele 2018). Those self-employed in regulated professions and associations are represented through structured professional associations, such as the Association of Literary Translators, while grassroots organizations and cooperatives are more common in the digital, knowledge and creative industries. For instance, freelance artists can join the cooperative SmartDE, which focuses on providing services to its members while also involving them in the management of the project (Borghi 2018). Finally, since the early 2000s new small occupational unions have established themselves as representative of high-skilled professional groups; for example, air pilots and doctors. Thanks to their strategic labour market position, they could bargain favourable agreements with employers, yet their influence was considerably scaled back as a law passed in 2015 prevented more than one collective agreement to be applied in the same enterprise (Keller & Kirsch 2020).

In the US, new forms of representation have emerged to fill the void, such as worker centres and broader social movements (Fine 2006; Lee & Tapia 2023a; Tapia, Lee & Filipovitch 2017). The "Fight for $15" movement, for example, while financially backed by the SEIU, has been instrumental in pushing employers as well as city councils and states to increase the minimum wage to $15 per hour. Worker centres focusing on servicing, advocacy and organizing have also been created across the nation, often representing migrant workers and/or workers from Black and brown communities that are not covered by the traditional unions. The Restaurant Opportunities Center (ROC), for example, is a worker centre that was created by and for restaurant workers in the aftermath of the terrorist attacks of 11 September 2001. Importantly, ROC uses an intersectional approach during their worker-centred campaign, framing their issues not just in economic ways but highlighting the systemic racism and discrimination that many restaurant workers face (Brady 2014; Jayaraman 2013; Tapia, Lee & Filipovitch 2017).

Failing to adapt

Finally, we also notice instances across our countries in which unions to a certain extent fail to adapt to the changing context, partially because of the way the unions are set up and structured, but also because of the disconnection between an increasingly diverse base and a union elite of functionaries reluctant to change (Lorens 2013). For instance, in Denmark most unions in the private sectors have a "youth problem", which could turn into a significant drop in union density as highly organized workers retire. Danish unions seem to have failed at adapting services to young workers and communicating about the benefits of unionization.

In Germany the limited authority of the general union confederation over its sectoral affiliates seems to have contributed to lower inclusiveness towards precarious workers of German unions until 2010 from a comparative European perspective (Benassi & Vlandas 2015). Furthermore, the decentralization of collective bargaining with an increasingly independent role of works councils has been argued to enable insider-focused strategies supporting labour market dualization (Benassi 2015; Benassi & Dorigatti 2015; Hassel 2007)

In their study of Italian and German metal unions in four matched manufacturing plants, Benassi and Dorigatti (2020) show that both unions failed to extend their representation to subcontracted workers performing "marginal" tasks (e.g. logistics and other industrial services) because they pursued traditional bargaining strategies that aimed at covering only workers employed in "core" occupations within the metal sector. Along similar lines, Durazzi (2017) argues that Italian unions initially did not prioritize the universalization of welfare and income protection to non-standard workers because they felt it was politically wrong to change their traditional bargaining strategies centred around the protection of (permanent) work.

Finally, in the US the unions have needed to adapt much earlier to respond to a changing context in comparison to many Continental European countries. They have done so partially through building coalitions with migrant, community, environmental and social justice organizations as well as by focusing on worker organizing campaigns. That said, arguably, unions in the US have yet to embrace a form that encourages minoritized workers to lead the movement, relying upon their full, lived experiences and centring demands of the most exploited (Lee & Tapia 2023b).

CONCLUSION

Union forms tend to vary along three dimensions: their internal governance structure, their constituency and their repertoires of representation. While union forms are shaped by the institutional setting, the form that unionism takes is destined to change over time, adapting to evolving labour markets and socio-economic contexts. The extent to which unions need to or want to change varies across countries – as well as sectors, even though this is not covered in this chapter – as our case studies suggest. The US stands out in this regard, having experienced the most radical liberalization and breakdown of traditional demarcations, making new social identities all the more important for new forms of unions (Givan 2007; Milkman & Ott 2014; Piore & Safford 2006). In contrast, Danish, German and Italian unions have largely maintained their forms even though they have expanded their membership pool and, in some cases, adapted their repertoire of representation for representing an increasingly diverse workforce (Arnholtz, Ibsen & Ibsen 2016; Benassi & Dorigatti 2015; Tapia & Turner 2018). Yet the emergence of "new" unions and the increased relevance of niche unions in the German and Italian case seem to suggest that new workforce segments, for example, in the gig economy or among creative professionals, do not feel that traditional unions are willing to include them or, if they were, that they can do so effectively.

REFERENCES

Alberti, G., J. Holgate & M. Tapia 2013. "Organising migrants as workers or as migrant workers? Intersectionality, trade unions and precarious work". *International Journal of Human Resource Management* 24: 4132–48.

Anders, C., H. Biebeler & H. Lesch 2015. "Mitgliederentwicklung und politische Einflussnahme: Die deutschen Gewerkschaften im Aufbruch?". *IW-Trends-Vierteljahresschrift zur empirischen Wirtschaftsforschung* 42: 21–36.

Andersen, S. & N. Wesley Hansen 2019. *Forligsmuligheder Og Knaster i Gode Tider Analyse Af Optakten Til OK20.* Copenhagen: FAOS.

Anner, M., M. Fischer-Daly & M. Maffie 2021. "Fissured employment and network bargaining: emerging employment relations dynamics in a contingent world of work". *ILR Review* 74: 689–714.

Arnholtz, J. & S. Andersen 2018. "Extra-institutional changes under pressure from posting". *British Journal of Industrial Relations* 56: 395–417.

Arnholtz, J., C. Ibsen & F. Ibsen 2016. "Importing low-density ideas to high-density revitalisation: the 'organising model' in Denmark". *Economic and Industrial Democracy* 37(2): 297–317.

Arnholtz, J. & S. Navrbjerg 2021. *Lønmodtageres Faglige Organisering 2000–2018*. Copenhagen.

Baccaro, L. & C. Benassi 2017. "Throwing out the ballast: growth models and the liberalization of German industrial relations". *Socio-Economic Review* 15: 85–115.

Baccaro, L., K. Hamann & L. Turner 2003. "The politics of labour movement revitalization: the need for a revitalized perspective". *European Journal of Industrial Relations* 9: 119–33.

Behrens, M. 2020. "Employment relations in national contexts: Germany". In *Comparative Employment Relations in the Global Economy*, edited by C. Frege & J. Kelly. Abingdon: Routledge.

Benassi, C. 2015. "From concession bargaining to broad workplace solidarity: the IG Metall response to agency work". In *The Outsourcing Challenge: Organizing Workers across Fragmented Production Networks*, edited by J. Drahokoupil, 237–54. Brussels: European Trade Union Institute.

Benassi, C. & L. Dorigatti 2015. "Straight to the core: the IG Metall campaign towards agency workers". *British Journal of Industrial Relations* 53(3): 533–55.

Benassi, C. & L. Dorigatti 2020. "Out of sight, out of mind: the challenge of external work arrangements for industrial manufacturing unions in Germany and Italy". *Work, Employment and Society* 34(6): 1027–44.

Benassi, C., L. Dorigatti & E. Pannini 2018. "Explaining divergent bargaining outcomes for agency workers: the role of labour divides and labour market reforms". *European Journal of Industrial Relations* 25(2): 163–79.

Benassi, C. & T. Vlandas 2015. "Union inclusiveness and temporary agency workers: the role of power resources and union ideology". *European Journal of Industrial Relations* 22(1): 5–22.

Benner, C. & F. Ghirmasion 2017. "Mitglieder mit Migrationshintergrund in der IG Metall–Gewerkschaften und Arbeitswelt als Wegbereiter für Integration". *WSI-Mitteilungen* 70: 296–300.

Benvegnú, C., B. Haidinger & D. Sacchetto 2018. "Restructuring labour relations and employment in the European logistics sector: unions' responses to a segmented workforce". In V. Doellgast, N. Lillie & V. Pulignano (eds), *Reconstructing Solidarity: Labour Unions, Precarious Work, and the Politics of Institutional Change in Europe*, 83–103. Oxford: Oxford University Press.

Bernhardt, A. & P. Osterman 2017. "Organizing for good jobs: recent developments and new challenges". *Work and Occupations* 33(1): 89–112.

BLS 2023. "Union Members Summary – 2022". Washington, DC: Bureau of Labor Statistics. www.bls.gov/news.release/archives/union2_01192023.htm

Borghi, P. 2018. "WP3. Country case study: Germany". In *Independent Workers and Industrial Relations*. Brussels: European Commission.

Brady, M. 2014. "An appetite for justice: the restaurant opportunities center of New York". In *New Labor in New York: Precarious Workers and the Future of the Labor Movement*, edited by R. Milkman & E. Ott, 229–45. Ithaca, NY: Cornell University Press.

Caraker, E. *et al.* 2015. *Fællesskabet Før Forskellene: Hovedrapport Fra APL III-Projektet Om Nye Lønmodtagerværdier Og Interesser*. Copenhagen: LO and FTF.

Carruthers, C. & M. Wanamaker 2017. "Separate and unequal in the labor market: human capital and the Jim Crow wage gap". *Journal of Labor Economics* 35(3): 655–96.

CEPR 2016. *Black Workers, Unions, and Inequality.* http://cepr.net/publications/reports/black-workers-unions-and-inequality

Crouch, C. 1993. *Industrial Relations and European State Traditions*. Oxford: Oxford University Press.

Dempsey, M. & D. McKevitt 2001. "Unison and the people side of mergers". *Human Resource Management Journal* 11(2): 4–16.

Doellgast, V. & I. Greer 2007. "Vertical disintegration and the disorganization of German industrial relations". *British Journal of Industrial Relations* 45(1): 55–76.

Donaghey, J. & J. Reinecke 2018. "When industrial democracy meets corporate social responsibility: a comparison of the Bangladesh accord and alliance as responses to the Rana Plasa disaster". *British Journal of Industrial Relations* 56(1): 14–42.

Dorigatti, L. & R. Pedersini 2021. "Employment relations in Italy". In *International and Comparative Employment Relations: Global Crises and Institutional Responses*, edited by G. Bamber *et al.*, 131. London: Sage.

Due, J. & J. Madsen 2006. *Fra Storkonflikt Til Barselsfond. Den Danske Model under Afvikling Eller Fornyelse*. Copenhagen: Jurist- og Økonomforbundets Forlag (DJØF).

Durazzi, N. 2017. "Inclusive unions in a dualized labour market? The challenge of organizing labour market policy and social protection for labour market outsiders". *Social Policy and Administration* 51(2): 265–85.

Fine, J. 2006. *Worker Centers: Organizing Communities at the Edge of the Dream*. Ithaca, NY: Cornell University Press.

Fiorito, J. & P. Jarley 2008. "Trade union morphology". In *Sage Handbook of Industrial Relations* edited by P. Blyton *et al.*, 189–208. London: Sage.

Frege, C. & J. Kelly 2003. "Union revitalization strategies in comparative perspective". *European Journal of Industrial Relations* 9(1): 7–24.

Friedman, G. & J. Godard 2020. "Employment relations in national contexts: the United States". In *Comparative Employment Relations in the Global Economy*, edited by C. Frege & J. Kelly. Abingdon: Routledge.

Givan, R. 2007. "Side by side we battle onward? Representing workers in contemporary America". *British Journal of Industrial Relations* 45(4): 829–55.

Givan, R. & A. Schrager Lang (eds) 2020. *Strike for the Common Good*. Ann Arbor, MI: University of Michigan Press.

Greer, I. 2008. "Social movement unionism and social partnership in Germany: the case of Hamburg's hospitals". *Industrial Relations: A Journal of Economy and Society* 47(4): 602–24.

Hassel, A. 2007. "The curse of institutional security: the erosion of German trade unionism". *Industrielle Beziehungen* 14: 176–91.

Healy, G., L. Hansen & S. Ledwith 2006. "Still uncovering gender in industrial relations". *Industrial Relations Journal* 37(4): 290–8.

Heinrich, S., K. Shire & H. Mottweiler 2020. "Fighting (for) the margins: trade union responses to the emergence of cross-border temporary agency work in the European Union". *Journal of Industrial Relations* 62(2): 210–34.

Hodder, A. 2014. "Organising young workers in the Public and Commercial Services union". *Industrial Relations Journal* 45(2): 153–68.

Holgate, J. 2005. "Organizing migrant workers: a case study of working conditions and unionization at a sandwich factory in London". *Work Employment and Society* 19: 463–80.

Holgate, J. & J. Wills 2007. "Organizing labor in London: lessons from the living wage campaign". In *Labor in the New Urban Battlefields* edited by L. Turner & D. Cornfield, 211–23. Ithaca, NY: Cornell University Press.

Hyman R. 1997. "Trade unions and interest representation in the context of globalisation". *Transfer* 3: 515–33.

Hyman, R. 2001. *Understanding European Trade Unionism: Between Market, Class and Society*. London: Sage.

Ibsen, C. 2016. "The role of mediation institutions in Sweden and Denmark after centralized bargaining". *British Journal of Industrial Relations* 54(2): 285–310.

Ibsen, C. 2021. "Social democratic trade unions in the knowledge economy: challenges, pathways and dilemmas". In *Social Democracy in the 21st Century*, edited by N. Brandal, Ø. Bratberg & D. Thorsen, 69–90. Bingley: Emerald.

Ibsen, C. & M. Tapia 2017. "Trade union revitalisation: where are we now? Where to next?" *Journal of Industrial Relations* 59(2): 170–91.

Ilsøe, A. 2020. "The Hilfr Agreement, Copenhagen". In *Collective Agreements for Platform Workers? Examples from the Nordic Countries*, edited by K. Jesnes, A. Ilsøe & M. Hotvedt. Oslo: FAFO.

Jaehrling, K., I. Wagner & C. Weinkopf 2016. "Reducing precarious work in Europe through social dialogue: the case of Germany". IAQ-Forschung 3.

Jaffe, S. 2019. "The Chicago teachers strike was a lesson in 21st-century organizing". *The Nation*, 16 November.

Jayaraman, S. 2013. *Behind the Kitchen Door*. Ithaca, NY: Cornell University Press.

Jesnes, K., A. Ilsøe & M. Hotvedt 2019. *Collective Agreements for Platform Workers? Examples from the Nordic Countries*. Oslo: FAFO.

Katz, H. & A. Colvin 2021. "Employment relations in the United States". In *International and Comparative Employment Relations: Global Crises and Institutional Responses*, edited by G. Bamber *et al.*, 55–79. London: Sage.

Keller, B. & A. Kirsch 2020. "Employment relations in Germany". In *International and Comparative Employment Relations*, edited by G. Bamber *et al.*, 179–207. London: Sage.

Kelly, J. 1998. *Rethinking Industrial Relations: Mobilisation, Collectivism and Long Waves*. London: Routledge.

Kjellberg, A. & C. Ibsen 2016. "Attacks on union organizing – reversible and irreversible changes to the Ghent-systems in Sweden and Denmark". In *Den Danske Model set udefra*, edited by T. Larsen, 279–302. Copenhagen: Jurist- og Økonomforbundets Forlag.

Lasarsfeld, P. 1937. "Some remarks on the typological procedures in social research." *Zeitschrift fuer Sozialforschung* 6: 119–39.

Lee, T. & M. Tapia 2023a. "Intersectional organizing: building solidarity through radical confrontation". *Industrial Relations* 62(1): 78–111.

Lee, T. & M. Tapia 2023b. "A critical industrial relations approach to understanding contemporary worker uprising". *Work and Occupations* 50(3): 393–9.

Lorens, R. 2013. *Gewerkschaftsdämmerung: Geschichte und perspektiven deutscher Gewerkschaften*. Bielefeld: Transcript Verlag.

Mara, C. & V. Pulignano 2020. "Collective voice for platform workers: riders' union struggles in Italy". Social Europe, 10 December. www.socialeurope.eu/collective-voice-for-platform-workers-riders-union-struggles-in-italy

Marchington, M. *et al.* 2005. *Fragmenting Work: Blurring Organizational Boundaries and Disordering Hierarchies*. Oxford: Oxford University Press.

Marino, S. 2015. "Trade unions, special structures and the inclusion of migrant workers: on the role of union democracy". *Work, Employment and Society* 29(5): 826–42.

Marino, S. *et al.* 2019. "Unions for whom? Union democracy and precarious workers in Poland and Italy". *Economic and Industrial Democracy* 40(1): 111–31.

Michels, R. 1962. *Political Parties: A Sociological Study of the Oligarchic Tendencies of Modern Democracy*. New York: The Free Press.

Milkman, R. & E. Ott (eds) 2014. *New Labor in New York: Precarious Workers and the Future of the Labor Movement*. Ithaca, NY: Cornell University Press.

Nicklich, M. & M. Helfen 2019. "Trade union renewal and 'organizing from below' in Germany: institutional constraints, strategic dilemmas and organizational tensions". *European Journal of Industrial Relations* 25(1): 57–73.

Oliver, R. 2010. "Insiders' institutions, outsiders' outcomes: Italian atypical workers in comparative perspective". Seventeenth International Conference of the Council for European Studies. Grand Plasa, Montreal.

Papadopoulos, O. 2014. "Youth unemployment discourses in Greece and Ireland before and during the economic crisis: moving from divergence to 'contingent convergence'". *Economic and Industrial Democracy* 37(3): 493–515.

Piore, M. & S. Safford 2006. "Changing regimes of workplace governance, shifting axes of social mobilization, and the challenge to industrial relations theory". *Industrial Relations* 45(3): 299–325.

Rinaldini, M. & S. Marino 2017. "Trade unions and migrant workers in Italy: between labour and social rights". In *Trade Unions and Migrant Workers*, edited by S. Marino, J. Roosblad & R. Penninx. Cheltenham: Edward Elgar.

Scharpf, F. 1999. *Governing in Europe: Effective and Democratic*? Oxford: Oxford University Press.

Semensa, R. & A. Mori 2018. "WP3. Country case study: Italy". In *Independent Workers and Industrial Relations*. Brussels: European Commission.

Simms, M. *et al.* 2018. "Organizing young workers under precarious conditions: what hinders or facilitates union success". *Work and Occupations* 45(4): 420–50.

Streeck, W. 2005. "The sociology of labor markets and trade unions". In *The Handbook of Economic Sociology*, edited by N. Smelser & R. Swedberg, 254–83. Princeton, NJ: Princeton University Press.

Tapia, M. & G. Alberti 2018. "Unpacking the category of migrant workers in trade union research: a multi-level approach to migrant intersectionalities". *Work, Employment and Society* 33(2): 314–25.

Tapia, M., T. Lee & M. Filipovitch 2017. "Supra-union and intersectional organizing: an examination of two prominent cases in the low-wage US restaurant industry". *Journal of Industrial Relations* 59(4): 487–509.

Tapia, M. & L. Turner 2013. "Union campaigns as countermovements: mobilizing immigrant workers in France and the United Kingdom". *British Journal of Industrial Relations* 51(3): 601–22.

Tapia, M. & L. Turner 2018. "Renewed activism for the labor movement: the urgency of young worker engagement". *Work and Occupations* 45(4): 391–419.

Tassinari, A. & V. Maccarrone 2020. "Riders on the storm: workplace solidarity among gig economy couriers in Italy and the UK". *Work, Employment and Society* 34(1): 35–54.

Tattersall, A. 2010. *Power in Coalition: Strategies for Strong Unions and Social Change*. Ithaca, NY: Cornell University Press.

Toubøl, J., C. Ibsen & A. Larsen 2015. *Udviklingen i Den Faglige Organisering På Det Segmenterede Danske Arbejdsmarked 1995–2012*. Copenhagen: FAOS.

Turner, L. 2003. "Reviving the labor movement: a comparative perspective". In *Labor Revitalization: Global Perspectives and New Initiatives*, edited by D. Cornfield & H. McCammon, 23–58. Bingley: Emerald.

Turner L. 2009. "Institutions and activism: crisis and opportunity for a German labor movement in decline". *ILR Review* 62(3): 294–312.

Turner, L., H. Katz & R. Hurd (eds) 2001. *Rekindling the Movement: Labor's Quest for Relevance in the 21st Century*. Ithaca, NY: Cornell University Press.

UAWD 2021. "About UAWD". https://uawd.org/about/

Vandaele, K. 2018. "Will trade unions survive in the platform economy?" In *Emerging Patterns of Platform Workers' Collective Voice and Representation in Europe* (19 June 2018). ETUI Research Paper-Working Paper. Brussels: ETUI.

Vandaele, K. & J. Leschke 2010. "Following the organising model of British unions? Organising non-standard workers in Germany and the Netherlands". Working Paper 2010.02. European Trade Union Institute.

Voss, K. & R. Sherman 2000. "Breaking the iron law of oligarchy: union revitalization in the American labor movement". *American Journal of Sociology* 106(2): 303–49.

Wagner, I. & B. Refslund 2016. "Understanding the diverging trajectories of slaughterhouse work in Denmark and Germany: a power resource approach". *European Journal of Industrial Relations* 22(4): 335–51.

Weil, D. 2014. *The Fissured Workplace*. Cambridge, MA: Harvard University Press.

CHAPTER 5

UNION GOVERNANCE: SOUTH AFRICA AND ITS LESSONS

Geoffrey Wood and Christine Bischoff

ABSTRACT

The South African labour movement's role in the collapse of apartheid in South Africa is renowned. Historically, the unions' opposition to apartheid can be attributed to their robust internal governance structures, which were premised on democratic workplace organizing rather than solely relying upon their leaders. In the post-apartheid era, workplace governance presented new challenges and, with the pressures to rapidly transform workplaces demographically, many firms targeted shop stewards for management roles, which in turn led to a significant loss of local leadership. Furthermore, the top union leadership structures had to contend with the demands placed upon them due to their participation in the African National Congress-led alliance. This chapter explores in detail how the model of successful union governance, which was centred on dynamic internal democracy, has substantially declined in three key areas of union strength.

Keywords: Union governance; South Africa; challenges for unions

INTRODUCTION

The mixed fortunes of unions around the world have led to increased interest in union governance, and the extent to which this is part of the recipe for success, supplementing earlier interest in union strategy and the organizing model (Visser 2019). The meteoric rise and decline of the South African labour unionism highlights some of the challenges in promoting effective governance of unions and sustaining it in the face of emerging challenges. The South African labour movement is well known to have played a leading role in the demise of apartheid, and was part and parcel of the negotiated transition, leading the promulgation of a highly progressive body of labour legislation. Part of the unions' success in opposing apartheid was highly effective internal governance structures, founded upon democratic workplace organizing, which reduced a reliance on key leaders, an experience relevant for unions in many countries under repressive rule. This focus made it much harder for the state to repress the labour movement.

However, effective workplace governance posed new challenges in the post-apartheid era. Under pressure to advance Blacks in the workplace, many firms took to poaching shop stewards, leading to the loss of a large strata of local leadership. If the democratic base of unions was weakened, a challenge to the top was operating within the constraints of the African National Congress (ANC) alliance, especially given the latter's neoliberal turn in the post-apartheid era. Again, this is relevant to unions in many parts of the world that face the challenges of reconciling strategic alliances with political parties with the need to fairly represent the interests of their members. This chapter highlights how a model of effective governance, based upon high levels of internal democracy, may become weakened on three levels: first, at the grassroots by shop stewards gaining new opportunities for career mobility; second, through new opportunities for union investment funds creating serious agency problems; and third, by new generations of leaders being less troubled by conventions, even if adhering to formal rules.

The history of South Africa's employment and labour relations are intertwined with the country's socio-political and economic conditions under which the unions played a significant role in the country's transition to democracy. The struggle for democracy engendered efforts to build internal democracy in the unions. Early efforts at organizing Blacks failed owing to a reliance on charismatic leaders at the expense of internal governance. The success and power of the unions at the close of apartheid was owing to a decision to prioritize internal governance. However, the type of internal governance, centring on shop-floor democracy, has faced very different challenges in the post-apartheid era, focusing on questions of unity and the extent of strategic compromise.

South African unions have drawn upon their collective bargaining power to strengthen the welfare of their members by influencing policy matters. However, South Africa's employment relations climate is now marked by volatility and fragmentation (Chinguno 2013). Industrial restructuring, increasing flexibility, workplace practice changes such as outsourcing and informalization and attendant job insecurity have all impacted upon unions' abilities to organize workers, and these changes have negatively affected union organizations, with membership numbers dropping and good union governance mattering little without members. More recently, new workplace practices such as remote working, automation and technological transformations have also undermined unions, as well as collective bargaining in the workplace, with a loss of membership being the outcome of these trends.

Since 1973 the union movement in South Africa has endeavoured to create a collectivist and democratic organizational culture but this culture has subsequently declined due to political and economic changes that have taken place in the past 20-odd years. Union democracy in the South African labour movement is taking strain. As convenient as the term oligarchy is for understanding the likely trajectories for union governance, it is not a given in these organizations. In the case of South African unions, the conundrum is not oligarchy but rather why the problems of union democracy have not brought about full-blown oligarchic rule in the unions (Buhlungu 2009). Thus, this chapter begins by examining the industrial relations system in South Africa after the democratic transition in 1994. Union membership figures and trends within this period are examined next. This is followed by a discussion of trade union organization and governance.

THE CONTEXT: SOUTH AFRICAN INDUSTRIAL RELATIONS SYSTEM

The post-1994 labour legislation upholds the rights of all to fair labour practices. Workers have the right to set up, join and partake in the activities of unions, to embark on strikes and to participate in collective bargaining. In post-apartheid South Africa, labour markets are centred around efforts to promote organized labour and corporatist institutions for social dialogue about a range of public policies. Specifically, the National Economic Development and Labour Council, a tripartite statutory body established in 1995, is charged with fostering dialogue between organized labour, business, community and government on public policy processes. These efforts had quite limited success, with limited relevance for those working for small business and the informal sector.

South African labour legislation is also favourable towards unions, and various strategies augment union power, such as catering for agency-shop and closed-shop agreements, validating legal strikes and supporting unions' rights to collect membership fees and to conduct shop steward elections. Section 66 of the Labour Relations Act (LRA) 1995 strives to encourage economic development, workplace harmony, social justice and, importantly, democracy in the workplace. This is accomplished by monitoring the organizational entitlements of unions in addition to supporting and expediting collective bargaining at the workplace and at a sectoral level. Furthermore, the LRA promotes the dispute resolution and labour peace with the establishment of the Commission for Conciliation, Mediation and Arbitration (CCMA), Labour Court, and Labour Appeal Court. Bargaining councils are voluntary entities set up by registered unions in partnership with employer organizations within a certain sector and area. Collective agreement on several matters, such as minimum wage as well as conditions of employment, is concluded during the collective bargaining process. Thus, the LRA provides directives for collective bargaining, provides a framework for managing conflict between employers and employees at the workplace, and it embeds and broadens the system of wage determination by means of centralized bargaining. There is a firm commitment to collective bargaining institutions and co-determination in the governance of workplaces is promoted. The CCMA was inaugurated and the introduction of the Skills Development Act (1998) was to fast-track skill development. On the one hand, it could be argued that all these initiatives place the unions in a very favourable position; on the other hand, these measures were only of relevance to those working in the formal sector. Even if the unions had perfect governance, they had to contend with a declining base. The shift towards an approach more associated with business unionism has led to internal debates within the labour movement, which, it is held, dilutes worker control and democracy.

The Employment Equity Act (1998) offers equal opportunities for the historically disadvantaged parts of the workforce. The institutional improvements that emanated from this new labour regime produced very favourable conditions for labour. The Basic

Conditions of Employment Act (1997) was passed to provide for the regulation of basic conditions of employment such as annual leave and hours of work. However, a large proportion of workers are not covered by collective bargaining. A national minimum wage has been introduced for workers who cannot improve their own situation through the bargaining processes, and it has been set at R3,500 (approximately US$207) per month. To address the high levels of unemployment in the country, the Basic Income Grant, a universal monthly payment, has been proposed. The government's Covid-19 Social Relief of Distress grant pays out R350 (approximately US$21) per month to those whose livelihoods have been affected by the Coronavirus pandemic, provided that they do not receive any other benefit, including support from the unemployment insurance fund.

MEMBERSHIP TRENDS

By late 2002 there were 225 registered unions in South Africa, representing almost 3 million workers. And while the number of unions has increased since 1994, union density has not in any comparable way. On the one hand, South African unionism is robust, as union membership numbers are high compared to developing countries and also are higher than the average international membership (Barker 2015). On the other hand, statistically there is a noted downward movement in union membership numbers, especially in the agricultural, construction, finance and manufacturing sectors. However, union density in the mining sector, which is traditionally the country's most highly unionized sector, increased from 71 per cent in 1997 to 80 per cent in 2013 (Uys & Holtshausen 2016), even as mining employment has declined. As Figure 5.1 shows, most people (67 per cent in 2021) in the formal labour force are not union members.

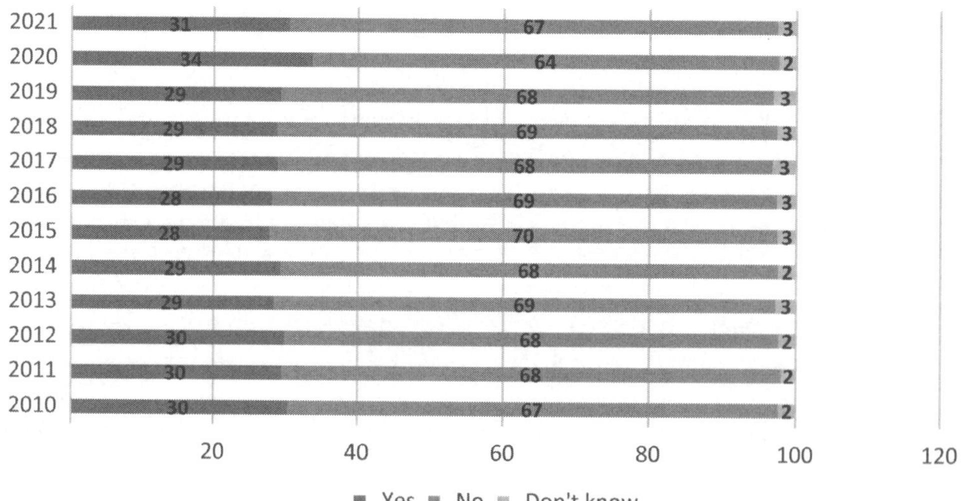

Figure 5.1 Union membership among the formal labour force (%)

Source: Statistics South Africa, Labour Force Survey, Q4 2021.

Presently, 31 per cent of employees in the formal, non-agricultural sector belong to unions, making unions an important partner in South Africa's labour relations system. There has been a steady increase in union membership in the public services, with public sector workers making up nearly half (49 per cent) of union membership in the formal sector (Forslund & Reddy 2015); that is, economic activity subject to state regulation and where there is formal employment, that being employment by a business or firm that is listed with the state. Due to South Africa's history of racial segregation and marginalization of the African workforce, legislative amendments included extending industrial democracy to the bulk of the country's workers, and this has been a factor in the increased public sector unionization, as well as its high levels of unionization, which have been consistent since 2010 onwards.

Union density is highest in the health and education sectors and, thus, there are as many high-skilled, white-collar workers among unionized workers as there are low-skilled, blue-collar workers. This means that the social configuration of the union movement is changing as the workforce includes more skilled and more educated employees (Webster 2013). However, the age of union members is also significant, as Mthethwa (2019) argues; younger workers are well aware of the issues in their workplaces but many do not know much about how unions could help them, or they think that unions are irrelevant or too politicized. As there are more non-unionized workers than there are unionized workers, this presents an opportunity to bolster the workers' movement. Moreover, the public sector unions operate under very different conditions to the private sector ones, and the shift in the internal balance in the Congress of South African Trade Unions (COSATU) towards the public sector meant that the relative power of some of the oldest and best governed unions was diluted.

Union density in the private sector has declined, and according to Cloete (2021), who drew upon internal union documentation, overall density currently is lower than one in three, with only one in four workers (around 23 per cent) employed in the economy and in the public services being union members, and under 10 per cent of workers receive a wage that is set by central collective bargaining bodies.

As Figure 5.2 shows, more than one in two workers in South Africa receives a wage dictated unilaterally by the employer. This crisis of internal representation among unions means that fewer workers are joining the unions and many state this is the case as they no longer trust them (Cloete 2021). Compounding this, what is lacking among workers is knowledge of what happens in the process of determination of working conditions (including wages) (Webster *et al.* 2009). Indeed, data in Figure 5.2 clearly indicates that over an 11-year period, it is the employer only that determines the salary (46 per cent in 2010, increasing to 52 per cent in 2021) of employees, with unions and employers setting the salaries only in around a quarter of cases (23 per cent in 2010 and 23 per cent in 2021).

In addition to tackling the challenge of recruiting younger employees and that employees possess a range of skill levels, unions also need to acknowledge that employees now have diverse demands and perspectives (Uys & Holthausen 2016). In studies on the issue of service delivery to members (see Masondo, Orkin & Webster 2015), shop stewards state that the quality of service delivery to members is of vital

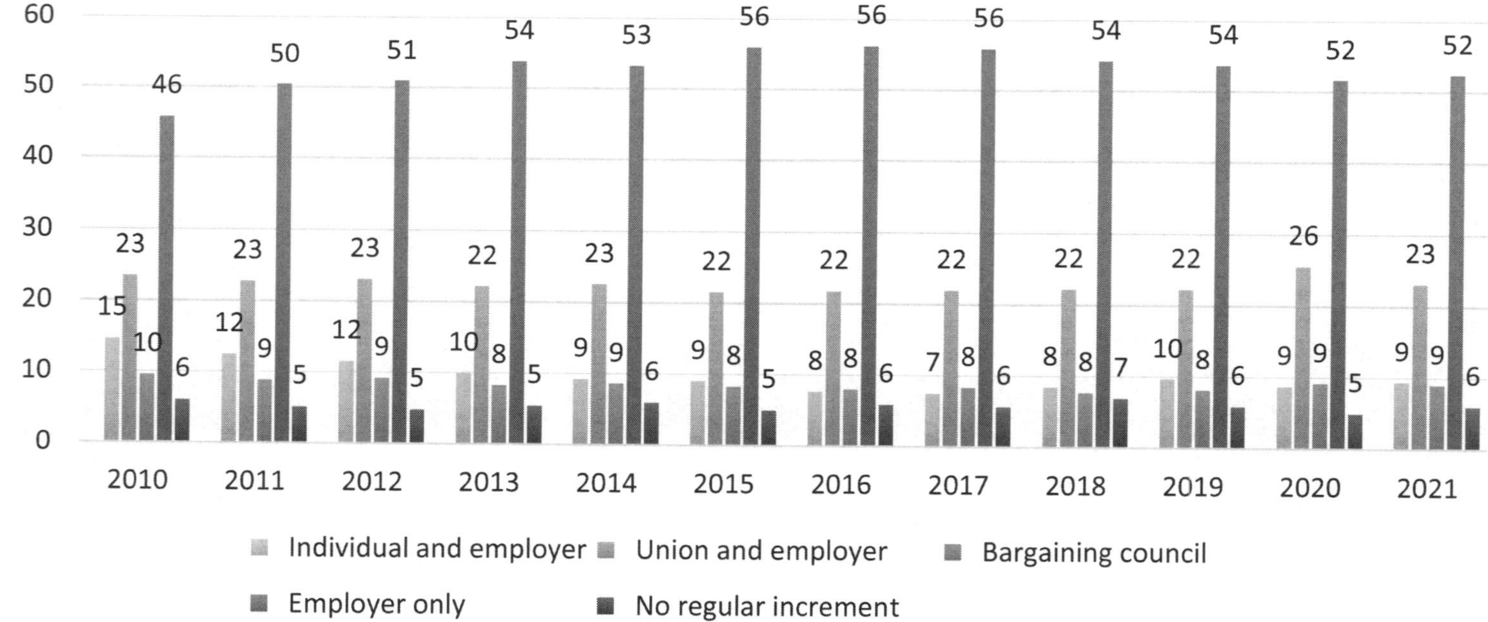

Figure 5.2 South Africa: where salaries are determined (%)

Source: Statistics South Africa, Labour Force Survey, Q4 2021.

importance and a significant issue concerning unions as it bolsters the probability that members will stay or employees will take up membership of the union. Thus, it appears that unions are organizations that have to shift and adapt in order to stay salient and they need to deal with the challenges facing them. Similarly, it would seem that unions need to introduce more services for members, beyond collective bargaining, and start to provide appropriate support to individual members.

UNION ORGANIZATION AND GOVERNANCE

In Africa, unions have a long tradition of political engagement, given their commitment to the anti-colonial movement. But after independence, unions were expected by their postcolonial governments to support the objectives of "national development", while they organized a minority – and so-called privileged section – of the labour force in Africa (Pillay & Webster 1991). By contrast, the liberation struggle in South Africa was against an internal opponent where the 1980s marked the high point of union mobilization, and it was at this point that the democratic traditions of the new unions were strengthened. Unions represented workers both in the workplace and community, this helping broaden the base of the unions and their scope.

As the occupational structure of the economy has shifted, the opportunity for young Black workers to move into the semi-skilled workforce has opened up. This has led to substantial changes in the make-up of union leadership, as the youth have joined the workforce and the unions. Younger union members co-joined with the older shop-floor activists and leaders who suffused the discourse of democratic unionism with content and meaning that resonated with their workplace experiences and community circumstances. Internal democracy, or the firmly held belief of worker control within COSATU of rank-and-file members filling the elected leadership positions, meant more than the prescribed processes of elections and meetings, for it gradually also became about empowerment and training and connecting the unions to the battle for democracy in the communities where the workers resided. Union democracy meant political democracy and working-class leadership in society (Buhlungu 2004).

The collectivist and democratic orientations were the predominant or hegemonic organizational culture in most unions. At a point, efforts were made to construct this collectivist and democratic organizational culture into the establishment of notions and practices of union democracy and worker solidarity. This culture was formed on notions of union democracy in the UK and the US but in South Africa its form was moulded by the experience of Black union mobilization from the early 1900s to the early 1970s, and the desire to evade state repression (Buhlungu 2004).

However, the labour regime in post-apartheid South Africa, often influenced by admiration for the West German model of co-determination, led to the institutionalization of industrial relations through corporatist-style negotiation between the actors, with external governance mechanisms sitting alongside, if not necessarily complementing, the internal governance mechanisms in the unions. When unions encounter management at the workplace, they are involved in bargaining councils and they are part and parcel of macroeconomic and political affairs. As union leaders have acquired the education, skills and knowledge to engage in institutionalized bargaining, and as unions have more members who need to be serviced, union officialdom has expanded. It can be argued that, in common with many unions around the world, the gap between workers and their leaders has increased and a shift from participatory democracy to representative democracy has taken place. Full-time positions for union office bearers and some shop stewards have also been implemented. However, as will be seen below, internal governance provided some checks to oligarchy.

Due to these organizational changes, hierarchical relations among leaders and between leaders and members emerged. Union leadership has become professionalized and is a full-time occupation with those in these positions receiving salaries and benefits (Masondo, Orkin & Webster 2015). Senior positions within the unions, such as president, general secretary, national and provincial organizers, have become more powerful sources of influence and authority. Their strategic location in unions provides them with opportunities to move onward and upward outside the union sphere into political positions, into the public service and into the private sector (see Bischoff 2015). These are the internal dynamics within all unions and federations.

The governing body of the union is the National Congress. Delegates are usually drawn from the branches of its affiliates, prescribed by the National Executive Committee (NEC). National congresses are usually preceded by provincial congresses. On the shop floor, shop stewards are the foundation of the organizational structure of the union. Shop stewards enlist new members and set up a structure of workplace representation that is intended to safeguard union democracy and ensure that the union is attentive to their members' needs and interests. The shop stewards jointly institute shop steward committees, which participate in plant-level negotiations over working conditions and wages (if there are no bargaining councils). The shop stewards also make sure that agreements are properly carried out. The shop stewards see their role as controlling and directing the union in the workplace.

Critically, shop stewards connect the union as an organization to the membership at the grassroots level. Members are dispersed over various workplaces, so the union officials rely on the shop stewards for this role. Shop stewards resolve grievances at the workplace and keep members informed of the union actions and policies of the shop steward committees – the bodies that engage and bargain with shop-floor management – in workplaces. Shop stewards eligible for election are those who have higher skill or education levels (Masondo, Orkin & Webster 2015).

Management has exerted great pressure on unions to have leaders who could make agreements without being constrained by the burdensome practices of worker control, which entail conferring with and checking constantly with members. Full-time officials have also been requested by management to take decisions for members, even when there is no mandate for this. Various shop stewards and full-time officials have capitulated when coerced, and this has generated friction within the unions. As unions in South Africa were fragmented racially, occupationally and politically, and differed in their organizational styles as well (Webster 2013), their responses to these challenges have differed and varieties of unionism have emerged (Bezuidenhout & Tshoaedi 2017).

Table 5.1 Varieties of unionism

Political unionism	*Business unionism*
COSATU	FEDUSA
NACTU	
SAFTU	
Solidarity	Solidarity

Business unionism, where unions limit themselves to workplace issues, is particularly represented by the Federation of Unions of South Africa (FEDUSA), Solidarity and other non-aligned unions (see Table 5.1). FEDUSA is the second-largest national union confederation and was founded in 1997. Its key leadership includes Godfrey Selamatsela as President, Riefdah Ajam as General Secretary and Martle Keyter as Deputy Secretary. FEDUSA has twenty affiliates. The top governing structure of FEDUSA is National Congress. According to FEDUSA's constitution, it holds a National Congress every three years. The purpose of the Congress is to report on the previous three years' work, make new resolutions for policy direction and elect new National Office Bearers. The highest governing structure of FEDUSA between the three-yearly National Congress is the NEC, which consists of representatives from each of the affiliated unions, and seats are allocated proportional to relative membership size. In a non-Congress year, the NEC meets once every quarter. The Management Committee's (ManCom) responsibility is governing the daily activities of the Federation. ManCom is made up of all the National Office Bearers, Senior Secretariat and additional co-opted representatives from various affiliated unions. The ManCom meets six times per year or as the need arises.

Solidarity, a Christian union, emerged in the post-apartheid era from the formerly Whites-only Mine Workers Union. Solidarity appears in both categories in Table 5.1 as in the early 1990s it started off as a business union but it has transformed itself into a political union as its institutions, such as the self-help fund, the Helping Hand, its technical training college and its university, Academia, predominantly benefit White Afrikaans-speaking members. Importantly, it has a civil rights arm called Afriforum. Solidarity's model of member representation is geared to servicing individual membership needs. Solidarity has about 140,000 members, all of who are compelled to adhere to the provisions of its constitution. Solidarity's governance structure is made up of a chairman, a chief executive and a general secretary, who is in charge of all labour relations and collective bargaining matters.

Political unionism, where unions are subservient to political parties (Pillay 2015) is represented by COSATU, its newly established offshoot and rival, the South African Federation of Trade Unions (SAFTU) and the National Council of Trade Unions (NACTU). NACTU is a national union federation set up in 1986. It is a creation of a merger between the Council of Unions of South Africa and the Azanian Confederation of Unions. Its leadership includes Narius Moloto, General Secretary, who is also the President of the Pan-Africanist Congress of Azania (a political party formed in 1959). NACTU has 22 affiliates and its constitution states that all committee and structures of the unions and the Federation need to be comprised of workers. In terms of the constitution's clause 3.1, a "bona fide Union means a union which has a constitution, and no less than 100 members, and is controlled by its members'.

COSATU was founded in 1985, being a popular democratic anti-system social movement union federation with some affiliates turning towards political unionism as many affiliates have always fought for more than just better working conditions.[1] COSATU is part of the Tripartite Alliance, with the ANC and the South African Communist Party (SACP). COSATU became a home of militant and democratic unions that re-emerged from the historic Durban strikes between 1971 and 1973. The absence of a political voice obliged unions to deviate from their traditional role of pursuing the "politics of production" to embracing the contestation of the politics of state power. COSATU's constitution was drawn up at its first congress. The constitution clarified that workers would govern the structures of the federation, that a national congress would be take place every two years and would be the premier decision-making body, that a central executive committee would convene a meeting every three months, that an executive committee would meet monthly and that the federation would elect its office bearers at the congresses.

Many of COSATU's affiliates stressed the importance of democratic workplace organization from the onset of their formation in the 1970s, and a resilient tradition of democracy in the unions, which entailed worker participation at the level of the shop floor with representative and accountable leadership at the higher levels, was created. Some of COSATU's affiliates were initially led by those who had a university education but they intentionally created democratic structures and practices to lay the foundation for democratic worker control of the unions (Buhlungu 2009). The unions started off with a small membership; however, as the membership increased, workers could not exercise direct participatory democracy but had to switch to representative democracy; that is, they elected or appointed leaders who would represent members' wishes and they would be held accountable. COSATU is, however, structured in a democratic centralist manner where majority decisions are binding. Its leadership also can take advantage of the organizational machinery to further their candidacies for leadership positions at election time.

Under the ANC-led government from 1994, COSATU struggled with its alliance partners to improve the lives of its members as well as the lives of the country's Black and working class. Union membership increased dramatically after 1994 and the gulf between shop-floor workers and shop stewards as well as between shop-floor workers and their branch, regional and national leadership became accentuated. During the 1990s some COSATU unionists were critical of government policies.

The 1996 Growth Employment and Redistribution (GEAR) programme, drafted to ostensibly bolster employment, led to job security loss among workers. GEAR transformed the relations within the Alliance, as COSATU was stripped of its power and the possibilities of redistributive social change were thwarted. This was because, with privatization and labour casualization combined with deindustrialization, rising unemployment and a drop in real wages, COSATU was disempowered to counter these changes. Its institutionalized and bureaucratized unions had failed to keep up with the

1. Social movement unionism is a "form of unionism that has a broad concept of the 'working-class'; has deep shop-floor democratic practices; engages in both workplace and broader struggles outside the workplace; actively forges alliances with organisations outside the workplace and maintains independence and organisational integrity" (Pillay 2015: 118).

changing nature of work. COSATU has battled to link up with and organize precarious labour, but also with community struggles and worker actions. Many South African workers demonstrate their militancy in wildcat strikes and community protests, but the unions have failed to capitalize on these regularly occurring events to secure fundamental material and ideological benefits for the working class (Buhlungu 2004).

COSATU remains the largest union federation with 21 affiliated unions. As the SACP is the vanguard of the working class, COSATU is under its influence and working-class support for the ANC is maintained.[2] COSATU's motto is "An injury to one is an injury to all", which signifies that COSATU's objective is the consolidation of working-class solidarity. However, the availability of posts in the public service, politics, business and the non-governmental sector have led to many senior unionists leaving their union positions (Buhlungu 1999). COSATU's key leadership includes Zingiswa Losi, who is the first female President.

In late 2013 COSATU expelled its largest affiliate, the National Union of Metalworkers of South Africa (NUMSA) for not supporting the ANC in the 2014 national elections in South Africa (Ashman 2015). After what some term the COSATU rupture (Bezuidenhout & Tshoaedi 2017), there are now two main strands of unionism in South Africa, as is depicted in Table 5.1. COSATU's weakness is particularly evident in its exhausted internal democracy. The distance between the leadership and its rank-and-file members is considerable, it does not take direction from workers and its leaders continue to enrich themselves.

The National Union of Mineworkers (NUM) is one of COSATU's largest affiliates. It was founded in 1985 and, from that time onwards, the rise of leadership at all levels has seen these leaders take over the union and reduce grassroots democracy to a mere ideal. According to Botiveau (2018), a special kind of leadership has emerged, being "labour aristocrats" who have been groomed by the union leadership to be capable organizers who can run a union. Power and decision-making are centralized at the top level and a type of organizational governance has been imposed on the NUM by its national office bearers (Buhlungu & Bezuidenhout 2008). This governance, in which democratic centralism increased and internal debate was gradually reduced, has muzzled dissent. A former General Secretary was known to have undermined democratic debate in the NUM.

The NUM became more inclined towards national politics as well as a career path for the NUM cadres and leaders. Many unionists moved into government, business or work for the mining companies, which has weakened the NUM. There has, however, been strong criticism of the NUM leadership, especially after the killing of 34 mineworkers in the area of Marikana in 2012, and a call for change – to go back to the grassroots to serve members. This serves as a reminder that the NUM is a democratic organization. Indeed, organizational agency was deployed to regenerate the unions and renew collective struggles. At the NUM's 15th Congress delegates had the right to choose new leaders and they did so. Even though NUM leadership was praised as a core union value,

2. The current general secretary of the South African Democratic Teachers Union, Maluleke Mugwena, was elected at the SACP's 14th National Congress for the 2017–22 political term as a Central Committee Member.

the NUM leader was voted out and replaced by his political rival. This does mean that the NUM is not a wholly oligarchic organization in which power cannot be contested internally. It has developed internal democratic safeguards, and bureaucratization is not necessarily to be viewed as the inevitable result of a degenerating organizational democracy: oligarchy is not inevitable but, of course, the gap between the NUM leaders and members did deepen (Botiveau 2018).

The hegemony of worker control has always been fragile as the generation of the culture and practice of worker control has been challenged and influenced by many bureaucratic and anti-democratic trends within a continuously fluid environment in which this hegemony had to be sustained. As a COSATU official remarked:

> I want to be guided by the very founding principle of COSATU and I know that was taking over from SACTU the principle of worker control. So, leadership in union movement is those comrades who are given responsibility to lead the union movement but acting on mandates that is collected and given by the general membership which means that those comrades are not implementing policies or desires of their own but they are leading because they have been given, they have been elected as leaders of the unions. And, of course, the principle of worker control reigns supreme. So that remains my understanding regardless of what the modern or latest issues should be or are, be it the issues to with the boards, because you find that some of these leaders are the ones who sit on SETA boards, on the investment companies, boards and now they get allowances which sometimes runs into millions. But still they are held accountable by the membership because they are there by virtue of being elected into those positions. So, they are not leaders because … in the first place were qualified for leadership. Leadership is earned through struggle and the trust that members put on you that you will be able to deliver. That remains my understanding on leadership.
>
> (Interview with COSATU official, 2015)

However, studies of COSATU affiliates have shown that shop-floor democracy is intact among its affiliates but that oligarchy occurs at the national level; that is, at the top the leaders operate without mandates from the workers and are not accountable to them. It is also difficult to vote leaders out of office after they have been elected to positions of power. Power in COSATU is firmly positioned at the top. COSATU is an oligarchic organization in that decision-making on some issues is the task of a few select leaders (Maree 1982).

The formation of SAFTU out of COSATU signalled the success of the discourse of democratic unionism and its related principles, namely leadership accountability with the full-time officials subservient to worker leaders. There has been a noted decline of the democratic culture in the unions that started in the early 1990s (Buhlungu 2009). The former General Secretary of COSATU, Zwelinzima Vavi, was dismissed by a COSATU special central executive committee meeting. Vavi was accused of playing a key role in trying to split COSATU from the ANC and the SACP (Satgar & Southall 2015).[3] Vavi set

3. In August 2013, Vavi was suspended from office over the sexual harassment of a COSATU staff member.

up and registered SAFTU in 2017 and became its General Secretary. Some COSATU affiliates, such as the Food and Allied Workers Union disaffiliated from COSATU and joined SAFTU. SAFTU's affiliates are general unions, which organize along value chains and in similar sectors. SAFTU has 21 affiliated unions, organizing 800,000 workers. At its founding congress, SAFTU affiliates agreed to a draft constitution and elected the first leadership structure, which consists of a president and two deputies, as well as a treasurer. Its key leadership included Mac Chavalala as the President and Nomvume Ralarala as the First Deputy President.

The NEC of SAFTU is attended by the SAFTU National Office Bearers, the affiliated unions' Presidents, Deputy Presidents, General Secretaries, Deputy General Secretaries as well as the provincial chairpersons and secretaries of the federation. The NEC deals with pressing organizational and socio-economic issues. SAFTU undertook to launch campaigns focusing upon matters such as poverty, unemployment, workers' right to strike and labour brokers. SAFTU committed itself to returning land to its rightful owners, attending to the nation's energy crisis and to paying attention to the unfolding catastrophe in the public health and education sectors (Luckett & Munshi 2017). It pledged to:

> ... give the South African working class a new home, a real revolutionary home with the objective of being a democratic, worker-controlled, militant, socialist-orientated, internationalist, Pan-Africanist from a Marxist perspective organisation that will struggle for the total emancipation of the working class from the chains of its capitalist oppressors.
>
> (Luckett & Munshi 2017)

SAFTU, thus, stresses building a democratic, worker-centred federation (Vavi 2020), but whether it represents an alternative unionism in South Africa remains to be seen. SAFTU's first congress offered some hope that it would develop a democratic culture. SAFTU's draft constitution set up a National Working Committee structure made up of various parts of the working class, including migrant workers. However, it has to be borne in mind that SAFTU was founded by leadership and not by the grassroots or worker militancy, as it emerged out of political battles within COSATU over its support for the ANC and SACP, as well as the internal union struggles that opposed corruption and the decline of internal democracy. It also remains to be seen whether SAFTU leaders will tolerate internal debate, whether workers will actually be in control of SAFTU, whether working-class struggles will be the focus and solid connections between unionized and non-unionized workers to community struggles will be forged (Luckett & Munshi 2017).

At SAFTU's founding congress, the commitment to the formation of a workers' party was made, yet SAFTU claims to be politically non-aligned. NUMSA, set up a "vanguard party" called the Socialist Revolutionary Workers Party (SRWP), which it expects SAFTU to support, in "defence of the working class". This is a current source of tension within SAFTU as it tries to resolve how to build the working-class movement politically (AIDC 2022). The SWRP is a predominantly small grouping of chiefly NUMSA worker leaders and staff that only managed to secure 24,439 votes nationally in the 2019 national elections. The question of support for the SWRP by SAFTU has led to in-fighting (AIDC 2022).

In March 2022, and prior to its second national conference in May 2022, the SAFTU President issued a letter of suspension to the General Secretary linked to purported financial misconduct charges, which the General Secretary challenged. The four national office bearers who issued the suspension were subsequently suspended by SAFTU's NEC. This angered NUMSA, SAFTU's largest affiliate, which considered the suspension of the four national office bearers as illegal (the SAFTU President, who comes from NUMSA, was one of the four suspended) (Feketha 2022). Vavi is seeking a second term but NUMSA and its General Secretary, Irvin Jim, do not support him. Four of SAFTU's affiliates are planning to launch a legal battle against the suspension of the four officials, as they contend that they operated in line with the federation's disciplinary code and procedures (AmaShabalala 2022a).

As SAFTU's second elective conference commenced, its delegates clashed over the union federation's constitution. In addition to this, a vote on the constitution and the constitutional powers of the NEC took place at this conference. It seems to be the case that SAFTU is battling to establish workers' control and democracy within all affiliates (AmaShabalala 2022c). So, SAFTU is considered to be in the formative stage, and it seems that the in-fighting indicates that the centre is not holding. SAFTU seems to have same practices that were endemic in COSATU (AmaShabalala 2022b). These practices include the self-enrichment of leaders, corrupt relationships with management and the suppression of internal democracy, as well as the problem of union investment funds.

FAILURES OF GOVERNANCE

Many unions have built up considerable retirement (pension and provident) funds, which have accrued to the affiliates. This is in addition to union membership dues. The income can be used to create jobs and for "socially responsible" investment. Even former socialist unionists have used union investment funds to invest in various sectors of the economy such as casinos, mobile phone and television deals (Bond 2000). Union investment companies sought to capitalize on the ANC's commitment to create Black Economic Empowerment as part of its non-racial nation-building plan, benefiting from privatizations, government contracts with the firms they own and through opportunities to gain stakes in White-owned firms seeking to prove their non-racial credentials (Iheduru 2004). Theron (2017) argues that COSATU money was siphoned off into investment companies without much debate and that the union investment companies enabled leaders to pursue their own interests, creating a class of "unionaires". If state policy and White businesses created many opportunities for union investments companies, this did not always mean that the interests of members were best serviced; in short, serious problems of achieving self-agency for and self-control by grassroots members have emerged.

Even though the investment appendages are theoretically detached from the unions, union office bearers are appointed to attend to the trusts. Unions and federations across the spectrum have investment companies: NACTU Investment Holdings, Kopano Ke Matla (COSATU's investment arm) and NUMSA's Investment Company (NIC) to name but a few. A closer examination of COSATU's investment companies has revealed that there is a substantial contradiction between the ideology of COSATU's commitment to socialism and the profitable business deals accomplished by its own union investment company, Kopano Ke Matla. This has contributed to the growing distance that has developed between the union leadership and its members as new "union entrepreneurs" have the space to commence their private investment undertakings. The era of "comrade capitalism" is brazenly part of COSATU and one of the reasons for the growing weakening of COSATU democratic governance, as the culture of self-enrichment among the union leaders who have extensive shares in union investment companies has grown. As COSATU and its unions, but especially the leaders, have profited from this arrangement, they are keen to preserve this status quo. Many can now afford to enrol their children in the former Whites-only schools, they can live in the formerly Whites-only residential areas and some drive luxury cars. Thus, many claim that union leaders are "sell-outs" as they lead lifestyles funded by the union investment company.

Nor are these governance problems confined to COSATU. NUMSA, a SAFTU affiliate, has the NIC and it is alleged that its life insurance arm, called 3Sixty Life, is so intertwined with the union that it has significant influence over how the union is governed – the NUMSA General Secretary's support of the NIC's CEO is evidence of this. The NIC sells insurance policies to NUMSA's 350,000 members and then reinvests that money in the union, presumably, one might assume, for members' benefit. 3SixtyLife was placed under curatorship in December 2021. In addition to this, there are leadership battles within SAFTU, which according to labour and policy analysts, are about the "outright struggle for control" of the federation, as the leadership is not concentrating on worker interests but rather on advancing themselves (Mahlaka 2022).

CONCLUSION

The rise of the South African labour movement was predicated upon good governance, centring on high levels of internal democracy. Although this tradition persists, decision-making in unions and federations has become more concentrated and bureaucratized (see also Chapters 4 and 15). In many unions a managerial outlook among shop stewards has become more dominant than a militant one, reflecting opportunities to move into managerial roles in their employing organization or to pursue lucrative careers in union investment companies (Masondo, Orkin & Webster 2015). The factional troubles within unions and federations are increasing, and in the current economic climate, with many more workers out of jobs, the loss of membership and of income intensifies these struggles and further fragments the labour movement.

These leadership struggles counter the tradition of worker control. Union general secretaries are now enormously powerful, and they dominate over the worker leaders, who are hypothetically their bosses. The authority of full-time officials has not been unopposed. There are still many union members of the federations who uphold the notion of worker control and support the democratic organizational culture in their unions. The rank and file have managed to maintain their control over leadership positions and decisions at the shop-floor level. However, good local governance is not matched by sound governance at the commanding heights of these labour union organizations. To date, the democratic union tradition in South African unions seems to be surprisingly robust, and unionists at different levels of union organization have opposed oligarchic tendencies. This demonstrates that the nature of union democracy is not predetermined nor is it constant but rather it is vibrant, and it is a process that is challenged as it depends on the balance of power between different layers and factions in unions.

What can unions around the word learn from the South African experience of unions and their governance? First, internal democracy is a powerful mechanism for defeating state repression where it is much easier for the state to remove a few union leaders than many, especially if these can be easily replaced. Second, and following the demise of apartheid, an essentially sympathetic government does not necessarily mean that it is easier to ensure sound internal governance. Progressive labour laws may simplify the job of unions in day-to-day organizing, but attempts to deracialize the South African economy have paradoxically opened up new challenges for unions.

Black empowerment measures have opened up new investment opportunities for unions, but with this has come something of a loss of innocence, as union officials and leaders have grasped opportunities for personal advancement and enrichment, leading to serious agency failings; union leaders have not always proved altruistic managers of the savings of ordinary members. It could be argued that this is a situation unique to South Africa. However, in any context where a sympathetic government is undertaking privatizations, there may be opportunities for union leaders to pursue opportunities that may lead to great wealth for themselves, if not their members.

Moreover, there are many instances around the world where union-administered pensions or saving schemes have made controversial investments that may not be in the interests of the rank and file, either on grounds of finance or principle. There is an extensive literature on how institutional investors often fail their proclaimed principles and ordinary savers, where union investment funds may be no exception.

Again, it is not just in South Africa where effective shop stewards may be tempted by managerial careers. This may again mean that sound principles of "governance" may be eroded by opportunism in practice. Finally, bitter personal struggles highlight how it is not just structures that make for good governance, but also principles and informal conventions. When individuals constantly challenge or seek to bypass the latter, unions may run into serious governance problems, rendering formal rules irrelevant.

REFERENCES

Alternative Information and Development Centre (AIDC) 2022. "Take back the union for its members". *Amandla* 32, March.

AmaShabalala, M. 2022a. "Irvin Jim draws first blood against Vavi in battle for soul of Saftu at congress. Numsa wins motion over unconstitutionality of acting president to chair the gathering". Times Live, 23 May. www.timeslive.co.sa/politics/2022-05-23-irvin-jim-draws-first-blood-against-vavi-in-battle-for-soul-of-saftu-at-congress/

AmaShabalala, M. 2022b. "Irvin Jim suffers major blow as suspended Saftu NOBs lose vote: after Numsa's dominance in debates for two days, voting favoured their nemesis, Zwelinzima Vavi". Times Live, 24 May. www.timeslive.co.sa/politics/2022-05-24-irvin-jim-suffers-major-blow-as-suspended-saftu-nobs-lose-vote

AmaShabalala, M. 2022c. "Vote on suspended office bearers will settle Vavi vs Jim tussle in tense Saftu congress". Times Live, 24 May. www.timeslive.co.sa/politics/2022-05-24-vote-on-suspended-office-bearers-will-settle-vavi-vs-jim-tussle-in-tense-saftu-congress/

Ashman, S. 2015. "The social crisis of labour and the crisis of labour politics in South Africa". *Dans Revue Tiers Monde* 224: 47–66.

Barker, F. 2015. *The South African Labour Market: Theory and Practice*. Fifth Edn. Pretoria: Van Schaik.

Bezuidenhout, A. & M. Tshoaedi 2017. *Labour Beyond COSATU: Mapping the Rupture in South Africa's Landscape*. Johannesburg: Wits University Press.

Bischoff, C. 2015. "COSATU's organisational decline and the erosion of the industrial order in South Africa". In *COSATU in Crisis: The Fragmentation of an African Trade Union Federation*, edited by V. Satgar & R. Southall, 217–46. Johannesburg: KMM Review Publishing.

Bond, P. 2000. *Elite Transition: From Apartheid to Neoliberalism in South Africa*. Pietermaritzburg: University of Natal Press.

Botiveau, R. 2018. *Organise or Die: Leadership of a Special Type in South Africa's National Union of Mineworkers*. Johannesburg: Wits University Press.

Buhlungu, S. 1999. "A question of power: co-determination and trade union capacity". *African Sociological Review* 3(1): 111–29.

Buhlungu, S. 2004. "The building of the democratic tradition in South Africa's trade unions after 1973". *Democratization* 11: 133–58.

Buhlungu, S. 2009. "The rise and decline of the democratic organizational culture in the South African labor movement, 1973 to 2000". *Labor Studies Journal* 34(1): 91–111.

Buhlungu, S. & A. Bezuidenhout 2008. "Union solidarity under stress: the case of the National Union of Mineworkers in South Africa". *Labor Studies Journal* 33(3): 262–87.

Chinguno, C. 2013. "Marikana: fragmentation, precariousness, strike violence and solidarity". *Review of African Political Economy* 40(138): 639–46.

Cloete, K. 2021. "Labour pains: trade union membership has declined badly and bosses are calling the shots". *Daily Maverick*, 2 March. www.dailymaverick.co.sa/opinionista/2021-03-02-labour-pains-trade-union-membership-has-declined-badly-and-bosses-are-calling-the-shots/

Feketha, S. 2022. "Fight for soul of Saftu is on as member unions threaten court action: suspension of 4 top leaders divides federation". *The Sowetan*, 28 March. www.sowetanlive.co.sa/news/south-africa/2022-03-28-fight-for-soul-of-saftu-is-on-as-member-unions-threaten-court-action/

Forslund, D. & N. Reddy 2015. "Wages and the struggle against income inequality". In *COSATU in Crisis: The Fragmentation of an African Trade Union Federation*, edited by V. Satgar & R. Southall, 83–114. Johannesburg: KMM Review Publishing.

Iheduru, O. 2004. "Black economic empowerment and nation building in post-apartheid South Africa". *Journal of Modern African Studies* 42(1): 1–30.

Luckett, T. & N. Munshi 2017. "Rebuilding a workers' movement". *Jacobin*, 15 May. https://jacobin.com/2017/05/south-africa-trade-unions-saftu-numsa-anc-suma

Mahlaka, R. 2022. "Leadership in South Africa's organised labour movement is imploding". *Daily Maverick*, 3 April. www.dailymaverick.co.sa/article/2022-04-03-leadership-in-south-africas-organised-labour-movement-is-imploding/

Maree, J. 1982. "Democracy and oligarchy in trade unions: the independent trade unions in the Transvaal and the Western province general workers' union in the 1970s". *Social Dynamics* 8(1): 41–52.

Masondo, T., M. Orkin & E. Webster 2015. "Militants or managers? COSATU and democracy in the workplace". In V. Satgar & R. Southall, 192–216. Sandton: KMM Review Publishing.

Mthethwa, G. 2019. "What have we learned about trade union organising in changing workplaces?" Labour Research Service, 24 January. www.lrs.org.sa/2020/02/26/what-have-we-learned-about-trade-union-organising-in-changing-workplaces-2

Pillay, D. 2015. "COSATU and the alliance: falling apart at the seams". In *Cosatu in Crisis: The Fragmentation of an African Trade Union Federation*, edited by V. Satgar & R. Southall, 115–33. Sandton: KMM Review Publishing.

Pillay, D. & E. Webster 1991. "COSATU, the party and the future state". *Work In Progress* 76: 31–7.

Satgar, V. & R. Southall 2015. "COSATU in crisis: analysis and prospects". In *Cosatu in Crisis: The Fragmentation of an African Trade Union Federation*, edited by V. Satgar & R. Southall, 1–34. Sandton: KMM Review Publishing.

Statistics South Africa 2021. *Quarterly Labour Force Survey*. Quarter 4, 2021.

Theron, J. 2017. "Solidarity for whom? How Cosatu became part of the establishment". *Global Labour Column* 280 (June).

Uys, M. & M. Holthausen 2016. "Factors that have an impact on the future of trade unions in South Africa". *Journal of Contemporary Management* 13: 1137–84.

Vavi, S. 2020. "Why are established trade unions comfortable with the status quo and threatened by the rise of Saftu?" *Daily Maverick*, 16 December. www.dailymaverick.co.za/article/2020-12-16-why-are-established-trade-unions-comfortable-with-the-status-quo-and-threatened-by-the-rise-of-saftu/

Visser, J. 2019. "Trade unions in the balance". ILO ACTRAV Working Paper. Geneva: International Labour Office, Bureau for Workers' Activities.

Webster, E. 2013. "The promise and the possibility: South Africa's contested industrial relations path". *Transformation: Critical Perspectives on Southern Africa* 81/82: 208–35.

Webster, E. *at al.* 2009. *NUMSA/FES/SWOP/DITSELA Research Changes in Production Systems and Work Methods: A Report on Work Restructuring in the Auto and Components Sector: The Benefits, Dilemmas and Opportunities Facing NUMSA*. Society, Work and Development Institute (SWOP), Wits University.

CHAPTER 6

UNION RELATIONS

Kurt Vandaele

ABSTRACT

This chapter examines and analyses the external relations of individual unions. Seeking to identify the interlinkages between power resources, union identities and union strategies towards external actors, the chapter makes three arguments. First, the interplay between the structural leverage and associational power of workers is linked to labour unions predominantly building relations in either the economic nexus or the state nexus, with each one marked by a prevailing union strategy, respectively mobilizing or political advocacy. Second, and although being dependent upon the institutional context, external relations within both nexuses have been deteriorating, relatively speaking, over time. Third, being associated with an organizing strategy, the development of coalitional power; that is, seeking alliances with non-labour actors in civil society, becomes consequently more desirable if unions want to retain or regain relevance in the twenty-first century. Opportunities lay in strengthening solidarities with, for instance, immigrant and ethnic communities and networks, on embarking on intersectional approaches to mobilizing and organizing, and union-environmental alliances.

Keywords: Union external relations; union strategy; challenges for unions

INTRODUCTION

Labour unions exist in largely predetermined worlds. Yet, in trying to change these worlds in order to attain their aims, they then must often necessarily choose to make decisions about what relationships they want to have with other actors with a view to achieving their objectives. This chapter examines and analyses the external relations of unionism; that is, the tactical and strategic behaviour of individual unions in building relations with their environment, this is other actors and forces outside the union movement, to pursue union goals.[1] Thus, three interrelated questions about union behaviour regarding external relations can be posed. First, why are labour unions building relations with external actors? In other words, what are the drivers for initiating and developing interactions with others?[2] Second, which other actors are potential candidates for establishing relations, and how do unions select them? Third, what is the nature of relation-building with those external actors? Put differently, is this relation-building temporal or permanent (structural), mutual or compatible, opportunistic or strategic, or materially underpinned or ideological mediated?

These questions practically direct us towards a study of the "essence of unionism" (Hodder & Edwards 2015) as it touches upon strategic choice and union strategies, union effectiveness and union goals (Frege & Kelly 2003; Gall & Fiorito 2016; Hyman 2007). While this "essence" calls for a nearly all-encompassing approach, this chapter builds upon a limited number of concepts for categorizing knowledge about the external relations of unionism. The concepts applied in this chapter are inspired and framed via the power resource approach (Refslund & Arnholtz 2021; Schmalz, Ludwig & Webster 2018; Silver 2003; Wright 2000; see also Chapter 3) and union strategies towards external actors (McAlevey 2016) in the arenas identified by Hyman's model of different engrained union identities formed on the collective interests of workers (Hyman 2001; see also Rhomberg & Lopez 2021). Based upon this triangular model, offering a dynamic typological framework, external actors are mapped in distinctive but interrelated sociological arenas: the "market", the "state" and "society" (Hyman 2001; see also Chapter 2). These three arenas or directions analytically and contextually align the research subject of this chapter.[3]

From unitary, pluralist or critical perspectives, offering "middle-range" theories, various academic disciplines have tried to understand the multifaceted relationships of labour unions and the dynamics and changes in one of the three arenas. The traditional subject of study for industrial relations scholars is the relationship between unions and management at workplace level or, beyond this, their relationship with employers and employer associations in the labour market (Gumbrell-McCormick & Hyman 2013). Political scientists and sociologists have mainly studied the (historical) relationships between interest groups such as unions and political parties in the context of the state (Ebbinghaus 1993; Haugsgjerd Allern & Bale 2017), state strategies towards unions, such as neo-corporatist arrangements (Berger & Compston 2002; Crouch 1993; Ebbinghaus & Weishaupt 2021), and the degree of embeddedness of unions within welfare institutions (Ebbinghaus 2021).

1. Intra-union relations and intra-confederation politics are, thus, largely disregarded here; see, for example, Chapters 4 and 5 on these issues.
2. This chapter starts in first instance from the perspective of workers and their representative organizations and puts far less emphasis on the motives and reasons of the other actors for building relationships with unions. Also, issues such as union representation, governance and leadership and the perception of members regarding unions' external relations are, as such, not studied here, although internal preconditions and union decision-making obviously influences priorities and strategies towards external actors.
3. Furthermore, this chapter is informed by union developments in the Global North, and Western Europe in particular, which is the region the author is most familiar with.

Researchers studying social movements have recently "rediscovered" unions as movements, and not solely as (bureaucratic and hierarchical) organizations, especially in the wake of anti-austerity discontent and resistance in various European countries in the aftermath of the collapse of the finance-led accumulation regime in 2007–08 (della Porta 2015; Grote & Wagemann 2019; Kriesi *et al.* 2020). Simultaneously, unions embracing a "whole-worker approach" (McAlevey 2016) and fostering coalition-building with equity-seeking identity groups and activist-led communities and social movements are increasingly considered as pathways towards union revitalization (Doellgast, Bidwell & Colvin 2021).

Seeking the interlinkages between power resources, union identities and union strategies towards external actors, underpinned by empirical studies, this chapter makes three arguments. First, the interplay between the structural leverage and associational power of workers is linked to labour unions predominantly building relations in either the economic nexus or the state nexus, with each one marked by a prevailing union strategy. Mobilizing for industrial action, or the credible threat of it, based upon structural leverage, is critical in the economic nexus of the labour market, while political advocacy addressing legal-based institutional resources is dominant in the political nexus of the state. Second, for various reasons, the external relations of unions within the economic nexus and the state nexus have been relatively deteriorating over time, although this is, for instance, dependent upon the institutional context. Third, from functional and normative perspectives, the development of coalitional power associated with an organizing strategy becomes consequently more desirable if unions want to retain or regain relevance in the twenty-first century (see also Hyman & Gumbrell-McCormick 2017; Kelly 1998; Robinson 2000; Tattersall 2010), albeit without being a panacea for (immediate) success (Heery, Williams & Abbott 2012). As such, the underlying rationale of this chapter is less about the environmental influences upon unions but more about them as possible active agents of change in the economic and state nexuses – unions have more than ever strategic choices to make.

INTERCONNECTING POWER RESOURCES AND NEXUS-DEFINED EXTERNAL ACTORS

The primary question that needs to be answered is why labour unions establish relations with external actors. In answering this question, it is useful to look first through the lens of the agency of the workers themselves, and not unions per se as just one of the organizational expressions of workers' associational power – albeit an important and still dominant one, yet increasingly marginalized in some economies of the Global North (Refslund & Arnholtz 2021). This is because it is assumed here that the power resources available to various types of workers both inform and influence – yet do not mechanically determine – the form that unions could take and their external relationships (see also Chapter 4). Although other factors might be of importance as well for analysing those relationships, power resources are a good starting point as they essentially put the emphasis on the structural position of workers in capitalism. Different types of workers can possess diverse power resources (Schmalz, Ludwig & Webster 2018; see also Chapter 3) among which structural power is the base. Varying within supply chains, across economic sectors and geographies over time, this power resource is inherently contingent on workers' material circumstances within capitalist employment relationships and labour markets.

In understanding those circumstances, a distinction is typically made between workplace bargaining power and marketplace bargaining power (Silver 2003; Wright 2000). Workplace bargaining power arises from the specific position of workers within distribution or production systems, influencing their ability to deploy their disruptive capacity via a withdrawal of their labour or any other manifestation of (everyday) individual or collective resistance. Marketplace bargaining power depends on workers' possession of certain qualifications or skills desired by management, the degree of unemployment or the extent to which workers can live from other non-market income sources. Structural power, whether based on marketplace or workplace bargaining power, can be conceived as latent. It only manifests itself and can gain momentum when workers collectively activate it through their associational power. This power resource – associational power – has therefore been defined as "power in action" to differentiate it conceptually from structural power, which is considered a "source of leverage" – hence, "structural leverage" (Rhomberg & Lopez 2021: 48). Even "spontaneous" walkouts by workers – without any union involvement – could be considered associational power (in the making), thereby converting structural leverage into collective action (Kelly 1998).

As a means to address power imbalances between labour and capital, associational power has, since the nineteenth century, formally coalesced into the formation of labour unions as voluntary, collective organizations (Offe & Wiesenthal 1980). Associational power facilitates the development of unions and explains why they are pursuing relations with external actors. After all, they are "intermediary organizations" (Müller-Jentsch 1985) combining the structural leverage of workers: unions are established after the creation of the capitalist employment relationship, signifying that their existence is contingent on the collective organization of workers by employers. This secondary character of unions implies that they are "fitting" into established interest and power structures even though they have a desire to change and amend these to varying degrees. Thus, for better or worse, their embeddedness in existing structures principally necessitates them building relationships at different levels with other actors such as management, employers and their associations, political parties, governments, public administrations and non-governmental organizations or other non-market actors for achieving their goals.

Apropos of Figure 6.1, solely theoretically focusing first upon workers' power resources, then dynamics and interconnections between them are of specific significance (Refslund & Arnholtz 2021; Rudman & Ellem 2022; Schmalz, Ludwig & Webster

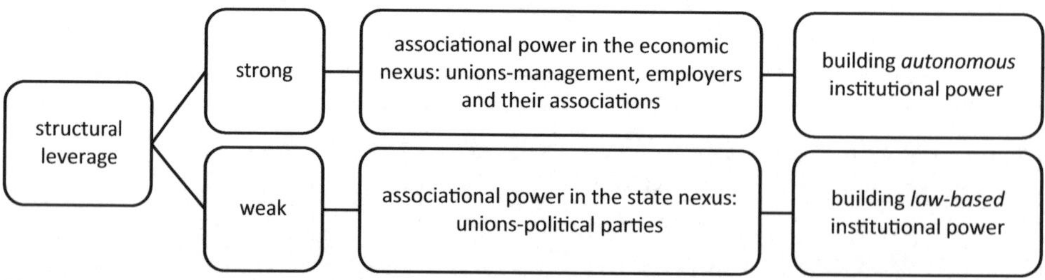

Figure 6.1 Workers' structural leverage and unions' default nexus in relation to external actors

2018). Essentially, associational power either "pools" structural power or partially compensates for the lack of it. If structural leverage is or remains strong, then workers, associated in labour unions, with sufficient financial and organizational resources, are likely to act collectively by drawing upon this power resource predominantly or almost exclusively. Conversely, it is expected that weak structural leverage entails workers aiming to strengthen associational power or rely on other power resources, or both. The external relations of individual unions and their preferences towards certain actors are, therefore, materially informed and influenced by the interplay between workers' structural leverage and associational power (see Figure 6.1).

Labour unions arising primarily from the pooling of strong structural leverage are likely to orient themselves first and foremost towards the economic nexus, emphasizing their pursuit of an understanding with management in the workplace and with employers and their associations in the labour market, and less so towards the state nexus. Apart from an asymmetric power relationship between labour and capital, mutual interdependency marks this relationship as well in the economic nexus. Fundamentally, workers need an income, and management and employers need them to perform tasks. To regulate this interdependency, if structural leverage for workers is strong, then unions tend to prefer or prioritize seeking institutional security from management, employers and their associations that is largely autonomous from the state (Crouch 1993; Streeck & Hassel 2003). For instance, in this way, the uncertainty of industrial action is channelled into an institutional compromise promising relative yet temporal stability. However, at the same time, the institutional realm is not necessarily superior compared to other power resources (Rudman & Ellem 2022). Institutional power can be Janus-faced as it can also be enchaining for workers or unions through, for instance, strict regulation on industrial action.

Labour unions predominantly composed of workers with relatively weak structural leverage will likely largely focus on the state nexus. They aim to compensate for this weakness in the economic nexus by strengthening workers' associational power via relationships with "union-friendly" political parties for swaying socio-economic policy-making and the law-making processes (Korpi 1983).[4] Organizational union-party ties can then assist the enactment, enforcement or expansion of workers' individual and collective (bargaining) rights in the economic nexus via procedural or substantive regulation. This brings an additional potential source of leverage to workers through emancipating law-based institution-building (see also Rhomberg & Lopez 2021), which transcends the economic nexus and state nexus.

All of this compels labour unions to combine multiple power resources and to build external relationships with respect to both nexuses as a means of trying to achieve their goals. While goals of labour unions could be of economic, political or social nature, the relation between those respective goals and the economic nexus or state nexus is not necessarily one-on-one. To illustrate this, the activation of structural leverage via industrial action in the economic nexus or via political mass strikes in the state nexus often tends to be blurred as both influence each other – a point that Rosa Luxembourg

4. Union–political relationships are here seen as being part of workers' associational power in the state nexus (Silver 2003; Wright 2000) and not as coalitional power, since political parties are not considered (fully) as a part of civil society, which is the arena of coalitional power.

already made (Gall 2013). Thus, industrial action is sometimes not only of purely economic nature but is also targeting the state nexus as aiming to generate pressure on political authorities, while political mass strikes can exert economic pressure as well in the economic nexus. Figure 6.1 therefore provides a too schematic picture of unions either building autonomous or law-based institutional power. In reality, the economic and state nexuses are not dichotomic and detached from each other but entwined and interweaved (Hyman 2008). Notable cross-national variation is, thereby, prevalent, as the specific articulation between both nexuses in terms of procedural and substantive regulation has given rise to and structured distinctive configurations of industrial relations system in which unions organize and operate.

UNION DIVISIONS: IDENTITIES, FORMS AND IDEOLOGIES

Transposing the above reasoning on the interaction between workers' power resources, in a "reifying way", helps to make an ideal distinction between labour unions oriented towards the "market" and others aligned to "agencies of class" in Hyman's (2001) framework on union identities. Put differently, it is suggested here that the conjuncture between structural leverage and associational power is *one of the sources* for shaping the very nature of unions; that is, their default dominant identity, which is linked to different nexuses in unions' external relations. Although other complex, interconnecting dynamics further shape union identities, in a dialectical way, and while internal tensions can arise between identities (Hodder & Edwards 2015; Hyman 2001), putting workers' power resources and their interplay – structural, associational and institutional – at the centre allows an assessment of union identities (see also Chapters 1 and 2) as diverse across industries and national contexts, and with the capacity to change over time. Yet, apart from their identities, unions can diverge as well in terms of, for instance, their ideology and form – both of which help further understanding the external relations of unionism.

Considering ideological beliefs, ideas and values (see also Chapter 2) as largely "external" to unions – yet concomitant internalized – enables the disentangling of such a "frame of reference" from the identity of unions, in the process linking the "relatively stable characteristics and orientations of an organization" with its external relations (Hodder & Edwards 2015: 4). Consequently, whereas the dominant identity of unions as either being "market" or "class"-oriented shapes the default nexus of external relationships; that is, the economic or state nexus, respectively, persistent ideologies rather more influence the union's overall meaning and purpose and tactics, the nature of those relationships and how they are contextually framed as threats or opportunities (Frege & Kelly 2003). The latter implies that the nature of relations of unions with external actors can change over time and across space. As a result, capturing and assessing this nature is essentially an empirical question. It is not only contingent upon union effectiveness and union goals, and the interaction with external actors and their respective strategies, but also upon ideologies. Ideology is, thus, one of the factors, besides material-based interests, affecting the closeness or distance between unions and external actors in both the economic and political nexuses. Relationships ideally flit then between being cooperative or accommodative or being adversarial or militantly advanced in both nexuses (see also Boxall & Haynes 1997; Tassinari, Donaghey & Galetto 2022; see also Chapter 22). This is respectively expressed by, for instance, integrationist ideas of "(social) partnership" or anarcho-syndicalist or anti-capitalist ideas.

Turning to the type of unionism, although idiosyncrasies in union morphology are, of course, possible, labour unions strongly emphasizing their identity as economic actors have historically been associated with specific characteristics and practices. Typical examples are unions bounded by distinct crafts, occupations or non-manual professions, where their well-defined organizing domain provides them with a clear identity to the outside world. Their members have common in-demand skills enabling unions to control their supply in the labour market. This high workplace bargaining power can, thus, explain why some such unions have been "surviving", whether affiliated to union confederations or not, especially in the case of sustained membership in stable product markets, even though the dominant union form has been shifted towards other forms such as industrial or general unions (Heery 2003; Smale 2020; Visser 2012). Often based upon exclusive membership, craft, occupational or professional, unions typically set common rules and standards via autonomous collective bargaining in the labour market, while they have often been reluctant to associate themselves with political parties (Hyman 2001). Therefore, such fairly homogenous and less politicized unions have been accused of having a "narrow perspective" as they "fail to develop the linkages, alliances, and broad reform vision required to build the necessary political power" (Turner 2004: 2). They have been pejoratively labelled as "business unions" for predominantly focusing upon the economic nexus.

Furthermore, some, but not all, labour unions organizing in manufacturing could also be strongly oriented towards immediate economic benefits for their members. This is the case if they possess considerable workplace bargaining power stemming from the disruptive capacity of strategic local stoppages within "highly integrated production processes or important export branches" (Schmalz, Ludwig & Webster 2018: 117). The same holds true for unions organizing in logistics and transport. As unions in those industries are often still operating in settings with largely entrenched labour rights (Silver 2003), then it comes as no surprise that a "labour repertoire" is prevalent within the action repertoire of logistical or transport workers, with grievances being framed in terms of the violation or expansion of workers' individual and collective rights (Vandaele 2016).

Importantly, political opportunity structures have influenced unions' choice as well for a particular nexus. Early labour unions could only engage as bargaining actors in the economic nexus if their legal basis was secured, whereas others considered targeting political parties in the state nexus more effective in case employers were unwilling to recognize them (as, in particular, in Southern Europe) (Hyman & Gumbrell-McCormick 2010; Lipset 1983). In fact, most if not all unions are almost inevitably also political actors, with those – par excellence – compensating for workers' weak structural leverage and, obviously, unions organizing in the public sector. In particular, unions primarily composed of ("less-skilled") workers have been aiming to strengthen the associational power of their workers via building structural relationships with "union-friendly" political parties. For explaining the selection of those parties, the structural embeddedness of unions in existing interest and power structures is again salient. This embeddedness also refers to associated phenomena such as nationalism and other aspects of consciousness and ideology.

Historical circumstances and the timing and sequence of events and processes such as industrialization and democratization are all important here, for the union landscape in (Western) Europe and elsewhere has been ideologically or otherwise divided in countries where societal divisions already existed prior to union formation or where such divisions became prevalent later (Ebbinghaus 1993; Lipset 1983; Streeck & Hassel 2003). Such divisions help to explain the choice of preferred actors within each nexus via unions either fostering intra-class or cross-class alliances. If political parties have ideologically been split prior to union formation, then political alignment has been sought within the same "ideological realm" of the labour union in question. As such, structural (organizational) links have been (historically) existing as well with communist or Christian democratic parties besides socialist or social democratic parties in the state nexus.

Although cross-national variation in union-party ties has featured from the start, labour unions and left-of-centre political parties have been labelled "Siamese twins" (Ebbinghaus 1993) based upon their shared ideologies and cultures and the close organizational links that have sometimes been institutionalized. Left-of-centre parties advancing the interests and needs of workers are considered a second type of workers' associational power besides unions. Socialist or social democratic parties have historically been predominantly the preferred political allies for many unions, in particular industrial ones, in most European countries (Visser 2012). Most of those unions have been "a synthesis between pragmatic collective bargaining and a politics of state directed social reform and economic management" (Hyman 2001: 55), which has been labelled "political economism". As a main form of unionism, unions being both economic and political actors is a development that in particular has become dominant in liberal democracies in Western Europe since the Second World War (Streeck & Hassel 2003). Relationships in the national economic and state nexus have then been crystallized into external institutional resources for workers and unions (Berger & Compston 2002; Crouch 1993). The considerable structural power of especially industrial unions and their mounting associational power has resulted in institutional resources in various fields and at various levels within nation states. Those resources largely reflect the balance in power at the time of the achieved postwar settlements, with compromises varying across industries and countries.

RELATIVE DEMISE OF UNIONS' EXTERNAL RELATIONS IN THE ECONOMIC AND STATE NEXUSES

Typical "political-industrial unionism" and its ideological legitimacy started to crumble after the 1970s (Hyman 2001; Visser 2012). Structural, interlinked dynamics in both nexuses have challenged and eroded workers' power resources, and this has been especially well documented and studied for the economic nexus. For instance, without being exhaustive, the managerial practice of mounting competition between plants to enforce local union concessions, so-called "whipsawing by management", has undermined workers' structural power in a context of financialization (Grady & Simms 2019; Kollmeyer & Peters 2019) and of European economic integration and globalization (Greer & Hauptmeier 2016). Furthermore, deindustrialization, the growing "demographic diversity" of labour markets (Kollmeyer 2022) and labour market "reforms" promoting flexibilization and "deregulation" (Baccaro & Howell 2017; Greer & Umney 2022; Simoni & Vlandas 2021) have all (indirectly) hampered and complicated labour unions' efforts in recruiting and organizing members in the increasingly "fissured workplace" (Weil 2014). All of this, among other factors, has generated an almost general decline in union density across Europe (Vandaele 2019), although workers in certain occupations or industries could still hold considerable associational power (and structural power) (Koçer 2018).

Organized or disorganized decentralization and fragmentation has marked nationally embedded multi-employer bargaining systems in most European countries over time. Both have accelerated and become more widespread in the aftermath of the collapse of the finance-led accumulation regime in 2007–08 (Marginson 2015; Müller, Vandaele & Waddington 2019). Consequently, where employee representation is present at company level, the external relationships of labour unions have been narrowed down to management and the individual employer in decentralized collective bargaining settings with company-level agreements. Relationship-building with employer associations at industrial level then becomes less vital in such settings. Those associations are simply less involved in multi-employer and centralized bargaining but instead prioritize other functions such as individual service provision and policy lobbying, while their role as negotiation actors nearly ceases to exist (Brandl & Lehr 2019; Demougin *et al.* 2019; Gooberman & Hauptmeier 2022).

Equally, arrangements for bipartite or tripartite social dialogue and political exchange in macroeconomic management at the national level in (Western) European countries have (incrementally) loosened and transformed over time. So, they have moved from "neo-corporatist" interest intermediation in the 1970s and the "functionalist fit" with demand-side Keynesian macroeconomics to often ad hoc and limited peak-level "social pacts" and supply-side "competitive corporatism" in the 1990s and early 2000s, and now to discretional "crisis corporatism" after the 2007–08 financial crisis (Ebbinghaus & Weishaupt 2021; Tassinari, Donaghey & Galetto 2022). The latter has, at least discursively, been more accommodating of labour demands since the Covid-19 pandemic (Meardi & Tassinari 2022).

The degree and depth of institutional change within those neo-corporatist arrangements, and its consequences for union effectiveness across countries, are debated, however. The most straightforward interpretation claims a converging trend towards a "common neoliberal trajectory" across countries (Baccaro & Howell 2017). A more moderated standpoint still sees room for cross-national variation in outcomes as being mediated by resilient complementary institutions underpinning collective bargaining and traditions in social dialogue (Hall & Soskice 2001), which, together with transformations in welfare state regimes, are associated with different "growth regimes" pursued by governments (Hassel & Pallier 2021). Indeed, it looks as if the relationships of labour unions can, nevertheless, be effective in the economic nexus in those countries where union density is relatively still high, and workers' interests are strongly "incorporated" in institutions of labour market and public policy governance (Crouch 2017). This "incorporation" in socio-economic policy-making is explained by traditions of "social partnership" and left-wing government participation, especially in small export-oriented economies (Ebbinghaus & Weishaupt 2021). Apart from this "political exchange" in neo-corporatists' arrangements in the state nexus, unions have been seeking privileged links with left-of-centre political parties, formalized functional representation in bipartite or tripartite quasi-public agencies and councils for administrating welfare provisions and controlling the implementation of public policies, and lobbying the parliament and government as a pressure group by offering expert knowledge (Streeck & Hassel 2003).

The main emphasis will be put here on the union-party links as an expression and extension of workers' associational power. A caveat should first be made, however, for the state nexus cannot be made synonymous solely with political parties or, indeed, any other actor within this nexus. Bringing in actors such as parliaments, governments, public administrations, courts or central banks simply does not fully capture the polymorphic nature of the state (Hyman 2008), whose role in employment relations has been shifting over time under capitalism (Howell 2021). Political parties are, nevertheless, crucial in the state apparatus in liberal democracies for they are still gatekeepers of the political system by delivering or appointing functionaries in the main executive and legislative bodies based upon electoral results. Political parties are, therefore, crucial for unions in terms of building relationships within the state nexus (Streeck & Hassel 2003). The crux of those relationships centres around unions delivering potential votes, financial donations or organizational support to the parties in question in exchange for favourable policies and rule-making limiting employer discretion (Haugsgjerd Allern & Bale 2017).

Traditional organizational union-party ties still exist today in (Western) Europe and they "are stronger where ... unions are larger, denser, and more unified, and where parties are less able to rely on the state to finance their organizational activities and electoral campaigns" (Haugsgjerd Allern, Bale & Otjes 2017: 337). Yet, beyond those organizational ties, relationships between labour unions and left-of-centre political parties have, for several reasons, generally become weaker and more pragmatic over time, although cross-national variation remains (Hyman & Gumbrell-McCormick 2010). Thus, union–party relationships have been diluted by cultural and ideological shifts influencing the political project of left-of-centre parties, making them less appealing to the working-class vote. Left-of-centre parties have started mobilizing voters based upon other identities than social class and "third way" ideas and policies have made them "more accommodative to market forces" so that "the social democratic ambition no longer consisted in developing counter-forces to the power of capital" (Rennwald 2020: 72–3). Cultural and ideological turns have often gone hand in hand with deindustrialization, causing the decline of blue-collar workers (and their communities) and the rise of white-collar workers and professional occupations in employment.

Structural long-term change in the workforce is important for the electoral outcomes of left-of-centre parties. Although they have never been purely working-class parties, left-of-centre parties have proportionally found more electoral support among the middle class in the long run, whereas working-class votes have been declining in the context of greater electoral competition (Gingrich & Häusermann 2015; Rennwald 2020).[5] Furthermore, political choices are of importance. While political attitudes among the working class have remained fairly similar, the socio-economic policies pursued by left-of-centre parties have distanced them from labour unions and their policy preferences (Rennwald 2020). Notably, external relations have become looser, too, in the state nexus regarding welfare state regimes. Unions have historically obtained functional representation in its administration in some countries. Yet this role has been diminished over the years (Ebbinghaus 2021), with union involvement in publicly subsidized unemployment insurance funds being undermined in Denmark, Norway and Sweden and weakening associational power in the economic nexus (Høgedahl & Kongshøj 2017).

Members of labour unions now perceive that they are less politically well-represented today than in the past (Rennwald & Pontusson 2021). Non-unionized as well as unionized workers and union activists might, therefore, associate mainstream unions with an "elite consensus" in a situation in which strong organizational ties are still present with left-of centre parties despite these pursuing detrimental policies of labour market deregulation and flexibilization, or being only able to mitigate them when in power (Rathgeb 2018). Consequently, left-of-centre parties, the traditional allies of (established) unions, have lost vote shares in many European countries since the 1990s (Rennwald 2020), while unions might also be simultaneously less appealing for them. Many unions providing traditional voters for left-of-centre parties have lost members, especially among blue-collar workers, while union members are not entirely resistant to the siren call of populist radical right parties (Oesch & Rennwald 2018), although generally less so in comparison with non-unionized voters (Mosimann, Rennwald & Simmermann 2019; Oesch 2008).

Waned associational power in the economic nexus and in the state nexus creates the space for rebuilding and broadening relationships with non-market actors for incorporating new, inspiring and imaginative ideas. Workers' power resources are instructive for identifying the external actors with whom labour unions aim to build relations within the economic and state nexuses. Deploying those resources and acting collectively is, however, contingent upon unions' capabilities and organizational learning (Hyman 2007; Lévesque & Murray 2010). What kind of strategies could unions then develop towards external actors in both nexuses?

PRIORITIZING UNION STRATEGIES TOWARDS EXTERNAL ACTORS AND COALITIONAL POWER

Whereas labour unions draw upon a wide range of union strategies, McAlevey (2016) makes a useful distinction between advocacy, mobilizing and (deep) organizing. Although not the focus of this chapter, such strategies also enable a consideration of

5. The mainstream right but especially the "radical left" and populist radical right parties are all able to capture an important share of the working-class vote (Rennwald 2020). In addition, participation in elections has declined among the working class; this has become a structural characteristic of working-class voting behaviour today.

the internal relations between union members and the union leadership. Each strategy can be associated with different degrees of union commitment and union participation, demonstrating interlinkages between unions' internal decision-making and strategies towards external actors (Snape & Redman 2004; Vandaele 2020; see also Chapters 16 and 17).

First, the mobilizing strategy advancing collective organization and action is oriented towards pre-existing networks of union activists, but largely depends upon professional staff for leadership. Mobilizing via (short-term) industrial action, or at least the credible threat of it, can be considered to feature primarily in the economic nexus, although it is highly contingent upon the (legal) regulation of industrial action and the degree of institutionalized neo-corporatist arrangements in general. Second, political advocacy is likely to be the preferred strategy of labour unions in the state nexus due to established union-party links. Individual instrumentalism and an economic relationship between mostly passive members and the union prevail in this strategy. Here, unions mainly rely on their professional union staff for providing services, while union commitment and participation is considered low and limited. Third, the more deep-rooted organizing strategy is primarily centred around unions' deepening attachment with members and (non-unionized) workers via engaging and activating them so that such well-connected social networks become aware of their own agency for pursuing social justice through collective action (Holgate, Simms & Tapia 2018). Allowing more room for (transformational) bottom-up processes, this empowering strategy seeks political congruence and ideational identification between (unionized) workers, grassroots activists and union leaders, mostly within the (wider) worker community – hence, the labels "community-based organizing" or "community unionism" (Holgate 2021; Tattersall 2010).

Importantly, the organizing strategy also encompasses an external dimension since potential members and activists could be sought not only within the workplace and the worker community but also in civil society at large for improving leverage in the economic or state nexus. As such, an organizing strategy facilitates a consideration of coalitional power in the further examination and analysis of the external relations of labour unions.[6] Coalitional power can be defined as the fostering of alliances or coalitions between unions and non-labour actors (i.e. those actors beyond the workplace and labour market or beyond the state) for voicing mutual support and achieving common goals (Frege, Heery & Turner 2004; Tattersall 2010). This power resource seems secondary to workers' structural and associational power as it aims to "boost" (Schmalz, Ludwig & Webster 2018: 122) those latter power resources for pursuing union goals by adding auxiliary resources. This "supplementary" but boosting nature implies that organizing interplays with the dominant union approaches in the economic and state nexuses. Consequently, this points to the importance of combining union strategies in terms of how they may either interact or overlap and how contingent they are upon historical legacies and the institutional context (Ellem, Goods & Todd 2020; Frege

6. Coalitional power can be considered part of societal power together with symbolic, discursive or ideational power aimed at influencing the public debate through mainstream press and social media via meta-narratives on, for instance, social injustice. The significance of this type of power is acknowledged, especially as it virtually transcends the other power resources, but is not considered here for reasons of simplicity.

& Kelly 2003). The secondary character of coalitional power explains why Figure 6.2 still maintains the binary distinction between the economic and state nexus. Depicting those nexuses together with "civil society" in a triangle would give the impression that all three sociological arenas for building external relations are considered of equal importance, with fostering coalitional power with actors in "civil society" being largely detached from the economic and state nexus.

Coalitional power affecting union strategies in the state nexus implies that the advocacy approach behind closed doors can then be accompanied by a mobilizing approach in the public realm, putting emphasis upon the social rights of workers as citizens and, thus, advancing a civil rights repertoire, with civil rights as the main framing logic for workers' demands and unions resorting to civil legal institutions for protection (Gentile & Tarrow 2009). Union alliances with other non-labour actors alternate then between "coalitions of protest" and "coalitions of influence" (Frege, Heery & Turner 2004). Such alliances have been conceived as labour unions turning (again) to "their role as protagonists in civil society" (Hyman 2001: 60). Unions are then seen as social movements addressing issues and expressing demands beyond the workplace and labour market, especially via mobilizing and campaigning. This has often been framed as unions shifting to protect precarious and vulnerable workers as well, instead of solely representing labour market "insiders" (Doellgast, Lillie & Pulignano 2018).

With mainly a focus on building workplace-level solidarity, strategies in the economic nexus oscillate then between mobilizing and organizing since coalitional power allows for the engagement and activation of under-represented groups within labour unions. Accordingly, a "civil rights repertoire" will be layered on to a labour repertoire in the economic nexus. Typical examples are grassroots unions that have been baptized as being "politically radical", and whose identities are marked by an outspoken critique of capitalism in general and not solely of neoliberalism as such (Upchurch, Taylor & Mathers 2009). Politically radical unionism is typically operating beyond the nation-state and oriented towards organizing precarious and diverse workers at the lower end of the labour market (Peròs 2020). Such grassroots unionism is in a minority position and limited to some European countries so far (Connolly, Kretsos & Phelan 2014), although such oppositional orientation could exist sectionally as well in mainstream unions. Also, anti-austerity movements and grassroots groups resisting labour market insecurity have emerged in the turmoil of the 2007–08 finance-led crisis in capitalism (Kriesi *et al.* 2020), especially in the most affected countries in Western Europe, although the rapprochement between them and mainstream unions has been somewhat mixed (Hyman & Gumbrell-McCormick 2017).

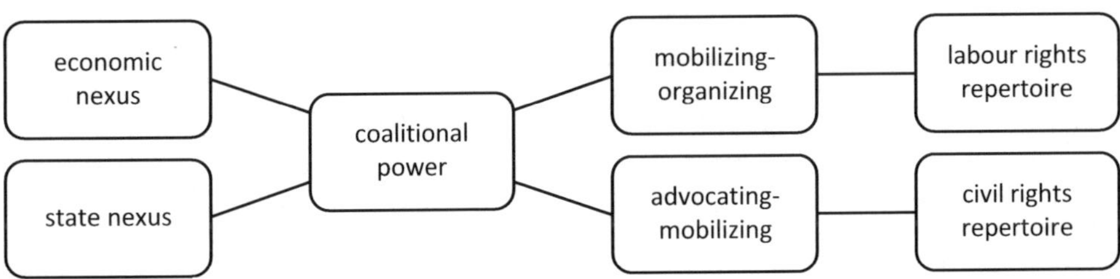

Figure 6.2 The influence of coalitional power upon union strategies and action repertoires

Coalitional power underlines how labour unions are also embedded and seeking alliances within civil society, revealing a union identity oriented towards "society" and a discursive framing beyond workplace and labour market issues. Union alliances with other non-labour actors can either be (opportunistically) based on ad hoc single issues or turned into longer-lasting and stronger alliances, resulting in different outcomes (Holgate 2015). Crafting coalitional power with equity-seeking identity groups and (other) social movements outside the traditional employment relationship has gained in importance for several reasons, not least because of workers' deteriorating power resources in the economic and state nexuses (see also Kelly 1998; Robinson 2000).

THE ADVENT AND PROMISES OF COALITIONAL POWER IN THE ECONOMIC AND STATE NEXUSES

Several macro-contextual reasons call for labour unions to build coalitional power with non-labour actors offering new opportunities for effectively overcoming weaknesses in the economic and state nexuses. Three main reasons are discussed here.

The first reason concerns the long-term demographic changes taking place in the labour market, such as feminization and increased labour immigration (including posted workers in the European Union), with especially foreign-born workers disproportionately employed in weakly unionized low-wage industries marked by small and fissured workplaces (Kollmeyer 2022).[7] Coalitional power might be beneficial here for constructing solidarities and strengthening workers' associational power, especially among a precarious or vulnerable workforce (displaying, nevertheless, favourable attitudes towards unions; see Gorodseisky & Richards 2020). For instance, dialogue and debate between labour unions and immigrant or ethnic communities and networks in the search for complementarities can support the former in better understanding the experiences, needs and issues of foreign-born workers in and outside the workplace and overcoming the possible cultural and linguistic barriers to organizing them – although unions strongly embedded in local communities might equally as well look to build trust directly among such workers in high-density countries, too (Refslund 2021).

Union relationships with such self-organized, bottom-up communities and networks are varied as union strategies and their discursive framing are contextually shaped by state policies towards immigration and social inclusion, internal union structures and union identities (Connolly, Marino & Martínez Lucio 2019). Thus, labour unions that recognize the importance of ethnicity or race tend to foster coalitions with immigrant and ethnic communities and networks; those that emphasize the common identity of ethnic minority workers or immigrants *as workers* directly engage them in the organization, preferably as activists, in the economic nexus; and, lastly, unions oriented towards the social rights of immigrants, also beyond the workplace, look rather to build relationships with counterparts in the state nexus. In practice, and not without tension, union strategies are dynamic, alternating mainly between the economic and state nexus

7. In many instances, horizontal structures are in place for internally representing (under-represented) worker categories such as migrants, young people and women, or other equity-seeking and minority groups in unions.

or between strengthening coalitional power in one of the two nexuses involving a mobilizing or organizing strategy. For being successful in terms of sustainable membership gains and organizing members, beyond-the-workplace approaches are also dependent on institutional support (James & Karmowska 2016).

Second, heterogeneous or multiple identities – worker identities beyond the employment relationship, such as gender, ethnicity, sexuality and disability – have become more salient in today's diverse societies calling for "inter-sectional approaches to organizing that move beyond seeing workers as a single class to recognize the amplified oppression workers face as a result of various marginalized identities" (Doellgast, Bidwell & Colvin 2021: 569). Mobilizing based on the intersectional nature of issues requires a "balance between universal and particularistic interests" and "politically managing differences in such a way as to build up *unity in diversity*" (Connolly, Marino & Martínez Lucio 2019: 164, emphasis in original).

This highlights the need to build "organic solidarity", where shared interests overlap, instead of the "mechanical solidarity" that has stereotypically been present in relative homogeneous worker communities in the past (Hyman 1999). Also, a "whole-worker approach" (McAlevey 2016) has been put forward for recruiting and organizing new union members, while the knowledge and experience from equity-seeking identity groups and community-oriented organizations can feed into such an approach. They can bring in extra-workplace issues such as discrimination in the rental housing market where it concerns, for instance, ethnic minority workers or immigrants. Union-community alliances and cooperation simultaneously then question the workplace as the sole locus for organizing workers. As a corollary, an intersectional lens on issues demands a "situated and contingent response of workers and organisers to the mix of subjective and structural gendered, racialised, juridical or class-based forms of oppression and exploitation, with the aim of uniting workers against the increasing differentiation of labour and the blurring of the employment relationship" (Alberti 2016: 89).

Third, there is the pressing and crucial issue of environmental destruction and crisis, and the need to transform the (capitalist) current mode of production given its diverse types of pollution and devastating impact upon the climate and biodiversity (Räthzel, Stevis & Ussell 2021).[8] Here, just as with other non-labour groups or movements, complexity and intersectionality can mark union-environmental alliances. Contrasting views and barriers for alliance-building are also frequently present, especially as job protection and promotion in certain strongly unionized industries, regions and countries could be at stake with decarbonization policies or other long-term environmental issues (Thomas & Pulignano 2021).[9] For instance, tensions between labour unions and community organizations has been diagnosed as "marching to different tunes" (Tapia 2013) when it comes to membership mobilization and organizational culture.

Union members have generally an instrumental commitment to their union, whereas a covenantal relationship with members prevails in community organizations. Such a finding might echo a dichotomous and stereotypical understanding of sterile unions

---

8. Environmental concerns and struggles at the workplace and beyond have historically been part of the labour movement from the very beginning.
9. Tensions arise also within unions and between unions at different levels and in different (world) regions, here, for example, between the Global North and Global South.

in contrast to dynamic community organizations. Even so, organizational culture in the latter might be equally marked by bureaucratization and formalization, especially if relationships in the state nexus have been matured and institutionalized (Hyman & Gumbrell-McCormick 2017).[10] Thus, relationships between unions and organizations in civil society are "multiform" (Heery, Williams & Abbott 2012) – not only cooperation and conflict are present but also indifference towards unions as civil society organizations themselves try to build relationships in the economic and state nexus.

However, fruitful union-community alliances are possible (Tapia 2019). "Bridge-builders" – individuals with ties to two or more organizations involved – have been identified as being instrumental in strengthening social ties between labour unions and non-labour movements (Heery, Williams & Abbott 2012). Issue-selection and its framing are equally important (Tattersall 2010). For instance, regarding union-environmental alliances, a more transformative approach to "just transition" policies might be better suited for overcoming potential tensions between unions and environmental movements in building coalitional power (Assi 2021; Dobrusin 2021).

In contrast to an accommodative approach that instrumentally focuses upon modifications within the existing capitalist framework, a transformative approach entails more profound socio-economic changes towards a low-carbon economy (Fraser 2005). The focus of a transformative approach is not only the most affected workers in high-carbon jobs in the transition towards a low-carbon society. Instead, it entails a broader just transition based on comprehensive alliance-building incorporating other groups of workers as well as relationship-building with actors in both the economic and state nexuses. A full circle would then be reached – coalitional power rooted in organizational diversity can enable the raising of workers' other power resources not only regarding climate change and environmental sustainability but also on the wider purpose of "developing a common agenda for a fundamental socio-ecological transformation" (Wissen & Brand 2021: 713).

CONCLUSION

Using the framework of the power resource approach, the starting point for examining and analysing the external relations of labour unions has not been the unions themselves but workers. This chapter has emphasized that the material interplay between workers' structural and associational power largely influences, although does not determine, the main nexus where unions, as "secondary organizations", will build external relations, whether or not institutionally supported, for achieving their goals. Dynamics between power resources have dialectically influenced union identities and forms, while ideologies have also mediated the selection of external actors and the nature of relationships built with them in the economic and state nexuses. The main actors are management at the workplace and employers and their associations in the labour market; political

10. Simultaneously, unions might see non-labour actors as rivals grasping political opportunities and envying their institutionalized access and positions.

parties within the state nexus; and equity-seeking identity groups, communities and social movements in civil society. Relationships between them and labour unions can vary between being adversarial and cooperative, and they can also be classified as either informal or more formal (via institutionalization), although in practice there is probably a blurring of such binary divisions along the continuum. Furthermore, dynamics within power resources, for instance, from shifts within employment structures, from a drift in union-political relationships or from an alteration of institutional arrangements within the employment relationship require unions to reconsider existing external relations and to foster relations with other or new external actors.

Further summarizing this chapter is barely possible given the complexity of its subject. Instead, attention is paid in this conclusion to its various limitations, which simultaneously allows for a highlighting of the main concepts and underlying ideas and for pointing to future research avenues. Thus, the focus has been laid upon actors and processes, and less so upon outcomes of the various external relations. Actors present at the scale of the nation-state have mainly been studied; less so actors at city-level, subnational actors or those operating beyond the national scale (on the latter, for example, see Brookes 2019). Furthermore, the dynamics within the conceptual framework linking the interaction between workers' power resources, union identities and union strategies have only been formulated in a sketchy way and stylized account. The framework suffers from a kind of structural functionalism as particular union forms have often been indirectly linked throughout this chapter to union purposes (via the interplay between workers' power resources).

Based on perhaps a too "mechanical" approach, there has been a rather unintentional prescriptive undertone running through the chapter, whereby building coalitional power is perceived as compensation for the deterioration of other power resources, which is introduced as an almost linear outcome. This is absolutely not the case in reality. Internal union debate and discussion matter for deciding on strategies in the economic and state nexuses, and labour unions operate in different spatial-temporal contexts while workers' power resources equally vary in time and between industries and countries. While cross-national variation has obviously been acknowledged, more detailed empirical studies can shed further light upon this. And the research focus has largely been on unions themselves, and not on "how the other half responds" in the economic and state nexuses. The motives and reasons of management, employers and their associations and (left-of-centre) political parties remain underexposed.[11]

REFERENCES

Alberti, G. 2016. "Moving beyond the dichotomy of workplace and community unionism: the challenges of organising migrant workers in London's hotels". *Economic and Industrial Democracy* 37(1): 73–94.

Assi, D. 2021. "Trade union policies for a just transition: towards consensus or dissensus?" In *The Palgrave Handbook of Environmental Labour Studies*, edited by N. Räthzel, D. Stevis & D. Ussell, 225–48. London: Palgrave Macmillan.

Baccaro, L. & C. Howell 2017. *Trajectories of Neoliberal Transformation: European Industrial Relations since the 1970s*. Cambridge: Cambridge University Press.

11. I am grateful for the comments of Gregor Gall and Bjarke Refslund.

Berger, S. & H. Compston (eds) 2002. *Policy Concertation and Social Partnership in Western Europe: Lessons for the Twenty-first Century*. New York: Berghahn.

Boxall, P. & P. Haynes 1997. "Strategy and trade union effectiveness in a neo-liberal environment". *British Journal of Industrial Relations* 35(4): 567–91.

Brandl, B. & A. Lehr 2019. "The strange non-death of employer and business associations: an analysis of their representativeness and activities in Western European countries". *Economic and Industrial Democracy* 40(4): 932–53.

Brookes, M. 2019. *The New Politics of Transnational Labor: Why Some Alliances Succeed*. Ithaca, NY: ILR Press.

Connolly, H., L. Kretsos & C. Phelan (eds) 2014. *Radical Unionism in Europe and the Future for Collective Interest Representation*. Bern: Peter Lang.

Connolly, H., S. Marino & M. Martínez Lucio 2019. *The Politics of Social Inclusion and Labor Representation: Immigrants and Trade Unions in the European Context*. Ithaca, NY: ILR Press.

Crouch, C. 1993. *Industrial Relations and European State Traditions*. Oxford: Clarendon Press.

Crouch, C. 2017. "Membership density and trade union power". *Transfer* 23(1): 47–61.

della Porta, D. 2015. *Social Movements in Times of Austerity: Bringing Capitalism Back into Protest Analysis*. Cambridge: Polity.

Demougin P. *et al.* 2019. "Employer organisations transformed". *Human Resource Management Journal* 29(1): 1–16.

Dobrusin, B. 2021. "A Just Transition for all? A debate on the limits and potentials of a Just Transition in Canada". In *The Palgrave Handbook of Environmental Labour Studies* edited by N. Räthzel, D. Stevis & D. Ussell, 295–316. London: Palgrave Macmillan.

Doellgast, V., M. Bidwell & A. Colvin 2021. "New directions in employment relations theory: understanding fragmentation, identity, and legitimacy". *ILR Review* 74(3): 555–79.

Doellgast, V., N. Lillie & V. Pulignano 2018. *Reconstructing Solidarity: Labour Unions, Precarious Work, and the Politics of Institutional Change in Europe*. Oxford: Oxford University Press.

Ebbinghaus, B. 1993. *Labour Unity in Diversity: Trade Unions and Social Cleavages in Western Europe, 1890–1989*. Florence: European University Institute.

Ebbinghaus, B. 2021. "Unions and employers". In *The Oxford Handbook of the Welfare State*, 2nd edn, edited by D. Béland *et al.*, 278–97. Oxford: Oxford University Press.

Ebbinghaus, B. & T. Weishaupt (eds) 2021. *The Role of Social Partners in Managing Europe's Great Recession: Crisis Corporatism or Corporatism in Crisis?* Abingdon: Routledge.

Ellem, B., C. Goods & P. Todd 2020. "Rethinking power, strategy and renewal: members and unions in crisis". *British Journal of Industrial Relations* 58(2): 424–46.

Fraser, N. 2005. "Reframing justice in a globalizing world". *New Left Review* 20(36): 69–88.

Frege, C., E. Heery & L. Turner 2004. "The new solidarity? Trade union coalition-building in five countries". In *Varieties of Unionism: Strategies for Union Revitalization in a Globalizing Economy*, edited by C. Frege & J. Kelly, 137–58. Oxford: Oxford University Press.

Frege, C. & J. Kelly 2003. "Union revitalization strategies in comparative perspective". *European Journal of Industrial Relations* 9(1): 7–24.

Gall, G. 2013. "Quiescence continued? Recent strike activity in nine Western European economies". *Economic and Industrial Democracy* 34(4): 667–91.

Gall, G. & J. Fiorito 2016. "Union effectiveness: in search of the Holy Grail". *Economic and Industrial Democracy* 37(1): 189–211.

Gentile, A. & S. Tarrow 2009. "Charles Tilly, globalization, and labor's citizen rights". *European Political Science Review* 1(3): 465–93.

Gingrich, J. & S. Häusermann 2015. "The decline of the working-class vote, the reconfiguration of the welfare support coalition and consequences for the welfare state". *Journal of European Social Policy* 25(1): 50–75.

Gooberman, L. & M. Hauptmeier (eds) 2022. *Contemporary Employers' Organizations: Adaptation and Resilience*. New York: Routledge.

Gorodseisky, A. & A. Richards 2019. "Do immigrants trust trade unions? A study of 18 European countries". *British Journal of Industrial Relations* 58(1): 3–26.

Grady, J. & M. Simms 2019. "Trade unions and the challenge of fostering solidarities in an era of financialisation". *Economic and Industrial Democracy* 40(3): 490–510.

Greer, I. & M. Hauptmeier 2016. "Management whipsawing: the staging of labor competition under globalization". *ILR Review* 69(1): 29–52.

Greer, I. & C. Umney 2022. *Marketization: How Capitalist Exchange Disciplines Workers and Subverts Democracy*. London: Bloomsbury Academic.

Grote, J. & C. Wagemann (eds) 2019. *Social Movements and Organized Labour: Passions and Interests*. Abingdon: Routledge.

Gumbrell-McCormick, R. & R. Hyman 2013. *Trade Unions in Western Europe: Hard Times, Hard Choices*. Oxford: Oxford University Press.

Hall, P. & D. Soskice (eds) 2001. *Varieties of Capitalism: The Institutional Foundations of Comparative Advantage*. Oxford: Oxford University Press.

Hassel, A. & B. Palier (eds) 2021. *Growth and Welfare in Advanced Capitalist Economies: How Have Growth Regimes Evolved?* Oxford: Oxford University Press.

Haugsgjerd Allern, E. & T. Bale (eds) 2017. *Left-of-Centre Parties and Trade Unions in the Twenty-First Century*. Oxford: Oxford University Press.

Haugsgjerd Allern, E., T. Bale & S. Otjes 2017. "Variations in party-union relationships: explanations and implications". In *Left-of-Centre Parties and Trade Unions in the Twenty-First Century*, edited by E. Haugsgjerd Allern & T. Bale, 310–41. Oxford: Oxford University Press.

Heery, E. 2003. "Trade unions and industrial relations". In *Understanding Work and Employment: Industrial Relations in Transition*, edited by P. Ackers & A. Wilkinson, 278–304. Oxford: Oxford University Press.

Heery, E., S. Williams & B. Abbott 2012. "Civil society organizations and trade unions: cooperation, conflict, indifference". *Work, Employment and* Society 26(1): 145–60.

Hodder, A. & P. Edwards 2015. "The essence of trade unions: understanding identity, ideology and purpose". *Work, Employment and Society* 29(5): 843–54.

Holgate, J. 2015. "An international study of trade union involvement in community organizing: same model, different outcomes". *British Journal of Industrial Relations* 53(3): 460–83.

Holgate, J. 2021. "Trade unions in the community: building broad spaces of solidarity". *Economic and Industrial Democracy* 42(2): 226–47.

Holgate, J., M. Simms & M. Tapia 2018. "The limitations of the theory and practice of mobilization in trade union organizing". *Economic and Industrial Democracy* 39(4): 599–616.

Høgedahl, L. & K. Kongshøj 2017. "New trajectories of unionization in the Nordic Ghent countries: changing labour market and welfare institutions". *European Journal of Industrial Relations* 23(4): 365–80.

Howell, C. 2021. "Rethinking the role of the state in employment relations for a neoliberal era". *ILR Review* 74(3): 739–72.

Hyman, R. 1999. "Imagined solidarities: can trade unions resist globalization?" In *Globalization and Labour Relations*, edited by P. Leisink, 94–115. Cheltenham: Edward Elgar.

Hyman, R. 2001. *Understanding European Trade Unionism*. London: Sage.

Hyman, R. 2007. "How can trade unions act strategically?" *Transfer: European Review of Labour and Research* 13(2): 193–210.

Hyman, R. 2008. "The state in industrial relations". In *The SAGE Handbook of Industrial Relations*, edited by P. Blyton *et al.*, 259–83. London: Sage.

Hyman, R. & R. Gumbrell-McCormick 2010. "Trade unions, politics and parties: is a new configuration possible?" *Transfer: European Review of Labour and Research* 16(3): 315–31.

Hyman, R. & R. Gumbrell-McCormick 2017. "Resisting labour market insecurity: old and new actors, rivals or allies?" *Journal of Industrial Relations* 59(4) 538–61.

James, P. & J. Karmowska 2016. "British union renewal: does salvation really lie beyond the workplace?" *Industrial Relations Journal* 47(2): 102–16.

Kelly, J. 1998. *Rethinking Industrial Relations: Mobilisation, Collectivism and Long Waves*. Abingdon: Routledge.

Koçer, G. 2018. "Measuring the strength of trade unions and identifying the privileged groups: a two-dimensional approach and its implementation". *Journal of Mathematical Sociology* 42(3): 152–82.

Kollmeyer, C. 2022. "Post-industrial capitalism and trade union decline in affluent democracies". *International Journal of Comparative Sociology* 62(6): 466–87.

Kollmeyer, C. & J. Peters 2019. "Financialization and the decline of organized labor: a study of 18 advanced capitalist countries, 1970 to 2012". *Social Forces* 98(1): 1–30.

Korpi, W. 1983. *The Democratic Class Struggle*. London: Routledge & Kegan Paul.

Kriesi H. *et al.* (eds) 2020. *Contention in Times of Crisis: Recession and Political Protest in Thirty European Countries*. Cambridge: Cambridge University Press.

Lévesque, C. & G. Murray 2010. "Understanding union power: resources and capabilities for renewing union capacity". *Transfer: European Review of Labour and Research* 16(3): 333–50.

Lipset, M. 1983. "Radicalism or reformism? The sources of working-class politics". *American Political Science Review* 77(1): 1–18.

Marginson, P. 2015. "Coordinated bargaining in Europe: from incremental corrosion to frontal assault?" *European Journal of Industrial Relations* 21(2): 97–114.

McAlevey, J. 2016. *No Shortcuts: Organizing for Power in the New Gilded Age*. Oxford: Oxford University Press.

Meardi, G. & A. Tassinari 2022. "Crisis corporatism 2.0? The role of social dialogue in the pandemic crisis in Europe". *Transfer: European Review of Labour and Research* 28(1): 83–100.

Mosimann, N., L. Rennwald & A. Simmermann 2019. "The radical right, the labour movement and the competition for the workers' vote". *Economic and Industrial Democracy* 40(1): 65–90.

Müller, T., K. Vandaele & J. Waddington (eds) 2019. *Collective Bargaining in Europe: Towards an Endgame*. Volume I, II, III and IV. Brussels: ETUI.

Müller-Jentsch, W. 1985. "Trade unions as intermediary organizations". *Economic and Industrial Democracy* 6(1): 3–33.

Oesch, D. 2008. "Explaining workers' support for right-wing populist parties in Western Europe: evidence from Austria, Belgium, France, Norway, and Switzerland". *International Political Science Review* 29(3): 349–73.

Oesch, D. & L. Rennwald 2018. "Electoral competition in Europe's new tripolar political space: class voting for the Left, Centre-Right and Radical Right". *European Journal of Political Research* 57(4): 783–807.

Offe, C. & H. Wiesenthal 1980. "Two logics of collective action: theoretical notes on social class and organizational form". *Political Power and Social Theory* 1(1): 67–115.

Però, D. 2020. "Indie unions, organizing and labour renewal: learning from precarious migrant workers". *Work, Employment and Society* 34(5): 900–18.

Rathgeb, P. 2018. *Strong Governments, Precarious Workers: Labor Market Policy in the Era of Liberalization*. Ithaca, NY: Cornell University Press.

Räthzel N., D. Stevis & D. Ussell 2021. "Introduction: expanding the boundaries of environmental labour studies". In *The Palgrave Handbook of Environmental Labour Studies*, edited by N. Räthzel, D. Stevis & D. Ussell, 1–31. London: Palgrave Macmillan.

Refslund, B. 2021. "When strong unions meet precarious migrants: building trustful relations to unionise labour migrants in a high union-density setting". *Economic and Industrial Democracy* 42(2): 314–35.

Refslund, B. & J. Arnholtz 2021. "Power resource theory revisited: the perils and promises for understanding contemporary labour politics". *Economic and Industrial Democracy* 43(4): 1958–79.

Rennwald, L. 2020. *Social Democratic Parties and the Working Class: New Voting Patterns*. London: Palgrave Macmillan

Rennwald, L. & J. Pontusson 2021. "Social class, union power and perceptions of political voice: liberal democracies, 1974–2016". Working paper No. 22. Geneva: Université de Genève.

Rhomberg, C. & S. Lopez 2021. "Understanding strikes in the 21st century: perspective from the United States". In *Power and Protest*, edited by L. Leitz, 37–62. Bingley: Emerald.

Robinson, I. 2000. "Neoliberal restructuring and US unions: towards social movement unionism". *Critical Sociology* 26(1/2): 109–38.

Rudman, A. & B. Ellem 2022. "Union purpose and power: regulation the fissured workplace". *Economic and Industrial Relations*. https://doi.org/10.1177/0143831X221139333

Schmalz, S., C. Ludwig & E. Webster 2018. "The power resources approach: developments and challenges". *Global Labour Journal* 9(2): 113–34.

Silver, B. 2003. *Forces of Labor: Workers' Movements and Globalization since 1870*. Cambridge: Cambridge University Press.

Simoni, M. & T. Vlandas 2021. "Labour market liberalization and the rise of dualism in Europe as the interplay between governments, trade unions and the economy". *Social Policy & Administration* 55(4): 637–58.

Smale, B. 2020. *Exploring Trade Union Identities: Union Identity, Niche Identity and the Problem of Organising the Unorganised*. Bristol: Bristol University Press.

Snape, E. & T. Redman 2004. "Exchange or convenient? The nature of the member–union relationship". *Industrial Relations* 43(4): 855–73.

Streeck, W. & A. Hassel 2003. "Trade unions as political actors". In *International Handbook of Trade Unions*, edited by J. Addisson & C. Schnabel, 335–65. Cheltenham: Edward Elgar.

Tapia, M. 2013. "Marching to different tunes: commitment and culture as mobilizing mechanisms for trade unions and community organizations". *British Journal of Industrial Relations* 51(4): 666–88.

Tapia, M. 2019. "'Not fissures but moments of crises that can be overcome': building a relational organizing culture in community organizations and trade unions". *Industrial Relations* 58(2): 229–50.

Tassinari, A., J. Donaghey & M. Galetto 2022. "Puzzling choices in hard times: union ideologies of social concertation in the Great Recession". *Industrial Relations* 61(1): 109–34.

Tattersall, A. 2010. *Power in Coalition: Strategies for Strong Unions and Social Change*. Ithaca, NY: Cornell University Press.

Thomas, A. & V. Pulignano 2021. "Challenges and prospects for trade union environmentalism". In *The Palgrave Handbook of Environmental Labour Studies*, edited by N. Räthzel, D. Stevis & D. Ussell, 517–38. London: Palgrave Macmillan.

Turner, L. 2004. "Why revitalize? Labour's urgent mission in a contested global economy". In *Varieties of Unionism: Strategies for Union Revitalization in a Globalizing Economy*, edited by C. Frege & J. Kelly, 1–10. Oxford: Oxford University Press.

Upchurch, M., G. Taylor & A. Mathers 2009. *The Crisis of Social Democratic Trade Unionism in Western Europe and the Search for Alternatives*. Aldershot: Ashgate.

Vandaele, K. 2016. "Interpreting strike activity in Western Europe in the past 20 years: the labour repertoire under pressure". *Transfer: European Review of Labour and Research* 22(3): 277–94.

Vandaele, K. 2019. *Bleak Prospects: Mapping Trade Union Membership in Europe Since 2000*. Brussels: ETUI.

Vandaele, K. 2020. "Newcomers as potential drivers of union revitalization: survey evidence from Belgium". *Relations Industrielles/Industrial Relations* 75(1): 351–75.

Visser, J. 2012. "The rise and fall of industrial unionism". *Transfer: European Review of Labour and Research* 18(2): 129–41.

Weil, D. 2014. *The Fissured Workplace: Why Work Became So Bad for So Many and What Can Be Done to Improve It*. Cambridge, MA: Harvard University Press.

Wissen, M. & U. Brand 2021. "Workers, trade unions, and the imperial mode of living: labour environmentalism from the perspective of hegemony theory". In *The Palgrave Handbook of Environmental Labour Studies*, edited by N. Räthzel, D. Stevis & D. Ussell, 699–720. London: Palgrave Macmillan.

Wright, E. 2000. "Working-class power, capitalist-class interests, and class compromise". *American Journal of Sociology* 105(4): 957–1002.

CHAPTER 7

UNION TERRAINS

Jamie Woodcock

ABSTRACT

Labour unions are institutions that develop to defend or advance the collective interests of groups of workers. When examining labour unions and movements over a longer historical scope, the terrain upon which they operate has been subjected to constant and far-reaching change. This chapter seeks to understand the current terrain in the UK by placing labour unions within their historical context. Drawing upon a critical account of the existing literature, it addresses the debates on union decline and renewal, different organizing models, and factors that shape the environment in which labour unions operate. This involves situating labour unions within a more expansive understanding of the terrain – both being changed by external factors and the ways in which labour unions and their allies have sought to change these. It also differentiates between the key present terrains for unions in the UK, summarizing the key differences and factors involved.

Keywords: External and environmental terrains; challenges for unions

INTRODUCTION

Labour unions are institutions that have been developed to defend or advance the collective interests of groups of workers. The origins of labour unions involve collective organizing and negotiation in opposition to employers, often within a hostile environment. At their roots then is a "unionateness" that Blackburn (1967: 19) outlines, which involves:

> ... collective bargaining and the protection of the interests of members, as employees, as its main function, rather than, say, professional activities or welfare schemes ... It is independent of employers for purposes of negotiation ... It is prepared to be militant, using all forms of industrial action which may be effective.

However, as labour unions develop and grow, they can become subjected to different pressures and interests, including the development of a layer of bureaucracy (see Chapter 15), institutionalization within bargaining procedures, amalgamation into larger formations, political engagement, regulation, and so on.

The current conjuncture for labour unions in the UK is one of longer-term secular decline, including membership limited within certain sectors of the economy and society (particularly the public sector), an ageing membership profile, a reliance upon a "service" unionism model, close links with a declining political party and operating in an increasingly hostile context – both in terms of employment relations and the relationship with the state and state regulation. Within this present context, many labour unions can be categorized as defensive and reactive. In part this can involve a loss of the "unionateness" that Blackburn (1967) identified. The terrain of work is constantly shifting, with new forms of employment and employment relations operating against the transformation of existing arrangements.

When examining labour unions and movements over a longer historical scope, the terrain upon which they operate has been subjected to constant and far-reaching change. This chapter seeks to understand the current terrain by placing labour unions within their historical context. Drawing upon a critical account of the existing literature, it addresses the debates on union decline and renewal, different organizing models, and factors that shape the environment in which labour unions operate. This involves situating labour unions within a more expansive understanding of the terrain – both being changed by external factors and the ways in which labour unions and their allies have sought to change these. It also differentiates between the key present terrains for labour unions, summarizing the key differences and factors involved.

The shape and nature of the terrain is something that labour unions – and researchers involved in the field – should have a deep and nuanced understanding of. It is the arena within which and on which labour unions organize, shaping the field of possibilities and the constraints under which they operate. An understanding of terrain has long been a focus of writings on tactics and strategy. For example, Sun Tzu (2007) argues that it was possible to distinguish between six different kinds of terrain: "accessible ground, entangling ground, temporising ground, narrow passes, precipitous heights, and positions at a great distance from the enemy". These operate at the level of local geography, determining when and where generals should engage in armed conflict. They emphasize considering the environment as a factor in tactics, understanding the balance of forces before committing to action.

Clausewitz (1984: 348) argues that terrain could have three affects: "as an obstacle to the approach, as an impediment to visibility, and as cover from fire. All other properties can be traced back to these three." Often, this has led to the tactic of trying to hold the high ground. However, there are also practical examples where unfavourable terrain can be overcome. For example, the Red Army demonstrated this during the battle of Seelow Heights, the last remaining major natural obstacle in the march on Berlin from the east. By following the shortest route to Berlin, albeit with overwhelming numbers, the Red Army overcame the Nazi advantage. Similarly, unfavourable terrain for unions, whether that be tight labour markets or particular forms of work organization, does not necessarily prevent an offensive.

Labour unions, as "intermediary organizations" (Müller-Jentsch 1985), are an institution that operate to mediate between labour and capital. However, as Draper argues (1970), using a similar metaphor:

> If you are in a revolutionary army, you have an officer corps ... you're fighting with them, while at the same time you may not like them ... I mean something similar when I emphasize that the trade union bureaucracy is the bureaucracy of our class.

The membership of labour unions, therefore, has to address the challenges of starting from the lower ground. The bureaucracy may be able to take the higher ground at various points, but the starting point is a structural disadvantage of capitalist social relations. Despite this, analysing the terrain for workers to advance on still matters. Whether that terrain is harder or softer, or offers "cover from fire" or other potential advantages is important. Drawing the metaphor out, there is also a much wider terrain that affects labour unions. There are features that could block the growth of union membership, prevent developing an understanding of the conditions or issues in a workplace or sector, or defensive bulwarks against employers.

There are, of course, limits to applying military analogies to the understanding of labour unions. Terrain, in a wider context, can refer to both the "political-strategic sense" as well as "political-economic sense" of territory, drawing out the complexity of both parts (Elden 2021: 174). It is also worth noting, as Woodward (2013) does, that the military had a significant impact on geography, not only taking it as an external factor to shape conflict. In the context of labour unions, clearly geography matters (Wills 1996), but so too does the workplace, industry, sector and wider economic, political and social context. There are also internal battles within labour unions that shape strategy or orientation, whether between the rank and file and the bureaucracy, or different factions within unions.

The military analogy also draws attention to the importance of tactics and strategies. Indeed, as Sun Tzu (2007) also argues: "Strategy without tactics is the slowest route to victory. Tactics without strategy is the noise before defeat." Unions are currently faced with a "fog of war" (Clausewitz 1984) that covers much of the terrain. Many labour unions are not equipped with a map of the terrain in which they are organizing, or if they do have a map it is outdated or inaccurate. Lifting the veil on the "fog of war" requires moving into unfamiliar terrain, something that is often considered much riskier than continuing to organize where established membership or institutional relationships exist.

Labour unions are not alone in this. In part this is a function of the kind of terrain that we are discussing in this chapter. This terrain is not primarily geographic, nor is it formed over millennia. Instead, this is a live terrain that is shaped by, and in turn shapes, people, both individuals and collective groups, workers and capitalists, the state and other organizations. People and their interests, experiences and practices are a key part of this. The terrain is therefore "formed and reformed as a dynamic process, rather than by processes" (Elden 2021: 182). The next part of this chapter will examine three terrains upon which labour unions operate: the economic, political and future terrains.

There is, of course, an interrelation between each of these three. However, drawing out the key dynamics from each provides one way to begin mapping the overall terrain. The concluding section of the chapter examines the future terrain of labour unions.

THE ECONOMIC TERRAIN

The economic terrain for labour unions is, at its core, the conflict between labour and capital at work. This means that the terrain scales from the individual workplace up to the national – and increasingly international and transnational – context. As intermediary organizations (Müller-Jentsch 1985), labour unions not only respond to the economic terrain but also become part of it, mediating the unequal balance between labour and capital. The economic terrain is, therefore, the battlefield upon which labour unions operate – whether they are taking strike action, organizing, negotiating, bargaining, accommodating or capitulating – as well as being a terrain upon which workers might fight for representation. The economic terrain, therefore, involves the composition of work, the shape and dynamics of industries, the nature of employment, introduction of technology, and strategies from management – to name but a few. It should, therefore, not come as a surprise that mapping this terrain is a challenging undertaking. In the time it would take to construct a perfect map, the conditions and dynamics would have changed. Despite this, it is still important to try to map out the key dynamics and drivers of change in order to sketch out the field of possibilities and constraints.

To make sense of contemporary dynamics, the end of the 1970s is often taken as a turning point – not only for labour unions but also for a wider transformation of economy and society. In particular, it marks a watershed moment of decline for labour unions. It is also often discussed as the start of neoliberalism. As Harvey (2007: 3) argues, this involves "a theory of political economic practices that propose that human well-being can best be advanced by liberating individual entrepreneurial freedoms and skills within an institutional framework characterised by strong private property rights, free markets, and free trade". Clearly, there is much here that involves the economic terrain. In particular, the state implementing programmes of "deregulation, privatisation, and withdrawal of the state from many areas of social provision" (Harvey 2007: 3).

There is a risk of just stating that neoliberalism has changed the economic terrain and that labour unions have been unable to adapt to this. However, neoliberalism should not be "a substitute of explanation", but instead a starting point from unpicking the changes that have taken place (Peck 2013: 153). Neoliberalism is not a singular phenomenon and has been articulated through a struggle between the interests of capital against labour, playing out not only on the economic terrain but more widely, especially in the ideological arena. This begins with the "structural crisis" of the 1970s, involving the falling rate of profit across the economy and a series of energy crises. Declining growth rates combined with unemployment and inflation. This new economic terrain provided the basis for the "new social order" of neoliberalism (Duménil & Lévy 2005).

This poses the question of what the transformation was away from. While there is a risk of viewing the pre-neoliberal period through rose-tinted glasses, there was a level of union strength that has not been repeated since (see Chapter 9). In the UK, union membership and density grew rapidly in the 1970s, with 85 per cent of workers being covered by "collective pay-setting mechanisms of some kind" (Baccaro & Howell 2017: 53). Labour union organization was decentralized down to the workplace level. This left little discretion for employers and "reflected a degree of routine local union influence over the conduct of work that is beyond the dreams of most twenty-first century trade union activists" (Brown & Edwards 2009: 7). This was noted in the Donovan Commission in 1968, which pointed out that there was a conflict between the formal and informal industrial relation systems. This was a terrain that labour unions and workers were successfully establishing and controlling significant parts of.

It is also worth noting that longer term changes in the composition of work had been unfolding alongside this. There is a widely held belief that the decline of labour unions can be blamed on these changes. In particular, the blame is put on the decline of employment in manufacturing and the creation of new kinds of service work. The argument is often not unpicked and gives the impression that there is no manufacturing left today in the UK. Freeman and Pelletier (1990: 3) note that "while the decreased share of employment in manufacturing undoubtedly hurt unions", they reject this change as a key driver of labour union decline for three reasons. First, density fell across most parts of the economy, including within heavy manufacturing. Indeed, their shift-share analysis showed that "changes in industrial composition explains a bare 0.4 points of the 1980–1986 8.6 point drop in union density". Second, the changes in industry mix in other countries such as Ireland did not lead to density falling. Third, the shift away from manufacturing employment was only slightly more in the 1980s than the 1970s when union density rose.

The complementary trend to the fall of manufacturing employment is the increasing share of service employment. The argument here is that the composition of these service workers must be less favourable to organizing for labour unions. However, Freeman and Pelletier (1990: 3) also found that the increasing share of employment among part-time, white collar or female workers cannot explain the decline in union density from the 1980s. Indeed, two-thirds of the growth in labour union membership in the 1970s was among white-collar workers (Price & Bain 1983: 5). Their conclusion is that the "something more" that must be responsible for the significant fall in density is "the legal environment for industrial relations, as reflected in laws regulating union and management behaviour in the area of union recognition and membership" (Freeman & Pelletier 1990: 6).

This is a factor that will be discussed in the following section of the chapter, but it is also worth noting that these models of union density decline often start from an understanding that density depends only on workers' decisions. However, in practice it involves three sets of behaviours: first, workers, including their evaluation of the benefits and costs of joining a labour union; second, labour unions, including tactical and strategic choices, as well as where resources are used; and third, employers, who may recognize unions and bargain or try to prevent unionization. There is, of course, an interaction between all three in the course of falling (or indeed increasing) union density (Freeman & Pelletier 1990: 7).

What is significant about the transformations of neoliberalism is that it involved more than these tendencies playing out as a backdrop. Instead, it involved a deliberate and planned reorganization of the balance of forces on the economic terrain in favour of the employers. This was, as Harvey (2007: 16) argues, part of a "project to achieve the restoration of class power". As Munck (2005: 62) argues, this involved reasserting "management's 'right to manage' … in all its splendour", introducing changes to ensure that the "market would not be allowed to suffer from 'political' constraints". The movement away from the state as an employer, whether through privatization or outsourcing, has the potential to undermine the strike proneness that could previously be found in the public sector (Dickerson & Stewart 2009). As a class project more widely, this meant shifting the balance of power away from workers and labour unions (Harvey 2007). Here the military analogies are more useful, as it comprised an offensive of ruling-class power.

This assault on the power of labour unions is, therefore, a key and defining feature of the economic terrain on which they operate today. During the postwar period, union density in the UK was stable at around 40–45 per cent, climbing sharply in the 1970s, with an even sharper fall from the late 1970s. From 1979 onwards, union membership has continued to trend downwards (Machin 2000: 632) falling to 22.3 per cent in 2022 (BEIS 2023). There have been varying claims and arguments about what happened from the 1980s onwards. For example, some point to changes in the composition of work, changes in employment law – discussed in the next section – or that the macroeconomic terrain had changed to become less favourable, drawing on the "business cycle model" (Bain & Elsheikh 1976; Booth 1983). Disney (1990) argues that while the "business cycle" can explain part of the membership trends in the postwar period, it fails to explain what happens from the 1970s. Freeman and Pelletier (1990: 5) also note that the "business cycle" alone cannot explain these changes.

Drawing on quantitative data, Machin (2000: 631) argues that across three measures – union density, union coverage and union recognition – a key driver of the decline is the failure of labour unions to organize in new workplaces that have been established since the 1980s. One reading of this is that "mechanisms that used to operate to enable trade unions to make employers concede recognition demands in new workplaces have ceased to exist", while, on the other hand, it could be that increased competitive pressures in new firms have made it much harder to win concessions (Machin 2000: 642). This shifts the terrain in an important way. Not only is this a decline of union membership, but the increasing "proportion of new union-free workplaces where unions are unable to even get a toe in the door, are trends that are unlikely to be reversed easily" (Machin 2000: 642). This represents a significant obstacle to fighting on large parts of the economic terrain.

The workplace is not of equally accessible ground (in Sun Tzu's terms). As Edwards (1979) argues, the capitalist workplace is a "contested terrain". Edwards uses the metaphor of terrain to understand how management attempts to gain and maintain control of the workforce and work in the workplace. What is useful about this is drawing attention to the ways in which managerial control attempts to reshape the terrain in the interest of capital. Therefore, current labour union activity builds upon histories of previous activities, being both facilitated and constrained by them.

The failure or inability to organize in new industries has shaped the ways in which these workplaces are formed and subsequently managed. For example, call centres have become symbolic of the new forms of service work that have increased during neoliberalism. In part the growth of call centres was facilitated by the deregulation of financial services and other neoliberal policies (Ellis & Taylor 2006), and became a rapidly growing area of employment, increasingly managed through electronic forms of surveillance and control (Taylor & Bain 1999). As noted elsewhere (Woodcock 2017: 32), the growth of call centres was not matched by unionization. Call centres, therefore, became increasingly shaped in the employers' interest, with high levels of stress, monitoring and burnout. In this light, Taylor and Bain (2004: 18) argue that "the future success of trade unions in call centres will depend in no small measure on their ability to contest and redefine the frontiers of control on terms desired by their members".

Beyond call centres, this has meant that an increasing number of workplaces have no labour union presence at all, instead replaced by human resources (HR) departments. It is worth noting here that Machin and Wood (2005) found that union decline in workplaces or industries that adopted HR management practices was no faster than those that did. However, the development of HR management practices in workplaces without any union density was undoubtedly different to those that had union (declining or not) presence.

These contemporary changes, therefore, involve the reorganizing of the economic terrain. The results of previous struggles over the terrain of the workplace reshape that terrain upon which new struggles about and over work are fought. While the key drivers for declining labour unions from the late 1970s may well not have been triggered or led by changes in composition of employment, this does not mean that the intervening 40 years has not seen widespread changes both in the nature and share of different kinds of employments, as well as changes in the labour process, technology and management techniques.

As Moody (2017) argues in *On New Terrain*, there have been substantial changes to capitalism since the 1980s. His usage of "terrain" refers to two key dynamics. First, what Moody (2017: 45) calls the "terrain of class struggle" and the transformation of capitalist production, within the US in particular, drawing out the increasing importance of logistics and the concentration of wealth and power with an increasingly small number of people. Second, the "changing political terrain" (Moody 2017: 91) examining the role of the state and electoral politics. Through both terrains Moody focuses upon changes in working-class composition. Rather than discussing these changes to "write off" the possibilities for labour union organizing or independent political action, Moody focuses upon the potential of these new compositions for new forms of struggle. In another vein, McAlevey (2016) argues that there is significant potential for rebuilding labour unions in the public sector, including workplaces such as schools and hospitals.

There are a variety of structural changes that have shaped the composition of work, including deindustrialization and changes in employment, increasing internationalization and globalization, demographic change and an increased push for flexibility at work (Baccaro & Howell 2017: 55). A key part of the debate has focused upon the growth of precarious work. At one end there are arguments that work has changed to such an extent that there is a new class called the "precariat" (Standing 2011). On the other side there are arguments that precarity is wildly overstated and that employment conditions have remained more or less stable (Doogan 2009). This is one of the difficulties of analysing a concept such as precarity. Workers do not just have to have shorter job tenure, as it can also involve a feeling that work is less secure, shaping the way that workers experience and respond to work.

In particular, in services such as hospitality, higher education, health and social care, and so on, there has been a growth of casualization. As Bourdieu (1998: 85) argues, "casualisation of employment is part of a mode of domination of a new kind, based on the creation of a generalised and permanent state of insecurity aimed at forcing workers into submission, into the acceptance of exploitation". This reshapes the economic terrain, seeking to undermine workers' collective power. However, there are notable examples of precarious workers organizing despite these conditions, including Latin American migrant cleaners or casualized academics in universities (Woodcock 2014). Similarly, there have been labour union responses to precarious work (Benassi & Dorigatti 2015). The terrain is, therefore, shaped by vulnerabilities that workers face, whether they be through low pay, lack of existing labour union organization or migration status (Pollert & Charlwood 2009).

Despite some examples that cut against the trend, the issue noted by Machin (2000) of a failure or inability for labour unions to organize effectively in the service sector remains. Labour unions "face considerable obstacles to extending their presence in private services, not least from hostile employers" (Williams & Adam-Smith 2009). However, as Walters (2002) found in a study of part-time women workers, there was not necessarily a reluctance to join: either workers simply had never been asked to join or they did not see that joining would achieve anything. It therefore appears that management has been able to utilize the "the more hostile political and economic climate for trade unionism to undermine their power and legitimacy" (Williams & Adam-Smith 2009).

This lack of labour union presence on the economic terrain has also left the battlefield open for more widespread changes to take place. Perhaps the most significant of these is the growth of the so-called "gig economy", with platform work at the sharpest end (Graham & Woodcock 2018). The growth of the "platform economy" (Srnicek 2017) has involved the use of bogus self-employment contracts (De Stefano & Aloisi 2019). This is an attempt to refigure the economic terrain away from employment and into a much more dispersed category of the "gig economy", imbued with the logic of entrepreneurialism (Duggan *et al.* 2020; Ravenelle 2019). This often denies labour unions a role in negotiation, whether through refusing to engage or because existing legislation prevents the self-employed from engaging in collective bargaining. However, over the past five years there have been waves of wildcat strike action in food delivery and private hire driver transportation work globally (Cant 2019; Joyce, Stuart & Forde 2020; Woodcock 2021). While there are some cases of labour union involvement – for example, the IWGB (Independent Workers Union of Great Britain) – in these struggles, many of them are taking place outside of these structures (Countouris & De Stefano 2019).

As Cant (2019) argues, platform work is acting as a laboratory for the testing of new technologies and management techniques. What happens on the economic terrain of platform work will matter much more widely than just for the sectors that are currently being transformed in this way. In particular, this work provides a battleground for the ways in which labour unions can be renewed. However, there is also an important reminder from Atzeni (2021) about the risk of fetishizing the labour union form. As Tassinari and Maccarrone (2020) have clearly demonstrated with Deliveroo riders, the contradictions of the labour process continue to generate conflict and solidarity in this kind of work. The absence of labour union organization does not mean there is an absence of worker struggle. A reorientation on the economic terrain involves starting not only from injustices at work, but from the labour process (Atzeni 2009). This constitutes the core of the economic terrain upon which labour unions operate.

THE POLITICAL TERRAIN

The political terrain is deeply connected to the economic terrain. In a Marxist sense, the political terrain is comprised of superstructural phenomena that are predominantly influenced by the economic base. The superstructure acts back upon the base, particularly providing opportunities or constraining actions for labour unions upon the economic terrain. The political connects to the micro-economic terrain of the workplace, as well as the macro-scale of the industry or economy and at multiple points. There are different actors involved in the political terrain, the most important of which is the state. This section of the chapter will review how the political terrain has shifted for labour unions, as well as considering some of the factors beyond the state that can play a role. Here, in addition to public policy, the political terrain is reified through legislative and regulatory action.

The state has undoubtedly played an important role historically in shaping industrial relations in the UK. It should perhaps be no surprise that the state has not played a singular role, given the division in the UK's parliament between the Labour Party (having grown out of the labour movement) and the Conservative Party. However, this has not involved a simple split between pro-labour union and anti-labour union governments. In the postwar period, there have been four important shifts in the political terrain that have had a significant impact on labour unions. During 1946–74, the Labour Party introduced a series of legal changes that strengthened labour unions. The first was the withdrawal of a series of restrictions, both previous legislation such as the Trade Disputes and Trade Unions Act 1927 and the removal of wartime restrictions on strikes. The Conservative Party also introduced changes in 1971 that including making collective bargaining legally binding and bringing in greater individual rights. While "unions opposed many of these provisions, they did nothing to weaken unionism and, arguably, strengthened collective bargaining" (Freeman & Pelletier 1990: 10).

The second phase was from 1974–79. The Labour Party introduced new legislation that strengthened labour unions. The Conservative Party's 1971 act was repealed with the introduction of the Trade Union and Labour Relations Act 1974. Further changes were brought in with the Employment Protection Act 1975, with rights to association, time off for union activity, rights to information disclosure from employers, as well as the creation of ACAS (Advisory, Conciliation and Arbitration Service). Further labour union rights were also introduced in the amendments to the act in 1976 (Freeman & Pelletier 1990: 11).

The third phase was from the 1980s onwards, where the legal balance was shifted away from labour unions and towards employers. As Baccaro and Howell (2017: 51) argue: "British industrial relations underwent de-collectivisation on a massive scale in a relatively short period of time." Or, to put it another way, it involved "the end of institutional industrial relations" (Purcell 1993). This change was driven by a shift in the political terrain: "What changed was the role of the British state, which ended its almost eighty-year support for collective bargaining, and the interests of employers and thus their investment in the collective regulation regime" (Baccaro & Howell 2017: 55).

This was a transformation of the terrain of labour unions that was deliberately planned by the Conservative Party. The Ridley Plan (Economic Reconstruction Group 1977), which was publicly leaked in 1978, detailed the plan for denationalizing industries. Industries are divided into three categories based on how many weeks of strike action "the nation could survive": zero weeks for Category I; between four to ten weeks for Category II; and "a long time" for Category III. There are considerations of making strike action unlawful in some industries, as well as a reflection that "as a long term policy government should seek to manoeuvre the nation out of a position where it is vulnerable to monopoly unions in vital industries". Rather than suggesting a "frontal attack", the plan prosed "preparing industries for partial return to the private sector, more or less by stealth" (Economic Reconstruction Group 1977: 15, 22). In the confidential annex of the report, a five-step plan is outlined for confronting and beating the labour movement. The significance of this report is that it represented a turning point ideologically for the Conservative Party. The "Thatcherite class political regime and the free-market authoritarian economic-political order began to emerge" (Gallas 2016: 96) from this point. The plan would later be put into action during the Conservative Party's defeat of the miners' strike of 1984–85.

The 1980s is often considered as a period that introduced anti-trade union laws in the UK. Driven at first by Margaret Thatcher's government, the Employment Act 1980 brought in a series of measures to curb labour union power: encouraging secret ballots for action and elections, limiting closed shops, restricting lawful picketing to workers' own workplaces, removing immunity from secondary action, and removing statutory recognition processes. The Employment Act 1982 went further with the removal of labour union immunities, including protection from actions in tort (meaning that employers could seek injunctions against labour unions, sue for damages, and it allowed courts to sequester union assets and funds), limiting immunities to trade disputes on industrial matters rather than political issues, reducing secondary industrial action, requiring secret ballots for closed shops, and prohibiting union labour only and union recognition clauses from commercial contracts. The Trade Union Act 1984 focused on elections, with the requirement of secret ballots before strikes, for the election of executive committee members, and for labour union political funds. Through all these changes in legislation, Dunn and Metcalf (1994: 22) argue that there was a "slow build-up in management confidence to resist unionisation".

From the end of the 1980s further changes were introduced. The Employment Act 1988 brought in a range of further limitations, as well as introducing postal ballots for union executive elections and political funds. The Employment Act 1990 aimed to limit unofficial industrial action. The Trade Union Reform and Employment Rights Act

1993 brought in a major change that built upon many of these prior changes, requiring all industrial action ballots to be conducted by post. At the same time, it repealed the Secretary of State's power to refund the costs of printing and postal costs for ballots. Taking industrial action became more restricted: labour unions had to give employers seven days' notice of intention to ballot, notify employers of the result, submit the ballot to independent scrutiny and provide seven days' notice of taking industrial action. Many other restrictions and requirements were introduced, particularly in relation to reporting and accounting. Most of the legislation from 1979 was consolidated in the Trade Union and Labour Relations (Consolidation) Act 1992 as amended (Pyper 2017).

The fourth phase, which starts with the Labour Party being elected to office in 1997, runs up until today. Looking at the previous phases, it may have seemed that Labour would repeal many of these anti-labour union restrictions. However, the Labour Party had promised that "key elements" of the Conservative reforms would remain in place. There were a series of changes introduced. The Employment Relations Act 1999 introduced procedures for union recognition where a majority of workers were in favour, the right for labour union representation during disciplinaries or grievances, and protections for union membership and when taking part in lawful industrial action. The Employment Relations Act 2004 clarified and furthered many of these changes, particularly relating to discrimination on the ground of labour union membership to bring it line with European legislation (Pyper 2017).

Most recently, the Trade Union Act 2016 introduced a series of significant restrictions on labour union activities. It introduced a 50 per cent turnout requirement for industrial action ballots, with "important public services" requiring a vote for of more than 40 per cent of all those entitled to vote, meaning that non-voters are effectively classified as voting against action. Labour unions were required to provide 14 days' notice for industrial action and postal ballot mandates expired after six months. The role of the Certification Officer was greatly strengthened, providing the ability to issue financial penalties up to £20,000, with the option to fund the Certification Officer in the future through a levy of labour unions (Pyper 2015).

As Bogg (2016: 299) argues, this represents "a historically significant realignment in the ideological politics of trade union regulation". Similarly, Ewing and Hendy (2016: 391) point out that the Bill "raised major concerns about freedom of association" and issues of human rights. Moore and Taylor (2016: 256) examined the proposals for restricting not only traditional forms of picketing, but other forms of "leverage", including social media. They conclude that "as with the Trade Union Bill in its entirety, government legislation fails to address the imbalance of power integral to employment relations". Indeed, the TUC (2017: 3) reported that "the Act represents the most serious attack on the rights of trade unions and their members in a generation".

This history of parliamentary or legislative struggle over the political terrain for labour unions has opened up possibilities for organizing and an action at some points, while severely constraining them at others. The transformations of neoliberalism have also involved a contradictory relationship with the state. Neoliberalism was driven by parts of the state, while also calling for the roll back of the state. The state plays a key role in providing essential services for capitalism to reproduce itself, as well as attempting to stabilize the business cycle and ensuring private markets operate through contracts and the legal system (Palley 2005: 27). As Munck (2005: 62) explains, "Government intervention was crucial to the making of markets, yet neoliberalism has as a central tenet the seemingly contradictory missions of 'driving back' state intervention." From the 1990s onwards there has also been a "second moment" of neoliberalism, which involved "a 'roll out' of new policies rather than just a 'roll back' of the state". This has involved new and aggressive policies to transform welfare, the justice system, urban regeneration, and migration.

This second moment of neoliberalism was driven forward after the election of Tony Blair in 1997. This had important implications for labour union struggle on the political terrain. While there have been contradictory moments of political representation for the labour movement in parliament through (parts) of the Labour Party, the election of Tony Blair marked a turning point for the party. Despite the movements around Jeremy Corbyn's leadership of the Labour Party, there seems to be very limited potential for labour unions to have an influence upon policies of the party, let alone having that party win an election and implement any of those policies.

THE FUTURE TERRAIN

Labour unions are not homogenous organizations, nor have they always followed the same tactics and strategies. Within labour unions there are significant differences and opposing ideas about how to operate on either the economic or political terrain. It is, therefore, worth focusing also upon the internal terrain of labour unions. In addition to the terrains discussed so far, there is also the broader neoliberal transformation of the ideological terrain (Harvey 2007). Part of this has been the increased individualization of society, the dismantling of collective organizations and institutions more widely, and the dominance of a "capitalist realism" that prevents the imagination of any alternatives (Fisher 2009).

The chapter started with a discussion of the "unionateness" (Blackburn 1967: 19), which suggested militancy in bargaining as the modus operandi. However, there have been significant shifts in how many labour unions operate in the UK. For example, the response from some unions to the decline after the 1970s was to adopt more of a servicing union approach, providing services to members instead of organizing, or even seeking partnership with employers (Bassett & Cave 1993). In one example, the TUC established the Organising Academy in 1998 in preparation for the new statutory recognition procedures to be brought in with the Employment Relations Act 1999. This was inspired by labour union programmes in the US, such as the Organizing Institute and Union Summer, which aimed to bring underrepresented groups into the labour movement. While this led to a large number of recognition deals in the years that followed the introduction of the Act, it was not able to renew the labour movement more widely. Simms and Holgate (2010: 157) argue that these approaches "tended to see organising as a 'toolbox' of practices rather than as having an underpinning political philosophy or objective". This runs the risk of launching "organising" campaigns without asking "the fundamental question of what are we organising 'for'?"

There have also been a series of mergers between labour unions that were designed to counter decline, as well as the establishment of new "community" membership models to bring new people into the labour movement (Holgate 2021b). As Holgate (2021a) argues, any overcoming of this involves taking seriously the issue of power. There are skirmishes on the economic terrain that show the potential for labour union renewal. The key problem is that "existing labor unions have proved incapable of mobilizing mass rank-and-file militancy to resist the ongoing deterioration in workplace conditions and the systematic erosion of workers' power" (Ness 2014: 1).

The history of labour unions highlights how important the economic and political terrain have both been in shaping the form and content of labour union activities. The economic terrain has been transformed through changing patterns of employment, shifts in the structure of capitalism, but also by the high points of labour unions. However, following the neoliberal transformations from the 1970s, the economic terrain is mainly one of non-unionized workplaces, precarious conditions of work, new technology being used to strengthen management and the rise of the so-called "gig economy". There has been a concerted reorganization of the economic terrain from capital and management, resulting in a terrain that is ever more challenging for labour unions to operate. This can be seen clearly with the "gig economy" and the use of bogus self-employment status that attempts to push labour unions out of the terrain.

Operating alongside this, the political terrain has opened opportunities and constrained the actions of labour unions. There is a history in the UK of shifts between pro-labour union and anti-labour union policies. However, following the 1970s there has been a notable shift towards both major parties introducing measures that constrain the legal powers and legitimacy of labour unions. Most recently this has involved introducing measures to make it much harder for labour unions to organize effective strikes, bringing in a series of thresholds required to legally take industrial action.

Against this current context of labour union power, the future terrain can be hard to imagine. However, through examining the economic terrain it is possible to see how labour unions have adapted and renewed through previous compositions of work (see Chapter 4). Work is never a stable phenomenon and is subject to constant change and reorganization under capitalism. This makes marking out and sketching out the limits and dynamics of the terrain challenging, but it also guarantees recognition that the terrain will change again in the future. In comparison to this, the political terrain has had a determining effect on labour unions over the past 40 years. It is worth remembering here that labour unions have not always been integrated into legal structures. Labour unions emerged before there was regulation and workers opposed legal restrictions that attempted to prevent them being established. The inability, even failure, of established labour unions to operate effectively in this context does not prevent class struggle from taking place. Instead, it can be deflected into informal struggles, conflict beyond the workplace or the formation of new organizations.

We can paraphrase Marx (1978) here: labour unions make their own history, but they do not make it just as they please for they do not make it under circumstances of their own choosing but, rather, under circumstances existing already, given and transmitted from the past. This is why critically analysing the different terrains of labour unions is important. This is "not a case of 'waiting for the fightback', romanticising the informal, or disregarding the capacity of unions to renew their own organisation and strategy", as Thompson and Ackroyd (1995: 629) put it, but instead "to put labour back in, by doing theory and research in such a way that it is possible to 'see' resistance and misbehaviour, and recognise that innovatory employee practices and informal organisation will continue to subvert managerial regimes" (Thompson & Ackroyd 1995: 629). Work is still a battleground between labour and capital, one that is continuing to be fought over whether labour unions operate on the terrain or not.

REFERENCES

Atzeni, M. 2009. "Searching for injustice and finding solidarity? A contribution to the mobilisation theory debate". *Industrial Relations Journal* 40(1): 5–16.

Atzeni, M. 2021. "Workers' organizations and the fetishism of the trade union form: toward new pathways for research on the labour movement?" *Globalizations* 18(8): 1349–62.

Baccaro, L. & C. Howell 2017. *Trajectories of Neoliberal Transformation: European Industrial Relations since the 1970s*. Cambridge: Cambridge University Press.

Bain, G. & F. Elsheikh 1976. *Union Growth and the Business Cycle: An Econometric Analysis*. Oxford: Blackwell.

Bassett, P. & A. Cave 1993. *All for One: The Future of the Unions*. London: Fabian Society.

BEIS 2023. *Trade Union Membership, UK 1995–2022: Statistical Bulletin*. London: Office for National Statistics.

Benassi, C. & L. Dorigatti 2015. "Straight to the core – explaining union responses to the casualization of work: the IG Metall campaign for agency workers". *British Journal of Industrial Relations* 53(3): 533–55.

Blackburn, R. 1967. *Union Character and Social Class*. London: Batsford.

Bogg, A. 2016. "Beyond neo-liberalism: the Trade Union Act 2016 and the authoritarian state". *Industrial Law Journal* 45(3): 299–336.

Booth, A. 1983. "A reconsideration of trade union growth in the United Kingdom". *British Journal of Industrial Relations* 21: 379–93.

Bourdieu, P. 1998. *Contre Feux*. Paris: Raisons d'agir.

Brown, W. & P. Edwards 2009. "Researching the changing workplace". In *The Evolution of the Modern Workplace*, edited by W. Brown *et al*. New York: Cambridge University Press.

Cant, C. 2019. *Riding for Deliveroo: Resistance in the New Economy*. Cambridge: Polity.

Clausewitz, C. von 1984. *On War*. Princeton, NJ: Princeton University Press.

Countouris, N. & V. De Stefano 2019. *New Trade Union Strategies for New Forms of Employment*. Brussels: European Trade Union Confederation.

De Stefano, V. & A. Aloisi 2019. "Fundamental labour rights, platform work and protection of non-standard workers". In *Labour, Business and Human Rights Law*, edited by J. Bellace & B. Haar, 359–79. Cheltenham: Edward Elgar.

Dickerson, A. & M. Stewart 2009. "Is the public sector strike prone?" *Oxford Bulletin of Economics and Statistics* 55(3): 253–84.

Disney, R. 1990. "Explanations of the decline in trade union density in Britain: an appraisal". *British Journal of Industrial Relations* 28(2): 165–77.

Doogan, K. 2009. *New Capitalism? The Transformation of Work*. Cambridge: Polity.

Draper, H. 1970. "Marxism and the trade unions (part 3)". Marxists.org. www.marxists.org/archive/draper/1970/tus/3-randforg.htm

Duggan, J. *et al*. 2020. "Algorithmic management and app-work in the gig economy: a research agenda for employment relations and HRM". *Human Resource Management Journal* 30(1): 114–32.

Duménil, G. & D. Lévy 2005. "The neoliberal (counter-)revolution". In *Neoliberalism: A Critical Reader*, edited by A. Saad-Filho & D. Johnston. London: Pluto.

Dunn, S. & D. Metcalf 1994. "Trade union law since 1978: ideology, intent, impact". Working paper, Centre for Economic Performance, London School of Economics.

Economic Reconstruction Group 1977. "Final report of the nationalised industries policy group". Economic Reconstruction Group.

Edwards, R. 1979. *Contested Terrain: The Transformation of the Workplace in the Twentieth Century*. New York: Basic Books.

Elden, S. 2021. "Terrain, politics, history". *Dialogues in Human Geography* 11(2): 170–89.

Ellis, V. & P. Taylor 2006. "'You don't know what you've got till it's gone': re-contextualising the origins, development and impact of the call centre". *New Technology, Work and Employment* 21(2): 107–22.

Ewing, K. & J. Hendy 2016. "The Trade Union Act 2016 and the failure of human rights". *Industrial Law Journal* 45(3): 391–422.

Fisher, M. 2009. *Capitalist Realism: Is There No Alternative?* Winchester: Zero Books.

Freeman, R. & J. Pelletier 1990. "The impact of industrial relations legislation on British union density". *British Journal of Industrial Relations* 28(2): 141–64.

Gallas, A. 2016. *The Thatcherite Offensive: A Neo-Poulantzasian Analysis*. Leiden: Brill.

Graham, M. & J. Woodcock 2018. "Towards a fairer platform economy: introducing the fairwork foundation". *Alternate Routes* 29: 242–53.

Harvey, D. 2007. *A Brief History of Neoliberalism*. Oxford: Oxford University Press.

Holgate, J. 2021a. *Arise: Power, Strategy and Union Resurgence*. London: Pluto.

Holgate, J. 2021b. "Trade unions in the community: building broad spaces of solidarity". *Economic and Industrial Democracy* 42(2): 226–47.

Joyce, S., M. Stuart & C. Forde 2022. "Theorising labour unrest and trade unionism in the platform economy". *New Technology, Work and Employment* 38(1): 21–40.

Machin, S. 2000. "Union decline in Britain". *British Journal of Industrial Relations* 38: 631–45.

Machin, S. & S. Wood 2005. "Human resource management as a substitute for trade unions in British workplaces". *ILR Review* 58(2): 201–18.

Marx, K. 1978. *The Eighteenth Brumaire of Louis Bonaparte*. Peking: Progress.

McAlevey, J. 2016. *No Shortcuts: Organizing for Power in the New Gilded Age*. New York: Oxford University Press.

Moody, K. 2017. *On New Terrain: How Capital Is Reshaping the Battleground of Class War*. Chicago: Haymarket.

Moore, S. & P. Taylor 2016. "'We planned a dispute by Blackberry': the implications of the trade union bill for union use of social media as suggested by the BA-BASSA dispute of 2009–11". *Industrial Law Journal* 45(2): 251–56.

Müller-Jentsch, W. 1985. "Trade unions as intermediary organizations". *Economic and Industrial Democracy* 6(1): 3–33.

Munck, R. 2005. "Neoliberalism and politics, and the politics of neoliberalism". In *Neoliberalism: A Critical Reader*, edited by A. Saad-Filho & D. Johnston. London: Pluto.

Ness, I. 2014. "Introduction". In *New Forms of Worker Organization: The Syndicalist and Autonomist Restoration of Class Struggle Unionism*, edited by I. Ness, 1–17. Oakland, CA: PM Press.

Palley, T. 2005. "From Keynesianism to neoliberalism: shifting paradigms in economics". In *Neoliberalism: A Critical Reader*, edited by A. Saad-Filho & D. Johnston. London: Pluto.

Peck, J. 2013. "Explaining (with) neoliberalism". *Territory, Politics, Governance* 1(2): 132–57.

Pollert, A. & A. Charlwood 2009. "The vulnerable worker in Britain and problems at work". *Work, Employment and Society* 23(2): 343–62.

Price, R. & G. Bain 1983. "Union growth in Britain: retrospect and prospect". *British Journal of Industrial Relations* 21(1): 46–68.

Purcell, J. 1993. "The end of institutional industrial relations". *Political Quarterly* 64(1): 6–23.

Pyper, D. 2015. "Trade Union Bill". House of Commons Library, Briefing Paper Number CBP 7295. https://commonslibrary.parliament.uk/research-briefings/cbp-7295/#fullreport

Pyper, D. 2017. "Trade union legislation 1979–2010". House of Commons Library, Briefing Paper Number CBP 7882. https://commonslibrary.parliament.uk/research-briefings/cbp-7882/

Ravenelle, A. 2019. *Hustle and Gig: Struggling and Surviving in the Sharing Economy*. Los Angeles, CA: University of California Press.

Simms, M. & J. Holgate 2010. "Organising for what? Where is the debate on the politics of organising?" *Work, Employment and Society* 24(1): 157–68.

Srnicek, N. 2017. *Platform Capitalism*. Cambridge: Polity.

Standing, G. 2011. *The Precariat: The New Dangerous Class*. London: Bloomsbury.

Tassinari, A. & V. Maccarrone 2020. "Riders on the storm: workplace solidarity among gig economy couriers in Italy and Britain". *Work, Employment and Society* 34(1): 35–54.

Taylor, P. & P. Bain 1999. "'An assembly line in the head': work and employee relations in the call centre". *Industrial Relations Journal* 30(2): 101–17.

Taylor, P. & P. Bain 2004. "Call centre offshoring to India: the revenge of history?" *Labour and Industry* 14(3): 15–38.

Thompson, P. & S. Ackroyd 1995. "All quiet on the workplace front? A critique of recent trends in British industrial sociology". *Sociology* 29: 615–33.

TUC 2017. *Trade Union Act 2016: A TUC Guide for Trade Union Reps*. London: TUC.

Tzu Sun 2007. *The Art of War*. Atlanta, GA: Dalmatian Press.

Walters, S. 2002. "Female part-time workers' attitudes to trade unions in Britain". *British Journal of Industrial Relations* 40(1): 49–68.

Williams, S. & D. Adam-Smith 2009. "Web case: trade unions and the prospects for unionisation in the service sector". In *Contemporary Employment Relations: A Critical Introduction*, edited by S. Williams & D. Adam-Smith, 2nd edn, 253–4. Oxford: Oxford University Press.

Wills, J. 1996. "Geographies of trade unionism: translating traditions across space and time". *Antipode* 28(4): 352–78.

Woodcock, J. 2014. "Precarious workers in London: new forms of organisation and the city". *City: Analysis of Urban Trends, Culture, Theory, Policy, Action* 18(6): 776–88.

Woodcock, J. 2017. *Working the Phones: Control and Resistance in Call Centres*. London: Pluto.

Woodcock, J. 2021. *The Fight Against Platform Capitalism: An Inquiry into the Global Struggles of the Gig Economy*. London: University of Westminster Press.

Woodward, R. 2013. "Military landscapes: agendas and approaches for future research". *Progress in Human Geography* 38(1): 40–61.

PART II

SPACE, POWER AND PERIODIZATION

CHAPTER 8

THE LIBERAL CAPITALIST STARTING POINT

Stefan Berger

ABSTRACT

This chapter traces the development of labour unions as a response to liberal capitalist structures between the late eighteenth and the middle of the twentieth century. It starts off by asking how the introduction of the factory system led to the formation of early unions and how these were different from the old guilds that had characterized the social organization of life in pre-industrial times. Thereafter, it traces their long struggle for recognition and the divisions within the union movement. The chapter discusses both the Global North and the Global South by looking at the development of labour unionism in the colonial and semi-colonial periphery before 1945.

Keywords: liberal capitalism; union genesis; union development

INTRODUCTION

There undoubtedly were a "variety of capitalisms" (Hall & Soskice 2001) that shaped industrialization processes in different parts of the world between the eighteenth century and the present day. And for much of the twentieth century, industrialization also occurred in communist regimes that postulated that they marked the transition from capitalism to communism (Stearns 2013). Yet, it still remains the case that labour unionism fundamentally began to emerge under conditions of industrialization in liberal capitalist regimes. This chapter traces their development as a response to liberal capitalist structures between the late eighteenth and the middle of the twentieth century. It starts off by asking how the introduction of the factory system led to the formation of early unions and how these were different from the old guilds that had characterized the social organization of life in pre-industrial times.

In many parts of the world, unions struggled for a long time to gain acceptance by employers and the state, often in the form of legal rights and policy. This chapter traces this struggle among some of these industrializing countries from the nineteenth century to the end of the Second World War. The fight of unions for legal recognition and legitimacy was accompanied by debates surrounding the best way to organize that came to focus upon two contrasting principles – organizing by craft or organizing by industry. At the same time, ideological divisions between unions emerged, with socialism, syndicalism, Christianity, liberalism and communism being the main contenders for unionized workers' loyalty. In the inter-war period unions benefited from democratizing impulses at the end of the First World War to which they actively contributed in their attempts to augment political with social and economic democracy.

However, this chapter also traces a right-wing backlash against these democratizing tendencies – with a range of right wing authoritarian and fascist regimes curbing union advances and the possibilities for workers to set up their independent interest organizations. Intriguingly, independent unions were also crushed in the communist Soviet Union during the 1920s, where unions then became more powerful bodies but completely dependent upon the governing Communist Party and subservient to the policies it devised (see Chapter 10).

The chapter will extend its coverage from the Global North to the Global South by looking at the development of labour unionism in the colonial and semi-colonial periphery before 1945. In the last section, the phenomenal success of the union movement in Sweden is accounted for, as its foundations were laid before 1945, but after 1945, as Chapter 9 will explicate, it became one of the prime examples of the social democratic high points of unions between the end of the Second World War and the 1970s.

UNIONISM AS A RESPONSE TO THE EMERGENCE OF INDUSTRIAL CAPITALISM

The most remarkable innovation of the economic life of the eighteenth century was the factory system. During the period of proto-industrialization, the so-called "putting-out" system decentralized production into cottage industries. In the linen industry, for example, the capitalist would purchase the raw materials and hand them on to the workers, who would produce the finished product in their own homes before handing it back to the capitalist, who paid for the service of these cottage workers and kept a healthy profit that they made from selling the finished product. Artisan families became cottage workers. The family in its own home was the central unit of production. When exploitation by the capitalists became too stark, cottage workers could rise in often violent protests, threatening the property and lives of the capitalists. The Silesian Weavers' Rising of 1844 was a typical form of workers' protest during proto-industrialization (von Hodenberg 1997; Ogilvie & Cerman 1996).

The factory system allowed the capitalist to centralize production in the factory (Chapman 1967). They bought and owned the necessary means of production, supplied the raw materials and needed only wage labour, which was in plentiful supply and, therefore, cheap. The more the capitalist could keep labour costs down, the bigger was their profit. As women and children could be paid even less than men, it was attractive to hire them as wage labourers where possible. Yet the more centralized forms of production brought workers closer together. Not only did they work next to each other in the factory, but they also often lived together closely in housing that was either supplied by the private housing market or by the employers, who offered housing both as an incentive to attract workers and their families as well as a means of social control. The closer links of workers on the shop floor and in neighbourhoods meant that they soon began to form associations to defend themselves against the harsh exploitation by the capitalists. These associations sought in particular to increase the wages of workers, improve safety in the factories and make working conditions better more generally. They called themselves various names but can be seen as early unions.

The first ones were founded in the motherland of the Industrial Revolution, in the UK, during the second half of the eighteenth century. In 1799 the Combination Act, passed in the UK parliament, banned unions. Later versions of the Act, such as the one from 1825, severely limited the actions of unions but they could not entirely prevent their formation, often under clandestine names and conditions. Thus, for example, the General Union of Trades, formed in Manchester in 1818, was also known as the Philanthropic Society. Unions only became legalized in 1871 with the Trade Union Act of the same year. The Trade Union Congress was already formed three years earlier in 1868 as an umbrella organization of all unions in the UK. By 1887 these unions had 675,000 members and were, thus, by far the biggest labour union organization of workers in the world (Laybourn 1997).

The factory system in Continental Europe was introduced slightly later than in the UK and pockets of industrial capitalism developed regionally rather than nationally (Pollard 1981). Thus, for example, in Belgium, which was the first Continental European country to follow the UK on the road to industrialization, the textile industry developed strongly in Flanders, while Wallonia became the centre of the early iron and metal industries. In Germany, Saxony, Upper Silesia and the Ruhr area were important centres of industrialization. Soon the map of Europe was dotted with early industrial centres, in which the factory system produced similar problems of the exploitation of labour against which workers formed their own interest organizations.

Everywhere the most powerful organizations were formed by highly skilled workers rather than by the unskilled, but in many places the early employers were powerful enough to enlist the state to repress workers' organizations. The less liberal the framework of the state, the more difficult it was to form early union organizations. Thus, the Romanov empire in Russia was a good example of how the absolutist and authoritarian regime allowed no open organization of early workers' interests (Thatcher 2005: 101–19). With the advances of the industrial revolution in North America, labour unions began to develop also there, especially in the post-Civil War era (Faue 2017).

THE STRUGGLE FOR ACCEPTANCE

While unions developed in many parts of the Global North under conditions of liberal capitalism in the nineteenth century, they became a widely recognized part of the industrial relations system only in the UK before the First World War. As mentioned above, they suffered heavy repression. However, from the mid-nineteenth century onwards, some of the unions successfully allied themselves to the Liberal Party. In a parliamentary political system in which the franchise was gradually extended to include at least some working people, it made sense for the political parties to woo workers as voters. The unions had been among those political forces campaigning heavily for an extension of the franchise already during the Chartist agitation of the 1830s and 1840s.

During the second half of the nineteenth century important union leaders managed to become Members of Parliament on the Liberal Party ticket. It allowed them to discuss welfare reforms in parliament and it brought a breakthrough in terms of the political recognition of unions as legitimate interest organizations of working people. This was most obviously embodied in the Trade Disputes Act 1906, which created a legal privilege for unions whereby they could not be sued for loss of business by the affected employer during a strike over a "trade dispute" between the union and employer over terms and condition of work. Employers had to accept, often reluctantly, unions as bargaining partners. Before the First World War around 25 per cent of workers in the UK were unionized and many of those were covered by collective wage agreements (Breuilly 1992; Geary 1981).

By comparison, in Germany, on the eve of the First World War, around 13,500 collective wage agreements covered around 2 million workers, among them around one-third of all union members in Imperial Germany. Less than ten years earlier, in 1906, only 3,000 collective wage agreements existed, showing the robust advances made by unions in the decade before the outbreak of the First World War (Welskopp 2009: 29–61). Nevertheless, it was still far behind the UK, being held back also by a more illiberal state that allowed less room for unions to develop as accepted interest organizations of workers in a system of industrial relations. Just before the First World War, Germany overtook the UK to become the primary industrial nation in Europe. Yet the situation of unions was much less secure than in the UK. Many employers refused to recognize them and to enter into negotiations with them. The so-called "master in one's own house" attitude dominated the approach of many employers. At best they developed a highly paternalist system of care for their core workers, whom they paid relatively well and provided with housing, pensions and sickness insurance. The Krupp family in Essen and the Stumm family in the Saar region are some of the best-known examples of such company paternalism that refused to accept unions as legitimate interest organizations of working people (Berger 2000; Stråth 1996: 33–7).

Imperial Germany also had a parliamentary system, albeit one tempered with strong doses of monarchical rule. Nationally, it had a fairly extensive male franchise after its foundation in 1871. Yet unlike in the UK, most unions did not ally themselves to the liberal parties in Germany but rather to the Catholic Centre Party, if they were Catholics or to the Social Democratic Party of Germany. The Catholic and Socialist unions were the biggest in the country, yet both struggled to gain acceptance, as did the parties they paid allegiance to. In the so-called Kulturkampf, Catholics were deemed anti-national because of their alleged loyalty to Rome, and the Socialists were persecuted under the Anti-Socialist Laws between 1878 and 1890. Even thereafter many thought of them as "fellows without a fatherland" who could not be trusted in terms of their loyalty to the new-found German nation-state. The union movement in Germany grew strongly, almost exclusively among skilled workers, from the second half of the 1890s onwards. By 1913, 2.5 million workers were unionized. Particular workers, such as metalworkers and building workers, had seen the strongest growth during this period. National federations appeared that were centrally organized and characterized by relative remoteness from the factory workshops (Schönhoven 1980).

The exclusion of workers from political, economic and social life was also common in other parts of the industrialized world before 1914. The strength of corporate capitalism in the US ensured that the main political parties, Republicans and Democrats, sided with liberal capitalism. Attempts to build up either unions or socialist parties were crushed, if necessary with extreme violence. Many companies hired security firms that were little more than contract killers, murdering union activists and socialists. Nevertheless, in some of the industrial heartlands, such as the Mid-West, union organizations were formed successfully and, from the 1880s onwards, the Knights of Labor developed into a powerful organization in defence of workers' rights (Weir 1996).

A pan-European strike wave between 1868 and 1873 spurred workers on to organize in many parts of industrial Europe. The end of the economic depression in the mid-1890s helped unionism further to build more powerful organizations. In the most prominent European republican nation, France, only 139,000 workers were unionized by the late 1880s. A national union federation, the General Confederation of Labour (CGT), was founded only in 1895. During the first decade of the twentieth century, mass suffrage movements in many parts of Europe coincided with the inspiration of the first socialist revolution in Russia in 1905 to give many union movements across the continent a major boost. Union campaigns for the eight-hour day now became a major focus for many union movements across the world. Melbourne in Australia had been the first place to introduce the eight-hour day for certain groups of workers, especially in the building trades, after stone masons downed tools in 1856. Many other workers in Victoria and New South Wales had to wait until 1916, when in these two Australian states the eight-hour day was introduced more generally (Scalmer 2021: 219–39).

New waves of mobilization brought national governments to introduce processes of conciliation between unions and employers that were meant to reduce the negative effects of strikes and violent labour protests. Thus, for example, the 1896 Conciliation Act in the UK brought in the Labour Department at the Board of Trade in order to mediate between the two sides. Arbitration legislation appeared in numerous other Continental European countries in the late 1890s and early 1900s, indicating that repression was giving way to attempts to incorporate labour into bargaining procedures. It reflected a growing acceptance of unions as legitimate interest organizations of workers, at least where we had relatively liberal state structures in Europe. In the more authoritarian Eastern European empires, policies of repression tended to continue (Eley 2002: 69–74).

CRAFT UNIONISM AND INDUSTRIAL UNIONISM

Despite a history of repression and a lack of acceptance of unions in many parts of the industrialized world before 1914, a great number of unions were formed, especially under liberal capitalist conditions. Soon they were to debate the best form of organizing workers into unions. Many of the older unions had been craft unions reflecting a strong pride of workers in whatever skills they possessed. They organized first and foremost to protect their skills, be they masons, glovemakers, shoemakers, carpenters, moulders or cutlery grinders, to give just a few examples of the many trades that existed across the landscape of industrialization. They attempted to negotiate with employers not on behalf of all workers in the factory but only on behalf of those who possessed the same skills and who shared a particular craft. A process whereby craft consciousness was translated into class consciousness needed active shaping on behalf of union activists (Haydu 1988).

Unions, it should be remembered, are not expressions of class or class conflict per se. They are primarily organizations to represent the interests of their members. As such, their interests and their solidarities are highly particularist and sectionalist. They are, above all, directed against a perceived adversary, the employer, factory owner and capitalist. Their strength depends on numbers but also on the internal homogeneity of the groups they seek to organize. This is also why early unions almost invariably sought to organize skilled workers and demarcate their interests from the unskilled and from women. Women's employment was a potential threat, thus early craft unions developed a highly masculinized identity with a clear separation of spheres for men and women. If women threatened deskilling, craft unions were concerned about controlling qualification processes and access to training, as they could only hope to maintain their relatively strong position via such controls. Unions also showed little solidarity towards the unemployed, as they were not seen as part of the in-group that needed protecting. Indeed, they were often seen as a threat to maintaining or advancing existing terms and conditions.

There is then, in early craft unionism, more than a hint of old guild thinking and, indeed, where guilds had been strong, unions developed. In the early metal industries in England and Germany, for example, the new industrial capitalism colonized earlier corporate work forms. The new wage workers incorporated both former master craftsmen and journeymen who formed unions that looked like guilds (Magnusson 1994). Like guilds, they were about maintaining forms of occupational distinctiveness. Yet there were also important differences between guilds and craft unions. Guilds were about the maintenance of an estates-based society and, as such, they were directed against socio-political change (Farr 2000). Yet, with the advances of industrial capitalism, precisely such change had occurred and the new unions were, therefore, directed against the adversaries identified with industrial capitalism. Craft unions were, therefore, not signs of a backward guild orientation of nineteenth-century workers, but rather signs of new forms of self-organization of groups of workers under conditions of industrial capitalism. Their associationalism had been voluntary in a way that guild membership never was. A class-based, not an estates-based, society determined the exclusiveness of early craft unions.

They were strongest where they organized workers that were non-expendable, irreplaceable and had considerable power over work processes in the factories. Their strength was further enhanced by developing relief and benevolent support systems that strengthened solidarity within the group. The more they could control the labour market, the more they could protect their own position within the labour market, which they sought to do with controls over labour exchanges, strikes, the founding of clubs and the sharing of information via a nascent union press. Collecting dues and recruiting more and more members willing to pay these dues also strengthened the early union movement (Boch 1985; Boll 1992; Eisenberg 1986; Welskopp 2019: 314–26).

This already indicates that a less particularist and more universalist class consciousness that sought to unite industrial workers as industrial workers was not something that came "naturally" or organically to workers. It had to be constructed, and one of the most powerful ideologies developing the idea that workers, regardless of skill, had much in common with other workers was socialism. Hence, it was the socialist unions that were often at the forefront of debates around the formation of industrial unions. Here the idea was that all workers, regardless of skill and profession, working in one industry, say the mining industry or the iron and metal industries, should join one union that could then represent all workers in that industry vis-à-vis the employers. This labour unionism was very much based from the outset on the principle that there was strength in numbers. Industrial unions were able to unite more workers in fewer unions than craft unions. On this aspect, they were, thus, potentially more powerful organizations.

Where the oldest unions had been able to establish themselves (i.e. in the UK), they often took the form of craft unions and here they survived and showed themselves resilient to change for a long time, not least because they were often successful as craft unions. This was the case especially in small- and medium-sized workshops. In Germany, by contrast, industrial unionism had made great strides already before 1914. Whereas in the 1880s the vast majority of union organizations in Imperial Germany were still craft-based, in the decades from the 1890s to the 1910s unions turned increasingly towards organizing the semi- and unskilled, in particular industries. During the 1890s, the General Commission of the Free Unions under the leadership of Carl Legien became a key motor for industrial unionism in Germany. The metalworkers and the woodworkers were the most outspoken in favour of the industrial principle and eventually, after considerable hesitation, other workers in other industries followed suit because these two unions had shown that organizing workers industry-wide did not lead to a weakening of union power.

On the eve of the First World War, industrial unions comprised 70 per cent of total union membership in Germany. And in the motherland of craft unionism, in the UK, the so-called "new unionism" that emerged in the 1890s also meant a significant push towards industrial unionism. In the three famous strikes in London of 1888 and 1889, the dockers, the gas workers and the matchgirls had demonstrated that the unskilled could be organized very effectively. The railway workers and the agricultural workers were among the first to build national industrial unions in the UK. General labourers' unions followed and many of the traditional craft unions that persisted became less exclusive about their membership (Mommsen & Husung 1985).

The movement from craft unions to industrial unions was a slow, uneven process but it occurred everywhere in the industrialized regions of the world. In Europe the new types of industrial unions were at the heart of new forms of labour militancy that characterized the European-wide strike waves of 1904 to 1907 and 1910 to 1913, as well as the massive strikes during and immediately after the First World War (Haimson & Sapelli 1992). In the US, the Knights of Labor were one of the earliest and most outspokenly programmatic industrial unions seeking to organize all industrial workers, regardless of ethnicity, gender and craft after 1869 (Fink 1985). The American Federation of Labor (AFL), which came into being in 1886, upheld for longer the principle of craft unionism and was far less universal in orientation than the Knights. Yet even the latter still predominantly organized the semi-skilled rather than the unskilled. Unlike many AFL unions, the Knights did not, however, explicitly exclude the unskilled and Black workers (Sieger 2007).

As in the UK, railwaymen in the US were at the forefront of industrial unionism. Eugene Debs organized the American Railway Union in 1894, which then undertook the immensely symbolical Pullman strike of the same year (Schneirov, Stromquist & Salvatore 1999). The Western Federation of Miners under Bill Haywood became another early champion of industrial unionism in the US. The revolutionary syndicalist Industrial Workers of the World, formed in Chicago in 1905, promoted the idea of the "one big union" (Dubowsky 2000). Revolutionary industrial unionism also took hold in Australia (Burgmann 1995). One of the major French union organizations, the CGT, was also anarcho-syndicalist in orientation before it turned communist in the inter-war period. It was an early champion of industrial unionism in France (Vigna 2021).

IDEOLOGICAL DIVISIONS

The issue of how best to organize labour unionism was intricately connected to ideological divisions in the union movement. More and more workers in the industrialized world flocked to the ideas of socialism, so that socialist and anarcho-syndicalist unions became prominent in many places. Before 1914 the socialist variants were strongest in Germany and Scandinavia, but socialism also made many converts among the new unionism in the UK from the 1890s onwards, which saw a slow but steady move away from Liberalism and towards the newly founded Labour Party after 1900. When the Miners' Federation of Great Britain joined the Labour Party in 1908, it was a highly symbolic move demonstrating that the political allegiances of unionism in the UK were shifting (Minkin 1991). Socialist trade unionists were often members of socialist parties that in turn tended to be committed to Marxism as their official ideology before 1914. Yet, their desire to see capitalism overthrown tended to be tempered with their concern to improve the conditions of workers under capitalism. Hence, socialist trade unionists often belonged to the reformist sections of socialist party politics seeking to achieve practical reforms under capitalism that might lead step by step to an overcoming of capitalism.

More orthodox Marxists and anarcho-syndicalists, by contrast, remained committed to the revolutionary overthrow of capitalism. Orthodox Marxism, represented before 1914 above all by Karl Kautsky, promoted a "revolutionary attentism" that did little to bring about the revolution and, instead, awaited the inevitable collapse of capitalism (Groh 1973). Left-wing socialists, such as Rosa Luxemburg, and anarcho-syndicalists preferred a much more activist stance promoting the general strike and mass strikes wherever possible to aid the collapse of capitalism. Especially among young militants, syndicalism seemed the best way forward to bring about a more just economic and social system. In France, the main union movement was syndicalist and syndicalism also was a considerable force in Spain and Italy as well as in the US and Argentina (Darlington 2013). The French CGT was often the inspiration behind syndicalist organizations elsewhere. Whereas the socialists put much emphasis on political democracy and sought to capture the state via the ballot box, in order to introduce social and economic reform, the anarcho-syndicalists put more emphasis on economic democracy and called on its supporters to mobilize in the factories to achieve concrete benefits for workers on the shop floor. Syndicalists opposed the principle of (indirect) representation and were in favour of local direct forms of democracy, wherever possible. They saw unions as important building blocks for a post-capitalist social order (Hirsch & van der Walt 2010; van der Linden & Thorpe 1990).

While syndicalists and socialists were opposed to liberal capitalism on principle and sought to overcome it, Catholic and Liberal unions accepted the liberal capitalist frame but sought to embed it in a way that made it look less unpleasant and more socially responsible. Catholic trade unionists often argued in terms of Christian ethics whereby capitalist greed had to be tempered by working-class organizations such as unions that could empower workers and enable them to get their fair share of the profits a company was making. The Papal Encyclical *Rerum Novarum* of 1891 indicted capitalism for its greed and laid the foundation for Catholic social teaching. Social justice, subsequent Catholic labour movement activists argued, could be brought about with the help of Catholic unions that developed strongly in many Catholic countries from the 1890s onwards. Unions could help to ensure the dignity of work and counter the sentiments of alienation that were identified with capitalist forms of production as well as bring about a more just distribution of income and wealth in society (van Voss, Pasture & de Maeyer 2005).

Liberals argued that interest organizations such as unions were part and parcel of liberal political systems that sought to organize different interests in society in a way that allowed for their peaceful regulation in diverse spheres of society. Unions and employers' federations represented different interests of workers and employers that needed to be brought into an equitable balance through a process of negotiation. From the last third of the nineteenth century onwards, a social liberalism developed in Europe that sought to align the classic concerns of liberalism with freedom and the new concerns surrounding the emerging "social question" in Europe. At the heart of this progressive liberalism stood the notion of a more interventionist state that would take proactive measures to guarantee a more socially just order, including a welfare state and legal protection for workers' organizations, such as unions. Social liberalism was about the reconciliation of interests rather than the socialist notion of the incompatibility of interests. Negotiation combined with arbitration was to replace the class struggle. While explicitly Liberal unions were a considerable force only in nineteenth-century the UK, social liberalism was an ideology that came to influence social democratic unions in the twentieth century and it also was an important influence upon the social reforms of the UK Liberal governments just before the First World War as well as the "New Deal" in the US that followed the Great Depression of the late 1920s (Freeden 1986; Rodrígues García 2010).

UNIONS AS DEMOCRATIZING FORCES IN THE INTER-WAR PERIOD

Unions had been in the forefront of those political forces seeking to extend the franchise as they recognized the vote as an important means of empowering working people. As part of broader labour movements, they sought, above all, to democratize both the political and the economic sphere. Franchise reform had taken place in many of the industrializing countries during the long nineteenth century, but a real burst of democratization coincided with the end of the First World War. In countries introducing greater political democracy, unions began to thrive. If we take the example of Germany, we have seen above how the unions were very much kept outside the company gates in Imperial Germany. When the revolution of 1918 swept away the edifice of this state and replaced it with the Weimar Republic, the new republican frame empowered the unions in major ways. Not only did employers now have to recognize unions as legitimate bargaining partners in a new industrial relations system, but the introduction of works councils after 1921 meant that workers inside the companies now had their own representatives who could speak on behalf of the workers to the company management. Economic democracy, indeed, now augmented political democracy. The extension of the welfare state was the logical conclusion of a greater empowerment of workers in the Weimar Republic (Milert & Tschirbs 2012).

With the collapse of the three Eastern European empires at the end of the First World War, a proliferation of new nation states came into being alongside a revolutionary wave that often sought to empower workers and their organizations. Where industrial workers were a prominent feature of regional economies, such as in parts of Austria, Czechoslovakia and Poland, labour unionism flourished and workers sought to extend their rights and to improve wages and working conditions. Like in the Weimar Republic, the Austrian republic saw a massive increase in union membership in the years after the First World War. The mobilization of workers made the Austrian unions a powerful instrument for the setting up of a strong welfare regime in the first Austrian republic (Klenner 1959). In Poland, unions had already formed in the late nineteenth century, when the country was still divided between Russia, Austria-Hungary and Prussia. They played an important part in the country's industrial centres, such as Łódź after 1918 (Sysiak *et al.* 2018). However, when Poland turned to right-wing authoritarianism after 1926 the government under Jósef Piłsudski did everything to weaken the nascent union movement. In the new-found Czechoslovakia after 1918, unions could also look back on a history commencing under the Habsburg empire. The revolutionary turmoil in its industrial centres led to a huge surge in union membership and the unions developed into an important bulwark for social reform in the successful democracy that was established in Czechoslovakia in the inter-war period. By 1938, 1.7 million workers were members in ideologically divided unions, the biggest of which were social democratic. Metalworkers, railroad workers and textile workers were well organized into powerful union movements. Their strong position in society in the inter-war period was not the least due to the fact that the state had given the unions the right to administer the payment of unemployment benefits (Čapka 2008).

In many parts of Western Europe, labour unionism also flourished in the 1920s and 1930s. In the UK, the unions remained powerful organizations that put their weight behind the Labour Party and contributed in no small part to the formation of the first two Labour governments in the UK in 1924 and 1929, respectively. Yet, despite its strength, the inter-war period was a time of disappointment for the UK's union movement. Its hopes for the nationalization of key industries at the end of the First World War had been dashed. Even where official state commissions, such as the Sankey Commission, had recommended nationalization, in this case of the coal industry, governments refused to act and liberal capitalism remained very much in place and in control of the economic system. The 1926 General Strike ended in failure and the Depression of the late 1920s and early 1930s significantly weakened the union movement. The inter-war years in the UK were dominated by Conservative governments hostile to trade unionism, and major social progress on the road to greater economic and social democracy was stifled (Reid 2005).

France and Italy saw the emergence of mass communist unions in the inter-war period. Communism was a new political force on the map of Europe that was closely aligned with the victory of Leninism in Russia and the formation of the Soviet Union. After the successful Bolshevik revolution, Lenin forced a split in the socialist movement between those willing to follow Leninism and those opposed to it. In the inter-war period, this translated into a split between social democratic and communist parties, and in the world of labour unionism a further ideological split (Tosstorff 2016). The "Biennio Rosso" in Italy at the end of the First World War saw factory occupations in which workers took control of the production sites and the management of companies in many parts of industrial Italy. It threatened the liberal capitalist order and was accompanied by a violent backlash of employers, in alliance with the state and the Catholic Church, which was eventually to result in the victory of fascism in 1923 (Pepe 1976). In France, the immediate postwar years saw massive economic unrest and several strike waves in different sectors of the industry. The general strike of 1920 was, however, a failure and in its wake many labour unionists were blacklisted by employers and fired. Union density, which had reached 25 per cent in 1919, declined sharply thereafter but recovered to reach another high of 19 per cent in 1930. A strong union movement was also a bulwark of a republican France electing a popular front government in 1935 and resisting the Europe-wide trend for right-wing dictatorships (Magraw 1992).

While in many parts of Europe liberal-democratic forms of capitalism seemed severely threatened in the inter-war period, in the US it continued to rule supreme. Yet the Depression resulted in a rethinking of liberal capitalism, which saw attempts to strengthen workers' rights and unions as part of the so-called "New Deal" that was to usher in a more embedded capitalism in the US. The National Labor Relations Act 1935, also known as the Wagner Act (after Senator Robert F. Wagner who wrote the Act), legalized independent unions and banned company unions. It gave unions the right to engage in collective bargaining and to strike. It, thereby, significantly extended political democracy to incorporate forms of economic democracy into the political system of the US. A National Labor Relations Board was set up that was to monitor violations of labour law and ban unfair labour practices as well as establish the rules for collective bargaining procedures (Morris 2005).

REPRESSION UNDER FASCISM AND RIGHT-WING AUTHORITARIANISM

If new forms of economic and social democracy came to be established under liberal capitalism in parts of the industrialized world during the inter-war period, many parts of Eastern Europe witnessed a marked turn to right-wing authoritarian rule in the 1920s. Fascism first made its appearance in Italy, where it gained power in 1923. While the leader of Italian fascism, Benito Mussolini, was a former socialist, and some of the anarcho-syndicalist supporters of Italian unionism became supporters of Italian fascism, the union movement had no particular say over economic and social relations under fascism. Independent labour unionism was soon persecuted by the fascist state and workers' independent interest organization was impossible. Yet the fascists wooed the workers with a mixture of nationalism, imperialism and the idea that they were a recognized party of the national community in Italy. Particular schemes, such as the National Afterwork Club (Opera Nasionale Dopolavoro), aimed to provide better leisure facilities for workers. Its sports activities proved especially popular and, by 1936, 80 per cent of salaried workers (but only 40 per cent of industrial workers) were members (de Grasia 1981). The National Socialists also implemented schemes, such as Strength Through Joy and Beauty of Labour, to win workers over to their cause. They provided cheap holidays and attempted to make workplaces better and more beautiful places of work (Baranowski 2004; Rabinbach 2018). Yet, ultimately, all of these schemes could not hide the simple fact that capitalists had once again taken full control of the economic system under fascism.

The repression of independent trade unionism was a hallmark of right-wing authoritarian and fascist regimes everywhere. In National Socialist Germany, the independent unions were crushed on 2 May 1933 and replaced with a National Socialist organization, the German Labour Front, which allegedly represented the interests of workers vis-à-vis employers (Smelser & Ley 1988). In reality, however, most employers were happy to take full control of the factories again and they were certainly not sorry to see independent labour unionism crushed, even if they rarely belonged to the most fervent supporters of National Socialism.

Francoism in Spain and the Estado Novo in Portugal also ended all independent labour unionism in their respective countries, persecuting socialists, communists and anarcho-syndicalists in equal measure (Balfour 1989; Borges Santos & Aronne de Abreu 2016). They either were murdered, disappeared behind prison walls or they went into exile. Right-wing authoritarian capitalism, whichever form it took, certainly meant serious setbacks for independent labour unionism when compared with liberal capitalist orders.

UNIONS UNDER COMMUNISM IN THE SOVIET UNION

Intriguingly, the fate of independent labour unionism was not much different in the Soviet Union (see also Chapter 10), the self-declared motherland of the international proletariat in which liberal capitalism had been abolished and socialism was allegedly under construction. Lenin always had a rather negative view of unions as reformist organizations that failed to work for the revolution and for the abolition of capitalism. However, during the Bolshevik revolution and its immediate aftermath, many unions were formed and took part in factory occupations promoting ideas of workers' self-management. When the Bolsheviks cemented their hold on power, the unions were transformed to become transmission belts of the decisions of the leadership of the Communist Party into the factories. The political system of democratic centralism de facto amounted to a Communist Party dictatorship in which there was no space for independent interest organizations of workers. The Communist Party represented the interests of workers. As such it had the right to dictate the general direction of political, social, economic and cultural developments. It was the task of the unions to ensure that the policies devised by the communist parties were communicated to the workers and put into practice (Filtser 1986).

Unions were given considerable power in the Soviet Union. In the factories they were a force to be reckoned with by management. They also began organizing social care and holidays for the workers and as such they became an important part of the social fabric of the Soviet Union. In the 1920s and 1930s, they organized literacy classes and evening schools for workers. A large part of the Soviet welfare state was administered through the unions (Hoffmann 2011). At the same time, unions sought to represent a heroic and romanticized image of workers that was meant partly to motivate workers. According to communist ideology, workers had brought about the fall of capitalism and they, therefore, deserved a special place in the communist state.

At the same time, work was a duty under socialism and part of a socialist citizenship. Workers, at least in theory, were no longer exploited at the workplace but fulfilled their obligations and exercised their rights as socialist citizens. Educating the workers ideologically became the prime task of unions in the Soviet Union. And the unions were also used by the secret police to spy on workers and report any behaviour that was critical of the Communist Party and its leadership (Chase 1987; Siegelbaum & Suny 1994). But the one thing they were not is independent interest organizations of workers who could decide to represent workers against the policies devised by the Communist Party.

It is quite a different matter if we talk about communist unions outside of the Soviet Union. Next to anarcho-syndicalist unions, they belonged to the most militant organizations of workers under liberal capitalism. They thrived where local traditions of union militancy persisted, such as in the mining pits of the South Wales valleys, or in the Nord Pas-de-Calais (Francis & Smith 1980; Kriegel 1968). Their locally rooted anti-capitalism and radicalism went hand in hand with a commitment to internationalism and a loyalty to the Soviet Union, but at the same time there were considerable tensions between the Soviet desire to direct a worldwide communist movement from Moscow and the fiercely independent local commitment to communism. Where mass communist parties emerged outside of the Soviet Union in the inter-war period, such as in Germany and France, communist unions underpinned communist workers' subcultures, propagating the fall of liberal capitalism. Where communist parties remained weak, such as in the UK, communists focused on their strength in the workplaces and the union movements (Fishman 2010). However, while much communist unionism remained committed to the overthrow of capitalism, in systems where collective bargaining procedures existed in industrial relations systems that recognized unions, they also had to engage on a day-to-day basis with negotiations and procedures that drew unions into the mechanisms of operation of liberal capitalist systems. This invariably produced tensions between their revolutionary ambitions and their functioning as unions.

UNIONS IN THE COLONIAL AND SEMI-COLONIAL PERIPHERY

So far, this chapter has almost exclusively concerned itself with the development of labour unionism under conditions of liberal capitalism in the industrialized Global North. Yet unions also developed in parts of the colonial or semi-colonial Global South before 1945, especially in some of the White settler colonies associated with the British Empire. In Australia, as mentioned above, unions had made advances towards achieving an eight-hour day well before 1914. The idea of labour unionism migrated to Australia within colonial contexts. The first unions were already founded in Sydney and Hobart in the 1820s and by mid-century several hundred union organizations were in place in the different states of Australia. Many were craft-based and focused on welfare assistance to their members. Unions were also very much behind the formation of Australian Labor Parties (ALP) in the 1890s, which united to form one federated ALP in 1901. In the early twentieth century, Australia developed into a laboratory for the welfare state. Compulsory arbitration procedures were introduced in the industrial relations system of the country in the inter-war years, and a central union confederation, the Australian Council of Unions, came into existence in 1927. Minimum wages were introduced and, although the unions suffered under the Great Depression of the 1930s, like they did elsewhere, Australia remained a bulwark of a liberal capitalist system that had accommodated labour unionism and allowed for the independent representation of workers through unions who had achieved considerable collective advances on behalf of workers (Scalmer 2006).

Within White-settler colonialism, unionism also thrived elsewhere, especially where there were centres of industrial work. In South Africa, many of the early unions that began to form from the 1880s onwards were for Whites only. They promoted employment policies based on racial discrimination. The Industrial Workers of Africa, founded in 1917, was the first Black union and it was inspired, like many Black unions in South Africa during the inter-war period, by syndicalist ideas organizing on industrial rather than craft principles. The South African Trades Union Council was formed in 1924, organizing both Black and White workers. Yet racism remained a strong dividing factor among workers in South Africa. During the so-called "Rand rebellion" of 1922, a strike that culminated in an armed uprising of White miners in the Witwatersrand region of South Africa, a slogan of the miners read: "Workers of the world unite and fight for a white South Africa." This had been inspired by the mining bosses' attempts to introduce cheaper Black labour into the mines of the Witwatersrand. Communist and syndicalist unions tried to counter the racist sentiments that ran high among White miners and led to several pogroms against Black workers during the rebellion (Krikler 2006).

The textile and jute mills of Bombay and Calcutta also saw the emergence of unions in India during the 1850s. The British colonial government set up a factory commission in 1879 to study the social problems accompanying industrialization in India. In 1891 the Indian Factory Act was passed but it turned out to be largely ineffective in protecting workers. There was an explosion of union formations in the 1890s, and the 1910s saw major strikes by printers, postal workers and textile workers. In 1926 a Unions Act was passed in the Central Legislative Assembly giving unions a minimum of legal protection. Already, in 1920, the All-India Trade Union Congress had been formed.

Like elsewhere, the union movement in India fought for recognition, better wages and working conditions and shorter working days (Sen 1977).

From the beginning of the twentieth century onwards labour unionism also developed strongly in Argentina. Continued violent state repression, for example during the "Tragic Week" in 1919, contributed to a strong revolutionary syndicalist orientation of a large part of the Argentinian union movement. During the right-wing dictatorship of the 1930s, sections of the syndicalist unions cooperated with the dictatorship in order to achieve concrete benefits for their members, but the socialist unions remained steadfastly opposed and suffered much persecution. In 1936 a general strike erupted and was again violently repressed. In the immediate post-Second World War years, Peronism in Argentina was based on a close alliance of Peron with the unions and a corporatist arrangement that allowed for major advances on behalf of Argentine workers (Alexander 2003).

In Brazil, early centres of a strong labour unionism were the industrial centres of Rio de Janeiro, Sao Paulo and Rio Grande do Sul, where strong union confederations developed that united to form the Brazilian Workers' Confederation in 1908. Socialists and syndicalists were the strongest factions among Brazilian labour unionists. The strike of textile workers in 1903 had a strong symbolic power of organizing the workforce, despite the fact that it was harshly repressed and none of its demands were met. In 1912 and 1913 a strike wave in several industries saw further advances of trade unionism in their struggles for higher wages and better working conditions (Alexander 2003).

In China, labour unionism emerged as a force to be reckoned with in the inter-war period. In early industrial centres, such as Canton, Hong Kong, Shanghai and Wuhan, workers formed unions that came together under the umbrella of the All-China Federation of Unions in 1925. The influence of the Chinese Communist Party over the federation was strong and it was increasingly restricted by the nationalist government of Chiang Kai-Shek (Chesneaux 1968).

In neighbouring Japan, a rapid industrialization process under the Meiji dynasty during the last third of the nineteenth century saw the emergence of union organizations, especially in mining, textile and metalworking. However, faced with state oppression and the frequent use of the military against striking workers, no permanent union organizations could be established and union density rates remained low. A more durable organization came into being only in 1921: the Japan Federation of Labour. It was ideologically divided between left- and right-wing labour unionists, leading to a split of the movement in 1925. The legal status of unions remained precarious and they never won the right to enter collective bargaining procedures with the powerful employers, who, in turn, had set up company unions very successfully. The Imperial government of Japan disbanded all independent unions in 1940 and brought in the Industrial Association for Serving the Nation, which was a government institution to optimize the functioning of the economy during the Second World War (Marsland 2019).

It has only been possible here to give a brief glimpse of the development of the union movement in the colonial and semi-colonial periphery of liberal capitalism, but it should have become clear that unions were by no means restricted to the Global North but followed a global pathway of industrialization that extended to many places in the Global South already in the nineteenth and first half of the twentieth century. The setting up of unions followed a complex pattern of adaptation of union models first developed in the European metropoles. Debates about craft versus industrial unionism and ideological splits can be found in the colonial and semi-colonial contexts as well, and in many cases what happened here in turn also influenced the development of labour unionism in the colonial centres. Here, as elsewhere in global history, the empire was "writing back", which makes it necessary to study labour unionism in its global entanglements rather than in its national specifications (Hall & Rose 2006).

PREPARING THE SOCIAL DEMOCRATIC HIGHPOINT: UNIONS IN SWEDEN BEFORE 1945

While there was undoubtedly a backlash of right-wing authoritarianism of global dimensions in the inter-war period that deleteriously affected unions, the one place where unionism continued to thrive significantly in the inter-war period was Scandinavia (see also Chapter 9). Its countries were overwhelmingly Protestant and its labour movements were thoroughly social democratic. State formations were liberal with a steady expansion of the franchise and representative forms of government in the course of the nineteenth and early twentieth centuries. Powerful union movements emerged, often alongside strong social democratic parties (Hilson, Neunsinger & Vyff 2017).

Nowhere was this more the case than in Sweden, where a political alliance between the Social Democrats and the Agrarian Party, looking after the interests of the farmers, paved the way for a long social democratic hegemony lasting from the 1930s into the 1980s. Political democracy here was accompanied by a steady expansion of economic and social democracy. A powerful corporatism emerged where the economic sphere was dominated by strong employers' federations and strong unions who, under the guidance of the state, regulated economic relations in a negotiated way that gave unions a powerful say over economic developments, welfare and workplace practices (Williamson 1989: 150). In a sense the kind of corporatism that emerged under liberal capitalist tutelage in Sweden was one in which the unions as part of a wider labour movement used the state and its political power to advance workers' interests (Korpi 1983: 46–50).

As far back as 1903, local labour exchanges in Sweden sought to regulate social problems created by industrialization – with worker and employer representatives equally represented here. They did not prevent relatively high levels of industrial conflict with employers frequently using the lockout to crush strikes. Highly centralized employer and union federations emerged in 1898 (Landsorganisationen, LO) and 1902 (Svenska Arbetsgivareföreningen, SAF) that moved towards conciliation procedures in a series of board-type arrangements that dealt with a variety of social problems, including work accident insurance and hours of work regulations. The Basic Agreement of 1938 between the LO and SAF was to regulate industrial relations and arbitration procedures in times of industrial conflict for much of the remainder of the twentieth century.

When the long period of Social Democratic government in Sweden began in 1932, unions became directly involved in policy-making and became a highly recognized pillar of Swedish society. Thus, the LO took responsibility for administering unemployment insurance, which led to high unionization rates, as workers saw manifest benefits in membership of unions. At the same time, the Social Democratic state in Sweden began to recognize that it would have to provide the frame for an internationally competitive Swedish capitalism. The labour movement had dropped plans for the comprehensive nationalization of industry already in 1932, and it began to work closely with the SAF's Industrial Research Institute in order to devise economic policies that would benefit Swedish capitalism. Under this system, the union movement was part and parcel of a system of liberal capitalism that would work on behalf of the many and not the few. It would provide the basis for the famous Swedish "people's home" (folkhem), an ideal popularized by the Social Democratic Prime Minister of Sweden between 1932 and 1946, Per Albin Hansson (Fulcher 2002: 279–94).

CONCLUSION

Under conditions of liberal capitalism between the middle of the eighteenth century and the middle of the twentieth century, a variety of factors determined to what extent unions could become effective interest organizations on behalf of workers in the political, social and economic lives of their respective societies (Berger 2019: 13–36). The first concerned the actual industrialization process and the extent to which older skills were still needed in the new factory production. Defending these skills was often the basis on which early craft unions organized. However, over the course of time, it also made increasing sense to organize workers by industry, and a definite trend from craft to industrial unionism can be identified – with major advances towards industrial unionism occurring in the first half of the twentieth century.

A second factor concerned the regional character of industrialization. The regional rather than national development of capitalist industrialization meant that unions formed regionally, and the regional conditions for workers to form interest organizations were, therefore, of crucial importance. A third factor concerned the strength of their opponents. Where employers' federations were strong and hostile to unionism, there they found it very difficult to gain acceptance and recognition as interest organizations of workers. A fourth factor concerned the character of the state. The industrial bourgeoisie was often influential in the state structures that formed a political context for industrialization. Unions often developed national umbrella organizations that could address the unions' concerns vis-à-vis national and imperial state structures.

To what extent they were successful in doing so depended upon how much influence they could take into these state structures, which in turn depended on the character of the state. Did it operate an extensive franchise that enfranchised working people? Did the unions manage to have political allies? To what extent were they able to make their voices heard in state institutions? Liberal state structures such as the ones in the UK and Sweden were far more capable of integrating unions and their concerns on behalf of workers than more authoritarian states.

Although the repressive potential of the liberal capitalist state should not be underestimated, it seems clear that the unions were able to benefit from liberal state structures and were also among the foremost organizations to work towards a greater democratization of these state structures and an extension of political democracy to the social and economic spheres. Yet, as we have seen, especially in the inter-war period, such democratization had its limits in the willingness of liberal capitalism to accommodate working-class interests. In many parts of Europe, right-wing authoritarian and fascist dictatorships curbed the freedom of independent labour unionism or ended it completely. Liberal capitalism turned illiberal in many places during the inter-war years, which had a deep impact on the development of unions. However, a strong social democratic labour movement in Sweden laid the groundwork for a form of embedded liberal capitalism in which workers and their organizations would have an important say in steering markets in a way that was to benefit the many and not the few. The promotion of a capitalism with a "human face" incorporated notions of social partnership between capital and labour that were significantly extended in the three decades following the Second World War (Berger 2002: 335–52; see also Chapter 22).

REFERENCES

Alexander, R. 2003. *A History of Organized Labor in Argentina*. Westport, CT: Praeger.

Balfour, S. 1989. *Dictatorship, Workers and the City: Labour in Greater Barcelona Since 1939*. Oxford: Oxford University Press.

Baranowski, S. 2004. *Strength Through Joy: Consumerism and Mass Tourism in the Third Reich*. Cambridge: Cambridge University Press.

Berger, S. 2002. "Social partnership, 1880–1989: the deep historical roots of diverse strategies". In *Policy Concertation and Social Partnership in Western Europe. Lessons for the 21st Century*, edited by S. Berger & H. Compston, 335–52. Oxford: Berghahn.

Berger, S. 2019. "Workers' participation: comparative historical perspectives from the nineteenth century to the end of the Cold War". In *The Palgrave Handbook of Workers' Participation at Plant Level*, edited by S. Berger, L. Pries & M. Wannöffe, 13–36. Basingstoke: Palgrave Macmillan.

Berger, S. 2000. *Social Democracy and the Working Class in Nineteenth and Twentieth-Century Germany*. Harlow: Longman.

Boch, R. 1985. *Handwerker-Sosialisten gegen Fabrikgesellschaft: lokale Fachvereine, Massengewerkschaft und industrielle Rationalisierung in Solingen 1870–1914*. Göttingen: Vandenhoeck & Ruprecht.

Boll, F. 1992. *Arbeitskämpfe und Gewerkschaften in Deutschland, England und Frankreich: ihre Entwicklung vom 19. zum 20. Jahrhundert*, Bonn: J. W. H. Diets.

Borges Santos, P. & L. Aronne de Abreu (eds) 2016. "Authoritarian states and corporatism in Portugal and Brazil". *Portuguese Studies* 32(2): 125–261.

Breuilly, J. 1992. *Labour and Liberalism in Nineteenth-Century Europe: Essays in Comparative History*. Manchester: Manchester University Press.

Burgmann, V. 1995. *Revolutionary Industrial Unionism: The Industrial Workers of the World in Australia*. Cambridge: Cambridge University Press.

Čapka, F. 2008. *Odbory v českých semích v letech 1918–1948*. Brno: Masarykova universita.

Chapman, S. 1967. *The Early Factory Masters: The Transition to the Factory System in the Midlands Textile Industry*. London: David & Charles.

Chase, W. 1987. *Workers, Society and the Soviet State: Labor and Life in Moscow, 1918–1929*. Urbana, IL: University of Illinois Press.

Chesneaux, J. 1968. *The Chinese Labour Movement*. Stanford, CA: Stanford University Press.

Darlington, R. 2013. *Radical Unionism: The Rise and Fall of Revolutionary Syndicalism*. New York: Haymarket Books.

Dubowsky, M. 2000. *We Shall Be All: A History of the Industrial Workers of the World*. Urbana, IL: University of Illinois Press.

Eisenberg, C. 1986. *Deutsche und englische Gewerkschaften. Entstehung und Entwicklung bis 1878 im Vergleich*. Göttingen: Vandenhoeck & Ruprecht.

Eley, G. 2002. *Forging Democracy: The History of the Left in Europe, 1850–2000*. Oxford: Oxford University Press.

Farr, J. 2000. *Artisans in Europe 1300–1914*. Cambridge: Cambridge University Press.

Faue, E. 2017. *Rethinking the American Labor Movement.* New York: Routledge.

Filtser, D. 1986. *Soviet Workers and Stalinist Industrialization: The Formation of Modern Soviet Production Relations, 1928–1941.* Armonk, NY: Sharpe.

Fink, L. 1985. *Workingmen's Democracy: The Knights of Labor and American Politics.* Urbana, IL: University of Illinois Press.

Fishman, N. 2010. *Arthur Horner: A Political Biography, 2 vols.* London: Lawrence & Wishart.

Francis, H. & D. Smith 1980. *The Fed: A History of the South Wales Miners in the Twentieth Century.* Cardiff: Cardiff University Press.

Freeden, M. 1986. *The New Liberalism: An Ideology of Social Reform.* Oxford: Oxford University Press.

Fulcher, J. 2002. "Sweden in historical perspective: the rise and fall of the Swedish model". In *In Policy Concertation and Social Partnership in Western Europe: Lessons for the 21st Century,* edited by S. Berger & H. Compston, 279–94. Oxford: Berghahn.

Geary, D. 1981. *European Labour Protest, 1848–1939.* London: Macmillan.

de Grasia, V. 1981. *The Culture of Consent: Mass Organisations of Leisure in Fascist Italy.* Cambridge: Cambridge University Press.

Groh, D. 1973. *Negative Integration und revolutionärer Attentismus: die deutsche Sosialdemokratie am Vorabend des ersten Weltkrieges.* Berlin: Propylaen.

Haimson, L. & G. Sapelli (eds) 1992. *Strikes, Social Conflict and the First World War: An International Perspective.* Milan: Feltrinelli.

Hall, C. & S. Rose (eds) 2006. *At Home with the Empire: Metropolitan Culture and the Imperial World.* Cambridge: Cambridge University Press.

Hall, P. & D. Soskice (eds) 2001. *Varieties of Capitalism: The Institutional Foundations of Comparative Advantage.* Oxford: Oxford University Press.

Haydu, J. 1988. *Between Craft and Class: Skilled Workers and Factory Politics in the United States and Britain, 1890–1922.* Berkeley, CA: University of California Press.

Hilson, M., S. Neunsinger & I. Vyff (eds) 2017. *Labour, Unions and Politics under the North Star, 1770–2000.* Oxford: Berghahn.

Hirsch, S. & L. van der Walt (eds) 2010. *Anarchism and Syndicalism in the Colonial and Postcolonial World: The Praxis of National Liberation, Internationalism and Social Revolution.* Leiden: Brill.

von Hodenberg, C. 1997. *Aufstand der Weber. Die Revolte von 1844 und ihr Aufstieg zum Mythos.* Bonn: J. W. H. Diets.

Hoffmann, D. 2011. *Cultivating the Masses: Modern State Practices and Soviet Socialism, 1914–1939.* Ithaca, NY: Cornell University Press.

Klenner, F .1959. *The Austrian Union Movement.* Brussels: ICFTU.

Korpi, W. 1983. *The Democratic Class Struggle.* London: Routledge.

Kriegel, A. 1968. *Les Communistes Français: Essai Ethnographie Politique.* Paris: Editions du Seuil.

Krikler, J. 2006. *Rand Revolt: The 1922 Insurrection and Racial Killings in South Africa.* Johannesburg: Jonathan Ball.

Laybourn, K. 1997. *A History of British Unionism, 1770–1990.* Stroud: Sutton.

van der Linden, M. & W. Thorpe (eds) 1990. *Revolutionary Syndicalism: An International Perspective.* Aldershot: Scholar Press.

Magnusson, L. 1994. *The Contest for Control: Metal Industries in Sheffield, Solingen, Remscheid and Ekilstuna During Industrialization.* Oxford: Berghahn.

Magraw, R. 1992. *A History of the French Working Class: Workers and the Bourgeois Republic.* London: Routledge.

Marsland, S. 2019. *The Birth of the Japanese Labor Movement: Takano Fusatarō and the Rōdō Kumiai Kiseikai.* Honolulu: University of Hawaii Press.

Milert, W. & R. Tschirbs 2012. *Die andere Demokratie. Betriebliche Interessenvertretung in Deutschland, 1848–2008.* Essen: Klartext.

Minkin, L. 1991. *The Contentious Alliance: Unions and the Labour Party.* Edinburgh: Edinburgh University Press.

Mommsen, W. & H. Husung (eds) 1985. *The Development of Unionism in Great Britain and Germany 1880–1914.* London: Allen & Unwin.

Morris, C. 2005. *The Blue Eagle at Work: Reclaiming Democratic Rights in the American Workplace.* Ithaca, NY: Cornell University Press.

Ogilvie, S. & M. Cerman (eds) 1996. *European Proto-Industrialization.* Cambridge: Cambridge University Press.

Pepe, A. 1976. *Movimento Operaio e Lotte Sindacali, 1880–1922.* Torino: Loescher.

Pollard, S. 1981. *Peaceful Conquest: The Industrialization of Europe, 1760–1970.* Oxford: Oxford University Press.

Rabinbach, A. 2018. *The Eclipse of the Utopias of Labor.* New York: Fordham University Press.

Reid, A. 2005. *United We Stand: A History of Britain's Unions.* London: Penguin.

Rodrígues García, M. 2010. *Liberal Workers of the World Unite? The ICFTU and the Defence of Labour Liberalism in Europe and Latin America, 1949–1969.* Bern: Peter Lang.

Scalmer, S. 2006. *The Little History of Australian Unionism*. Melbourne: Vulgar Press.

Scalmer, S. 2021. "Remembering the movement for eight hours: commemoration and mobilization in Australia". In *Remembering Social Movements: Activism and Memory*, edited by S. Berger, S. Scalmer & C. Wicke, 219–39. Abingdon: Routledge.

Schneirov, R., S. Stromquist & N. Salvatore 1999. *The Pullman Strike and the Crisis of the 1890s: Essays on Labor and Politics*. Urbana, IL: University of Illinois Press.

Schönhoven, K. 1980. *Expansion und Konsentration. Studien sur Entwicklung der Freien Gewerkschaften im wilhelminischen Deutschland, 1890–1914*. Stuttgart: Klett-Kotta.

Sen, S. 1977. *Working Class of India: History of Emergence and Movement, 1830–1970*. Calcutta: Bagchi.

Siegelbaum, L. & R. Suny (eds) 1994. *Making Workers Soviet: Power, Class and Identity*. Ithaca, NY: Cornell University Press.

Sieger, R. 2007. *For Jobs and Freedom: Race and Labor in America since 1865*. Lexington, KY: University Press of Kentucky.

Smelser, R. & R. Ley 1988. *Hitler's Labor Leader*. Oxford: Berg.

Stearns, P. 2013. *The Industrial Revolution in World History*. Boulder, CO: Westview.

Stråth, B. 1996. *The Organization of Labour Markets: Modernity, Culture and Governance in Germany, Sweden, Britain and Japan*. London: Routledge.

Sysiak, A. *et al.* 2018. *From Cotton and Smoke. Łódź – Industrial City and Discourses of Asynchronous Modernity, 1897–1994*. Łódź: Łódź University Press.

Thatcher, I. 2005. "Late imperial urban workers". In *Late Imperial Russia: Problems and Prospects*, edited by I. Thatcher, 101–19. Manchester: Manchester University Press.

Tosstorff, R. 2016. *The Red International of Labour Unions, 1920–1937*. Leiden: Brill.

Vigna, X. 2021. *Histoire des Ouvriers en France au XXe Siècle*. Paris: Perrin.

van Voss, L., P. Pasture & J. de Maeyer (eds) 2005. *Between Cross and Class: Comparative Histories of Christian Labour in Europe, 1840–2000*. Bern: Peter Lang.

Weir, R. 1996. *Beyond Labor's Veil: The Culture of the Knights of Labor*. Pennsylvania, PA: Pennsylvania State University Press.

Welskopp, T. 2019. "German unions in the nineteenth century: sociological foundations in comparative perspective". *German History* 37(3): 314–26.

Welskopp, T. 2009. "Transatlantische Bande: eine vergleichende Geschichte der Gewerkschaften in Deutschland und den USA im 19. und 20. Jahrhundert". In *Solidargemeinschaft und Erinnerungskultur im 20. Jahrhundert. Beiträge su Gewerkschaften, Nationalsosialismus und Geschichtspolitik*, edited by U. Bitsegeio, A. Kruke & M. Woyke, 29–61. Bonn: J. W. H. Diets.

Williamson, P. 1989. *Corporatism in Perspective: An Introductory Guide to Corporatist Theory*. London: Sage.

CHAPTER 9

THE SOCIAL DEMOCRAT HIGH POINT, 1945–79

Greg Patmore

ABSTRACT

This chapter will explore the performance of social democratic parties in Sweden, the UK and Australia during the period of prosperity that followed the Second World War, highlighting their general approach to social and economic policy and their specific relationship to unions. The parties were closely linked to their national union movements and placed an emphasis upon protecting and improving workers' living standards. While the Swedish labour movement made major gains with SAP, the relationship between unions and their respective labour parties in Australia and the UK were tense by the end of the 1970s as they tried to deal with the economic and social fallout from the end of the postwar economic boom. There was a growing recognition, particularly in the UK and Sweden, of need for worker voice at the enterprise level.

Keywords: social democratic parties; union ascendancy; union influence

INTRODUCTION

This chapter will explore the performance of social democratic parties during the period of prosperity that followed the Second World War. It commences with a review of the concept of social democracy and provides an overview of the electoral fortunes and policy initiatives of social democratic parties in Sweden, the UK and Australia, with a particular focus on their relationship with unions. The Swedish Social Democratic Party (SAP) dominated politics in Sweden during the postwar period, while the UK's Labour Party (UKLP) held power from 1945–51, 1964–70 and 1974–79. Australia had only relatively short periods of national Australian Labor Party (ALP) governments, 1945–49 and 1972–75, with state ALP governments being at the forefront of industrial and social reforms.

While the Swedish labour movement made major gains with the SAP, the relationship between unions and their respective labour parties in Australia and the UK were frayed by the end of the 1970s as they tried to deal with the economic and social fallout from the end of the postwar economic boom. Unions in the UK then faced the neoliberalism of Thatcherism, while the ALP adopted more neoliberal economic policies to restore their credentials as economic managers in the wake of the apparent economic mismanagement of the Whitlam government.

SOCIAL DEMOCRACY

Social democracy and democratic socialism are vague terms. Social democracy emerged with industrial capitalism and the working class, which came with it. It chose labour unionism over syndicalism, parliamentary democracy over the dictatorship of the proletariat, and reform over revolution. It began as part of the movement for socialism to become a left-of-centre party for political, social and economic reform. Social democrats built their philosophy of the need to capture the state through parliament, which was viewed as a class-neutral institutional device that once captured by electoral means could be used to "civilize capitalism" by reducing the exploitation of workers and improving their lives. That definition combined a developing faith in the capacity of the state as an administrative instrument with a refusal to countenance the use of the industrial wing of the movement to strengthen parliamentary initiatives.

The social democrats emphasized nationalization of key industries to coordinate the economy and reduce exploitation of consumers and workers, economic planning, the welfare state and macroeconomic levers to generate growth and employment. This parliamentary approach contrasted sharply with the ideas of anarchists and syndicalists, and at times the communist parties, who called for workers' control of industry, through state ownership of the means of production or worker owned cooperatives (Gallop & Patmore 2010: 1; Macintyre 1986: 3; Patmore & Coates 2005: 121–3). The social democrats' outlook was optimistic and even idealistic, believing that "human beings acting collectively, could use the power of the democratic state to create a better world" (Berman & Snegovaya 2019: 7).

Giddens (2013: 7–11) has defined "classical social democracy" as possessing several features. Social democrats, while viewing free-market capitalism as creating inequality and injustice, believed these problems could be overcome through state intervention in the processes and outcomes of the market to ameliorate their effects. There is an emphasis on the partial replacement of markets through corporatist collective decision-making involving business, unions and government. Social democrats emphasise the need for a "welfare state" with a social safety net to rescue families in need and help those individuals who are unable to help themselves. There was a strong concern with social equality through progressive taxation, for example. While there was a Keynesian influence with a focus upon the mixed economy and demand management, some social democrats went beyond Keynesianism, as in the UK, and favoured nationalization for some sectors because of the deficiencies of markets and the idea that it was not in the "national interest" that certain key industries be in private hands.

There is some confusion over what a social democratic party is. A distinction is made between the "labourist" parties such as in the UK and Australia and European social democratic parties such as in Germany and Sweden (Macintyre 1986). These labourist parties, however, also came under the label of social democracy. They, like their social democratic party counterparts, have links with organized labour and, according to Lavelle (2008: 7), "are reformist because they see reforms within capitalism as ends in themselves". There were differences among these social democratic parties. While the ALP and the UKLP had a commitment to the socialization of industry, the SAP placed an emphasis on expanding the welfare state (Lavelle 2008: 8).

The period of prosperity after the Second World War saw the heyday of support for most social democratic parties in Europe. They generally attempted to reach out to middle-class voters, recognizing that industrial workers did not provide sufficient votes to form a majority government. This strategy started in Scandinavia in the 1930s and expanded to over European countries, including the UK in the 1940s and Germany in the 1950s. The social democratic parties generally downgraded traditionally classed based policies in favour of policies that gained the support of public sector employees, urban professionals and agricultural workers. Their focus included the construction of a welfare state and the nationalization of monopolies and later liberal social policies in areas such as gender equality (Benedetto, Hix & Mastrorocco 2020: 930–31). Similar patterns can be seen outside Europe with the ALP, through federal leader, Gough Whitlam, and state leaders such as Don Dunstan, broadening its appeal beyond its traditional working-class base to embrace liberal social policies, providing the foundations for its 1972 federal election win after 23 years of conservative rule (Beilhars 1988: 17).

The considerable wealth generated by the postwar boom allowed for increased government expenditure on social welfare and other social democratic initiatives. As Lavelle (2010: 56) noted during the postwar boom, "governments of varying political persuasions were able to use the bounty afforded by historically high and consistent rates of economic growth to oversee socially progressive measures". Most European social democratic parties saw their popularity peak in the 1960s and 1970s but decline with the challenges at of the end of the postwar boom, which included the 1970s oil crisis, inflation, unemployment and changing trade patterns. There were also ideological challenges with the rise of neoliberalism and the growing environmental movements (Benedetto, Hix & Mastrorocco 2020: 930–31). The chapter now proceeds to examine three countries during this high point of social democracy: Sweden, the UK and Australia.

SWEDEN

The classic example of a social democracy is Sweden. The foundation of the Swedish SAP in 1898, unlike the UKLP and ALP, largely predated union activity. The SAP's early focus was initially upon union organization rather than political activity, which shaped the character of the Swedish unions away from craft unionism and towards centralization with the formation of the national union federation (the LO) in 1898. Further, the character of Swedish industrialization, which was centred on industries such as iron ore mining and timber, which were geographically dispersed and dominated by large companies, favoured centralization. The SAP also saw tensions between reformists, largely driven by their leader, Hjalmar Branting, and revolutionaries, culminating in a split in 1917 when the revolutionists left to form the Social Democratic Left Party. The SAP was also limited initially by the political system, with universal suffrage for men in 1909 and women in 1918 (Epsing-Anderson 1990: 38; Hilson 2006: 36–7, 51; Sejersted 2021: 40, 141).

While the party entered into a coalition government with the Liberals from 1917 to 1920, during the 1920s the SAP was unable to secure a majority in its own right and could only briefly form minority governments. Sweden faced political instability with ten governments between 1919 and 1932, with none of the non-socialist parties able to win a majority in their own right. While Sweden remained neutral during the First World War, it was vulnerable to disruptions in international trade with a British blockade on shipping. During the war there was a tripling of wages, prices and the cost of living, and there was a postwar economic downturn that hit Sweden hard with an interest rate policy that increased unemployment and decreased industrial production.

While by mid-1920s the economy improved, unemployment remained high and there was major industrial conflict. These political and economic problems fuelled disillusionment with both democracy and the capitalist economy. The SAP also broadened its electoral appeal beyond the working class by embracing ideals such as the "people" and the "nation" and rejecting the nationalization of industry (Berman 2006: 130, 161–3; Patmore & Balnave 2018: 118; Sejersted 2021: 156). Sejersted (2021: 40) argues that anti-nationalization reflects the SAP's view that state initiatives undertaken during the First World War were ineffective compared to private sector initiatives and pushed the SAP to support not only coexistence with capital but also its effective development.

All these developments laid the basis for the SAP winning office in 1932 and gaining the support in 1933 of the Farmers' Union with higher agricultural tariffs. The SAP, which briefly lost office for three months in 1936, enacted economic stimulus policies that reduced the unemployment arising from the Great Depression and a range of social reforms during 1937–39 that included a pension indexed to the cost of living, child support and a housing loan fund (Epsing-Anderson 1990: 41; Pontussen 1990: 58, 67; Sejersted 2021: 84, 121, 166–70).

The SAP dominated Swedish postwar politics, retaining power until 1976. Swedish industrial relations during the postwar period were built upon the Saltsjöbaden Agreement of 1938, negotiated between employers and unions, which included an informal understanding that labour and capital would cooperate to generate economic growth. The Swedish labour movement was left free to pursue the Rehn-Meidner model, which was adopted by the LO in 1951, and promoted the labour movement's idea of wage egalitarianism through wage solidarity – "equal pay for equal work" across various industries. The approach forced low productivity firms to upgrade or exit the market and provided for unemployed workers through an active labour market programme that allowed them to move to high-productivity firms and jobs. LO membership expanded from 1,107,000 in 1945 to 2,089,000 by 1979. Besides the close links with the LO, through which members could collectively join the SAP, social democracy permeated Swedish society through cooperatives, education and leisure societies, tenant associations and pensioner organizations. When its alliance with the farmers, which was a declining sector of the Swedish population, weakened over pension reform in the late 1950s, the SAP made a successful appeal to the growing middle class that helped underpin its outright victory in the 1968 elections (Coates 2000: 94–7; Epsing-Anderson 1990: 43, 46–7; Johnston 1981: 100–1; Sejersted 2021: 221).

The SAP also took an active interventionist role in the labour market with the exception of pay, which it left to the peak organizations of labour and employers to negotiate (Coates 2000: 96–7). Unlike the UKLP, the SAP continued to not place an emphasis upon nationalization, with Berman (2006: 184) arguing that it viewed nationalization "as both economically unnecessarily and politically unwise". In 1948 the SAP established the National Labour Market Board (AMS), which retained supervision of the pre-existing Swedish labour exchanges. While there was union and employer representation on the AMS, the union representatives held the majority of seats. The AMS was also given independence from the government, the civil service and parliament. It subsequently pursued the policy of full employment and reinforced the Rehn-Meidner model through a range of measures that included the relocation and retaining of workers to allow the strengthening and restructuring of Swedish industry. It also introduced reforms such as the three-week summer holiday and the five-day week (Coates 2000: 97; Hilson *et al.* 2017: 19; Rothstein 1985; Sejersted 2021: 221–3).

The SAP was not the principal architect of the welfare state, but was the inheritor of a long tradition of tolerance for Swedish state intervention in welfare policy dating from at least the late nineteenth century. Postwar welfare, with one major exception, was introduced through political compromise with the bourgeoise parties. From the late 1950s, the SAP expanded the welfare provision so that by 1980 Sweden was at the top of several international measures of equality, including the progressiveness of the tax system, the generosity of welfare benefits, the relative absence of poverty and overall income equality.

A significant feature of this expansion was the General Supplementary Pension (ATP), which supplemented the old age pension and was funded by payroll taxes paid by employers. The ATP was opposed by the bourgeoise parties, with it passing through parliament by only one vote in 1959. The amount paid to workers on retirement depended upon their earnings and the number of years employed. The SAP also provided publicly subsidized and regulated social housing that made it possible for low-income groups to obtain good housing, largely through municipal rental accommodation (Coates 2000: 95; Hilson *et al.* 2017: 20; Sejersted 2021: 251–2, 262). Hilson *et al.* (2017: 20) argue that the Swedish welfare policy was designed largely for male, waged and industrial workers, "the breadwinner", and not marginalized groups such as unpaid women working in the home and the Indigenous Sami.

SAP hegemony came under challenge in the 1970s. The OPEC oil crisis of 1973 and the subsequent economic downturn was dealt with by an expansive fiscal policy that fuelled a "profits explosion" in 1973–1974 and a "wages explosion" in 1975. The latter highlighted the failure of centralized wage negotiations to adapt to changing economic circumstances. "New left" issues such as the environment and nuclear power, which the SAP supported, siphoned off middle-class support to the small Communist Party and the Centre Party (previously the Agrarian Party). Middle-class workers were critical of the SAP policy of decreasing wage differentials by both tax and wage policies, which culminated in a 1971 public sector strike (Epsing-Anderson 1990: 45–7, 53–4; Hilson *et al.* 2017: 21; von Otter 1983: 207–8).

There was also a push towards economic and industrial democracy that heightened class tensions. There were underlying concerns with the failure of centralized bargaining to address workplace issues, highlighted by a major strike at the Kiruna iron ore mine in northern Sweden in 1969–70, which was against a background of a surge in wildcat strikes in 1969–71 nationally. This strike was also a challenge to the Stockholm-based SAP leadership and the traditional workplace and union hierarchies. Volvo also took a lead among Swedish employers to highlight the pursuit of alternative work practices. The LO responded to the workplace challenges by rejecting paragraph 32 of the Saltsjöbaden Agreement of 1938, which left production decisions in the hands of management, and going further than the systems of consultative works councils that had been built up in agreements with employers since 1946. While the SAP had shown an interest in the idea of industrial democracy in the 1920s, it was overshadowed by the SAP's political setbacks in the 1920s, the economic problems of the Great Depression and the LO's preference for a combination of its wage policies and SAP legislation to protect workers. The LO had also been concerned about being compromised through being involved in production decisions and the possibility that social democracy could create a rival organization competing for worker loyalty. The SAP in 1972 introduced legislation allowing two union representatives on management boards for companies with more than 100 employees. There was also the Joint Regulation of Working Life Act 1976, which introduced the core concept of participation agreements and required employers to provide a regular flow of information to employees (Gourevitch *et al.* 1984: 254–5; Hilson *et al.* 2017: 20–1; Johnston 1981: 117–8; Sejersted 2021: 367–72).

While these changes challenged employer power, class conflict was particularly heightened by the LO proposals in 1971 to establish wage earner funds, whereby profits would be converted directly into shares for employee investment funds that would increasingly gain control of the company. The proposal was a major incentive for the Swedish bourgeoise parties to forget their longstanding differences, with encouragement of Swedish employers, and form a non-socialist bloc. While the SAP moved towards the idea in its policy platform, the wording was vague and without obligation. The SAP Prime Minister, Olof Palme, distanced himself from any idea that sanctioned wage earner funds as a means of transferring dominant economic power to workers. However, the SAP lost power at the 1976 elections when a coalition of bourgeoise parties formed a government (Epsing-Anderson 1990: 54–5; Johnston 1981: 117–8; Markey, Balnave & Patmore 2010: 240; Sejersted 2021: 372–4).

While the SAP did regain office as a minority government in 1982, Lavelle (2008: 143–8) argues that it began to shift away from social democracy after its period of opposition. There were in the 1980s and early 1990s regressive tax reforms modelled upon those introduced by President Ronald Reagan in the US. After the SAP regained office in 1994 after three years out of power, there was a shift towards public spending cuts and "sound public finance". There was a growing tolerance of social inequality if it increased investment and spurred economic growth. There was also a weakening of the relationship between the Swedish unions and the SAP, which embraced more neoliberal economic policy. The unions were particularly concerned with the replacement of full employment by low inflation as a major macroeconomic concern and the weakening of the welfare state. The SAP did introduce a watered-down version of wage earner funds, where shares were purchased without coercion on the market and shareholdings were limited to 8 per cent, but these were phased out after the conservative Bildt government took office in 1991 (Epsing-Anderson 1990: 56: Sejersted 2021: 374–6).

THE UK

In the first 50 years of its existence, the UKLP learned to be a strong state party, particularly in relation to the economy. This was genuinely a learning process, with three recognizable stages. There was the initial positioning of the UKLP in the wider debate about the potentialities of the democratic state as a tool of social reform. The UKLP operated within a tradition that looked to parliament and legislation to advance labour interests, drawing upon Fabianism with influences including George Bernard Shaw and the Webbs. Then there was the UKLP's incorporation during the 1920s into the established practices of UK parliamentarianism, experiencing minority government in 1924 and 1929–31, where the party leadership's associated acceptance of a particularly narrow definition of the scope of legitimate political action furthered the focus on the state as a means of reform. This focus developed alongside a refusal to countenance the use of the industrial wing of the movement to strengthen parliamentary initiatives.

The party's learning process was then topped up in the 1930s by the adoption of public ownership and planning as key policies to control the excesses of market capitalism. The culmination of this learning process was the willingness shown by postwar UKLP governments to use the existing machinery of the state as an instrument of economic reconstruction, as a vehicle of economic management and electoral manipulation, and as a means of social engineering and social reform. (Jones & Keating 1985; Miliband 1961; Panitch 1971; Patmore & Coates 2005: 123; Shepherd & Davis 2011; Thompson 2006: 7–83; Tomlinson 2002).

Unlike the SAP, the UKLP did not hold continuous office in the immediate postwar period. While it won a landside victory in 1945, broadening its appeal to the middle-classes, especially in the south of England, the Conservatives were returned to office in 1951. The UKLP then held office from 1964–70 and 1974–79, forming a minority government after the February 1974 elections and having a narrow majority after the October 1974 elections. As in Sweden and Australia, the UK economy grew after the Second World War with per capita income growing by 2.2 per cent annually in the period 1950–64 compared to 1.2 per cent per annum for the period 1870–1964. There were, however, issues with the UK economy, such as balance of payments problems, the stability of the pound and the relatively booming economies of competitors such as Germany, Italy, France and Japan. This constrained reform led the Wilson Labour government to call upon its supporters to make sacrifices to allow the modernization of the UK economy and prevent the UK falling behind its international competitors (Gourevitch *et al.* 1984: 25, 30; Kirk 2011: 164, 219; Pontusson 1990: 59, 65; Siegler 2010: 96).

While the chosen methods of economic management by the postwar UKLP governments did change, they were wedded to the view that the state could buy or direct industries and companies, build and run large welfare bureaucracies, and pull and push economic levers to generate employment and growth. It was in that sense a "big state" party, but it was also a party that took as axiomatic the existence of a huge private sector. The UKLP was committed to the public provision of education, health care and welfare, but it was also a party that left unchallenged the governing assumptions and institutions of a civil society within which life chances were still overwhelmingly fixed by inequalities of class, gender and ethnicity. In the triangular relationship of state, economy and society that governed its internal policies, the UKLP, like the SAP, left large swathes of the economy subject to market forces and private control. The principle of voluntarism, collective bargaining without state intervention, still dominated industrial relations. Except for a growing sensitivity in the 1970s to the need for devolution, and an ongoing desire to reform the House of Lords, the UKLP did not view constitutional change as a key to the modernization of the economy and society that its other policies perennially sought (Howell 2005: 107; Patmore & Coates 2005: 124–5, 129).

The Attlee governments (1945–50, 1950–51) inherited both wartime direct controls and postwar resource shortages. Unlike the SAP, they nationalized heavily, taking 20 per cent of the economy into the state sector, including railways, the Bank of England and steel, but did not challenge the pre-war managerial prerogatives in the private sector and appointed very few worker directors in the nationalized sector, with a union preference being for strengthening collective bargaining in these industries and class sentiment undermining any further push for industrial democracy. They initially exercised direct financial and fiscal controls over the remaining 80 per cent, and were heavy purchasers of military equipment under the Ministry of Supply. They developed a central economic planning staff, under whose guidance they progressively abandoned direct controls in favour of Keynesian-inspired aggregate demand management (Edgerton 1992; Gourevitch *et al.* 1984: 24; Phillips 2019: 190–1; James & Markey 2006: 33; Patmore & Coates 2005: 126–7).

A generation later, the first Wilson government experimented with centralized national planning in relation to prices, investment, regional policy and the application of science and technology to overcome the balance of payment problem. While Wilson renationalized the steel industry in 1967, as Byrne, Randall & Theakston (2020: 65) noted, Harold Wilson "was consistently committed over his political career to a mixed economy with a role for interventionist and modernising state". His government used the instruments of national planning created by the Conservatives, such as the tripartite National Economic Development Committee (NEDC) and liaison with a series of industries through the local level economic development councils ("little Neddies"), in a more aggressive manner to modernize Britain's manufacturing base.

The first Wilson government created a Ministry of Technology (Mintech) to sponsor new industries and the Industrial Reorganisation Corporation (IRC) in January 1966 to promote the rationalization and reorganization of industries to gain the advantages of increased economies of scale. It was also given the power to establish and assist the establishment and development of any industrial enterprise. There were tensions between the UKLP's traditional desire to curb monopoly power and its promotion of increased industrial concentration, and Mintech's efforts to restructure industries such as shipbuilding were not very successful due to insufficient resources, the lack of powers for effective intervention and hostility from the private sector. UKLP governments in the 1970s did even more: creating the National Enterprise Board (NEB) to extend the IRC role (to include the public ownership of failing companies and industries) and using the powers of the Industry Act 1975 to seek planning agreements with major companies. Unsuccessful there, the Wilson and Callaghan governments of the 1970s then established a string of sectoral working parties under the NEDC, used procurement policies to reconfigure the military wing of the industrial base, and generously supported private investment with public funds. The UKLP saw its job as picking and developing national winners and it used public ownership and aid to private industry extensively for that purpose (Coates 1980: 86–146; Goodman 1983: 49; Gourevitch *et al.* 1984: 24; Kirk 2011: 219; Patmore & Coates 2005: 127; Thompson 2006: 180–82).

The Labour government in the 1940s created a universal welfare provision through child allowances in 1945, the National Health Service (NHS) in 1948, unemployment assistance and pensions, and extended the quantity and length of state education provision. It actively pursued policies of full employment and industrial relocation, encouraged the building of publicly provided housing for rent and created planning structures to contain urban sprawl. As in Sweden, the male breadwinner model was heavily embedded in the initial UK welfare state. A generation later, UKLP governments recommitted themselves to the maintenance of the NHS and its associated welfare bureaucracies, actively engaged in urban renewal and focused its reforming seal both upon the restructuring of higher education (polytechnics and the Open University) and on the removal of selection from the secondary education sector.

The first Wilson governments were also notable for the liberalism of their legislation in the area of sexual orientation and abortion, a liberalism that in the second set of Wilson governments was reinforced by legislation outlawing sexual discrimination. The record of the Wilson governments on issues of immigration and race relations was noticeably patchier: the liberal legislation of 1949 being replaced by restrictive immigration legislation in the 1960s, but then balanced in the 1970s by major initiatives to outlaw acts of racial discrimination.

The UKLP was committed across the entire social agenda to the establishment and improvement of basic standards and rights, and to the development of individual opportunities for self-improvement and social mobility. Comparing the social sphere to the economic, the UKLP was less ubiquitously active, but where it was, its initiatives tended to be universal rather than targeted, and to be designed directly to shape specified sets of social outcomes. The cumulative impact of its programmes was the creation of a large and labour-intensive publicly funded set of welfare institutions, which by 1979 absorbed 28 per cent of gross domestic product (GDP) (the figure for the much smaller GDP of 1938 had been 11.3 per cent), and distributed pensions, unemployment and disability benefits and child allowances to nearly half the population (Clarke & Langan 1993: 32, 38; Patmore & Coates 2005: 128–9; Thompson 2006: 143).

The UKLP was progressively more and more corporatist as it sought to strengthen the competitive position of UK-based industry while seeking full employment and price stability. Union access to ministers was extensive and frequent during the Attlee governments, lubricated by the presence of former union leaders in cabinet posts and as heads of nationalized industries. The Attlee governments also negotiated and sustained the first set of postwar incomes policies. The later Wilson governments of the 1960s created new corporatist bodies (the National Board for Prices and Incomes and the IRC) and settled into a pattern of national bargaining with unions on wages and productivity.

Between 1966 and 1970, however, Wilsons's determination to maintain the UK's status as a great power led to major strains on the economy, with Wilson forced to devalue the pound in 1967. There was also conflict with the Trades Union Congress (TUC) over a Prices and Incomes Bill in 1968 and proposals for legal constraints to be placed on unions in exchange for union recognition. This had negative effects on labour movement unity and turned many union members against the government, with the UKLP losing the June 1970 elections to the Conservatives, led by Ted Heath. There was a 4.7 per cent swing against the government, the greatest since 1949, with the UKLP's share of the working-class vote declining by 12.4 per cent compared to the previous election (Howell 2005: 112; Kirk 2011: 221; Patmore & Coates 2005: 127).

Following its return to office in 1974, against the background of the end of the postwar boom and serious stagflation, UKLP governments, led again by Wilson and later James Callaghan, who became Prime Minister after Wilson's resignation in 1976, inherited and created yet more tripartite bodies (the Manpower Services Commission, Advisory Conciliation and Arbitration Service and NEB) to which executive powers were delegated, and introduced new and favourable sets of labour legislation. There was a surge in union membership, fuelled by a growth of white-collar unionism, with union density increasing from 44.2 per cent in 1969 to a peak of 53.4 per cent in 1979. The UKLP negotiated with union leaders a voluntary "Social Contract", whereby unions cooperated on wage restraint in exchange for the removal of the Conservative's Industrial Relations Act 1971 – which was a fundamental departure from many aspects of the largely voluntary industrial relations system and strongly opposed by the TUC – and action against rising prices, high rents and social inequality.

The UKLP followed through with its commitments after its election by repealing the contentious legislation, granting the miners a 29 per cent wage increase, raising old age pensions, freezing rent increases on council houses and increasing food subsidies. The "Social Contract" by 1977 limited wage increases, reduced living standards and cut inflation, falling from 25 per cent in 1975 to 8 per cent in 1978 (Byrne, Randall & Theakston 2020: 73; Crook 2019: 92–3; Goodman 1983: 65–7, 69; Gourevitch *et al.* 1984: 42; Kirk 2011: 222, 226; Patmore & Coates 2005: 127; Thompson 2006: 224; Waddington 1992: 289–97). Goodman (1983: 67) notes that the 1975–77 incomes policy was certainly "one of the most effective of the many repeated government attempts during the 1960s and 1970s to moderate the rate of inflation by direct action".

There was also some interest in industrial democracy. Unions gained parity representation on the board of the nationalized British Steel Corporation in 1967. The TUC proposed union representation on company boards in its submission to the Donovan Commission on Industrial Relations in 1968, and in 1974 adopted a policy for parity representation of union-nominated employee representatives with shareholder representatives on company boards, although there was considerable union scepticism over the desirability of such involvement.

The Callaghan government set up the Bullock Committee to examine a form of employee participation at the board level. It recommended in early 1977 the continuation of unitary boards with equal numbers of shareholder and employee directors, with the balance of directors being held by independent individuals acceptable to both groups. The committee endorsed a TUC proposal that workers be union representatives but subject to approval by all employees irrespective of union membership. The proposals would apply to companies with more than 2,000 employees. The three employer members of the committee did not sign the Majority Report and the recommendations were strongly opposed by employer organizations and the Conservative Party, claiming that union directors represented a major challenge to managerial prerogative and shareholders.

This reinforced Conservative Party opposition to social democracy and unions. The Labour government retreated in the face of disinvestment threats if the legislation was proceeded with. It did issue a consultative White Paper on Industrial Democracy in May 1978 that reduced the employee representation to one-third, but proposed that unions be given statutory rights to discuss with companies major aspects of corporate strategy, such as takeovers and closures, irrespective of their forms of employee representation. Subsequently, no legislation arising from the Bullock Committee was passed into law. The Labour government provided financial assistance in several cases to workers who wanted to organize worker cooperatives in the face of closures and maintain employment, but few survived for very long (Goodman 1983: 44, 70–1; Howell 2005: 118; Markey & Patmore 2009: 38–9; Phillips 2019: 196–7).

The Callaghan government ran into further difficulties. There was a movement away from Keynesian policies and deficit financing towards deflationary budgets, with International Monetary Fund (IMF) support for the pound being underpinned by massive reductions in public expenditure and £500 million of shares in (state-owned) British Petroleum being offered for sale in December 1976. Unemployment doubled between 1974 and 1978; however, the UKLP maintained a commitment to the welfare state, with the cuts to public expenditure almost entirely related to capital expenditure rather than current expenditure and welfare spending cuts a "no-go area". The return to free collective bargaining came with the ending of the "Social Contract", when the TUC rejected a proposed 5 per cent ceiling at its annual conference in September 1978, which led to the "Winter of Discontent" in 1978–79, with major industrial and political unrest.

Rising wages and prices brought the Labour government into conflict with sections of the UKLP, which also rejected any form of wage restraint at its October 1978 annual conference, and unions who supported collective bargaining. Divisions appeared with some on the left, both within the UKLP and Communist Party, criticizing militant labourism for damaging the cause of socialism and the UKLP's electoral prospects. They called for more economic planning, industrial democracy and public ownership of key firms to counter the growing power of multinationals.

While there was an agreement in February 1979 between the TUC and the UKLP covering grievance handling, picketing and the conduct of strikes, there was no firm joint commitment on wage increases. These divisions contributed to the UKLP's defeat in the May 1979 election to the Conservatives led by Margaret Thatcher, whose neoliberal policies were to gain ascendency, with the UKLP losing support among skilled workers, where there was discontent with the narrowing of wage relativities under the "Social Contract", and the broader population (Byrne, Randall & Theakston 2020: 70–71; Crook 2019: 93, 101; Dorey 2016: 109: Goodman 1983: 48, 68; Gourevitch *et al.* 1984: 58; Howell 2005: 124; Kirk 2011: 226–8; Thompson 2006: 224). As Crook (2019: 100) argues, the UKLP's "stress on working with the trade unions to deal with inflation turned from a potential vote winner into a massive electoral liability".

The UKLP next won office in 1997 after 18 years in opposition, during which Conservative governments led the way in the adoption of neoliberal policies. Even before taking office, future Prime Minister Tony Blair and future Chancellor Gordon Brown saw the state's economic role as giving both space and legitimacy to market forces. They saw a need to distance the UKLP from "Old Labour" to overcome electoral suspicions that the party could not manage the economy. They lacked their UKLP predecessors' enthusiasm for "big government". The state would be the "lubricator" of the provision of social service by private institutions and as the "enabler" of self-development and self-help by a modern individualized citizenry (Patmore 2019: 135).

AUSTRALIA

The ALP enjoyed earlier electoral success than either the UKLP or SAP with earlier moves for full adult suffrage, with the exception of Indigenous people, who were excluded until 1966. The various state Labor parties in Australia had mixed success at first. The ALP found it necessary to broaden its base beyond labour unions and appeal to Catholics and small farmers to gain electoral support. Despite this, unions still saw their ALP affiliation as an important means for improving wages and conditions. In 1910 the ALP won majority government both in New South Wales (NSW), the largest state, and federally. It subsequently formed federal governments before the outbreak of the First World War until 1913, 1914–16 and 1929–31. There were constitutional constraints on what a federal government could do, for example, in the arena of industrial relations or with regard to nationalization, and there were legislative problems if it did not control the senate, where there was an equal number of representatives elected from each state. There were also significant splits that brought down ALP governments in 1916 and 1931 over military conscription and the handling of the Great Depression. The ALP was particularly successful at the state level, where unions won major improvements in working conditions through legislation rather than industrial action. In contrast to the UK, the state played a growing role in industrial relations through wages boards and industrial tribunals (James & Markey 2006: 24–5; Patmore & Coates 2005: 124; Patmore & Balnave 2018: 7).

While socialist ideas and movements abounded within the ALP, it did not fundamentally challenge capitalism. There have been many attempts to characterize the ALP. It has been perceived as "labourist", "nationalist", "populist", "socialist", "liberal" and "social democratic". The varied perceptions reflect the wide range of influences on the ALP. While there have been attempts to play down the impact of ideas on ALP activists and politicians and emphasize political pragmatism, the ALP thought was complex (Battin 1994: 41; Laurent 1995; Scott 2001: 44). Laurent (1995) argues, for example, that ALP heroes such as Second World War prime ministers, John Curtin and Ben Chifley, had a wide range of influences, including Karl Marx, Henry George, Charles Darwin, Tom Mann, HG Wells and John Maynard Keynes. Whatever ideas influenced them, ALP activists, like their counterparts in Sweden and the UK, accepted the need to capture the "neutral" state through parliament to improve workers' lives. While the major thrust of ALP thinking favoured the nationalization and state intervention, it focused upon monopolies or the "money power" rather than the complete overthrow of the capitalist system (Patmore 1991: 80–81).

The ALP regained federal office during the Second World War in October 1941 when two independents switched their support from a non-labour government. The ALP won a majority of seats in the House of Representatives in the 1943 and 1946 federal elections with control of the senate from 1944–49. The economy boomed with the wartime expansion of industry, and unemployment fell from 9 per cent in 1939 to 1 per cent in 1945. While the economy contracted briefly in 1944–45, GDP grew by 52 per cent by 1949. The government led by John Curtin gained union support for the war effort. It initiated regular meetings with officers of the national peak union council, the ACTU (Australian Council of Trade Unions). While the ACTU advised on manpower and industrial issues, Curtin would not allow union representation on the Economic Planning Committee, which planned essential war production.

The direct threat to Australia following the Japanese attack on Pearl Harbor in December 1941 removed any remaining official union doubts about the war. In 1942 the government, through the defence power of the constitution issued National Security Regulations that established extensive controls over wages and prices. These regulations forbade for three years industrial tribunals, including those in state jurisdictions, from awarding pay increases unless existing rates were found to be anomalous and consent was obtained from the federal Minister for Labour and National Service. The cost-of-living adjustments in the basic wage were permitted. The regulations prohibited employers granting and employees accepting wage increases outside statutory industrial tribunal awards. There was some industrial unrest among rank-and-file workers as the war continued. Curtin issued regulations to prevent strikes in mining and deployed naval personnel to fill the places of striking seafarers. While Curtin was committed to economic planning, he promised not to nationalize any industries during the war (Butlin 1987: 139; Kirk 2011: 169; Patmore 1991: 88–9, 150; Patmore 2003: 8–9).

The Curtin government proceeded with an unsuccessful referendum on 19 August 1944 to provide the Commonwealth with the constitutional power to legislate on 14 points until the war ended and for five years afterwards to assist with postwar reconstruction. The 14 points included powers relating to prices, national health, repatriation of return military personnel, and Indigenous Australians. Voters were also requested to approve the insertion into the constitution of clauses protecting the freedoms of speech, expression and religion. There are numerous reasons for its defeat, including concerns about the extension of wartime controls into peacetime, divisions within the ALP over the diminution of states' rights, particularly in NSW, and the tactical blunder of including multiple measures in the referendum (Fox 2008; Kirk 2011: 169).

Curtin died in July 1945, one month before the end of the war. Ben Chifley, his treasurer and successor, enacted the ALP's plan for postwar reconstruction, which was strongly influenced by Keynesian ideas of state intervention and would modernize Australian capitalism. The success of wartime planning contributed to a distrust of unregulated market forces. ALP governments also began to rely on experts recruited to the growing federal public service from universities and business, who pushed the ALP down the road to "a new order" based upon Keynesianism. The focus of the ALP's plan was full employment – the 1930s Depression exacerbated the ALP's fear of unemployment if the wartime prosperity ended. Although Chifley was thwarted by the constitution in his attempts to nationalize the banks, these plans were based upon a desire to maintain a tight control of the economy rather than an ideological commitment to state socialism.

Other sections of private enterprise were seen as socially beneficial and integral to developing an independent Australian economy. While Chifley generally supported the interests of domestic capital against those of international capital, he faced a caucus revolt over his plans to ratify the Bretton Woods Agreement of 1944, which established the IMF and the World Bank. Chifley was concerned about the possibilities of international intervention in domestic policy, but believed that international collaboration was essential to revive capitalist production and trade in the postwar world. There were fears within the labour movement that ratification would lead to the dismantling of tariff barriers and a "world dictatorship of private finance". He used the authority of the party federal executive to end caucus dissent. Chifley in the meantime continued the wartime price and wage controls to curb inflation (Crisp 1961: 210–11; Day 2001: 398–401, 457; Freudenberg 2001: 84–5; Johnson 1986; Massey 1994: 55–7; Scalmer 1998: 54–7).

The federal ALP governments during and after the Second World War did introduce a wide range of social welfare programmes, such as unemployment benefits, child endowment and widows' pensions, underpinned by a successful constitutional referendum in 1946 that extended federal powers over social security. It failed in its efforts to establish a health benefits scheme, and its two attempts at pharmaceutical legislation were both declared unconstitutional by the High Court. However, all these schemes were a "safety net" if the primary objective of full employment failed. Further, Australia lagged behind the rest of the world in regard to social security and had one of the lowest expenditures in this area of GDP during the 1950s compared to other countries. Welfare was not a top priority for Chifley, with welfare being residual to wages provided through the compulsory arbitration system and focusing on those who could not undertake paid work. Consolidated revenue rather than contributory insurance schemes funded this residual social welfare system (Castles 1985: 28–31; Garton & McCallum 1996: 124, 128–9; Patmore & Coates 2005: 128; Watts 1987: 61–127).

Chifley's economic plans clashed with the rising expectations of workers, with union density reaching an all-time peak of 64.9 per cent in 1948 (Bowden 2011: 63). One major union demand was the 40-hour week. Although an ALP federal conference in 1943 called for legislation to introduce shorter hours within six months of the end of the war, Chifley stalled and denied that the federal government had the power to do so. He was concerned with the inflationary impact of the reduced hours and persuaded the ACTU to put the case before the Commonwealth Arbitration Court. This process delayed the introduction of the reduced hours in the federal arbitration jurisdiction until January 1948. Chifley lifted wage controls in mid-1947 only after a major dispute in the Victorian metal trades from November 1946 to May 1947 forced him to do so. There were also notable disputes in the coal, steel and transport industries. Chifley and his colleagues increasingly dismissed these strikes as the work of communist agitators.

During the 1949 coal strike, the Chifley government mobilized the state against the strikers by freezing union funds, gaoling union officials and even sending the army to work open-cut mines. The Chifley government lost office in the 1949 federal elections (Patmore 1991: 80–81).

While the ALP was out of office federally, for most of the period of postwar prosperity and with unemployment remaining minimal, it still held office in the states for the varying periods. This was despite a split arising in the mid-1950s from a conflict between Catholics and communists in the Australian labour movement that led to the formation of the Democratic Labor Party. This channelled away votes from the ALP to the Conservatives through the preferential voting system. Structural changes contributed to a decline in union density from 62 per cent in 1954 to 49 per cent in 1970.

The NSW ALP held government continuously from 1941 to 1965. Its leadership shared Chifley's views on postwar reconstruction and embarked upon an urban planning scheme for Sydney and major public works programmes such as dams, public housing and hospitals. A close relationship developed between the ALP government and the Labor Council of NSW, the peak NSW union body, which brought major legislative benefits for unions and their members, such as annual leave, long service leave and a 40-hour week (Bowden 2011: 66; Markey 1994: 336–7; McDougall 2015: 32; Patmore & Coates 2005: 126).

The ALP held state government office in South Australia from 1965 to 1968 and 1970 to 1979, promoting a range of progressive policies, particularly through the influence of lawyer, Don Dunstan, who was leader after 1967. The Dunstan approach represented a "technocratic thrust, consistent with the social democratic attachment to advanced planning techniques" (Parkin 1986, 302). There was a focus upon the expansion of public enterprise through the public service and statutory authorities, such as the Development Commission, and increased regulation of the private sector in areas such as consumer affairs. As Minister for Aboriginal Affairs, Dunstan introduced pathbreaking legislation that recognized Indigenous self-determination and land rights. The Prohibition of Discrimination Act 1966, the first anti-discrimination legislation in Australia, prohibited discrimination on the basis of race, national origin or skin colour in the provision of goods and services. Dunstan also promoted industrial democracy through the establishment of the Unit for the Quality of Working Life, later the Industrial Democracy Unit, in 1974, but faced hostility from both the public and private sectors (Markey & Patmore 2009; Parkin 1986: 294–5, 302–3).

The Whitlam government, which held power from 1972 to 1975, coincided with the end of the postwar economic boom and the rise of stagflation. As in the UK, there was a surge in union membership fuelled by white-collar unionism with union density growing to 56 per cent by 1975. Whitlam had diminished the influence of the ALP on its parliamentarians by obtaining the direct representation of politicians in 1967 on the federal ALP conference and executive. ALP parliamentarians also began to dominate the federal policy committees. This allowed Whitlam to ensure that his social and urban policies were adopted by the ALP that, for instance, allowed the extension of sewerage services to Australia's expanding suburbs. It expanded public sector expenditure, with the budget deficit as a proportion of GDP increasing from 1.6 per cent to 4.7 per cent. The government, however, became more fiscally conservative as unemployment and inflation both rose. The 1975 budget emphasized restraint on "excessive" government spending. The government also became embroiled in scandals, notably the "loans affair" and faced a hostile senate, which frustrated many of its legislative reforms and played a crucial role in its dismissal by the governor-general in November 1975 (Bowden 2011: 66; Faulkner 2001: 215–6; Johnson 1989: 63–75; McDougall 2015: 34–5).

Despite its social radicalism, the Whitlam government commenced a move towards to market liberalization. Whitlam did share the Curtin and Chifley governments' pre-occupation with the establishment of independent and modern Australian economy, industry and culture through Keynesian intervention. Whitlam believed that strong economic growth would provide a platform for new expenditure programmes in areas such as social welfare, which in turn would provide the basis for future economic growth. Like Curtin and Chifley, he favoured domestic capital over multinational capital in developing the Australian economy. There was, however, a weakening of the traditional suspicions within the ALP of market outcomes and a desire for more competitive market economy. His government established an Industries Assistance Commission to review industry protection, slashed tariffs by 25 per cent in July 1973 and moved against restrictive trade practices (Johnson 1989: 51–62; Leigh 2002: 491–4; Massey 1994: 59–61; Scott 2001: 244–5; Whitlam 1985: 185).

The Whitlam government attracted high expectations of social reform after more than two decades of conservative rule. It introduced Medibank, which provided for universal health benefits. This was not nationalized medicine and involved subsidizing private sector medicine and health insurance companies. The Whitlam government did raise pensions, abolished means tests for pensioners over 70 and ended fees for tertiary education. While it supported the formal adoption of equal pay in the Commonwealth Arbitration Commission, it reinforced the concept of the "male breadwinner" model by introducing supporting mothers' benefit.

The Whitlam government formally ended the White Australia Policy, which discriminated against non-European immigrants, and there was a movement away from migrant assimilation towards multi-culturalism, which meant more respect for immigrant cultures. His government introduced the Racial Discrimination Act to challenge domestic racism and ended the policy of assimilation for Indigenous Australians. It provided more support for Indigenous land rights, particularly in the Northern Territory, and against the background of the 1967 constitutional referendum that extended federal powers to cover Indigenous Australians, established a new Department of Aboriginal Affairs to provide medical and legal services to Indigenous Australians and set up the first national Indigenous advisory body (Johnson 1989: 64–71; Kirk 2011: 217; Lee 2021: 58; McDougall 2015: 33–4).

The relationship between the unions and the Whitlam government was strained. Whitlam avoided a close dialogue with the unions and even considered them to be an electoral liability. He failed to consult the unions or his own caucus over the tariff cuts. Despite union objections concerning controls over wages, the Whitlam government tried unsuccessfully in December 1973 to amend the Commonwealth constitution to allow it control over prices and income in its fight against rising inflation.

As the economy deteriorated, the government turned to the Commonwealth Arbitration Commission to restrain a "wages explosion" through the indexing of wages to price increases. Unions accepted wage indexation as they were concerned at the growth of unemployment, the impact of the wage increases on relativities and the election of a conservative government. Tensions, however, remained, with the ACTU insisting on the right to bargain outside the Commission. Like its counterparts in Sweden and the UK, the Whitlam government showed an in interest in industrial democracy. Clyde Cameron, the Minister for Labour in the Whitlam government, urged management in 1973 to share decision-making processes with labour, as did state and national leaders of the Liberal Party at this time. The Whitlam government encouraged this by introducing limited representation of employees or union officials on the boards of federal government agencies such as the Australian Broadcasting Commission and Australia Post. In 1975 the federal ALP platform called for promotion of industrial democracy. In the same year the Jackson Report on policies for the development of manufacturing industry recommended policy to improve the working environment and increase employee participation. Despite their disagreements, unions supported the Whitlam government following its dismissal and during the ensuing unsuccessful election campaign, which resulted in a massive swing to the non-ALP coalition (Hagan 1981: 414–25; Markey & Patmore 2009: 46; Singleton 1990: 10–49). As Kirk (2011: 227) notes, Whitlam, like his UKLP counterparts, Wilson and Callaghan, "had been hit particularly hard by their failure to tackle the growing demands of their workers in the face of worsening socio-economic conditions".

The ALP regained federal government office in 1983, retaining it until 1996, the longest in Australian political history. The Hawke-Keating ALP governments were willing to dump Keynesianism and social democratic policies, which were viewed as discredited by the collapse of the postwar boom, and needed to establish their economic credentials as a responsible party of change and modernization following the apparent economic mismanagement of the Whitlam government. They embraced pre-Keynesian "neo-classical" economics as part of their drive to modernize Australia. The consequences of this thinking were the dismantling of state regulation, state enterprise and public investment. The tense relationship between the ACTU and the Whitlam government was not to be repeated by the Hawke-Keating governments through the development of Prices and Incomes Accord, which traded off wage restraint for concessions such as tax reform and a universal health insurance scheme (Patmore 2019: 134).

CONCLUSION

While noting their differences on issues such as nationalization, the SAP, the UKLP and the ALP are all considered to be social democratic parties because of their commitment to parliamentary democracy and willingness to accept the continued operation of the private sector in a sizeable proportion of their economies. They were closely linked to their national union movements and placed an emphasis on protecting and improving workers' living standards. There were differences between the SAP, UKLP and ALP over their emphasis on nationalization. They enjoyed varying periods of government during the postwar boom, with the SAP enjoying an uninterrupted period of government. While the UKLP held national government for a substantial part of 1945–79, the ALP only held national government for periods at the beginning and end of this period. The inability of the ALP to hold national government for a substantial part of this period was offset by its ability to hold state government offices, which exercised considerable powers under Australia's federal constitution.

All these parties when in office saw their role as modernizing their economies through planning and collaboration with both capital and labour, particularly in Sweden and the UK. The Whitlam government showed an interest in opening up the economy to international markets by slashing tariffs without union consultation. On the basis of growing economic prosperity and social change, there was an expansion of the social welfare safety net and liberal reforms, particularly in Australia and the UK, to challenge phenomena such as racism. There was, however, a continued commitment to the ideal of the "male breadwinner" when it came to wages policy.

The end of postwar prosperity created issues for all three parties with stagflation and wage explosion. Their relationship with unions was important for the development of policies to contain inflation such as the UK's "Social Contract" and Australian wage indexation. There was, however, labour unrest over these efforts, as highlighted by the 1978–79 "Winter of Discontent" in the UK. Despite the pressures to reduce government expenditure as the economies deteriorated, there was a continued commitment to social welfare and reform. The period also highlighted the need of workers to have a greater voice in their workplace with varying degrees of interest in industrial democracy, which challenged traditional notions of managerial prerogative and faced strong opposition from employers and their political allies, particularly in Sweden and the UK. The Swedish wage earner fund proposal went the furthest as it also involved worker ownership as well as worker voice and contributed to the formation of a coalition of the bourgeois parties that won the 1976 Swedish elections. The electoral defeats in the 1970s of the SAP and the UKLP governments and the dismissal of the Whitlam government in Australia marked a turning point in the history of social democracy. The loss of office shifted the social democratic parties in all three countries towards neoliberal policies as a solution to the economic problems that followed the end of the postwar boom.

REFERENCES

Battin, T. 1994. "Keynesianism, socialism and labourism, and the role of ideas in labour ideology". *Labour History* 66 (May): 33–44.

Beilhars, P. 1988. "A hundred flowers faded". In *Staining the Wattle: A People's History of Australia since 1788*, edited by V. Burgmann & V. Lee, 164–70. Melbourne: McPhee Gribble/Penguin Books.

Bendetto, G., S. Hix & N. Mastrorocco 2020. "The rise and fall of social democracy, 1918–2017". *American Political Science Review* 114(3): 928–39.

Berman, S. 2006. *The Primacy of Politics: Social Democracy and the Making of Europe's Twentieth Century*. Cambridge: Cambridge University Press.

Berman, S. & M. Snegovaya 2019. "Populism and the decline of social democracy". *Journal of Democracy* 30(1): 5–19.

Bowden, B. 2011. "The rise and decline of Australian unionism: a history of industrial labour from the 1820s to 2010". *Labour History* 100: 51–82.

Butlin, N. 1987. "Australian national accounts". In *Australian Historical Statistics*, edited by W. Vamplew, 126–44. Sydney: Fairfax, Syme & Weldon.

Byrne, C., N. Randall & K. Theakston 2020. *Disjunctive Prime Ministerial Leadership in British Leadership: From Baldwin to Brexit*. Cham: Palgrave Macmillan.

Castles, F. 1985. *The Working Class and Welfare: Reflections on the Political Development of the Welfare State in Australia and New Zealand, 1890–1980*. Sydney: Allen & Unwin.

Clarke, J. & M. Langan 1993. "The British welfare state: foundation and modernisation". In *Comparing Welfare States: Britain in International Context*, edited by A. Cochrane & J. Clarke, 49–76. London: Sage.

Coates, D. 1980. *Labour in Power?* London: Heinemann.

Coates, D. 2000. *Models of Capitalism: Growth and Stagnation in the Modern Era*. Cambridge: Polity.

Crisp, L. 1961. *Ben Chifley: A Political Biography*. Sydney: Angus & Robertson.

Crook, M. 2019. "The Labour governments 1974–1979: social democracy abandoned?" *British Politics* 14(1): 86–105.

Day, D. 2001. *Chifley*. Sydney: HarperCollins.

Dorey, P. 2016. "'Should I stay or should I go?': James Callaghan's decision not to call an autumn 1978 general election". *British Politics* 11(1): 95–118.

Edgerton, D. 1992. "What ever happened to the British warfare state? The ministry of supply 1945–51". In *Labour Governments and Private Industry: The Experience of 1945–51*, edited by H. Mercer, N. Rollings & J. Tomlinson, 91–116. Edinburgh: Edinburgh University Press.

Epsing-Andersen, G. 1990. "Single-party dominance in Sweden: the saga of social democracy". In *Uncommon Democracies: The One-Party Dominant Regimes*, edited by T. Pempel, 33–57. Ithaca, NY: Cornell University Press.

Faulkner, J. 2001. "Splits: consequences and lessons". In *True Believers: The Story of the Federal Parliamentary Labor Party*, edited by J. Faulkner & S. Macintyre, 203–18. Sydney: Allen & Unwin.

Fox, C. 2008. "The fourteen powers referendum of 1944 and the federalisation of Aboriginal affairs". *Aboriginal History* 32: 27–48.

Freudenberg, G. 2001. "Victory to defeat: 1941–1949". In *True Believers: The Story of the Federal Parliamentary Labor Party*, edited by J. Faulkner & S. Macintyre, 76–89. Sydney: Allen & Unwin.

Gallop, G. & G. Patmore 2010. "Social democratic governments and business". *Labour History* 98: 1–5.

Garton, S. & M. McCallum 1996. "Workers' welfare: Labour and the welfare state in 20th century Australia and Canada". *Labour History* 71: 116–41.

Giddens, A. 2013. *The Third Way: The Renewal of Social Democracy*. Cambridge: Polity.

Goodman, J. 1983. "Great Britain. Labour moves from power to constraint". In *Worker Militancy and its Consequences: The Changing Climate of Western Industrial Relations*, edited by S. Barkin, 39–80. New York: Praeger.

Gourevitch, P. *et al.* 1984. *Unions and Economic Crisis: Britain, West Germany and Sweden*. London: Allen & Unwin.

Hagan, J. 1981. *The History of the ACTU*. Melbourne: Longman Cheshire.

Hilson, M. 2006. *Political Change and the Rise of Labour in Comparative Perspective: Britain and Sweden 1890–1920*. Lund: Nordic Academic Press.

Hilson, M. *et al.* 2017. "Labour, unions and politics in the Nordic countries, c. 1700–2000: introduction". In *Labour, Unions and Politics under the North Star: The Nordic Countries, 1700–2000* edited by M. Hilson, S. Neunsinger & I. Vyff, 1–70. New York: Berghahn.

Howell, C. 2005. *Trade Unions and the State: The Construction of Industrial Relations Institutions in Britain*, 1890–2000. Princeton, NJ: Princeton University Press.

James, L. & R. Markey 2006. "Class and Labour: the British Labour Party and the Australian Labor Party compared". *Labour History* 90: 23–42.

Johnson, C. 1986. "Social harmony and Australian labor: the role of private industry in Curtin and Chifley governments' plans for Australian economic development". *Australian Journal of Politics and History* 32(1): 38–51.

Johnson, C. 1989. *The Labor Legacy: Curtin, Chifley, Whitlam, Hawke*. Sydney: Allen & Unwin.

Johnston, T. 1981. "Sweden". In *Trade Unions in the Developed Economies*, edited by E. Owen Smith, 97–122. London: Croom Helm.

Jones, B. & M. Keating 1985. *Labour and the British State*. Oxford: Clarendon Press.

Kirk, N. 2011. *Labour and the Politics of Empire: Britain and Australia 1900 to the Present*. Manchester: Manchester University Press.

Laurent, J. 1995. "'That old treasure-house of constructive suggestion': Australian Labor ideology and war organisation of industry". *Labour History* 68: 46–62.

Lavelle, A. 2008. *The Death of Social Democracy: Political Consequences in the 21st Century*. Aldershot: Ashgate.

Lavelle, A. 2010. "The ties that unwind? Social democratic parties and unions Australia and Britain". *Labour History* 98: 55–75.

Lee, D. 2021. "Labor, the external affairs power and the rights of Aborigines". *Labour History* 120: 49–68.

Leigh, A. 2002. "Trade liberalisation and the Australian Labor Party". *Australian Journal of Politics and History* 48(4): 487–508.

Macintyre, S. 1986. "The short history of social democracy in Australia". *Thesis Eleven* 15(1): 3–14.

Markey, R. 1994. *In Case of Oppression: The Life and Times of the Labor Council of New South Wales*. Sydney: Pluto Press.

Markey, R., N. Balnave & G. Patmore 2010. "Worker directors and worker ownership/cooperatives". In *Oxford Handbook of Participation in Organizations*, edited by A. Wilkinson *et al.*, 237–57. Oxford: Oxford University Press.

Markey, R. & G. Patmore 2009. "The role of the state in the diffusion of industrial democracy: South Australia, 1972–9". *Economic and Industrial Democracy* 30(1): 37–66.

Massey, N. 1994. "A century of labourism, 1891–1993: an historical interpretation". *Labour History* 66: 45–71.

McDougall, D. 2015. "Edward Gough Whitlam, 1916–2014: an assessment of his political significance". *The Round Table* 104(1): 31–40.

Miliband, R. 1961. *Parliamentary Socialism*. London: Merlin.

Panitch, L. 1971. "Ideology and integration: the case of the British Labour Party". *Political Studies* 19(2): 184–200.

Parkin, A. 1986. "Transition, innovation, consolidation, readjustment: the political history of South Australia since 1965". In *The Flinders History of South Australia*, edited by D. Jaensch, 292–338. Adelaide: Wakefield Press.

Patmore, G. 1991. *Australian Labour History*. Melbourne: Longman Cheshire.

Patmore, G. 2003. "Industrial conciliation and arbitration in NSW before 1998". In *Laying the Foundations of Industrial Justice: The Presidents of the Industrial Relations Commission of NSW: 1902–1998*, edited by G. Patmore, 5–66. Sydney: Federation Press.

Patmore, G. 2019. "Different types of societal regulation – co-ordinated market economy, social democracy, aspiration of worker control". In *Handbook of the Politics of Labour, Work and Employment*, edited by G. Gall, 127–43. Cheltenham: Edward Elgar.

Patmore, G. & N. Balnave 2018. *A Global History of Co-operative Business*. New York: Routledge.

Patmore, G. & D. Coates 2005. "Labor parties and the state in Australia and the UK". *Labour History* 88: 121–41.

Phillips, J. 2019. "Participation and nationalisation: the case of British coal from the 1940s to the 1980s". In *The Palgrave Handbook of Workers' Participation at Plant Level*, edited by S. Berger, L. Pries & M. Wannöffel, 187–204. Cham: Palgrave Macmillan.

Pontussen, J. 1990. "Sweden and Britain compared". In *Uncommon Democracies: The One-Party Dominant Regimes*, edited by T. Pempel, 58–81. Ithaca, NY: Cornell University Press.

Rothstein, B. 1985. "The success of the Swedish labour market policy: the organizational connection to policy". *European Journal of Political Research* 13: 153–65.

Scalmer, S. 1998. "Labor's golden age and changing forms of workers' representation in Australia". *Journal of the Royal Australian Historical Society* 84(2):180–97.

Scott, A. 2001. *Running on Empty: Modernising the British and Australian Labor Parties*. Sydney: Pluto Press.

Sejersted, F. 2021. *The Age of Social Democracy: Norway and Sweden in the Twentieth Century*. Princeton, NJ: Princeton University Press.

Shepherd, J. & J. Davis 2011. "Britain's second Labour government, 1929–1931: an introduction". In *Britain's Second Labour Government, 1929–1931: A Reappraisal*, edited by J. Shepherd, J. Davis & C. Wrigley, 1–15. Manchester: Manchester University Press.

Siegler, P. 2010. "Harold Wilson, 1964–1970, 1974–1976". In *From New Jerusalem to New Labour: British Prime Ministers from Attlee to Blair*, edited by V. Bogdanor, 93–106. London: Palgrave Macmillan.

Singleton, G. 1990. *The Accord and the Australian Labour Movement*. Melbourne: Melbourne University Press.

Thompson, N. 2006. *Political Economy and the Labour Party: The Economics of Democratic Socialism, 1884–1995*. Second edn. London: UCL Press.

Tomlinson, J. 2002. "The limits of Tawney's ethical socialism". *Contemporary British History* 16(4): 1–16.

von Otter, C. 1983. "Sweden. Labor reformism reshapes the system". In *Worker Militancy and its Consequences: The Changing Climate of Western Industrial Relations*, edited by S. Barkin, 187–228. New York: Praeger.

Waddington, J. 1992. "Trade union membership in Britain, 1980–1987: unemployment and restructuring". *British Journal of Industrial Relations* 30(2): 287–324.

Watts, R. 1987. *The Foundations of the National Welfare State*. Sydney: Allen & Unwin.

Whitlam, G. 1985. *The Whitlam Government 1972–1975*. Ringwood: Penguin.

CHAPTER 10

THE "SOCIALIST" EXPERIMENT

Jeremy Morris

ABSTRACT

This chapter discusses the quasi-corporatist labour relations in the USSR. Soviet Taylorism necessitated the absorption of unions into the apparatus of state from the very beginning. The unions' role became one of assessing the value of workers as applied to production processes. However, Taylorism in the USSR never fully developed, and unions became important in implementing paternalistic workfarism and welfarism. This chapter then analyses so-called "transition" in socialist countries to marketized capitalism and in China's case to a so-called "socialist market economy". Path dependency from the socialist period led in the same direction – quasi-corporatism, with co-option by the ruling elite of federated unions and so-called tripartite structures, with China diverging somewhat in its search for a legalist solution to labour unrest. In conclusion, the demise of legacy labour representation leaves open the question of the "traditional" unions' continuing role in authoritarian states that lack the welfare state histories of their Western European counterparts.

Keywords: Unions under socialism; state-controlled unionism; Russia; China

INTRODUCTION

While the meaning of "actually existing socialism" in terms of typologizing the political economy of communist and socialist countries (state capitalism, bureaucratic collectivism, etc.) is beyond the scope of the chapter, the first section discusses in some depth the substantive character of "quasi-corporatist" (Pravda 1983) labour-enterprise-state relations in these countries in the period roughly between Stalinism in the USSR in the 1930s to the transition of China towards some form of hybrid market socialism in the 1990s and beyond. The point of this section is to argue that such corporatist dependency left an indelible mark on labour subjectivity and relations, organizational forms and resources of unionism. This focus also allows reflection on the degree to which the high socialist period (1945–89) offers points of similarity between Eastern Bloc countries and social democratic ones. While China, Vietnam, Eastern Europe and the USSR – the places with the most wide-ranging extant academic literature – were all different, their common dependency upon the party-state nexus shaped the existential terrain in profound ways. This comprised, among other things, the distance between union leadership and activism, a hidebound conservative ideology, and, relatedly, a paternalist, quasi-welfarist sense of role and identity.

The second part of this chapter analyses so-called "transition" in these countries to marketized capitalism and, in China's case, to a so-called "socialist market economy", with the latter characterized by significant political economic diversity affecting organized labour since the 1990s, as evidenced by the Chongqing model. In this section, emphasis is put upon the spatially comparative dimension. Despite the widespread perception that China was successful in avoiding the "Shock Therapy" of the 1990s that resulted in massive strikes in Russia (Mandel 2004), the path dependency created from the socialist period led in the same direction for both countries – corporatism one more time, with co-option by the ruling elite of federated unions and so-called tripartite structures. The chapter reflects on this period as offering some signs of divergence – with Russia and China again the examples. While in the former, this period is seen as a low point – the complete defeat of organized labour, enforced quiescence (Ashwin 1999), with workers sacrificed on the altar of rapid privatization, China in the late 1990s to early 2000s sees experimentation with legal arbitration (Pringle & Clarke 2011) – a false class compromise with a limited capacity to sustain itself given the upheavals of the global financial crisis of 2007–08.

This path dependency, and partial divergence in the Chinese case, leads to a paradox of organized labour in the present moment of crisis. This is the weakness in ideology and the existential terrain of what can be called the still dominant legacy forms of "transmission belt" relations (Metcalf & Li 2006) obscuring emergent forms of politicized grassroots labour organization and activism. In the final part of the chapter, future trends are examined by reflecting upon key, representative examples of labour conflict in the present (via case studies of medics and utility workers from Ukraine, and autoworkers from Russia and China). Despite China's attempt to substitute legalism for institutional development and Russia's renewed and yet stunted corporatism, a striking similarity can be observed – politicized "new" labour movements on the rise while decaying "paternalist" organizations continue to play a dominant, and detrimental, role to the cause of labour.

Indeed, one question for the future, posed by Friedman (2014) on China – but equally applicable elsewhere – is why does the particular constellation of labour unrest, failed class compromise, tactical victories and (relatively strong) structural power of labour not lead to substantive institutional change? The answer is that, given the rapidity of economic change, authoritarian states are particularly weak in substantively co-opting labour beyond phoney-corporatism, evidenced by their frequent need to resort to nationalist populism (Kalb 2011). The chapter ends, not by celebrating the demise of legacy labour representation, but by leaving open the question of the "traditional" unions' continuing role in authoritarian states that lack the welfare state histories of Western European ones.

QUASI-CORPORATIST UNIONS UNDER SOCIALISM

While historical research on workers and unions, particularly in the Soviet Union, might seem a dry exercise in scholastics, it is increasingly relevant to contemporary struggles across the globe. After all, authoritarianism and atomization in the sphere of work – two key terms relating to Stalinist labour discipline – are making something of a global comeback, in part due to the advent of platform capitalism, which itself can be seen as a novel reinvention of Taylorism. Taylorism is understood as a set of techniques to get more production out of workers though individualized specialization of work, organization, monitoring and incentivizing. Why is Taylorism important? Because right from the beginning of the "workers' state" in 1917 it had already made inroads into communists' thinking about the scientific subordination of labour and the centralized management of production that would characterize the whole socialist period (Peci 2009). A dominant view has been that Soviet industrialization essentially "imitated" the capitalist model, with all its "stigmata" (Braverman 1974, cited in Van Atta 1986).

The "militarization of labour" in the context of world war, civil war and then Stalinism, finds in Taylorism a mechanism to discipline and motivate workers, particularly via piece work in the context of a low-skills and a developmental model. This logic also necessitates the absorption of unions into the apparatus of state from the very beginning. The unions' role then becomes one of assessing the value and skills of workers as applied to production processes (in a system where productivity of output cannot be measured adequately in aggregate because of the absence of a market for goods or for labour). In unions' monitoring role and as an appendage of what we now call "management", they can both offer monetary incentives for the achievement of production goals, and allocate and distribute wider social benefits (Bailes 1977). Both have the advantage of distancing the real holders of power – the party *nomenklatura* and industrial bureaucrats – from the unseemly process of authoritarian labour relations in a so-called "workers' state". That at least was the theoretical logic of the early Soviet state. The reality was quite different, and neither Taylorist methods, nor union as ersatz management were ever effectively dominant in the Soviet system.

Despite the absence of manifest union independence after the Russian civil war, the unions' role in defending workers against abusive management was still acknowledged. Strikes remained legal until the 1930s and there are plenty of documented examples of labour strife and even union activism through the 1930s (Lebskii 2021: 100). The Labour Code of 1922 was liberal, protecting against dismissal and channelling discontent into arbitration (Mandel 2004).

The industrial history of Stalinism (1928–53) paints a model of coercion based upon the complete domination of labour by a totalitarian dictatorship, but this is simplistic at best. The standard view is that company towns are built around raw materials sites and suck in from the countryside traumatized peasants who are quickly moulded into atomized, submissive and disposable human material in Stalin's drive to industrialize before the looming world war. Company towns far from "civilization", facilitate labour discipline – physical movement within and between such industrial sites can be controlled. Unions take on an unambiguous role through rationing benefits such as housing and food. Despite the appearance of "total" control, a particular contractual model starts to emerge: workers trade freedom for at least guaranteed physical survival in a system where concentration camps are visible to all social strata.

A long-planned "autocracy in the factories", based upon Lenin's own assessment of worker control as a disaster, is complete as a system by 1929 (Lewin 1985: 196), but immediately creates the need for delegated control further down the chain – foremen, nominated by the unions, but who are also responsible to the lowest rung in the union ladder – the local workshop *savkom*. The role of unions in shaping the forms of compensation is reduced, and collective agreements abolished. Infractions of labour discipline and even things beyond worker direct control such as the breakdown of machines are criminalized, sometimes with capital penalties. At the same time, gradually, a crudely paternalistic workfarism develops – workers go to the front of the queue for provisions and shelter and maybe one day will inherit a welfare state. This is bolstered by a more consistent attempt to impose widened differentials in wages, bonuses and preferential treatments (Lewin 1985: 253).

A landmark work in this vein is Kotkin (1997) on the iron ore company town, Magnitogorsk, in the 1930s. Kotkin's approach is based on a Foucauldian analysis of power relations – where mutual surveillance and subjective incorporation of discipline and working-class "culturedness" are just as important as the machinery of the state. Power is not just about what it oppresses, but what it produces, and Kotkin's task is to answer the question: "How did it come about that within a period of less than twenty years, the revolutionary proletariat of Europe's first self-proclaimed workers and peasants state were turned into Europe's most quiescent working class?" (Kotkin 1997: 198). Kotkin's approach is aimed at correcting – via an argument about internalization by workers of aspects the disciplining ideology of the regime – what he sees as overly simplistic or incomplete explanations.

These are, in turn, long-held truisms in historical scholarship about repression and class disintegration (peasantization) in the 1930s, and then, to explain supposed support among workers for Stalinism, co-option or false consciousness. While Kotkin contrasts his approach to another prominent historian of Soviet labour, namely, Donald Filtzer (e.g. Filtzer 1986), they both focus on micro-tactics of resistance that could never become collective forms of labour organization – for a critique of Kotkin as Eurocentrically liberal, see Krylova (2000).

Kotkin's (1997: 201) addition to the analysis of "little tactics of habit" is to propose a meagre form of rather schizoid agency for workers – as both acceptance (of the subjectivizing demands of regime) and dissembling and private doubt. He also argues that unions as agents of social insurance gain prominence after 1935 in a nexus of heightened ideological attention to labour. Production becomes a source of heroic "deeds" and self-sacrifice, but also as a site of conflict about differentiated roles and rewards, and experimentation with conditions of sweating, rationalization and intensification (Kotkin 1997: 213). Unions are institutionally reincorporated, but hardly respected or trusted by either state or workers (Kotkin 1997: 369).

After Stalin, coercion slackens but does not disappear. Under his successor, Khrushchev, in 1962 a decision to raise food prices led to widespread discontent among factory workers across the USSR. While possible strike action was a subject of shop floor and even factory-wide discussion, only in the city of Novocherkassk, in an electric locomotive factory, did a labour stoppage and protest break out. When public protests threatened to spiral out of control and spread beyond the city, interior ministry troops shot dead dozens of people. While harsh, targeted repression replaced mass terror, recent archival work by historians showed discontent and the existence of the ever-present threat of labour unrest beneath the surface of Soviet society (Hornsby 2013). Unions could not respond to worker demands that the regime should live up to its own worker-orientated rhetoric.

Similarly, the repressive state apparatus had no answer to demands from below to make good on the promises of communism (which Khrushchev had promised by 1980). This situation repeated itself across the Eastern Bloc and was investigated by sociologists such as Burawoy in Hungary in the 1980s. He even coined the term "negative class consciousness" to describe this phenomenon (Burawoy & Lukács 1992). In a similar vein, Ticktin (1992: 46) argues that workers were profoundly atomized because the system was "transparently unequal and exploitative".

In this period between the 1960s and 1990s, regimes exhibit with varying degrees the willingness to make at least token concessions to workers, or at least not overstep certain bounds (Mandel 2004: 3). Union functions are threefold but with a focus on the administration of social benefits in what were now all more developed and wealthier societies. They retain (at least in the eyes of workers) the secondary role of assisting management. Defending workers is a distant third function. The regime focused on the problems of increasing productivity and compliance where there were widespread labour shortages. However, a lack of meaningful wage differentiation, and no weapons of unemployment to discipline labour, meant that unions were rarely called upon to undertake direct "defence" of members' interests. Instead, the legitimizing and broad socializing role of the union as the agent of the distribution of the so-called "social wage" should not be underestimated. While social benefit administration also fell to unions – including sick and maternity leave, pensions and vacation benefits – the so-called "social wage" was much broader, including faster access to better housing conditions, superior food supplies and more fast-tracked childcare and medical care than that accessible to many white-collar workers. This was even more important when recalling the low value of money itself in a society of shortages and, by the 1980s, rationing.

Mandel (2004) provides some much-needed nuance on what is called the "late socialist" period. He draws attention to the fact that while the regimes saw union work as primarily directed towards "productivity-orientated activity", after 1958 union agreement was required for any dismissals to occur and, at least theoretically, they could exercise power in terms of intervening in the production process where health and safety issues were a concern. In reality, most were content to subordinate themselves to management. Leaders were appointed by the party, and union officials did not emerge from the shop floor. It remained a serious political, and therefore severely punished, crime to try to "mobilise workers or assume the leadership of their spontaneous protests" (Mandel 2004: 7). At the same time, workers developed structural power despite the toothlessness of organized labour – they were able to vote with their feet, go slow (at least most of the time), and even absenteeism became difficult to effectively punish. Clarke *et al.* (1993: 16) undertook "plant sociology" just before the end of the Soviet Union, finding that "Soviet workers are powerful, in that managers are unable to impose labour discipline, and have to make concessions to enlist their co-operation, but they are weak in that they are atomised and have no means of collective resistance".

Beyond the USSR, in both Yugoslavia and Hungary, economic reforms allowed a less centralized system and unions were able to make limited interventions, especially in the area of wages. In China, from the 1950s up to the end of the 1980s, we can speak about a Soviet-type of labour unionism with partial cessation of union activities during the "Great Leap Forward", and in Romania, for example, extreme forms of control of labour representation by the Communist Party (Littler & Lockett 1983).

Between the 1950s and 1980s more than half of Africa's newly independent countries at one point or another considered themselves "socialist". Even though the demise of the USSR led to regime change in numerous self-described Marxist-Leninist states such as Ethiopia, or the dumping of most socialist political baggage in Mozambique and Zambia, socialist ideology endured in some locales, such as Tanzania in a populist variation, leading some researchers to assert that in the twentieth century the world's most dynamic socialist movements were to be found in Africa (Pitcher & Askew 2006: 1). As Pitcher (2006) showed in the case of Mozambique, the USSR was very much the model for the total incorporation of the industrial worker, even if by comparison the industrial workforce in African developmental states was much smaller. This incorporation included social benefits via union membership with unions also fulfilling a disciplining, ideological and mediating role (Pitcher 2006: 93).

Overall, the late socialist period saw the strengthening of workers' structural position, albeit in a system of decaying paternalism and economic decline. Attempts to improve productivity could not be effective in a system without unemployment and where money had little value. As Ticktin (1992: 12) memorably puts it, the Soviet worker "has to alienate his product to the management. He cannot choose not to work, but he can choose not to work as management would prefer he work." Ticktin proposes a strong form of worker "atomization" that is not fully borne out by the sociological evidence. Important legacies that impacted upon labour relations in the post-socialist period appeared, such as the delegation of production autonomy to the lowest unit of organization with inconsistent oversight, a form of genuine collective identity and even loyalty among workers towards "their" enterprise. Certainly, debate still continues as to the nature of class power and workplace relations during this period.

These are summarized by Haynes (2006: 6), who discusses the degree to which we can consider the Soviet workshop as a unique form that was then "decisive in determining the overall character of the USSR" as a mode of production. Haynes argues that a focus on workplaces is limited in what it can tell us. It might be more important to look at workers in the context of a society where consumption was suppressed and the productivity pressure upon workers was considerable. From the 1950s, serious attempts were made to develop a management ideology and normative forms of compliance. At the same time, Haynes is less convincing when he questions the importance of paternalistic relations and a web of informal and instrumental deals substituting for meaningful formal labour relations. Paternalism (and loyalty) was, and to some extent remains, visible as a legacy of the lack of labour unionism as it is understood in the social democratic period in the West as mediating relations within enterprises.

THE FALL OF THE USSR: THE ROLE OF WORKERS AND UNIONS

Under Gorbachev (1985–91), a key aspect of a general philosophy of democratization and transparency as an instrument in raising productivity and commitment saw the institution of work-collective councils (STKs) and election of plant managers. This was hardly "autonomism". Gorbachev's overall labour policy (although not worthy of that name) was a confused mix of Leninist voluntarism and renewed collectivism on the one hand, and a groping towards market mechanisms – making the threat of unemployment manifest – on the other. But 1989 was far too late to be tinkering to create market socialism, and mass labour unrest arose in Russia for the first time since 1917. The wider economic reforms of Gorbachev only undermined labour further by freeing prices too quickly (inflation), promoting private ownership in consumer goods areas in ways that disadvantaged the working class (cooperatives) and widening the "brigade" system of labour organization where manifest incentives and punishments were to be introduced.

In 1988 a conservative and official assessment of the overemployment in Soviet enterprises was 20 per cent of the workforce (Lane 1992). Inequalities between groups in society became more visible as the economy shrank and reforms were inconsistently and half-heartedly carried out. The work-collective councils were not a meaningful concession to labour – the election of managers was subject to veto by the authorities and the councils' powers were vague. More importantly, a genuine commitment to market socialism, or indeed socialism, would have meant giving workers a broader political role, but self-management was only ever envisaged at the level of the plant, not the wider economy (Mandel 2004: 11). Nonetheless, workers responded to the greater openness ("glasnost") and discussion of reform by releasing a long bottled-up militancy from 1987. This period also saw the emergence of sympathy strikes, the rise of informal leaders and the appearance of more democratic political demands. In summer 1989, 300,000 coal miners went on strike across the USSR over shortages and low wages. The government conceded to the economic demands of the strikers. By September a Russian United Workers' Front had formed. In October 1989 strikes became legal but these rights were still quite limited.

From October 1990, Mandel argues, new alternative unions were emerging spontaneously in response to demands from below for militant and active representation at the very same time as the Soviet leadership reversed course on self-management, abolishing the election of managers. He also argues that, despite success in creating new organizational forms to adapt to the new economic circumstances, labour activists were unable to think through the relationship between state and enterprise – that enterprise "independence" was meaningless without a wider societal consensus on labour policy – and a lack of mobilization from the base to articulate alternatives to privatization.

By spring 1991 the USSR was gripped by a wave of strikes, which played an indirect role in the collapse of the federal state. Workers supported Yeltsin over Gorbachev and then conceded power to leaders of the soon-to-be-independent Soviet Republics, which promptly adopted further policies that excluded workers from the reform process (Mandel 2004: 15–16). A cycle of conflict similar to that which had taken place in Poland ten years earlier now took place, with economic and organizational demands that could not but help have political implications and that precipitated wholesale systemic change. In Poland in 1980–81, the Solidarity – Solidarność – movement opened up alternatives to traditional labour unionism, including adversary unionism and neo-corporatism. In Hungary, unions operated within a system of state corporatism weakly institutionalized but more effective than in the USSR, leading Pravda (1983) to argue that unionism in communist Eastern Europe could be characterized as developing into a form of state corporatism.

Whereas the strength of the Polish Solidarność movement was enduring and brought about a broad anti-Soviet social movement, its outcome was hardly different than in Russia. The overturning of conservative communist elites merely hastened marketization, mass unemployment and deindustrialization. Summarizing this period, it is hard to disagree with Mandel (2004) that labour organization, even at its most militant, was easy prey to a hegemonic liberal narrative already well established in the 1980s and to Soviet Republic political entrepreneurs who promised more paternalism, but whose real motives were to increase their own power by destroying the federal state and accelerating marketization in order to transform themselves into capitalists.

POST-1989–91 ECONOMIC TRANSITION AND TRANSFORMATION

The rapid transition to a market economy has been dubbed "shock therapy", and it is hard to describe different Russian and East European countries' sudden encounters with unemployment, high inflation, a massive reduction in state capacity and destruction of social safety nets at this time. The new reality of overemployment in woefully low-productive and uncompetitive enterprises could only result in large-scale layoffs in the early 1990s across the former communist countries. Militancy in places such as Romania persisted longer than in other places throughout the 1990s and included periodic national strikes, but even there it could not defend standards of living (Kideckel 2008), despite maintaining high protest capacity to this day (Varga & Freyberg-Inan 2015).

In the "brave new world" of transitioning Eastern Europe, where all social strata were under severe economic pressure, militancy by working classes contributed to the emergence of a strong anti-collectivist discourse and even stigmatization of the broader working class, who were seen as retarding reform and change (Morris 2017). Even in the "poster-boy" country of disciplined transition – Poland – unemployment still stood at 17 per cent in 2001. The rise of new enterprises and transnational corporations could not compensate for the loss of state jobs in large industrial enterprises that were the mainstay of employment in communist-type economies. Nonetheless, in Russia in particular, transition was a drawn-out affair with governments throughout the 1990s forced to maintain support for at least some loss-making large-scale enterprises because the social costs of closing them would have led to mass unrest. This was compounded in many places, not just in Russia, by the legacy of urban planning under communism, which created many "company towns" – sometimes called "monotowns" – whose employment was concentrated in just one factory, which made it difficult for politicians to allow sudden closure (Crowley 2020).

While the social infrastructure such as kindergartens and canteens continued to decay or was privatized, what remained of it served as a form of paternalistic glue, cementing the worker to the enterprise into the 2000s and even substituting for welfare provision from the state itself. Thus, the 1990s and even after can be referred to as a period of decaying enterprise paternalism, fear of unemployment, demographic decline and a drawn-out period of deindustrialization hardly conducive to labour organization.

Nonetheless, from 1990 onwards there were serious attempts to build alternative unions, suggesting that transformation of the established and inflexible industrial relations system might have been possible (Gerasimaov & Bisyukov 2018; Greene & Robertson 2009). However, these attempts failed, partly because of internal organizational conflicts as well as continued support for traditional "yellow" unions from the state. While alternative unions were able to demonstrate structural power in several sectors, especially manufacturing and transport, they experienced significant difficulties in gaining stable associational power or meaningful institutional power. In Russia, the most significant achievement was the emergence of an alternative federation in 1995, KTR (Confederation of Russian Labour), which was able to carve out its own space legally as a more activist-focused organization and gain about two million members. Nonetheless, employment relations in Russia are still dominated by the traditional unions.

An end to the inertia in organized labour came about because of the growing significance of foreign firms in Russia in the 2000s. This heralded a new area of alternative unionism (Chetvernina 2009; Olimpieva 2012). Newly emerging unions, independent of the traditional system of employment relations, quickly emerged. Like the alternative unions of the 1990s, small unions at company level grew out of conflicts not about wage arrears – still a problem in "Soviet-style" factories dominated by the old unions – but in new and foreign-owned corporations. These multinationals often sold jobs as "modern", yet made deals with regional politicians to keep wages low where automation became less necessary or was forgone, as labour costs were so much lower.

A pattern emerges across Eastern Europe where various forms of manufacturing relocated from Western Europe during this period to exploit low wages. In Poland, for example, the bargaining power of UK workers was undermined when chocolate maker, Cadbury, was able to relocate there and close the UK plant (Meek 2017). Many of these transnational corporations were also given preferential legal and tax treatment because they operated in Special Economic Zones. In many zones, unions are restricted because the zones are under the authority of subnational agencies not subject to national labour laws (Neveling 2014). In the Polish case, the weakness of unions is a major attraction to the parent company. In a Russian case in the Volkswagen car plant in Kaluga, there are a more encouraging signs, which are discussed below. There, unions made use of workers' strong marketplace and workplace bargaining power in the automotive industry, mobilizing large groups of workers in the production process to achieve demands relating to wages, but also working time and workplace health and safety. Most of these small local unions are affiliated to the Interregional Trade Union of Autoworkers (Russian abbreviation: MPRA).

THE STATE OF POST-SOCIALIST WORKER POWER

As Crowley (2004: 394) observes: "Labor is indeed a weak social and political actor in post-communist societies, especially when compared to labor in Western Europe", and this is best explained by the institutional and ideological legacies of the previous period. The watchword became "quiescent" labour and unions. Crowley shows that all post-communist societies attempted some form of social compact at this time to maintain social peace, but that, as Ost memorably puts it (2000), the result was "illusory corporatism". Crowley (2004) argues that labour weakness had five key causes. First, corporatist co-option constituted by the decaying legacy unions of the communist period continuing their existence through inertia and continuing their role in acting against the interests of workers. Here, Russia's umbrella union organization was still the largest in Europe with approximately 20 million members but it was anti-militant and membership was involuntary, a legacy of compulsory enrolment in the USSR, and since corporatist unions' founding in 1990, these organizations have mainly defended their institutional "partnership" position, at the expense of their members' pay, conditions, and security (Vinogradova, Kosina & Cook 2012). Second, the relative lack of competing labour organizations or other civil society outlets for worker grievances. This will be returned to below. Third, worker organization was more effective during periods of economic growth and demand for workers. Fourth, the large size of the informal and, sometimes, subsistence economies in the 1990s in these states provided an alternative to collective action for workers. Consequently, where in-work insecurity was encountered, many choose to "exit" into the informal economy. Fifth, the general legacy of communism – both as the devaluation of class as a rallying point for organization, and the inability of unions to rethink their purpose beyond paternalistic buffers between capital and worker (Crowley 2004: 398).

Despite agreeing with the broad thrust of the argument that labour was cowed and organization severely undermined by the 1990s transition period in post-communist states, one can point to a process of diffusion and learning among new alternative unions in parallel to the decaying paternalist corporatism of the traditional unions. Ironically, new and more militant unions have tended to arise in the factories of global capital, with spontaneous and less organized forms of labour protest continuing with varying intensity in the more traditional factories remaining after the Soviet period because of these enterprises' inability to pay wages on time (Gerasimova & Bisyukov 2018). Indeed, at political and economic crisis points such as in 1998, mass labour unrest again forced temporary concessions. However, as Greene and Robertson (2009) point out, these cycles of intense and desperate protest action take place beyond the organizational structures of traditional unions in Russia.

In the less authoritarian context of Ukraine, Gorbach (2020a) found paternalistic relations more enduring and, therefore, a continuing challenge to new unions. For Ukraine coal miners, transport and utility workers, Gorbach documents examples of disorganized strikes and "spontaneous outbursts of collective protest", showing the similar informal tactics as in Russia, but in the context of a more gradual transition away from paternalism towards militant unionism (2020b: 7). Similarly in the Belarus context, Artiukh (2021: 53) draws attention to the relative strength of post-socialist workers at "moments of struggle".

Returning to Russia, in this period what is also noteworthy is the low level of solidarity between sectors; there is no history of coordination between unions and political parties, although there are trilateral agreements and institutions. Since the early 2000s an increasingly repressive labour code backed by a resurgent security apparatus has been ready to pre-empt industrial conflict by directly targeting activists. At key moments, Putin has made political interventions that combine appeals to "authoritarian" order, paternalistic rhetoric directed at workers couched in the language of social conservatism, and concessions (usually indefinitely deferred) regarding better pay and conditions. At the same time, even the stifling atmosphere of Putin's Russia cannot completely extinguish resistance on the part of labour. Labour protests elude the repressive code, or utilize the informal or indirect forms of resistance, individualized tactics or online campaigns. Key sectors such as the automotive and service industries, with intense exploitation and a field less dominated by traditional unions, represent niches for new activist labour organizers to colonize. As the nature of work changes and new forms of employment relations arise and union legacies recede, labour organization is increasingly possible, despite, or even because of, the many other obstacles in its way. This can be highlighted by the case of the MPRA union (see below).

RUSSIA'S MILITANT MPRA UNION

The most important independent union association, the MPRA, emerged in 2007 out of a local union at St Petersburg's Ford plant, after an intense, year-long labour conflict (Olimpieva 2012). The MPRA brought together members from across 40 regions. However, at both the plant and sector level, union organizations are still learning and struggle to stabilize their resources within the MPRA. The difficulty of uniting the varying interests of members and at the same time informing the workforce about current negotiations with management was a crucial task for obtaining lasting bargaining power. Similarly, bargaining power at the sector level is fragile because collective agreements, where they exist, are limited to the plant level. This keeps the unions' actions primarily local, limiting their influence. Unionists view any attempt to reach binding agreements beyond the plant as beyond their remit, and prioritize improving basic working conditions in their own factory.

The state's hostility towards union action is visible in the drastically reformed labour code of 2001. This restricted the right to action, especially for smaller, alternative unions (Greene & Robertson 2009; Olimpieva 2012). Their aims of directly affecting policy-making and influencing labour markets and social politics, or even of attaining the capacity to provoke forms of social unrest, posed enough of a threat to the government to justify these changes early in Putin's rule. Any transformation of the established system would give alternative unions opportunities to gain leverage. Therefore, the government was eager to support traditional unions as dominant actors, despite the fact that their level of approval in society is continually eroding.

So far, as a small union association mostly in transnational and automotive companies within a fragmented system of employment relations, the MPRA's scope to expand to the broader working class or society is limited. Overall, this new movement is marked by the difficulties in transforming its successes at the plant level, and even along value chains, into lasting organizational power and meaningful influence in institutions and politics, with the latter being particularly restrained by authoritarianism. It remains an open question whether new unions will be able to not only survive but evolve under these hostile circumstances.

Hins and Morris (2017) undertook an in-depth case study of Volkswagen in Russia from 2009–13. The new German plant was successfully unionized by the MPRA with significant outside support, shortly after the factory opened. By 2012, and with some 1,200 workers organized, the union gained formal and legal recognition from management, becoming the dominant union at the plant, although membership was relatively low, around 20 per cent of production operatives. "Traditional" union representation was also present, which MPRA activists described as "Soviet-minded" (using the "yellow" label). There was obvious significant cooperation with management. Relations between management and the MPRA were tense and difficult from the very beginning. The German management underestimated the MPRA's ability to gain access to the plant and "salt" the workers with well-trained activists.

In 2012 the union entered into collective bargaining with the plant owners – significant because of the high threshold of membership legally required for recognition of the union – and in 2013 was effective in calling for strikes (the legality of which was ruled unlawful by a court) and protests. A mark of the "novelty" of this kind of labour activism is the use of unconventional and hybrid tactics – such as protest actions, political pickets of non-VW car dealerships, sabotage of production, as well as working-to-rule, known in Russian as the "Italian strike". The union was relatively successful, winning concessions on pay and reductions in agency labour contracts. Spurred on, the union's second campaign sought to reduce shift lengths and limit the hours in the working week. At the time of writing in 2022, the union remains the key player at the plant but less able to mobilize than before.

CHINA: FROM THE PEOPLE'S REPUBLIC TO MARKET SOCIALISM

As in Soviet Russia, after 1949 the Chinese Communist Party's core concern was "work stability". In the initial period, industrial relations were characterized by *danwei*, a collective institution similar to the social contract forms of paternalist for workers in the Soviet system, but also containing instruments to ensure low labour turnover, centralized decisions on wages and ideological efforts to integrate managers and workers (Pringle 2011). The infamous *hukou* regulations limiting labour migration underpinned the *danwei* from the 1950s. Pringle sees this party-led administrative system of industrial relations as persisting into the reform period and its withdrawal as presenting unions with a crisis in legitimacy. There is some evidence that Chinese workers benefited more overall from this system than their Soviet counterparts: lower wage differentials between workers, forced inclusion of managers on the shop floor and also more militancy (Lee Ching-Kwan 2000, cited in Pringle 2011).

While during the "Cultural Revolution" of 1966–76, union activity was virtually prohibited, this period was ambiguous in terms of working-class opportunity and identity. The industrial reform period can be dated from 1978 where enterprise autonomy focused on profit retention and a small bonus paid to the workforce. Further elements included production planning, marketing and capital utilization (Pringle 2011).[1] In 1982 the right to strike was removed from the constitution. One effect of later 1980s reforms was the removal of the system of guaranteed permanent employment that a small number of urban workers had enjoyed and the state's control over job allocation. Like in Soviet Russia, the removal of price controls and rises in inflation were a source of discontent for workers. In 1989 students protested in Tiananmen Square against corruption and for democracy. The Beijing Workers Autonomous Federation (BWAF) showed solidarity with the students and was critical of official labour unionism – the fifth example in modern China's existence of workers asserting political demands (Chan 1993). As a result, the Chinese Trade Union Federation called for the banning of the BWAF and conflict and repression ensued. Another important factor in the reform process was that by the late 1980s the roll out of Special Economic Zones was well underway in more than a dozen coastal cities. Pringle (2011) examined whether workers themselves exercised agency in winning the benefits of *danwei*, or whether the system was really just a form of control over the working class. The former interpretation then sees events in the late 1980s as a rearguard action by workers to defend the social and economic gains from globalization and reform.

While, in a sense, China's reform path was taken more slowly and conservatively, it entailed massive layoffs in a short period from 1995 to 2000 – state employment fell by 48 million in that period (Cai 2005: 1). Thus, it is the violation of the "subsistence ethic", known as the "iron rice bowl", and economic grievances that primarily drove labour protest in this period (Chen 2000). In a wide-ranging comparison of Russia and China in 2005, Clarke (2005: 4) draws attention to further commonalities during the periods of marketization, such as the attempts in the 1990s to exert more control over workers, and the attraction to reforming elites of the role of unions in ameliorating tensions rather than restraining managerial ambitions. In characterizing the 1980s to the present as a drawn-out period of the dismantling of all social and economic guarantees by (former communist) states towards workers, Clarke emphasizes not the conservatism or indolence of union leaders, but the structural limitations legacy unions are subject to. Unlike in Russia, Clarke sees the outcome of reform in China as characterized by a paradox of both stricter subordination of unions (because of Tiananmen) but also an enhancement of their status and ability to promote unions' own position – within certain limits (Clark 2005: 10).

In this sense, the Chinese Communist Party's chosen path has been more effective in stifling the opportunities for independent unions to emerge. Nonetheless, strikes and protests, particularly in southern China, came to a head in 2002 – largely outside union structures and challenging their legitimacy. Clarke and Pringle (2011: 64–6) provide evidence of a similar arc of labour militancy in Vietnam rising in the 2000s reaching its height in 2008 and mainly in private and foreign-owned enterprises. Unions there showed little appetite for militancy and conflict took the form of direct action because

1. The rest of this paragraph's historical summary is mainly taken from Pringle (2011).

of the failure of the dispute resolution system – an overly complex legalistic procedure (Clarke, Lee & Chi 2007).

Scholarship on the renewed labour conflicts in China in the twenty-first century discusses the challenges to cooperation and collective resistance. The main targets of action are local governments and state-owned enterprise management where workers aim for concrete economic goals in a context where there exists a political space for collective action to be exploited, and local, peaceful, non-political action is non-contentious but still a channel to pursue workers' interests (Cai 2005). The solution from the state's perspective has been the development of collective consultation that only involves the top hierarchy of unions. A second aspect is contractual regulation that contains little detail and only minimum standards in wages and safety, essentially just confirming national and local laws. This legalistic tripartite solution was inadequate to stop exploitative forms of labour relations in China in the early 2000s where there were over 100,000 workplace deaths due to accidents (Metcalf & Li 2005), and ineffective and far short of any genuine collective bargaining (Clarke, Lee & Li 2004).

The legal regime in labour relations has not prevented growing unrest and protest, particularly after the 2008–09 global financial crisis. Chinese migrant workers have become more able to challenge the state's regulatory regime but are as yet constrained judicially and, therefore, a breakthrough to introduce the recognition of collective rights is unlikely because of both the continuing manipulation of unions by the state and the influence of global capital (Chan 2014). It appears that China will continue to try to impede the emergence of collective contestation by unions by pursuing corporate social responsibility programmes, but this will hardly prevent future conflict.

In a case study on Honda, Chan (2014) found strikes in 2010 heralded a new stage of labour resistance because of their success, duration, level of organization and political demands (for democratic union reform). Unlike in Russia, these strikes also acted as an impetus for other autoworkers in China. Unlike in Russia, these actions also forced the issue on to the agenda of the Chinese Federation of Unions. According to Chan, only the opposition of overseas business interests halted the process of transformation of unionism in China into a system of collective bargaining. Since those 2010 strikes, Zhang and Yang (2019) are more guardedly optimistic – challenging the view that labour unionism is a dead end but sensitive to the possibility of a more repressive turn. They find that democratically elected unions exploit a limited space of opportunity in the system to carry out genuine collective bargaining and incrementalism, which have improved workers' lives.

Friedman (2014) proposes that a key problem in China is the "insurgency trap" of labour politics – cycles of disorganized violent confrontation over labour conditions, the formation of informal associations that try to resolve grievances, and vigilante and police responses. Insurgent workers are dispersed in so many different enterprises that unions struggle to build meaningful connections with them. Therefore, labour unrest is more like a social movement than the classical framing of labour organization and contention. Thus, the density of informal interpersonal networks engenders trust, solidarity and reliability in protest mobilization (Liu 2016), yet fails to translate into coordination and solidarity between different workers (Cai 2005: 6).

Finally, turning again to Africa in the contemporary period, one can observe similar legacies to those that exist in the Global North post-socialist states – widespread yet decaying membership, organized and less organized strikes in response to privatization and other common grievances, along with co-opted and demoralized unions. Writing on Tanzania railway privatization and strikes in Mozambique, Pitcher and Askew (2006) argue that the utopian imaginings of the socialist period continue to inspire overt resistance in the present. Collective identities opposed to privatization in Tanzania were partly facilitated by union work. More broadly, research indicates that the notion of unions as engines of social movements is widely accepted in the African cases. Waterman (1993) argues, in the case of South Africa in the 1980s, that mineworkers unions were not "traditional" but inflected by cultural affinities and community loyalties.

While South Africa was never a socialist state, the post-apartheid country has been shaped by the African National Congress, a member of the Socialist International and an uneasy alliance of nationalists and socialists. South Africa's transition can, therefore, be compared to those in post-socialist countries in terms of its neoliberal turn and privatization (Peet 2002). Further, unions in post-socialist countries from Africa to Eurasia similarly face the challenge of the growing salience of how to connect to informal workers, something African unions, with their experience of alliance with broader societal militancy, are arguably better equipped to do (Ness 2016).

CONCLUSION

Tictkin (1992), one of few to predict the USSR's demise, focused upon workers' structural power under communism as a key factor in the system's inability to reform itself. The USSR after Stalin was fundamentally unable to discipline workers. Harrison (2002) argues that the incentive system failed to raise productivity and, under Gorbachev, workers lost their fear of penalization and were rewarded with concessions, accelerating a breakdown in Soviet institutions. Ironically, the Soviet project partly failed because the structural strength of the working class prevented rapid and full marketization, so to "succeed" the system had to destroy itself, and the lives of countless workers and their families in the process (Ticktin 1992).

Scholarship since 2010 on unions, and labour power more generally after communism, can be characterized as moving away from an interpretation of quiescence and decaying paternalism towards a greater appreciation of three factors. First, an understanding that the potential power of workers' structural position has manifest effects – in the Russian case, the latent threat of labour-related unrest was enough to worry even the authoritarian Russian state. Second, researchers pay attention to how new independent unions, even if only occasionally successful, provide models of learning and action for others; these forms of action are hybrid forms of labour agency rather than traditional balloted strikes such as informational pickets. In the words on Wada (2012), strikes have become "modular". In the West the "strike" has been transferred to non-traditional labour contexts (sex strikes, student strikes, climate strikes). But we can equally talk about modular labour protest in post-communist countries that use repertoires of contention borrowed, transferred and sometimes adapted from other actors, issues and locations. Third, and relatedly, scholars acknowledge the influence of social movement theory on labour organization, agency and action – although this position remains subject to critique (Neary 2002).

Varga (2014), in Romanian and Ukrainian contexts, draws specific conclusions about the difference between labour strength and representation. Representation requires unions that are autonomous, legitimate and effective. This is simply not possible consistently in places such as Ukraine and Russia because most union activity is co-opted or controlled by management. Varga also reminds us that despite continued paternalism and legal constraints, there are specific strategies unions can use to succeed in protecting workers, including making their threat-potential known, sometimes in a political manner. So, it is important to recognize that nationalist challenging of grievances by elites has been successful and that the decay of legacy unions continues to be a distracting, and challenging elite strategizing, even in the most authoritarian regimes such as Belarus and Russia, can be countered by workers if they are able to articulate labour interests.

Varga (2014: 7–8) argues that "loyalty" was a barrier to contention in the socialist period and still has echoes today. Attachment via feelings of loyalty prevents unrest – an argument of relevance too beyond post-communist countries expressed in the phrase attributed to Gramsci that "hegemony is born in the factory" (Varga 2014: 189). Varga does not argue that this is a legacy of communism, but this is worth considering, not necessarily as part of the cultural transference of labour relations from that period, but as a structuring constraint given the insights of various scholars about the continuing relevance of paternalistic relationships. It might not be the case that employers can meaningfully offer workers anything comparable to the social wage they received during communism, but we should not discount how the language of paternalism affects workers living in extremely difficult material circumstances in states where authoritarian populists can blame others and simultaneously prevent any circulation of discourses of worker solidarity or leftist ideology. The shadow of the discrediting of socialism is a long and dark one for workers in Eastern Europe. However, paternalism is still a meaningful language of communication and may yet provide some common terrain for workers and unions together to transform corporatism into contention and credible collective action. Numerous scholars draw on the concept of "moral economy", arguing that the articulation of entitlements represents a meaningful resource to unions today and a positive legacy of the Soviet period (e.g. Morrison, Croucher & Cretu 2012).

Nonetheless, the appearance of new independent and democratic unions is possible, even in increasingly authoritarian post-communist states. New unions such as the one in Kaluga use unconventional methods of protest to promote worker interests. Entangled interconnections and dependencies of transnational firms along the value chain, as well as the differing national, political and economic determinants of former socialist countries, make an appraisal of the situation of workers and their unions challenging. New unions successfully represent workers and challenge legacy systems by comprehensively organizing members, often in foreign-owned firms. Nevertheless, the prospects for lasting consolidation are not overly positive. Workers still have high primary bargaining power in markets where demographic change and pauses in productivity gains from automation constrain firms' actions.

The Russian case of the MPRA shows a notable drop in members due to the progressive deterioration in automotive employment associated with ongoing economic problems in Russia, leading to stagnation in the development of associational power. The unions' exclusive focus on the local level, while successful, precludes pursuing sector-wide and regional agreements. This obstacle continues up to institutional level, where those new union formations have practically no way of overcoming the stalled institutions of employment relations marked by a continuing monopoly of traditional unions and a pseudo-paternalist state. Thus, a shift in the power balance of this established system is a long-term prospect.

Ironically, it is the actions of authoritarian states that have the potential to accelerate matters. Continuing austerity policies in the public sector in Russia have led to more grassroots labour organization among public sector workers – in 2019, 20 per cent of labour protests were by medical workers protesting low wages (TsSTP 2020). If activist unions are to regain the initiative, they need to transition from the locales of material production and enter the fray where neoliberalization is now at its most disruptive in Russia – in the public sector and among the new service sectors such as Uber and food delivery (Morris 2023).

REFERENCES

Artiukh, V. 2021. "The anatomy of impatience: exploring factors behind 2020 labor unrest in Belarus". *Slavic Review* 80(1): 52–60.

Ashwin, S. 1999. *Russian Workers: The Anatomy of Patience*. Manchester: Manchester University Press.

Bailes, K. 1977. "Alexei Gastev and the soviet controversy over Taylorism, 1918–24". *Soviet Studies* 29(3): 373–94.

Braverman, H. 1974. *Labor and Monopoly Capital: The Degradation of Work in the Twentieth Century*. New York: Monthly Review Press.

Burawoy, M. & J. Lukács 1992. *The Radiant Past: Ideology and Reality in Hungary's Road to Capitalism*. Chicago, IL: University of Chicago Press.

Cai, Y. 2005. *State and Laid-Off Workers in Reform China: The Silence and Collective Action of the Retrenched*. London: Routledge.

Chan, A. 1993. "Revolution or corporatism? Workers and trade unions in post-Mao China". *Australian Journal of Chinese Affairs* 29: 31–61.

Chan, C. 2014. "Constrained labour agency and the changing regulatory regime in China". *Development and Change* 45(4): 685–709.

Chen, F. 2000. "Subsistence crises, managerial corruption and labor protests in China". *China Journal* 44: 41–63.

Chetvernina, T. 2009. *Trade Unions in Transitional Russia: Peculiarities, Current Status and New Challenges*. Russian Research Center Working Paper Series 16. Kunitachi, Tokyo.

Clarke, S. 2005. "Post-socialist trade unions: China and Russia". *Industrial Relations Journal* 36(1): 2–18.

Clarke, S., C. Lee & D. Chi 2007. "From rights to interests: the challenge of industrial relations in Vietnam". *Journal of Industrial Relations* 49(4): 545–68.

Clarke, S., C. Lee & Q. Li 2004. "Collective consultation and industrial relations in China". *British Journal of Industrial Relations* 42(2): 235–54.

Clarke, S. *et al.* 1993. *What About the Workers? Workers and the Transition to Capitalism in Russia*. London: Verso.

Crowley, S. 2004. "Explaining labour weakness in post-communist Europe: historical legacies and comparative perspective". *East European Politics and Societies* 18(3): 394–429.

Crowley, S. 2020. "Global cities versus rustbelt realities: the dilemmas of urban development in Russia". *Slavic Review* 79(2): 365–89.

Filtzer, D. 1986. *Soviet Workers and Stalinist Industrialization: The Formation of Modern Soviet Production Relations, 1928–1941*. London: Pluto.

Friedman, E. 2014. *Insurgency Trap: Labor Politics in Postsocialist China*. Ithaca, NY: Cornell University Press.

Gerasimova, E. & P. Bisyukov 2018. "The strike movement and labour protests in Russia". In *Workers' Movements and Strikes in the Twenty-First Century: A Global Perspective*, edited by J. Nowak, M. Dutta & P. Birke, 289–306. London: Rowman & Littlefield.

Gorbach, D. 2020a. "Industrial hegemony and moral economy in a Ukrainian metalworking city". *Politix* 132(4): 49–72.

Gorbach, D. 2020b. "Changing patronage and informality configurations in Ukraine: from the shop floor upwards". *Studies of Transition States and Societies* 1: 3–15.

Greene, S. & G. Robertson 2009. "Politics, justice and the new Russian strike". *Journal of Communist and Post-Communist Studies* 43(1): 33–54.

Harrison, M. 2002. "Coercion, compliance, and the collapse of the Soviet command economy". *Economic History Review* 55 (3): 397–433.

Haynes, M. 2006. "Rethinking class power in the Russian factory 1929–1991". Working Paper Series, Management Research Centre, Wolverhampton University Business School.

Hins, S. & J. Morris 2017. "Trade unions in transnational automotive companies in Russia and Slovakia: prospects for working-class power". *European Journal of Industrial Relations* 23(1): 97–112.

Hornsby, R. 2013. *Protest, Reform and Repression in Khrushchev's Soviet Union*. Cambridge: Cambridge University Press.

Kalb, D. 2011. "Introduction". In *Headlines of Nation, Subtexts of Class: Working-class Populism and the Return of the Repressed in Neoliberal Europe* edited by D. Kalb & G. Halmai, 1–56. New York: Berghahn.

Kideckel, D. 2008. *Getting by in Postsocialist Romania: Labor, the Body, and Working-Class Culture*. Bloomington, IN: Indiana University Press.

Kotkin, S. 1997. *Magnetic Mountain: Stalinism as a Civilization*. Berkeley, CA: University of California Press.

Krylova, A. 2000. "The tenacious liberal subject in Soviet studies". *Kritika: Explorations in Russian and Eurasian history* 1(1): 119–46.

Lane, D. 1992. *Soviet Society Under Perestroika*. Revised edn. London: Routledge.

Lebskii, M. 2021. *Rabochii klass SSSR: shisn' v usloviiakh promyshlennogo paternalisma*. Moscow: Gorisontal.

Lewin, M. 1985. *The Making of the Soviet System*. Methuen: London.

Littler, C. & M. Lockett 1983. "The significance of trade unions in China". *Industrial Relations* 14(4): 31–42.

Liu, J. 2016. "Credibility, reliability, and reciprocity: mobile communication, guanxi, and protest mobilization in contemporary China". In *Asian Perspectives on Digital Culture: Emerging Phenomena, Enduring Concepts*, edited by S. Sun, R. Cheryll & R. Soriano, 69–84. New York: Routledge.

Mandel, D. 2004. *Labour After Communism: Auto Workers and Their Unions in Russia, Belarus and Ukraine*. Montreal, ON: Black Rose Books.

Meek, J. 2017. "Somerdale to Skarbimierz". *London Review of Books*, 20 April.

Metcalf, D. & J. Li 2005. *Chinese Unions: Nugatory or Transforming. An Alice Analysis*. London: Centre for Economic Performance.

Morris, J. 2017. "An agenda for research on work and class in the postsocialist world". *Sociology Compass* 11(5).

Morris, J. 2023. "Activists and experiential entanglement in Russian labor organizing". In *Varieties of Russian Activism: State-Society Contestation in Everyday Life*, edited by J. Morris, A. Semenov & R. Smyth. Bloomington, IN: Indiana University Press.

Morrison, C., R. Croucher & O. Cretu 2012. "Legacies, conflict and 'path dependence' in the former Soviet Union". *British Journal of Industrial Relations* 50(2): 329–51.

Neary, M. 2002. "Labour moves: a critique of the concept of social movement unionism". In *The Labour Debate: An Investigation Into the Theory and Reality of Capitalist Work*, edited by A. Dinerstein & M. Neary, 149–78. Aldershot: Ashgate.

Ness, I. 2016. *Southern Insurgency: The Coming of the Global Working Class*. London: Pluto.

Neveling, P. 2014. "Structural contingencies and untimely coincidences in the making of neoliberal India: the Kandla foreign trade zone, 1965–1991". *Contributions to Indian Sociology* 48(1): 17–43.

Olimpieva, I. 2012. "Labour unions in contemporary Russia: an assessment of contrasting forms of organization and representation". *Journal of Labor and Society* 15(2): 267–83.

Ost, D. 2000. "Illusory corporatism: tripartism in the service of neoliberalism". *Politics and Society* 28: 4.

Peci, A. 2009. "Taylorism in the socialism that really existed". *Organization* 16(2): 289–301.

Peet, R. 2002. "Ideology, discourse, and the geography of hegemony: from socialist to neoliberal development in post-apartheid South Africa". *Antipode* 34(1): 54–84.

Pitcher, M. 2006. "Forgetting from above and memory from below: strategies of legitimation and struggle in postsocialist Mozambique". *Africa* 76(1): 88–112.

Pitcher, M. & K. Askew 2006. "African socialisms and postsocialisms". *Africa* 76(1): 1–14.

Pravda, A. 1983. "Trade unions in east European communist systems: toward corporatism?" *International Political Science Review* 4(2): 241–60.

Pringle, T. 2011. *Trade Unions in China: The Challenge of Labour Unrest*. Abingdon: Routledge.

Pringle, T. & S. Clarke 2011. *The Challenge of Transition: Trade Unions in Russia, China and Vietnam*. Basingstoke: Palgrave Macmillan.

Ticktin, H. 1992. *Origins of the Crisis in the USSR: Essays on the Political Economy of a Disintegrating System*. London: Routledge.

TsSTP 2020. *Kak protestuiut rossiiane: resultaty monitoring protestnoi aktivnosti*. http://trudprava.ru/expert/article/protestart/2184

Van Atta, D. 1986. "Why is there no Taylorism in the Soviet Union?" *Comparative Politics* 18(3): 327–37.

Varga, M. 2014. *Worker Protests in Post-Communist Romania and Ukraine: Striking with Tied Hands*. Manchester: Manchester University Press.

Varga, M. & A. Freyberg-Inan 2015. "Post-communist state measures to thwart organized labor: the case of Romania". *Economic and Industrial Democracy* 36(4): 677–99.

Vinogradova, E., I. Kosina & L. Cook 2012. "Russian labour: quiescence and conflict". *Communist and Post-Communist Studies* 45(3/4): 219–31.

Wada, T. 2012. "Modularity and transferability of repertoires of contention". *Social Problems* 59(4): 544–71.

Waterman, P. 1993. "Social-movement unionism: a new union model for a new world order?" *Review* (Fernand Braudel Center) 16(3): 245–78.

Zhang, L. & T. Yang 2019. "Worker activism and enterprise union reform in China: a case study of grassroots union agency in the auto parts industry". *Development and Change* 53(2): 396–423.

CHAPTER 11

THE NEOLIBERAL LOW POINT

Chris Howell

ABSTRACT

The chapter is concerned with the 40-year crisis of organized labour in the advanced capitalist world since the end of the 1970s. While that crisis has had an uneven spatial spread and temporal evolution, the overall trajectory of organized labour has been, by all conventional measures, one of unrelenting secular decline. The chapter argues that labour strength and influence have been shaped by the underlying growth model and the extent to which unions and collective regulation contribute to or conflict with the dominant form of growth. As such, the collapse of Fordism and its replacement by a range of accumulation models not dependent upon wage-led growth have been central to the story of decline. The chapter also argues that foregrounding neoliberalism as a form of regulation of contemporary accumulation helps explain how and why labour movements have experienced such reversals and declines over the last four decades.

Keywords: Union decline; neoliberalism; challenges for unions

INTRODUCTION

Previous chapters have traced the rise of organized labour in both Western countries (Chapter 9) and state socialist ones (Chapter 10) to their highpoints in the 1970s. These highpoints were characterized by organizational strength, political influence and economic power. Indeed, it is often forgotten that the 1970s was widely seen as a moment when the long postwar class compromise disintegrated, not because of the strength of capital, but rather that of labour (Crouch & Pizzorno 1978). By the end of the 1960s, in country after country, labour militancy reflected a growing self-confidence on the part of workers and a willingness to challenge the Fordist bargains of social democratic countries, the Keynesian-welfarist compromises of most capitalist democracies, the indicative planning model of developmental states and the forms of labour paternalism in state socialist countries. In many, if not most cases, this newfound militancy involved challenging union bureaucracies (see Chapter 15) themselves through unofficial or wildcat industrial action. The response of governments, particularly those on the left, was to move left to accommodate this challenge, through new rights to unions, an extension of co-determination, a further decommodification of the labour market, and experiments in fund socialism, worker self-management and industrial planning. Few observers in 1975 would have then anticipated the long 40-year crisis of organized labour that ensued and that is the concern of this chapter.

One observer who did was the historian Eric Hobsbawm, who in 1978 gave the Marx Memorial Lecture with the remarkably prescient title, "The Forward March of Labour Halted?". Its focus was the UK and its analysis tentative, but in retrospect it understood very well the changes to the composition of the working class and the constitution of postwar capitalism that have accumulated and contributed to the crisis and decline experienced by unions, a decline that has only deepened and become well-nigh universal in the intervening period (Baccaro & Howell 2017). We have now entered the fifth decade of neoliberalism, and while this chapter will chart its uneven spatial spread and temporal evolution across Western Europe, North America and Australasia over that time, the trajectory of organized labour has been, by any conventional measures, one of unrelenting secular decline.

The decline in union membership across the advanced capitalist world has been to some degree hidden by its different timing in different countries. For the first two decades of decline, it was always possible to point to countries where union density was still growing or at least was not in retreat. Pontusson (2013: table 1) had identified three distinct periods of decline from peak union membership. The peak was reached in the 1960s in several countries, including the US, France, Japan, Norway and Austria. Another larger group of countries saw peak union density in the period between 1976 and 1989, including Australia, the UK, Italy, Spain, Denmark and Canada. For a final small group, which included Germany and Sweden, the peak came in the first half of the 1990s. And yet decline has been universal, regardless of the start date. As Avdagic and Baccaro (2014: 706) put it in their survey of employment relations: "The data suggest that union decline is a truly general phenomenon, which is not just limited to the CEE [Central and Eastern Europe] and LME [Liberal Market Economy] countries, but involves the CME [Coordinated Market Economy] countries as well."

It should be noted that collective bargaining coverage has been more resilient, although even here a clear trend towards decentralization is evident (Avdagic & Baccaro 2014; Baccaro & Howell 2017; Visser 2019), but this reflects not union strength but rather state extension mechanisms and the existence of employers' associations. Similarly – and discussed below – the revival of social pacts and other forms of concertation from the 1980s onwards was a result of state efforts to manage liberalization rather than labour strength. The other unambiguous measure of the decline of organized labour is the near universal collapse of strikes (Avdagic & Baccaro 2014: 709). While one could at one point during the social democratic highpoint have made the case that a low strike rate reflected a labour movement so strong it did not need to exercise its power, that case is no longer tenable.

If the dawn of the neoliberal era, at least with regard to organized labour, can be dated to the near simultaneous assaults on union power by the Thatcher government and Reagan administrations. These are neatly symbolized by the state response to the miners' strike in the UK in the mid-1980s and the air traffic controllers' strike in the US in the early 1980s. Thus, in the ensuring decades, unions everywhere have experienced neoliberalism as a 40-year class war, sometimes cold, sometimes hot, but with no return to the compromises and bargains of the first 35 years of the postwar era.

The chapter is divided into four parts. The first gives the political-economic context of the period, with an emphasis upon the changing capitalist growth model after Fordism. The second part examines the concept of neoliberalism and develops an argument to help explain why its spread has been so destructive of the institutions and resources of organized labour. The third part is comparative, examining varieties of neoliberalism and varieties of union experience across the advanced capitalist world. The final part is both conclusion and a look forward to how organized labour has adapted, whether there are reasons to believe that union decline is at an end, and whether the Covid-19 pandemic presages a retardation or acceleration of the trends of the last four decades.

CHANGED CONTEXT FOR ORGANIZED LABOUR

The next section will more directly examine neoliberalism as the mode of regulation of contemporary capitalism, while this section is primarily concerned with what changed in the context facing unions in the advanced capitalist world from the early 1980s onwards. That is too broad a task for one section of one chapter of this handbook. There is voluminous material on changes in class structure, economic conditions, political situation, and so broad are the combined changes as to make the equivalent context of the highpoints of the 1970s almost unrecognizable from the current period (Martin & Ross 1999). Rather, this section will identify some of the most significant changes but focus upon the evolution in capitalist growth models. The main claim here will be that labour strength and influence are shaped by the underlying growth model and the extent to which unions and collective regulation contribute to or conflict with the dominant form of growth. As such this section is heavily indebted to the "Regulation Theory" approach to political economy (see Boyer 1990).

The social democratic highpoint was marked by different national versions of a shared growth model that has generally been described as Fordism. What was distinctive to Fordist growth was the linkage of mass production and mass consumption through a series of institutional mechanisms designed to solve the endemic over-accumulation problem faced by capitalist economies (Harvey 1989). In its ideal-typical form, mass production technologies in oligopolistic sectors of manufacturing produced high levels of long-term productivity growth. Some portion of productivity gains was shared with workers to ensure adequate demand creating – again in ideal-typical form – a virtuous cycle of productivity growth, wage growth, demand growth and profit growth (Glyn *et al.* 1990). The primary institutional mechanism for tying productivity growth to wage growth was collective bargaining, usually on the sectoral or industrial level where wages could be taken out of competition. In country after country, some version of industry bargaining, pattern bargaining or corporatist bargaining not only offered workers job security and a certain level of material comfort, but also benefited capitalist growth.

Whether described as the "golden age of capitalism", or the "*trente glorieuses*", it carved out a class compromise that, while fragile and constantly subject to challenge and industrial conflict, nonetheless lasted until the late 1970s. While the core elements of Fordist growth only applied to countries with a substantial mass production sector, large domestic consumption and unions strong enough to take advantage, its influence affected a much broader swathe of countries with small export-led economies dependent upon mass consumer markets in the Fordist core, and an assortment of institutional forms of state substitution for the effects of Fordist collective bargaining, such as Keynesian demand management, aggressive use of a statutory minimum wage and extension mechanisms or awards that applied Fordist agreements reached in one part of the economy more broadly. But whatever the specific mechanism, the key point from the perspective of organized labour was that Fordist growth made unions and collective bargaining a public good, a positive benefit for capitalist growth (Baccaro & Howell 2018). Thus, what was good for General Motors might be good for America, but what was good for its production line workers was also good for capitalism. This partially insulated unions from the sort of industrial and political attacks that came later.

For reasons beyond the scope of this chapter, the Fordist growth model appeared exhausted by the early 1980s and came under sustained challenge from capital and its political allies. Categorizing and explaining what has replaced it – "post-Fordism" – has been something of a growth industry in itself with a cascading set of models finding academic favour. That search has had two main tracks. One sought to expand upon evidence of an alternative form of growth alongside the dominant Fordist one and which instead rested upon high skill, high quality, "flexible specialization", emblemized by the German Mittelstand firms (Piore & Sabel 1984). By the end of the century this approach had become embedded in the influential "Varieties of Capitalism" (Hall & Soskice 2001) approach to comparative political economy that posited two possible production models: a liberal market economy based around low wage, low skill and low value production, which best operated with largely unregulated labour and capital markets, and a coordinated market economy in which bargaining and deliberation between class actors served to regulate economic activity in service to high wage, high skill and high value "diversified quality production" (Streeck 2009: 110). There was a role for organized labour in determining wages, ensuring worker voice and solving collective action problems in the provision of training in the second model but none in the first.

The second track, more indebted to a regulationist approach, sought to identify emerging features of contemporary capitalism and their contribution to a new growth model or regime of accumulation. The question for this set of scholars was whether an alternative to Fordist growth had emerged, and particularly one able to solve its chronic demand problem or, rather, whether capitalism had entered a stage of endemic instability, crisis and short-term fixes. Again, this literature is too large to do justice to here, but three elements of it are worth highlighting. First, a form of growth that is the inversion of Fordism, what Harvey (1989) characterized as flexible accumulation and Jessop (1993) labelled "Schumpeterian workfare" where constant innovation and revolutionizing of production is accompanied by a redisciplining of the labour market, driving down wages. Both the expansion of the post-industrial service sector through a ruthless anti-Fordist emphasis upon precarity, surveillance and low wages exemplified by Walmart in the US (Lichtenstein 2006), and the platform economy that accompanies digital capitalism (Srnicek 2017) are consistent with this use of new forms of extraction (the exploitation of data, for example) and production forming the basis of growth without an accompanying source of demand.

Second, financialization as a distinct growth model of its own serving, both to discipline labour markets through the imposition of shareholder value as an imperative and to incorporate debt as substitute source of consumption, whether through credit card debt, housing bubbles, student debt or the myriad other ways in which the lifecycle has been cannibalized by financial markets (Durand 2017; van der Swan 2014; Umney *et al.* 2017). Finally, much of this research has been incorporated into the emergence of a new growth models approach that identifies a range of growth models, including consumption-led and export-led growth (Baccaro, Blyth & Pontusson 2022). For the purposes of this handbook, the broad conclusion of this line of research is that none of these growth models rests upon a functional role for organized labour or wage-led demand. As a result, and perhaps unsurprisingly, they are all highly unstable and subject to crisis (Baccaro & Howell 2018).

While always an exaggeration, nonetheless, one could make a plausible case that in many countries unions did not have to work very hard to thrive during the period from the end of the Second World War until the end of the 1970s. Certainly strikes occurred, although more where labour was less strongly institutionalized (France, Japan and Italy, for example), but strikes were less part of an existential struggle for survival on the part of unions than they have become since the early 1980s. As long as they remained within the broad parameters of Fordist bargaining – avoiding significant incursions into managerial prerogative and keeping wage demands below productivity gains – unions contributed to and benefited from the growth model.

As a result, business either explicitly through closed shops, or tacitly through dues check-off systems, organized workers on behalf of unions. The structure of mass production workplaces, with the preponderance of semi-skilled workers and high levels of job security, encouraged a certain union consciousness that made membership a norm. Governments either provided positive legal protections for organizing and industrial action or carved out immunity for unions. Governments across the political spectrum tolerated unions and collective regulation because of their role in the dominant growth model, but ties were especially close between social democratic and labour parties and organized labour in numerous countries (Allern & Bale 2017; see also Chapter 9). Unions in several countries, Australia, the UK and Sweden, for example, had formal constitutional roles in their respective centre-left parties borne of a shared history, whereas unions elsewhere had informal but no less important roles in influencing party policy and leadership selection (see also Chapter 9). A dual bargain informed this party-union relationship. A political bargain traded the votes of blue-collar union members for a role in setting party policy, while an economic bargain traded "responsible" behaviour in the labour market, in some cases institutionalized in corporatist incomes policies (Panitch 1986), for full employment policies and an expansion of the social wage (Howell 2001).

While not mono-causal, all these elements of the postwar model of labour regulation became subject to erosion and reversal once the Fordist growth model appeared no longer able to provide the basis of a positive and productive class compromise (Harvey 1989). To the extent that the earlier era incorporated a functional role for labour in the growth model, that had ceased to be the case by the end of the 1990s. The 1980s and 1990s saw capital become more hostile to organized labour: strikes were more likely to be provoked by employers than be the product of worker militancy in manufacturing; as deindustrialization reduced the share of employment in manufacturing, firms in the service sector, and especially the emerging platform economy, were far less likely to see potential benefits from union organization; and business associations became more militantly anti-union, more willing to deploy political resources to challenge collective regulation and more likely to withdraw from tripartite arrangements.

Simultaneously, governments, first of the right but increasingly of the left, saw less public good and economic benefit from their relationship with organized labour. On the right this occasionally translated into a direct assault on unions and the legislation or immunity that protected them. On the left it led first to a distancing of party and unions (Allern & Bale 2017; see also Chapter 9), and then to a harder turn towards neoliberal economic policy (Mudge 2018). The degree and scale of organized labour's isolation has varied across the advanced capitalist world, as a later section will describe, but everywhere capital and the state have become more distant and less supportive of unions and collective regulation of the economy. This has not meant that unions everywhere were subject to the same degree of attack and erosion, but it does mean that they have had to rely on their own resources to survive rather than benefit from a contributory role in the functioning of a dominant model of economic growth.

NEOLIBERALISM AND LABOUR REGULATION

The use of the term neoliberal to describe the current period, and to do so by pairing the term to an historic low point for labour movements, is a choice not without its challenges because it may obscure as much as it reveals. Neoliberal is now so widely used as to have lost analytical purchase, to have devolved into an epithet and general term of opprobrium rather than a concept that helps us make sense of the world (Brenner, Peck & Theodore 2010; Cahill *et al.* 2018). Nonetheless, it is used here because, suitably defined and explicated, it does still have value. Thus, the concept of neoliberalism has three distinct advantages. First, it directs us to the ways in which the current era is not simply a cyclical return to previous periods of laissez faire. Second, it identifies two distinct parts or faces of the manner in which contemporary capitalism is regulated. And, third, it centres and foregrounds the state as an actor. All three are relevant to understanding how and why labour movements have experienced such reversals and declines over the last four decades.

The term neoliberalism is used here to describe the institutional, political and ideational architecture of contemporary capitalism in the period from the end of the Fordist era to the present (Howell 2021). It is the manner in which post-Fordist capitalism is regulated in the regulation theory sense of the term, where regulation is understood more expansively than simply what governments do. Neoliberalism is, then, "the contemporary mode of existence of capitalism" (Ayers & Saad-Filho 2014: 603). Within this broad conception of neoliberalism, it should be understood as simultaneously a response to the crisis of the earlier Fordist growth model in the 1970s and 1980s and a response to the failure of classical liberalism in the inter-war era. Thus, one can distinguish two distinct elements or faces of a neoliberal mode of regulation. The first is neoliberalism in support of the growth model, or a family of growth models, which have replaced Fordism and require a new set of institutions to stabilize growth and permit markets to function. This is, to borrow from Habermas (1973) and Offe (1984), a response to the accumulation imperative of post-Fordist capitalism. The second face of neoliberalism is a constructivist project that involves reshaping society to allow a market order to function. As discussed below, this second role is a response to the legitimation imperative of contemporary capitalism in the sense that it attempts to discursively reshape the boundary between the state and market and the citizen and the market.

The elements of neoliberalism that have supported and served to regulate post-Fordist capitalism are probably familiar to most and follow in part from the discussion in the last section that outlined how the evolution of capitalist growth models has been theorized in academic circles. Streeck (2014) has nicely described a series of regulatory "fixes" that managed the increasing gap between productivity growth and wage stagnation dating in the US from the mid-1970s, and in Western European from the end of the 1980s. First, public debt in service of broadly Keynesian goals propped up consumption and, after mounting government debt became politically untenable, it was replaced by private debt, what Crouch (2009) termed "privatised Keynesianism". Each had institutional needs with financial deregulation serving to permit the massive expansion of consumer debt, and then the socialization of losses by financial institutions to protect the financial system after the inevitable bubbles burst (Blyth 2013). At the same time, changes to the institutions regulating industrial relations and labour markets, which will be the subject of the next section, recommodified and disciplined labour. While the Fordist era was characterized by collective regulation of the labour market and regulatory institutions that constrained and disciplined financial markets, the neoliberal era has seen the de-collectivization and disciplining of the labour market, while financial markets have been subject to deregulation. One further feature of post-Fordist growth models is important here. As noted above, they have proved much less stable and more crisis-ridden than their Fordist predecessor (Vidal 2014). As a result, they demand more regulatory intervention to manage.

The second face of neoliberal regulation will be less familiar, and follows from an understanding of neoliberalism as quite distinct from earlier periods of laissez faire and market expansion. As Dardot and Laval (2014) argue, classical liberalism, as both economic and political project, collapsed in the inter-war era through a combination of depression, fascism and war. The effort to create a market order (in Polanyi's sense of the commodification of labour, nature and money, accompanied by self-regulating markets) turned out not to emerge spontaneously and naturally but rather to generate opposition and resistance. It could not be the basis for a mass democratic politics. The nineteenth-century liberal expectation of the market order's natural legitimacy and self-evident efficiency proved false. In this context the neoliberal project, as conceived, for example, by the Mont Pelerin Society (Peck 2010), involved a reshaping of a wide array of institutions, practices and modes of life to better permit a market order to function. Foucault (2008) in his elaboration of biopolitics described elements of this process, including atomization and the transformation of the citizen into an economic subject. It is in this sense that this second face of neoliberalism can be understood as a constructivist project. As Cahill and Konings (2017: 14) put it, neoliberalism recognizes that a market order "needs to be actively constructed, institutionally and politically". It is also why neoliberalism can be described as a terraforming project: making contemporary society safe for a market order (Howell 2021).

The last reason for analysing the low point of organized labour through the framework of neoliberalism is that it highlights the central role of the state in the regulation of labour in the contemporary period. Neoliberalism is not laissez faire. Unlike classical liberalism, neoliberalism is not about limiting state intervention, but rather about using state power in new ways to institutionalize a market order. As Davies (2018: 273) has put it: "The critical distinction between liberalism and neoliberalism is that the latter abandons the vision of market and state as independent and ontologically distinct entities." Dating from at least Gamble's analysis of Thatcherism has come a recognition of the general affinity in the current period between "the Free Economy and the Strong State", and particularly in the realm of class relations where state action was required "in order to unwind the coils of social democracy" (Gamble 1988: 32), which had piecemeal become layered on to the economy in the postwar period. Gamble was discussing the UK economy, but his argument can be generalized beyond it. These coils sought to partially decommodify the labour market, strengthen unions and build a welfare state. These coils could not be hacked away by business alone; rather it required a more interventionist state.

This disciplinary role for the state intersects with an emerging scholarly literature on "political capitalism" (Riley & Brenner 2022) or "neo-feudalism" (Dean 2020), which points to the expanded role that the state plays in the creation of value for capital under neoliberalism. This includes the privatization of public assets, the opening up of public services, particularly in the care sector, to for-profit providers, the entrenchment of monopoly rents in the platform economy, and so on. The state can be said to not only be providing regulatory fixes for post-Fordist growth models but also contributing directly to capitalist accumulation. Neoliberalism is, therefore, not hostile to the state, although a strong case can be made that it is hostile to democratic institutions of the state.

It is this feature of neoliberalism that helps to make sense of the central role played by states in the rolling back of labour gains and reconfiguration of class compromises that has taken place across the advanced capitalist world. To be sure, as the next section will illustrate, that role has varied from country to country, as has the scale and timing of state intervention. Nonetheless, in no case has the reshaping of industrial relations taken place without a substantial state role. A return to the two faces of neoliberalism should indicate why this has been the case. Both the economic project of regulating emerging growth models, providing the necessary institutional fixes and supports, and the political and ideational project of reshaping institutions, behaviours and modes of life to better embed a market order in society, are almost impossible without state action. States are crucial actors because they retain legitimacy, material resources and institutional capacity, which are far less available to even the most powerful non-state actors. Further, class actors are often cautious, unable to see beyond short-term interests and face collective action problems in pursuing their goals. Finally, the scale of the transformation of industrial relations over the last four decades produced higher levels of class conflict as employers strove for change and unions resisted. In this context the state was bound to be drawn into managing the conflict.

Summarizing the argument of this chapter so far, the period from the end of the 1970s onwards in the advanced capitalist world was one in which each element that had contributed to the strength of unions during the previous 30 years was challenged, weakened and eroded. With deindustrialization and the proliferation of insecure work, the resulting industrial and workplace structures and labour markets were less conducive to union organizing. With Fordist growth faltering, the interest of employers in collective regulation weakened. With changes in class structure and increased capital mobility, both the interest and the ability of governments to enter into partnerships with union movements around a Keynesian-welfare bargain were undercut. At the ideational level, broadly neoliberal diagnosis, prognosis and prescriptions became hegemonic, on the left as much as on the right. This chapter is agnostic as to the precise direction of causality, added to which the outcome seems to have been overdetermined with all the causal drivers pushing in the same direction. Whether Fordist growth became exhausted and only then did employers and states turn upon labour movements, or whether attacks on unions pre-dated the weakening of Fordism but made it unstainable is hard to untangle, and almost certainly varied by country. In some cases the move to transform industrial relations in a more neoliberal direction was early and eager, while in others it came later and reluctantly.

What we can say with some confidence is that by the end of the 1990s it was almost universal: national business organizations were everywhere urging greater decentralization and individualization of bargaining and weaker labour market regulation. This was in stark contrast to the expectations of the "Varieties of Capitalism" approach, which anticipated that industrial actors would defend existing institutions. Governments everywhere acceded to or were actively driving institutional change, and unions were on the defensive and dependent upon their own diminished mobilizational capacity rather than external supports. It was in this context that the neoliberalization of industrial relations played out.

VARIETIES OF LABOUR TRANSFORMATION

Neoliberalism is not reducible to a single blueprint or institutional architecture, or achievable by a single pathway. Neoliberalism is too often conceptualized in the academic literature in Anglo-American terms, as taking the form associated with liberal market economies and especially the experience of the UK and the US. As a result, the scale of the marketization that has taken place over the past three decades in countries widely considered to be hostile terrain for a neoliberal ideology and traditionally neoliberal political parties (northern Europe as well as France, for example) is widely missed.

The more recent literature on neoliberalism (Ban 2016) suggests that its emergence in any given country is not a top-down process nor a simple process of exportation of liberal market ideas and then their reception in native culture. More often it involved a hybrid relationship of market ideology with distinct indigenous political cultures. What might be better described as neoliberalisation – in order to emphasize the process rather than the outcome – was a reinterpretation of domestic political cultures and economic traditions around a set of universal economic and social objectives. As such, not just capital and ideologically militant governments, but also social democratic and other centre-left governments, public sector actors and even unions themselves could play a role in its emergence.

A strong case can be made, for example, for a distinct Nordic neoliberalism or "neoliberalization through progressivism", which has taken place not outside of the way the Nordic model has been traditionally understood, but rather inside and through the model.[1] The image of Nordic countries as resistant to liberalization remains largely true for the sphere of industrial relations, even as it is often acknowledged that change has taken place. In the Anglo-American imagery of neoliberalism, one would need to observe a direct assault upon the collective power of organized labour and upon collective regulation itself to label the institutional transformation of industrial relations as neoliberal, and for the most part that has not happened, and yet the liberalization of Nordic industrial relations has, in fact, been profound.

In what follows, no effort will be made to offer a comprehensive account of the state of labour movements across the advanced capitalist world, or of the transformation of industrial relations. Rather, this section will highlight a set of themes that help make sense of the different national experiences and yet common trajectory during the neoliberal era. In each country case, the institutional forms, the actors and the processes have been somewhat different, but they all share the fundamentally neoliberal goals: expanding employer discretion over wage determination, work organization and hiring and firing; individualizing and differentiating labour market outcomes; rewarding market power rather than collective power; and transforming public sector industrial relations to mirror those in the private sector. These elements are in contrast to industrial relations organized around principles, outcomes and institutions of solidarity, collective determination and the countervailing class power of unions.

Institutional pathways

The different institutional pathways towards the liberalization of industrial relations that we have seen across the advanced capitalist world primarily reflect the different obstacles to liberalization in each country. In the UK, for example, the obstacle was the system of collective regulation put in place from the 1890s onwards and then decentralized to the firm level from the late 1950s onwards. Labour law played a relatively light role in directly regulating the labour market. The mechanism for liberalizing

1. See the research programme "Neo-liberalism in the Nordics: Developing and Absent Theme", led by Jenny Andersson; https://nyliberalisminorden.se/

industrial relations was an active dismantling of the institutions of collective regulation and the means of their enforcement. This involved above all de-collectivization – weakening unions themselves – and an individualization of relations between employers and employees (Davies & Freedland 1993; Howell 2005).

In France, by contrast, the main obstacle to liberalization was the state in the form of legal regulation of the labour market and legal support for collective bargaining. But fears of labour militancy dating to 1968 and reappearing periodically throughout the 1980s, 1990s and 2000s persuaded governments of both left and right that a direct assault on unions and labour market protections would be dangerous. Thus, the mechanism of institutional change used was to encourage a decentralization of bargaining to the firm, both by offering greater flexibility in return for negotiated change and by the creation of legal obligations inside the firm. Given the endemic weakness of French unions at the workplace level, the state was forced to create new collective actors who would negotiate and legitimize workplace change (Amable 2016; Howell 2009). In Germany, the obstacle to a liberalization of industrial relations was partly legislative, in the form of employment protections, but primarily the system of collective regulation, with sectoral agreements at its core. The mechanism for institutional change was, thus, not a frontal assault, or even the construction of new institutions, so much as erosion, deregulation, conversion and the creation of escape hatches for firms to opt out of sectoral bargaining (Baccaro & Benassi 2017; Hassel 2002).

The obstacles to a liberalization of industrial relations in Sweden lay both in restrictive labour market regulation and a form of collective regulation that put severe limits upon employer discretion in wage determination and work organization. As a result, institutional change took place through a recentralization of the collective bargaining system in the late 1990s, the construction of new institutions for mediating industrial conflicts, and some limited legislative deregulation of the labour market. The renaissance of coordinated multi-industry bargaining featured a minimalist role for the centre – setting a wage ceiling and imposing a peace obligation upon local bargaining – while permitting a much greater role for decentralized and even individualized bargaining. This was the very opposite of solidarism, as wages became more and more determined by local conditions (Baccaro & Howell 2017; Elvander 2002).

Australia and the US are mixed cases where the obstacles to liberalization came from both sectoral bargaining and a heavily statist and juridically structured industrial relations system. In the American case, sectoral bargaining was largely limited by the end of the 1970s to the mass production industries, although it was also extensive and protected in the public sector, and while federal and state legislation only lightly regulated the labour market, they provided crucial protections for collective action (Lichtenstein 2002). In Australia, the award system operated in symbiosis with union strength in selected firms and sectors. Gains in those areas could then be transmitted more broadly through the economy through compulsory awards. Institutional change in both cases involved an important role for the respective state actors (Cooper & Ellem 2008). In the US, an outsize role was played by the judiciary and its reinterpretation of labour law at both federal and state level, while in Australia, after a flirtation with concertation in the 1980s, a series of comprehensive industrial relations reforms reduced the role of the award system, cut the link between it and union bargaining and encouraged the decentralization of bargaining.

One can extract broadly three reform pathways from these cases. The first involved deregulation through changes to legislation that had once supported collective regulation and limited employer discretion. The obvious examples are the UK during the Thatcherite period, Australia from the mid-1990s through the mid-2000s, and the US from 1980 onwards, although here it was less legislation and more judicial reinterpretation of labour rights that served the goal of de-collectivization. But one can include the weakening of the Ghent system in Sweden, deregulation of the strict labour market protections that had protected French workers, and similar Harts IV reforms in Germany. It is also important to note the decision by centre-left governments – "new" Labour in the UK, the Democrats in the US, the Australian Labor Party – not to re-legislate protections for collective organization and industrial action when they re-entered government.

A second approach was the use of derogation, whereby legal or contractual constraints remain, but states permit industrial actors to bypass or ignore them, to permit a liberalization of industrial relations without having to formally end or replace existing institutions. It can be thought of as a form of neoliberalism as exception (Ong 2006). Derogation permits liberalization without having to frontally challenge industrial actors, and as such was more palatable in countries with either stronger unions or forms of industrial partnership. It could be justified as an emergency measure, or as institutional change under carefully controlled conditions. Examples include the introduction of opening clauses in Germany, the linking of flexibility, achieved via exemption from labour law, to workplace social dialogue on France, and the ability of sectoral and eventually firm level agreements in Sweden to derogate from legal limits on atypical employment. They all involved expanding employer discretion without the need to formally reconstruct industrial relations institutions.

A third approach involved what Streeck and Thelen (2005) have referred to as "institutional conversion", whereby formal institutional continuity masks a change in the function of institutions so that they become more discretion-enhancing for employers. While much of comparative industrial relations research has rightly focused attention upon institutions, it is important to recognize that there is a plasticity to institutions such that how they function is heavily determined by the forcefield of power resources within which they operate. Institutions that benefited unions in a period of full employment, labour strength and state and business support can serve to undermine collective action when all those conditions changed, as they did in the neoliberal era.

For example, works councils in Germany and France, which were once subordinate institutions, supportive of the dominant role of unions in collective regulation, under conditions of weakened unions and changes to labour law, increasingly served to detach firms from the wider industrial relations system and tie worker interests more closely to those of the firm, encouraging de facto enterprise unionism. Streeck (1984) has labelled this micro-corporatism, in that, just as the macro-corporatist institutions of the postwar era encouraged powerful labour movements to internalize the needs of the broader economy, so now firm-specific forms of worker representation encourage an identification with the needs of the firm. Indeed, turning to those forms of macro-corporatism, peak-level concertation, once a mechanism for solidarism and achieving worker gains, came in the current era to encourage a decentralization of bargaining in Sweden, and to legitimize austerity and deregulation in Italy and Ireland. In each case, peak-level bargaining became discretion-enhancing for employers, a mechanism for overcoming obstacles to liberalization rather than one for strengthening labour solidarism (Howell 2021).

Of these three pathways, it is important to note that only the first incorporated a direct assault upon unions, either by the state or business. For the other two, the liberalization of industrial relations could be just as profound, but less conflictual and less easily recognized as liberalization.

Business and the state

This chapter has argued that the exhaustion of the Fordist growth model, with its imperative for wage-led growth, and replacement by a range of post-Fordist growth models, which share only that they do not rest upon the wages of the great mass of working people to sustain demand, undercut postwar systems of collective regulation and with them the strength and influence of labour movements. But capitalist growth models do not change industrial relations institutions on their own, independently of actors. In the neoliberal era, both business and the state defected from support of collective regulation and sought to transform them in a liberalizing direction. The state in particular mattered, as noted above, but business interests too had to decide that maintaining support for existing institutions no longer served its interests.

This period saw a general tendency towards greater politicization and a greater willingness to challenge industrial relations institutions inherited from the past on the part of business organizations. Streeck notes (2014: 18) that, while social scientists were quick to recognize labour as a political and strategic actor, as well as an economic one, that same recognition for employers has been slower, not least because collective organization has always been more important for labour than capital as workers require collective action and collective organization not only to sanction employers but also to define a labour interest in the first place (Offe & Wiesenthal 1985). For employers, on the other hand, collective organization has secondary benefits but interests are fed back to employers through the market and the simple act of not hiring or not investing is sufficient to sanction workers.

This period has seen, in almost every case, the emergence of a more self-confident, more politicized employer class willing to seek substantial change to national industrial relations systems, and always in a more liberalizing direction. In France, main employer confederation adopted a new name and an increasingly neoliberal discourse and sought collective agreements with smaller reformist unions in order to liberalize the labour market and industrial relations (Woll 2006). Sweden is the clearest case of employers defecting from the postwar model of industrial relations and then constructing a new model (Johansson 2005). Beginning with large engineering firms seeking separate sectoral agreements in the early 1980s, through the withdrawal of the Svenska Arbetsgivareföreningen (SAF) from corporatist institutions in 1990, to its more politicized and radical role as the rebranded Svenskt Näringsliv (SN), employers led and unions followed. In both countries the main business organization shifted focus away from a primary function as employer representative and collective bargaining agent towards an organization emphasizing the entrepreneurial function of business and a role of lobbying the state.

But even where a formal organizational change did not take place, employer organizations adopted a more overt neoliberal discourse, proved much more willing to revisit and challenge longstanding elements of the industrial relations landscape and, where unable to negotiate the changes that they wanted with unions, sought state support for liberalization. After initial hesitation at the radicalism of the Thatcher industrial relations project, the Confederation of British Industry also underwent a radicalization, calling for further limits on strike action and the reach of European Union directives, along with emphasizing individualization of industrial relations as the preferred trajectory of change. The German metalworking employer association launched a political and public relations campaign to deregulate the labour market through the New Social Market Initiative (Kinderman 2014), an effort that bore fruit in the Harts reforms. Even Confindustria in Italy proved willing to abandon a bargaining route to industrial relations reform in the 2000s when it appeared possible that a friendly government might unilaterally deregulate the labour market.

The evidence from business behaviour is consistent with the argument that the first-order preference of employers is usually a liberalization of industrial relations institutions. This was the "dormant wish" of employers (Ibsen *et al.* 2011: 336), and once the political opportunity structure and the ability of labour organizations to resist changed, that wish rose to the surface. The political opportunity structure and capacity of unions to resist was, in turn, heavily shaped by the role of the state. As noted above, different obstacles to liberalization produced different pathways to liberalization and different state strategies but, across Western Europe, North America and Australasia, states became more interventionist in class relations.

Major projects of industrial relations reform in liberal market economies were state-led. the UK in the 1980s, when the Thatcher government passed six substantial legislative packages of reform and endured a year-long strike in the mining industry, was a prime example, but the same process took place in the US, although it was more protracted and involved a larger role for the judicial branch. In France, New Zealand and Australia, where industrial relations systems had been heavily statist from earlier in the twentieth century, were dismantled by those same states. Among coordinated market economies, the state role was more indirect but no less essential. The Swedish state was a central actor in the transition from the solidaristic regime of the Saltsjöbaden era to the coordinated but decentralized regime today, and in Germany the state unilaterally deregulated the labour market when the social partners failed. There was also the emergence of a new form of neoliberal concertation, state-initiated and state-led, in countries without earlier corporatist traditions such as Italy and Ireland, to overcome local labour resistance to industrial relation and labour market reform and to legitimize austerity.

State intervention in industrial relations certainly existed before the neoliberal era. Nonetheless, the past 40 years have seen remarkably expansive state-led projects of industrial relations reconstruction, and these projects have differed in important ways from the collectivist and corporatist past in that they contributed to the liberalization of industrial relations. In comparison to the earlier era, more recent state interventionism has also been more ambitious, in that it has involved efforts to transform and reconstruct industrial relations systems along quite different lines from before (Howell 2021). The role played by states has in part been one of institutional deconstruction and then reconstruction, but it has also involved taming labour movements, whether through legislative changes that weaken their capacity to launch collective action and engage in solidarism across sectors, firms and groups of workers, through direct intervention into industrial disputes, or through new forms of social pact that incorporate peak union organizations into reform efforts.

UNION FUTURES

The chapters in this middle part of the handbook form a temporal bridge between the early chapters that speak to the structures, ideologies, resources and internal dynamics of labour unions, and the later chapters that assess the evolving strategies and democratic potential of unions in the contemporary period. This concluding section will look backwards as well as forwards in evaluating how the neoliberal era has shaped labour movements.

Unions entered the 1980s shaped by a set of social, economic, political and ideological resources inherited from the social democratic period, resources that shaped their organizational structures, strategies and practices. These resources reflected the predominance of manufacturing and mass production technologies, organizing semi-skilled, mostly male workers in large industrial unions, the functional role played by labour in the Fordist regime of accumulation, the shared interests and close ties between social democratic and labour parties and unions, and an ideological commitment to broadly respecting capitalist ownership and managerial prerogative in return for a share in the value added.

Two decades into the twenty-first century, all of those resources have been challenged, weakened and undermined, with resulting changes to union structures, strategies and practices. Deindustrialization and the gig economy have reshaped the labour force and the spaces that unions try to organize. New growth models have no need of collective regulation. Governing parties of the left and the right see little benefit from rewarding unions and in many cases run for office explicitly against unions. With the collapse of the grand Fordist bargain, strategies premised upon partnership or class compromise with capital have equally collapsed and neoliberal ideologies have chipped away even at labour movement worldviews. One by one, social, economic, political and ideological resources have been rendered far less valuable.

This chapter has focused upon the shift in growth models as the linchpin of the transformation of labour movements during the neoliberal era, arguing that neoliberalism itself forms the superstructure of the growth models that have replaced Fordism. This emphasis upon the central role played by the transformation of capitalist growth helps to explain both the timing and the universality of the trajectory of and for labour movements and industrial relations. Greater contingency and a focus upon the more recent past cannot explain why liberalization both began in the 1980s and has been generalized across so much of the advanced capitalist world. But it is also worth noting three other factors, which will be only very briefly described here, all of which have pointed in the same direction of liberalization.

The first is the most political. Within Europe the acceleration of economic integration from the end of the 1980s, with the implementation of the Single European Market, through Economic and Monetary Union, the primacy of the Court of Justice in determining the balance between social goal and liberalization, and the reaction of European institutions to the public debt and deficits following the 2008–09 global financial crisis and the sovereign debt crisis, have all had the effect of closing down the space for different, non-liberal, economic models and macroeconomic policies, as well as encouraging (or in some cases mandating) the liberalization of industrial relations institutions (Erne 2018; Scharpf 2010; Zhang & Lillie 2015). The European project over the past three decades has operated to deepen and institutionalize broader neoliberal projects and the forces of liberalization that shaped the context within which European employment relations systems have been transformed. As such, the political context at the supranational level has strongly reinforced liberalization from below.

The second factor is the long-term effects of the financial and economic crisis of 2008–09 and after (Blyth 2013). Despite a brief flirtation with a Keynesian response, the long-run response was both austerity and a renewed assault upon unions and industrial relations. Public debt became the justification for reforms to public sector industrial relations, once relatively insulated from liberalizing trends. In the US, for example, a combination of more aggressive anti-unionism at the state level and Supreme Court interventions have created legislative and judicial obstacles to public sector unionism. And as noted above, the sovereign debt crisis has also served to justify externally enforced liberalization of the industrial relations institutions in several southern European members of the European Union (Koukiadaki & Kokkinou 2016; Marginson 2015). What is important about the response to economic crisis is that it did not reverse the liberalizing course of the previous three decades but rather exacerbated it.

Finally, as the world moves on from the Covid pandemic, we can begin to assess its longer-term implications for the future of labour. The sudden confrontation of consumers in affluent Western countries with their dependence upon a poorly paid and precarious service class, the reaction of the members of that service class to working during a pandemic, and the adoption of more or less generous forms of income and job support at the height of the pandemic, do not for the most part seem to have left a lasting legacy beyond some evidence of increased labour militancy (for a survey of the pandemic's impact on collective bargaining, see ILO 2021). To do so would have required a political force to ally and build upon nascent and scattered industrial action. A partial repudiation of the neoliberalization of the European and American left from Syriza's initial radicalism in 2015 and Podemos's rise, through the Corbyn period of the Labour Party and the Sanders insurgency (Watkins 2016), was reversed prior to the pandemic and little suggests that Covid itself has created political space for its renewal (Winant 2021).

Later chapters assess the prospects for labour unions in this epoch. Labour movements in advanced capitalist countries are now in their third decade of strategic reorientation away from the strategies (and to a lesser extent the structures) of the social democratic era. The widespread although uneven shifts towards organizing and some version of social movement organizing have in recent years further shifted towards protest around issues of work outside of unions, or outside of the workplace and conventional collective bargaining (Béroud 2018), perhaps most advanced in France where the Gilet Jaunes has posed the greatest threat to neoliberalization in recent years. From the perspective of this chapter and the argument it has made, resistance to liberalization alone is unlikely to be a viable long-term strategy, especially as unions continue to see membership and organizational decline. However, strategies that target the growth model itself may have more success. Where the wage-led components of growth can be strengthened, or the liberalizing and class disciplining implications of financialized growth can be limited, for example, there will be more space to struggle for collective regulation and to defend the urgent need for collective organization and action.

REFERENCES

Allern, E. & T. Bale (eds) 2017. *Left-of-Centre Parties and Trade Unions in the Twenty-First Century*. Oxford: Oxford University Press.

Amable, B. 2016. "The political economy of the neoliberal transformation of French industrial relations". *ILR Review* 69(3): 523–50.

Avdagic, S. & L. Baccaro 2014. "The future of employment relations in advanced capitalism: inexorable decline?" In *The Oxford Handbook of Employment Relations: Comparative Employment Systems*, edited by R. Deeg, A. Wilkinson & G. Wood. Oxford: Oxford University Press.

Ayers, A. & A. Saad-Filho 2014. "Democracy against neoliberalism: paradoxes, limitations, transcendence". *Critical Sociology* 41(4/5): 597–618.

Baccaro, L. & C. Benassi 2017. "Softening institutions: the liberalization of German industrial relations". In *European Industrial Relations Since the 1970s: Trajectories of Neoliberal Transformation*, edited by L. Baccaro & C. Howell, 97–120. Cambridge: Cambridge University Press.

Baccaro, L., M. Blyth & J. Pontusson (eds) 2022. *Rethinking Comparative Capitalism: The New Politics of Growth and Stagnation*. Oxford: Oxford University Press.

Baccaro, L. & C. Howell 2017. *European Industrial Relations Since the 1970s: Trajectories of Neoliberal Transformation*. Cambridge: Cambridge University Press.

Baccaro, L. & C. Howell 2018. "Unhinged: industrial relations liberalization and capitalist instability". MPIfG Discussion Paper 17/19. Cologne: Max Planck Institute for the Study of Societies.

Ban, C. 2016. *Ruling Ideas: How Global Neoliberalism Goes Local*. Oxford: Oxford University Press.

Béroud, S. 2018. "French trade unions and the mobilisation against the El Khomri law in 2016: a reconfiguration of strategies and alliances". *Transfer: European Review of Labour and Research* 24(2): 1–15.

Blyth, M. 2013. *Austerity: The History of a Dangerous Idea*. Oxford: Oxford University Press.

Boyer, R. 1990. *The Regulation School: A Critical Introduction*. New York: Columbia University Press.

Brenner, N., J. Peck & N. Theodore 2010. "Variegated neoliberalization: geographies, modalities, pathways". *Global Networks* 10(2): 182–222.

Cahill, D. & M. Konings 2017. *Neoliberalism*. Cambridge: Polity.

Cahill, D. *et al.* (eds) 2018. *The Sage Handbook of Neoliberalism*. Thousand Oaks, CA: Sage.

Cooper, R. & B. Ellem 2008. "The neoliberal state, trade unions and collective bargaining in Australia". *British Journal of Industrial Relations* 46(3): 532–54.

Crouch, C. 2009. "Privatised Keynesianism: an unacknowledged policy regime". *British Journal of Politics and International Relations* 11(3): 382–99.

Crouch, C. & A. Pizzorno (eds) 1978. *Resurgence of Class Conflict in Western Europe Since 1968* (vols. I and II). New York: Holmes & Meier.

Dardot, P. & C. Laval 2014. *The New Way of the World: On Neoliberal Society.* New York: Verso.

Davies, P. & M. Freedland 1993. *Labour Legislation and Public Policy: A Contemporary History.* Oxford: Clarendon Press.

Davies, W. 2018. "The neoliberal state: power against 'politics'". In *The Sage Handbook of Neoliberalism*, edited by D. Cahill *et al.*, 273–83. Thousand Oaks, CA: Sage.

Dean, J. 2020. "Neofeudalism: the end of capitalism?" *Los Angeles Review of Books*, 12 May.

Durand, C. 2017. *Fictitious Capital.* London: Verso.

Elvander, N. 2002. "The new Swedish regime for collective bargaining and conflict resolution: a comparative perspective". *European Journal of Industrial Relations* 8(2): 197–216.

Erne, R. 2018. "Labour politics and the EU's new economic governance regime (European Unions): a new European Research Council project". *Transfer: European Review of Labour and Research* 24(2): 237–47.

Foucault, M. 2008. *The Birth of Biopolitics: Lectures at the College de France, 1978–79.* Basingstoke: Palgrave Macmillan.

Gamble, A. 1988. *The Free Economy and the Strong State: The Politics of Thatcherism.* Basingstoke: Macmillan.

Glyn, A. *et al.* 1990. "The rise and fall of the golden age". In *Golden Age of Capitalism: Reinterpreting the Postwar Experience*, edited by S. Marglin & J. Schor, 39–125. Oxford: Clarendon Press.

Habermas, J. 1973. *Legitimation Crisis.* Boston, MA: Beacon Press.

Hall, P. & D. Soskice (eds) 2001. *Varieties of Capitalism: The Institutional Foundations of Comparative Advantage.* Oxford: Oxford University Press.

Harvey, D. 1989. *The Condition of Postmodernity: An Enquiry into the Origins of Cultural Change.* Oxford: Blackwell.

Hassel, A. 2002. "The erosion continues: a reply". *British Journal of Industrial Relations* 40(2): 309–17.

Hobsbawm, E. 1978. "The forward march of Labour halted?" *Marxism Today*, September, 279–86.

Howell, C. 2001. "The end of the relationship between social democratic parties and trade unions?" *Studies in Political Economy* 65: 7–37.

Howell, C. 2005. *Trade Unions and the State: The Construction of Industrial Relations Institutions in Britain, 1890–2000.* Princeton, NJ: Princeton University Press.

Howell, C. 2009. "The transformation of French industrial relations: labor representation and the state in a post-dirigiste era". *Politics & Society* 37(2): 229–56.

Howell, C. 2021. "Rethinking the role of the state in employment relations for a neoliberal era". *ILR Review* 74(3): 739–72.

Ibsen, C. *et al.* 2011. "Bargaining in the crisis – a comparison of the 2010 collective bargaining round in the Danish and Swedish manufacturing sectors". *Transfer: European Review of Labour and Research* 17(3): 323–39.

International Labour Organization 2021. *Covid-19, Collective Bargaining and Social Dialogue: A Report on Behalf of ILO ACTRAV.* Geneva: ILO.

Jessop, B. 1993. "Towards a Schumpeterian workfare state? Preliminary remarks on post-Fordist political economy". *Studies in Political Economy* 40(1): 7–39.

Johansson, J. 2005. "Undermining corporatism". In *Power and Institutions in Industrial Relations Regime*, edited by P. Oberg & T. Svensson, 77–106. Stockholm: Arbetslivsinstiutet.

Kinderman, D. 2014. "Challenging varieties of capitalism's account of business interests: the new social market initiative and German employers' quest for liberalization, 2000–2014". Max-Planck-Institute for the Study of Societies Discussion Paper 14/16.

Koukiadaki, A. & C. Kokkinou 2016. "Deconstructing the Greek system of industrial relations". *European Journal of Industrial Relations* 22(3): 205–19.

Lichtenstein, N. 2002. *State of the Union: A Century of American Labor.* Princeton, NJ: Princeton University Pres.

Lichtenstein, N. 2006. *Wal-Mart: The Face of Twenty-First-Century Capitalism.* New York: The New Press.

Marginson, P. 2015. "Coordinated bargaining in Europe: from incremental corrosion to frontal assault?" *European Journal of Industrial Relations* 21(2): 97–114.

Martin, A. & G. Ross (eds) 1999. *The Brave New World of European Labor.* New York: Berghahn.

Mudge, S. 2018. *Leftism Reinvented: Western Parties from Socialism to Neoliberalism.* Cambridge, MA: Harvard University Press.

Offe, C. 1984. *Contradictions of the Welfare State.* Cambridge, MA: MIT Press.

Offe, C. & H. Wiesenthal 1985. "Two logics of collective action". In *Disorganized Capitalism: Contemporary Transformations of Work and Politics*, edited by C. Offe, 170–220. Cambridge: Polity.

Ong, A. 2006. *Neoliberalism as Exception: Mutations in Citizenship and Sovereignty.* Durham, NC: Duke University Press.

Panitch, L. 1986. *Working-Class Politics in Crisis: Essays on Labour and the State.* London: Verso.

Peck, J. 2010. *Constructions of Neoliberal Reason.* Oxford: Oxford University Press.

Piore, M. & C. Sabel 1984. *The Second Industrial Divide.* New York: Basic Books.

Pontusson, J. 2013. "Unionization, inequality and redistribution". *British Journal of Industrial Relations* 51(4): 797–825.

Riley, D. & R. Brenner 2022. "Seven theses on American politics". *New Left Review* 138: 5–27.

Scharpf, F. 2010. "The asymmetry of European integration, or why the EU cannot be a 'social market economy'". *Socio-Economic Review* 8(2): 211–50.

Srnicek, N. 2017. *Platform Capitalism*. Cambridge: Polity.

Streeck, W. 1984. "Neo-corporatist industrial relations and the economic crisis in West Germany". In *Order and Conflict in Contemporary Capitalism*, edited by J. Goldthorpe, 291–314. Oxford: Clarendon Press.

Streeck, W. 2009. *Re-forming Capitalism: Institutional Change in the German Political Economy*. Oxford: Oxford University Press.

Streeck, W. 2014. *Buying Time: The Delayed Crisis of Democratic Capitalism*. London: Verso.

Streeck, W. & K. Thelen (eds) 2005. *Beyond Continuity: Institutional Change in Advanced Political Economies*. Oxford: Oxford University Press.

Swan, van der N. 2014. "Making sense of financialization". *Socio-Economic Review* 12(1): 99–129.

Umney, C. *et al.* 2017. "The state and class discipline: European labour market policy after the financial crisis". *Capital & Class* 42(2): 333–51.

Vidal, M. 2014. "Incoherence and dysfunctionality in the institutional regulation of capitalism". In *Comparative Political Economy of Work*, edited by M. Hauptmeier & M. Vidal, 73–97. Basingstoke: Palgrave Macmillan.

Visser, J. 2019. "Trade unions in the balance". ILO ACTRAV Working Paper. Geneva: ILO.

Watkins, S. 2016. "Oppositions". *New Left Review* 98: 5–30.

Winant, G. 2021. "Strike wave: a new chapter in US labour militancy?" *New Left Review*, Sidecar blog, 25 November.

Woll, C. 2006. "National business associations under stress: lessons from the French case". *West European Politics* 29(3): 489–512.

Zhang, C. & N. Lillie 2015. "Industrial citizenship, cosmopolitanism and European integration". *European Journal of Social Theory* 18(1): 93–111.

PART III

THE PRACTICE OF BUILDING PRESENCE AND POWER

CHAPTER 12

UNIONS AND THE AGENDA OF JOINT REGULATION

Miguel Martínez Lucio

ABSTRACT

This chapter focuses, first, upon debates around collective bargaining and in relation to union strategy. Second, it examines how collective bargaining structures and practices develop and operate in distinct national and economic contexts. Third, it looks at rethinking the role of joint-regulation in terms of its dynamics, practices, aims and actors whereby a widening of the ways in which we view collective bargaining is required. So, collective bargaining and new variations of it need to be placed within a broader approach to the dynamics and politics of work since it has become an even more contested. Approaches linked to such a broadening allow a better understanding of the contexts that are beyond the "developed" economic ones.

Keywords: Collective bargaining; character and form of collective bargaining; challenges for unions

INTRODUCTION

In an episode of the American cartoon comedy series, *The Simpsons*, the young Bart Simpson has broken his leg and is resigned to spending summer in his room, unable to engage in fun activities with his friends. He turns to the television set in his room to take his mind off his predicament only to find that very old programmes are being broadcast due to staff summer holidays. To Bart's horror, they are reruns of black-and-white television interviews with union leaders about the crisis of collective bargaining. This clearly exacerbates the situation for Bart, but also serves as a gentle and albeit peculiar example of the way collective bargaining and related union issues are viewed in some popular circles. Even within various academic circles of a more sociological orientation – and even, increasingly, in some high-profile work-related journals – the question of collective bargaining is increasingly seen as a particular and limited feature of work and discussions pertaining to it. Hence, its exclusion as an acceptable theme on which to publish. Collective bargaining is seen as a narrow and minimalist dimension of the regulation and politics of work. Radical academics – a broad term that would include the present author – are also particularly worried about the way collective bargaining can sometimes be seen as being overstated, especially within the mainstream history of industrial relations debates.

However, collective bargaining as a concept within the study of work and employment remains a central feature of industrial relations. Approaches to collective bargaining – and changes in it – have been evolving in ways that reflect the manner in which work, and employment itself, have been changing. In many ways, collective bargaining encapsulates how working conditions are improved in formalistic and textual artefacts such as collective agreements (and informal processes such as verbal and custom-based approaches), but such formalistic and textual artefacts also present a dilemma in that they can confine or limit debates on the way work is organized and how the determination of working conditions is understood. These differences have defined the variations within the tradition of industrial relations, as collective bargaining is seen as not only a way of constraining management action and committing it to some type of social orientation, but also in some cases as restricting further debate regarding broader improvements at work. Yet, alongside this ongoing discussion – which has mutated, as the chapter will explain – there is an interest in the way collective bargaining varies across national and social contexts while also being subject to significant changes and "stretched" in the way it operates, and how it is operated by an even broader set of actors.

This chapter will, therefore, outline the traditions and changes related to collective bargaining. First, it will start with a focus upon some aspects of the more established debates around collective bargaining and in relation to union strategy generally. The way the debate has been framed during the twentieth century has been pivotal in the establishment of academic disciplinary boundaries and in the practices of various organizations in relation to it. Second, it will focus upon how the understanding of what collective bargaining structures and practices develop and operate in distinct national and economic contexts. This forms part of the comparative industrial relations tradition that has grown in prominence in recent decades and brought a new range of insights into the role of collective bargaining in different contexts. Third, the chapter takes on another genre of debates that, curiously, has become embedded within the study of industrial relations, namely, that the crisis and fragmentation of collective bargaining and worker representation is an important part of the way we have begun to rethink the role of joint regulation, and, at times, this has to some extent diluted some of the boundaries between pluralist and radical schools of thought. It has also led to a growing defence of its role in a neoliberal context and alternative ways of researching the concept. However, and fourth, these dynamics of change have also brought new forms and practices within the processes of joint-regulation as we see new actors, activities and forms of representation linked to the demands for changes in working conditions. To some extent, this has led to a form of widening of the ways in which we view collective bargaining and related activities – this has begun to some extent – and, broadly speaking to de-institutionalize certain aspects of the way collective bargaining is understood.

In this respect, the deregulation – or, to some extent, crisis – of collective bargaining and industrial relations has forced the need for new critical approaches and has questioned some traditional ontological boundaries. The chapter ends with outlines of discussions as to how we need to "place" collective bargaining and new variations of it within a broader approach to the dynamics and politics of work since it has become an even more contested and engaging space and arena. This may, indeed, allow for a broader approach that includes other contexts beyond the "developed" contexts so beloved of mainstream industrial relations scholars.

REPOSITIONING CONFLICT AND THE POLITICAL IN RELATION TO COLLECTIVE BARGAINING

The study of industrial relations has been viewed from various perspectives and has included a range of thematic issues linked to the actors of industrial relations, the nature of regulation, and the forms that conflict and conflict management take. It is an area that covers a broad set of themes but seems to be especially focused historically on the questions of the institutions of work and employment. Although there are competing views as to the nature of and the problems associated with such institutions, the focus of study has tended to be clear. Within this curiosity for institutions and forms of regulation at various levels of the employment relationship, collective bargaining has been a pivotal focus of key parts of the discipline. The issue becomes one of how the rules and regulatory processes are developed in relation to the main actors of industrial relations (Goodman 1984).

Much depends on how one views collective bargaining: whether it is perceived fundamentally as a joint outcome of specific arrangements between actors or as a hidden form of incorporation that conceals the uneven balance of force between actors (as will be discussed later in this section). However, collective bargaining has become synonymous with being the core dimension of the interface between different actors and interests. To this extent, it is either a strategic "compromise" within the socio-economic system of capitalism – allowing for those other than employers to partake in establishing some aspects of their working conditions – or a significant feature of the way a democratic logic is extended into the realm of the economic (Slichter 1941).[1]

This tension in viewing the role of collective bargaining varies over time, but its centrality to union strategy and activity can be seen to evolve across the twentieth century – especially as unions consolidated their institutional roles (generally speaking), thus, to a degree, relegating social and mobilization-based approaches to become a supplementary feature of these roles.[2] The focusing of mobilization to specific working

1. For some this "democratic function", which was associated with collective bargaining, had varying outcomes. For sociologists, such as Dahrendorf (1959), collective bargaining could ameliorate the negative aspects of class conflict in modern society. However, others during a similar period were concerned with the negative effects such forms of institutionalization would have on the independence and political influence for change within unions, especially their leadership (Wright Mills & Schneider 1948; see also Geary 2001 for a detailed discussion).
2. See Atzeni (2010) for a discussion of this relation in contemporary Argentina.

conditions over time not only facilitated a set of clearer exchanges and transactions across the employment relation between actors, but also influenced and led to the increasingly instrumentalized nature of organizational conflict in many cases: in some cases, "bargaining power is seen as the ability strike or take a strike" (Flanagan 2008: 408).

To that extent, the nature of economic transactions and relations are key to understanding the development of collective bargaining (Flanagan 2008). Nevertheless, over time the nature of collective bargaining relations may become embedded across a set of particular and mutual expectations between actors and social forms that determine its scope and reach. Unions may steadily rethink certain aspects of their purpose – knowingly or unknowingly – and orient themselves to a "market" view of transactions, as opposed to a social view of one (Hyman 2000). Pohler (2018), in a significant overview of collective bargaining, draws attention to the role of cognitive institutions – especially referencing the work of Scott (2001) – where rules are often sustained by virtue of the fact that they may be taken for granted. This is something that more critical scholars also tend to point to, in that collective bargaining – regardless of how it varies across contexts –in some instances takes a "common sense" stance, with its processes not always being questioned.

Within industrial relations, the debate in the immediate post-Second World War years up until the 1970s, broadly speaking, therefore began to focus on issues of form. Clegg (1976) was important in attempting to show how collective bargaining was not only key in shaping labour unionism but also in the way in which it operated. Issues of structure became important to the discussion in terms of the extent of collective bargaining, its level, depth of union engagement, security afforded by employers to this process, and ways and extent collective bargaining is controlled. Across these dimensions, there was an attempt to explain differences across contexts and, in turn, the way in which these shaped unions. Such an approach remains implicitly embedded in many studies and approaches – and, in some cases, even explicitly, as a major recent 28 country comparative study of collective bargaining in the European Union reveals (Waddington, Müller & Vandaele 2019). This parallels an important issue in relation to collective bargaining in terms of how we need to appreciate the way different levels and combinations of structures are developed and sustained (for an example, see Traxler 1996).[3] Curiously, especially since the global financial crisis of 2008–09, as the narrative and policies of deregulation have developed

3. The issues of collective bargaining in terms of its more complex institutional dynamics have also been seen in the work of Walton and McKersie (1965). Burchell (1999) pointed to the need to return to this work more rigorously due to the importance of thinking beyond just levels or questions of formality, especially in the way the dynamics of negotiation develop. The importance of separating out distributive and zero-sum approaches to specific issues (such as pay rates, typically) to integrative and potentially mutually beneficial aspects of work and employment (one such being training, in some cases) is important if the dynamics of conflict and consensus are to be appreciated to their fullest extent. However, intra-organizational bargaining, a further way in which the dynamics of collective bargaining must be viewed as consensus generation within each "actor" is as important as those between them – as the fascination with the internal tensions in unions and, increasingly, employers illustrate. Finally, attitudinal structuring should also be seen as a significant aspect of negotiation relations as expectations and attitudes across themes and interests need to be engaged with for longer-term forms of consensus and agreement to be sustained. This attitudinal dimension has, in some respects, become important due to the increasing inability in some contexts – especially more marketized and neoliberal ones – to generate forms of consensus and shared narratives regarding the process of collective bargaining.

within policy circles and employer lobbies, the study of formal structures of collective bargaining and their role has to some extent returned to the centre of debates, especially within European industrial relations, as the need to outline their changes becomes a focus. Thus, in some senses as seen in various comparative studies on Europe, this has brought a counter point of looking at union-form and sector-level regulation, for example, as an exclusive counterpoint to deregulation (which is an ambivalent development in how to view industrial relations responses).

The contributions of industrial relations as a discipline are, therefore, significant not only in terms of how they highlight different dimensions of collective bargaining, but also how they highlight the more informal and discreet forms of relations that existed. In basic terms, much of industrial relations between the 1960s and the 1980s was increasingly concerned with the way informal bargaining relations developed, not only in the case of the UK, but also in other instances such as Italy, where workplace-based labour unionists negotiated continuously beyond formal agreements on matters related to the patterns and processes of work on the ground, using forms of collective action of an unofficial or direct nature to sustain such negotiations (Terry 1993).

The discipline of industrial relations was engaged with explaining these informal and unofficial spaces (something that current commentators on new forms of worker action in the gig economy seem to ignore). Terry (1977) outlined the importance of this informal dimension, which paralleled, supplemented and, in some cases, undermined more formal mechanisms, as in the case of the UK – see also Joyce (2016), for a more contemporary intervention. Much of the labour process debate within the sociology of work has mirrored these concerns with the way workplace struggle and both indirect and direct negotiations with local management can be seen to be a central part of the ongoing tensions between workers and management (see Martínez Lucio 2010). This "informal" aspect was also common in other less-institutionalized contexts, such as authoritarian Spain during the early 1960s and 1970s, especially where the official means afforded to workers to improve the terms and conditions of their employment were limited (Martínez Lucio 1988).[4]

In a review of the debate on reforming industrial relations within Britain in the 1960s and the perceived problems of informality, local conflict and the fragmentation of bargaining, among other matters, Goldthorpe (1974) raises a series of critical issues with regards to the deeper problems and injustices of the context of regulation. According to Goldthorpe (1974: 448), when referring to the official government interventions and studies regarding these informal or "unorganized" dimensions, argue:

> Nowhere in the Donovan Report, or in the entire tradition of academic industrial relations writing on which it drew so heavily, is there to be found any systematic consideration of how the functioning of the economic system as a whole and of its constituent units of production is founded upon, and sustains, vast differences in social power and advantage; nor of how these are

4. Poole's (1981) mapping of the broader cultural and environmental factors that shape collective bargaining and labour unionism is an example of an intervention that sought to locate a broader approach to the issue of "institutionalization". Yet, much of the debate has curiously continued to focus on form, levels and reach whether formal or "informal".

> then generated – in undoubtedly complex ways – on the one hand, objective oppositions of interest and, on the other, subjective responses of frustration, resentment and antagonism, and also in some degree aspirations and movements towards an alternative dispensation.

To this extent, the patterns of negotiation are often much broader and more critical, consequently, as formal spaces have been challenged and, in some cases, undermined in recent years by a range of factors.

The radical critique of collective bargaining has taken various forms. There has always been a concern that collective bargaining generates or reinforces a type of apolitical "wage consciousness" that instrumentalizes the exchange between workers and their employers. Hyman (1975a) outlined this issue when discussing the stance held by Lenin and other pessimistic approaches to unions in the Marxist tradition that reference the need for external, political leadership if unions and their struggles are not to be institutionalized and if broader political agendas are to be sustained. The critique of collective bargaining as a process that narrows and disconnects struggles is one that has been prevalent and that has taken on various forms. The context of struggles and conflict – whether related to specific sets of bargaining scenarios or not – vary and their outcomes are sometimes better understood by examining the practices and experiences of workers that exist on the ground (Hyman 1975a). To that extent, the arena of action may be greater than one may anticipate. Much depends on how such institutions as collective bargaining are framed across various organizations, especially in terms of the role of the state and broader questions of ideology (Hyman 1975b).

The roles of mobilization and counter-strategies in relation to collective bargaining are, therefore, not aberrations as some in the pluralist arena would have us believe but, rather, form an important part of the relationships that constitute the nature of regulation and negotiation. In some senses it depends upon the location of such practices along a wider spectrum of union strategy and practice. Increasingly, the academic widening of the understanding of mobilization in terms of its dynamics and stages within industrial relations suggests that we cannot start from some institutionalist privileging of collective bargaining in order to read off substantive outcomes from the act of workers negotiating with employers.[5]

Heery (2016: 86) points to the way Kelly's work is an important attempt to broaden the sources of radical analysis in social scientific terms by looking at mobilization across its different stages and processes of politicization and not just as an adjunct to "institutional" activities. There is, therefore, greater sensitivity within radical, and especially neo-Marxist or Marxist traditions, as to how to locate the practice and outcomes of collective bargaining in broader political terms. The argument that radical traditions negate collective bargaining ignores the meditations within these traditions (although there are clear anarcho-syndicalist positions that continue to sustain that engagement through such mechanisms as collective bargaining are problematic in the long term and limit the link between unions and the workforce). There is also, as Heery (2016: 89)

5. See Kelly (1998) on the development of mobilization theory within industrial relations, and Holgate, Simms & Tapia (2018).

observes in the work of the early Hyman (1975b), a call for analysts and activists to note the fragility of such institutions of job regulation: in effect, their reach and effect are never fully conclusive and able to close off the tensions the employment relation generates between workers and management.

Curiously, in an attempt to deepen and distance pluralism from the "institutionalist" tendencies within it, Dobbins, Hughes & Dundon (2021) call for a need for issues related to inequality and power to be more central to understanding the dynamics and limits of industrial relations analysis. Much of this renewed radicalism within the pluralist tradition draws upon the concept of there being a "structured antagonism" within the employment relation (Edwards 1986), but that worker interests and motives are more complex and contradictory (Dundon & Dobbins 2015).

The relation with the "institutional" is, therefore, more complex, although, within this radical pluralist school of thought, the institutional remains a key point of reference – unlike other radical perspectives that prefer to locate it in the broader spaces of the "social" and "political". Be this as it may, the need to take a broader view of radical perspectives towards collective bargaining is required if we are not to fall into the frequent mistake of assuming a simple binary of positions with regard to it. Furthermore, in some senses, the issue is one of how the boundaries between the different spheres of activity are constructed within unions both internally and in relation to other social actors (see, for example, Chapters 6 and 7).[6] In effect, what is important to understand is that, due to the range of strategic and political factors to which they are subject, specific institutions in themselves need to be understood in terms of how they develop, consolidate and sustain themselves. The issue of collective bargaining is important because not only does it have substantive effects – positively and, in some cases, negatively, economically, both in the short to long term – but collective bargaining also has negative and ambivalent institutional effects on the actors involved, not least unions. Collective bargaining is subject to a range of competing interventions. This ambivalence in the role and the position of collective bargaining across the array of other relations that exist means that greater attention is required, even from a critical perspective. The next section will widen the analysis of the debate to include political economy, gender and welfare, and ideological factors.

LOCATING COLLECTIVE BARGAINING IN THE CONTEXT OF THE SOCIAL AND POLITICAL

The contemporary study of industrial relations, and the more specific issue of collective bargaining, has been further widened and approached from various perspectives, typically by academics located in the heterodox economic tradition, organizational theory

6. These boundaries and the role of different voices in the process of collective bargaining or bargaining more generally have increasingly become the focus of debate where the issues of spheres of activity and their relation within discourses and narratives, and practices, is not insignificant, as collective bargaining is one aspect of a union's mobilizing or negotiating function alongside social, political and other forms of engagement (Martínez Lucio 1988; Martínez Lucio & Mustchin 2019). Radical debates increasingly relay a broader view of the position of collective bargaining and do not necessarily start from it as the main point of departure.

and labour sociology. This shift further seems to suggest that specific institutions can only be understood in terms of their effects by looking at their "location" within such broader relations and not just their "processes". Three such issues are outlined below.

First, collective bargaining exists in relation to a range of other state practices and relations that are relevant to the nature of work and the structure of the labour market. The role of the socio-economic system within which collective bargaining exists, and to which it contributes, should be central to any analysis. The nature of the national business system, as Whitley (2007) describes it, situates industrial relations systems and their processes alongside the nature of ownership and non-ownership structures. Central to why such systems vary is the pattern of hierarchies and networks across employers, with some systems being more hierarchically closed and hierarchical than others. Moreover, the role of a range of public and private actors within industrial relations structures can vary contributing to the nature of organizational practices across time and their inclusive nature or not. Furthermore, the vogue for using the "varieties of capitalism" approach has emerged as a way of trying to establish a sensitivity to the role of the cultures and practices of regulation in different contexts in terms of how they are coordinated, developed and linked (or not) (Hall & Soskice 2001)

Collective bargaining can reflect the extent of economic centralization and coordination – and it can also contribute to these across time and sustain them. To that extent, positioning collective bargaining in an economic systems bureaucratic complexity is one way of determining its structure and processes. Questions of legislative expectations and requirements, along with normative expectations, vary across systems, leading to different depths, forms and levels of coordination (see Pohler & Riddell 2019).[7] Hence, the broadening out of the study of collective bargaining into a deeper analysis of socio-economic systems has been prevalent.

Second, part of the problem of the study of collective bargaining has been its tendency to focus on particular structures and processes in isolation from the social relations and structures that encompass the employment relationship. Hence, the context and impact of collective bargaining is related to the diverse nature of the workforce, and its distinct access to social rights and services or support should be given more attention. This has been a central feature of a broad body of heterodox economists and labour market specialists who have attempted to widen the focus of analysis that has been derived from the more institutionalist views just described.

Bosch, Lehndorff & Rubery (2009: 15–20) argue that we need to study the process and nature of work by locating our interest in production regimes (e.g. patterns of ownership or industrial organization) and employment regimes (e.g. labour market regulation and industrial relations systems) alongside welfare regimes (e.g. in relation to gender, the role of family and social services). In effect, for example, how women have been included within or excluded from the tapestry of regulation that exists, and whether their specific circumstances have been recognized. This interest in sensitizing any analysis to the position of specific groups of workers and their specific social context has implications for understanding the way collective bargaining reinforces or challenges the hierarchies within workforces.

7. For a general discussion on the varieties of capitalism approach and different critiques, see Pohler (2018).

Collective bargaining may reflect certain specific aspects – for example, gender pay gaps within work (a matter in which it has not typically been effective or to which it has not been sensitive) – but without an awareness of the nature of social protection or care systems in relation to the state where the impact joint-regulation has on women's working conditions may be limited (Rubery & Fagan 1995; Távora 2012; Távora & Rubery 2013). That the subject of equality has emerged as an important part of collective bargaining and union strategies in various countries – due to a range of political, legal or social pressures, as well as the steady renewal of union priorities – should not be discounted (Kirton & Greene 2005). However, the way collective bargaining sits within a broader network of regulatory practices requires the strategies of actors such as unions and employers to be understood and approached more broadly (Connolly, Marino & Martínez Lucio 2019; Greene, Kirton & Wrench 2005).

This issue of appreciating the boundaries of regulation has emerged in a much stronger form in recent decades, as has the realization that collective bargaining has also acted as a form of exclusion or segregation in certain contexts, reflecting workforce hierarchies related to gender, race, and the like. To that extent, collective bargaining is more dependent on social and political hierarchies – and changes within these – in terms of its form and its content. These factors will also vary across national and regional contexts.

Third, we increasingly need to locate collective bargaining and joint-regulation within a cultural and ideational landscape. It is all too easy to forget that collective agreements and forms of bargaining may have a different meaning across distinct contexts, rather than just national ones. The way the status of an agreement varies can be the outcome of different institutional approaches, which can also mean that it has a variable effect. In more judicial cultures, written collective agreements are referenced in different ways in terms of their details and legitimacy than, for example, those cultures in which they are not published and circulated regularly among unionist activists or human resources (HR) managers, for example.

The way specific terms and conditions of employment are referenced in detail within the workforce (e.g. working hours) can also vary, not solely because of the higher visibility of agreements within certain contexts, but also because of the legalistic traditions that may exist. Kirk (2018) speaks of the importance of a legal consciousness around specific forms of worker rights in certain instances, and there is no reason not to presume this is not relevant to the position of collective agreements as well. The political visibility of collective agreements and their broader social status can also be important in framing this awareness of the content of collective agreements, and may also play a role in securing the centrality of its role and limit attempts to undermine its legitimacy. Moreover, the cultural and broader media context of industrial relations and the way it is framed is important, as well as visible, in the way bargaining processes and actors associated therewith (such as unions) may be highlighted or even stigmatized, as can be the case in some contexts where anti-union narratives have been mobilized through cinema or television drama or comedy as in the UK in the 1960s and 1970s (MacKenzie & Martínez Lucio 2014).

In addition, the broader "ideational environment" is also important in the way broader political paradigms are underpinned by policy ideas and discourses concerning work-related issues such as collective bargaining by a range of actors (Morgan & Hauptmeier

2021). The way expertise is mobilized, and the way in which ideas are represented and mobilized by a range of actors, such as thinktanks, can shape political and popular understandings of aspects of industrial relations processes and institutions, as in the focus on deregulation and its depiction (Morgan & Hauptmeier 2021). This can also vary across national and even industrial contexts, thus limiting or propelling different trajectories of reform; for example, as in the stigmatizing of the supposed "rigidities" of employment regulation in Spain since the 1980s (Fernándes Gonsález & Martínez Lucio 2013). Increasingly, we need to appreciate these cultural and ideological contexts of industrial relations, as Hyman (1975a, 1975b) reminds us, in terms of the use and abuse of certain terminology and concepts that can either legitimate or stigmatize systems of regulation.

Increasingly what we have seen is an interest in the context of collective bargaining. Contemporary studies have pointed to the need to understand the position of bargaining structures and how these are articulated and combined. In addition, we have also noted how the socio-economic context has emerged as a significant aspect of the study of collective bargaining, looking at how it relates to other forms of state intervention in relation to the economy and key labour market relations such as training regimes. These have been further complemented by the equality and feminist-related aspects of academic research that have aligned the form and content of collective bargaining to questions of gender and other societal aspects and relations, especially the role of the state in relation to welfare. To that extent, we have seen a greater tendency to place the development and impact of joint-regulation in a broader regulatory and political context.

Finally, the concern with ideas and ideology further add to the political understanding of collective bargaining in terms of how its processes and actors are represented more broadly in either positive or, as seems to be the case, increasingly negative terms. To that extent, we have seen a significant move in the way we appreciate the position and role of collective bargaining. This is significant for the study of such forms of joint-regulation because approaching it from such a set of perspectives does not assume that it is inevitably central to the process of industrial relations or HR management. It allows for different contexts to be appreciated beyond the colonial norms of European and North American industrial relations. This then permits a greater sensitivity of a decolonization nature to be developed in the way dynamics of representation and regulation can be approached in terms of different sets of economic systems, broader social forms of regulation within the private and public spheres, and the varied roles of political and national level discourses and narratives.[8] It is one set of steps that allows us to situate such processes in terms of the way we can study them in a much broader comparative manner.

DECLINE AND CRISIS: LOCATING COLLECTIVE BARGAINING IN RELATION TO THE CONTRADICTORY NATURE OF CHANGE

The issue of the contemporary political challenge to the role and centrality of collective bargaining has become a significant field of discussion, especially given that such developments substantively undermine working conditions and limit the reach

8. For a discussion of the importance of such developments in the study of organizations more broadly, see Ibarra-Colado (2006).

and role of the workforce through its representatives. One cannot ignore the reality of how significant the debate on decline is (including that related to unions). It has been extensive to some extent, especially since the 1980s – given the emergence into the political field of the "New Right" at that time with its antagonistic approach to unions – and has tended to be a part of various debates. It parallels the concern with the decline of traditional industrial cultures and societies (see Gorz 1987) and it also parallels the concern with greater levels of individualization and fragmentation socially and within the employment relation (Purcell 1991). These are debates that mark a particular moment where scholars and practitioners were alerted to the onset of change.

However, substantive changes in relation to collective bargaining in terms of its supposed decline or weakening are complex, as are the outcomes of a range of contradictory factors concerning forms of renewal that have subsequently emerged.[9] Such changes may be an outcome of deliberate strategies to undermine collective bargaining, but there are also developments that increasingly "stretch" this form of regulation in terms of having to deal with an ever more complex employment environment. This forces us to understand the issues of crisis and change as being the outcome of a broad range of factors and actors. Let us start with the first of these two situations in terms of the "neoliberal" challenge.

Mustchin (2019) alerts us to the need to locate the creeping narrative and practices of anti-unionism in a broader historical context through a range of political and ideational actors such as right-wing thinktanks. The 1970s appear to be a key turning point in the European context, especially in the UK where the political and economic crisis seems to have generated new political options intent upon undermining the role of joint-regulation and collective representation. However, in the case of the UK, Mustchin (2019) points to the way right-wing thinktanks developed practices for management in their attempt to weaken unions. For this reason, the supposed Copernican shift (or reverse shift, depending on your point of view) of the 1970s was based upon a growing critique of the perceived negative impact of collective bargaining in market terms.

The role of the Thatcher and Reagan administrations in the 1980s are considered key – and remain key – to the way unions and collective bargaining processes have been undermined. The focus of policy reforms was on the debilitating of collective worker rights in relation to the right to collective action, the remits of such action and the general activities of trade unions (see McIlroy 1991). The re-emergence of political hostility towards unions and related activities in terms of collective bargaining was clear from the anti-collectivist rhetoric of the political Right and its use of individualism as a way of paving the way towards a greater emphasis upon privatization and markets as a way of allocating resources – it was also a way of undermining unions and allowing employers increasingly to dominate the sphere of worker representation (MacKenzie & Martínez Lucio 2014). This was a form of "authoritarian populism" that critiqued established collectivist economic and labour market regulation while lauding the role of "individual" market-related rights (Hall 1985).

However, this political framing of collective regulation also began to influence social democratic agendas within North America and parts of the European continent.

9. See Martínez Lucio (2006) for an outline of different economic, social, and political factors related to the crisis of modern industrial relations and labour unionism.

Aspects of social democratic or centrist politics further engaged with the view that collective bargaining was an antiquated or problematic way of setting terms and conditions of employment. The 1990s onwards saw a shift towards a more market-oriented social democracy that sustained a less engaged relation with organized labour and collective forms of regulation (see Smith 2009 on the UK) with the view that such regulation was excluding certain "outsider" interests from being mobilized through various political discourses for primarily ideological reasons (see Fernández González & Martínez Lucio 2013 on Spain). This shift has been clear within the European context in general, with the push to deregulation as a feature of economic policy becoming for some a hegemonic feature of public policy (Baccaro & Howell 2017; see also Chapter 11).

Within this context, the perceived positive experiences of countries, such as the UK, politically assisted various institutional interests in the development of a desire to restrain the impact of collective bargaining, especially after the 2008–09 global financial crisis, given the belief that deregulating collective bargaining and labour markets would facilitate economic stability (Martínez Lucio, Koukiadaki & Távora 2019). While all-out and substantive change has not been that visible in many parts of Europe (see Chapter 11), there appears to have been an erosion of multi-employer bargaining in some parts since 2009 and a challenge to unions in their having to deal with more fragmented forms of bargaining (Marginson 2015).

This push towards greater deregulation and decentralization has been paralleled with greater emphasis upon individual forms of wage setting and participation at work. The emergence of forms of management practices that emphasise individual channels of communication and systems of performance management have been seen to undermine the previous balance of power between management and unions (Bacon & Storey 1993), although to what extent this is the main driver of union decline is a matter for debate (Machin & Wood 2005; see also Chapter 7). Furthermore, the emergence of new forms of greater levels of subcontracting at work and self-employment during the current context of the gig and platform economies have generated a challenge to the way working conditions are set (Inversi 2019; Stewart & Stanford 2017; see also Chapter 7). Such developments can fragment collective bargaining structures as in subcontracting or obviate the need for bargaining as workers are classified as self-employed.

Alongside these developments, there has also been a greater emphasis on anti-union practices that, while always present to varying degrees within industrial relations, have experienced increasing intensification and use through specialist consultancy and law firms by employers (Dundon & Gall 2013). The management consultancy industry, which increasingly advises firms regarding how to undermine or constrain collective bargaining practices and unions, has been deepening its role even in contexts where joint-regulation is regarded as being central to democratic processes. The role of blacklisting union members and activists is an established practice that only recently has been the focus of broader and systematic study (see Drucker 2016). This question of the coercive aspects of industrial relations in terms of employers and the state has steadily gained ground in the way the politics of work develops, even in the context of the gig economy, and new forms of work where anti-unionism is a commonly embedded feature.[10]

However, a "crisis" is not solely related to issues of decline in, impoverishment of, or increasing challenges to collective bargaining. A crisis is also an "overdetermined" moment[11] when forms of institutional and established behaviour and activities are subject to competing imperatives for change and a range of contradictory pressures. In the case of collective bargaining, the pressures emerging from specific sets of employers and state policies has, in certain contexts, focused on limiting its remit and even its existence in some cases at the same time as having to contend with the increasing demands on collective bargaining actors and processes to extend their remit and reach towards newly emerging or increasingly vocal demands on a range of social issues from a wide variety of social groups and interests. That is to say, the second set of pressures contributing to the crisis, shall we say, of collective bargaining was the fact that demands related to issues such as the need for greater health and safety, intra-workforce equality, and developmental aspects of work were increasing. The changing nature of work and the workforce has brought forth a need to ensure new ways of sustaining and enhancing working conditions beyond a traditional narrow remit.

To this extent, collective bargaining – and various parallel activities of an informal nature related to it – have had to respond increasingly to pressures from the workforce: pressures that have been placed upon employers, the state and internal union structures (in terms of intra-bargaining relations and the role of diverse internal voices). Following on from key moments when these issues came to the fore – the 1970s being an example in certain national contexts such as the UK – and when pressures at work coincided with social pressures more broadly, there was a steady widening of the content of collective bargaining. In some respects this reflects a moment that could be considered a shift or "overloading" of new rights issues in terms of work (Martínez Lucio & Simpson 1992) – the question becoming who is it that is capable and strategic enough to be the relayers and agents for such new rights (Martínez Lucio & Simpson 1992)?

Tailby and Moore (2014) argue that collective bargaining has often been insensitive to such development as equality-related workplace issues, for example, and, if anything, has often acted as a barrier to the development of those issues – something that

10. However, there is another set of challenges to collective bargaining that emerge from the way academics beyond the realm of industrial or labour relations have tended themselves to associate collective bargaining with antiquated forms of worker activity. Be it within broader realms of the sociology of work through to the more managerialist approaches to the issue of HR management, the role and reach of collective bargaining is often seen as passé and irrelevant. This may be due to a range of new industrial sectors where its presence is uneven (IT, the gig economy, and so on). However, on occasions there seems to be a residual pleasure in such circles in discounting such forms of regulation and falling into stereotypes that have emerged, ironically, from the right of the political spectrum.

11. This issue of how different factors play themselves out can be seen in terms of the nature of struggles and the competing patterns of dominance and subordination in any given context. It requires an understanding of change that is based on an awareness of the contradictions and tensions that exist within capitalism, although what the dominant strand and underlying drivers have been is a subject of discussion (see Park 2013).

is increasingly well documented in the industrial relations literature (see also Blackett & Sheppard 2002, as quoted in Tailby & Moore 2014). Yet, the changing nature of social and political dynamics has meant a growing sensitizing of collective bargaining content to the question of equality. In cases such as Spain, the equalities legislation paralleled by broader union framing of social issues contributed to a social politics that also reshapes joint-regulation and other areas of union engagement more broadly (see Barranco & Molina 2019).[12] In some cases, this has led to specialist union offices, or departments specializing in such issues providing advice to the internal bargaining functions of the unions and even specialist forms of representation regarding equality within workplaces, as in the UK (Moore & Wright 2012). This mirrors the increasing specialist representation around matters of health and safety, as well where we see an expansion and widening of representation and negotiation roles.

To a great extent, one could argue that the broadening of social struggle and new voices and modes of activism within the workforce have played an important role in reshaping the contours of union agendas and priorities in relation to equality or training (see, e.g., Moore 2010). This widening of the agenda may have different push factors involved but the emergence of new constituencies in migrant communities and others are important in shaping the way collective bargaining or union agendas evolve (Moore 2010). Such a widening of joint-regulation is, therefore, the outcome of new forms of protest, mobilization, activism and social agendas, and even social movements. This is something that occurs at various levels of the labour movement, too, as Wever (1998), points out in terms of the organizing of new worker constituencies). In effect, the social and political context of industrial relations intrudes into its forms and reshapes it in different ways at different times.[13]

However, and returning to Tailby and Moore (2014), this widening, to call it that, can be ambivalent. The case of the UK, they point out, is telling, as progressive developments such as the minimum wage and other forms of progressive legislation can hinder and obviate the role of unions more generally in the establishment of wage levels (especially where collective bargaining is underdeveloped or non-existent). In addition, the emergence of new forms of civil society organizations that include various other bodies, including employers, can limit the role of unions and collective bargaining mechanisms. The role of the living wage as a form of voluntary and presumably "socially responsible" way of setting a "fair" wage level can, ironically, undermine the processes of collective bargaining and its role (Tailby & Moore 2014). Furthermore, the attempt by employers to "colonize" the space of representation through various forms of communication may

12. The role of power relations and how they frame the content of collective bargaining is becoming an increasingly important feature of the study of collective bargaining and its content (see Bourguignon & Coron 2021). Moreover, the role of unions and how they influence the state and other actors with regard to questions such as health and safety can also be significant in the widening the remit of collective bargaining (Martínez Lucio 2020).

13. This has implications for unions as well as others involved in the processes of collective bargaining as they must sustain and support information systems and research related to the ever-widening nature of collective bargaining in terms of a range of themes. This places new pressures on unions and their bureaucratic structures, which either enlarge them further or force a need to generate new informational networks.

also inhibit democratic forms of negotiation and the role of unions (Edwards 2009; MacKenzie & Martínez Lucio 2014). The presence of a range of new regulatory actors may, thus, expand the field or regulatory space but it need not necessarily democratize it or deepen worker influence (for a discussion of the politics of the regulatory space, see MacKenzie & Martínez Lucio 2005, 2014).

The importance of such approaches is that they point to the social dynamics and tensions facing collective bargaining. It may be challenged at one level by management and employer strategies that seek to limit or even eliminate its role while the economy becomes increasingly fragmented in neoliberal terms, but it also must face a set of social challenges that provide opportunities for its democratic and social expansion thematically. However, various developments and the role of new actors and institutions can, ironically, displace the role of traditional or core actors and lead to a decentred – and what Tailby and Moore (2014) consider to be voluntarist – context. This tension within the evolution of collective bargaining suggests we must widen our views of how it operates and impacts. We need to adopt an awareness of the complex exchanges that exist in relation to collective bargaining and how it is positioned along a range of different developments. It forms part of a tapestry of political and social relations with its centrality and form being, in part, dependant on a range of factors and contradictory dynamics.

RETHINKING BARGAINING IN AN AGE OF ORGANIZATIONAL "DISTANCE"

How collective bargaining and related changes are represented by a range of interests is also key, as is where it is made to "fit" in the spectrum of regulation.[14] It could be that we need to draw attention to the way competing projects attempt to reframe the role and purpose of collective bargaining. Using Hyman (2000) for understanding union identity (see also Chapter 1), it could be that we need to see how collective bargaining is situated in different ways. There may be market-oriented approaches that privilege forms of mutual interests between workers and unions or other worker representatives in the attempt to link productivity and worker rights (making them conditional upon each other); or it may be that collective bargaining is driven and remade around more explicit class-related politics that seek to enhance worker conditions as a demand in themselves and sustain the autonomy of unions in relation to employers. Finally, it could be that collective bargaining is linked to broader social dynamics and issues with a strong societal dimension and sets of alliances. These three approaches are not precise ways of conceptualizing such developments and strategic turns, but they suggest that we need to extend how we understand the context and location of collective bargaining. The future of collective bargaining rests across different sets of scenarios depending

14. It is interesting that, in the European Union, many labour-related research bodies have been working on the issue of collective bargaining as a central part of the European social project. The role of the European Trade Union Institute has been important in not only studying the contours and changes of collective bargaining across Europe, but also its role as a central feature of the social dialogue agenda along with national forms of tripartite policies. This broad and varied "project" is an example of how the positioning of collective bargaining politically can develop within research and policy agendas.

on the balance of forces, political context and the nature of the regulatory actors and resources that exist. The "expansion" of collective bargaining is contingent on how these factors play themselves out.[15]

A further set of developments is also clear in the strategic use of collective bargaining relations. One notices in many contexts new forms of "negotiation" that are more akin to an arms-length relation between the parties due to the role of new employers with less interest in collective industrial relations, and partly due to the emergence of new collective worker organizations that are increasingly weary of such firms and their agendas, as in the case of the platform economy (Martínez Lucio *et al.* 2023; see also Chapters 7 and 21). Batstone (2015), in describing post-Second World War industrial relations in key parts of the French system, points to the way paternalistic or anti-union strategies led to arms-length approaches within union strategies (and what were, in effect, low-trust relations) (Jefferys 2015). This is then a way of understanding such strategies that is also useful for explaining the way collectives of workers negotiate at a distance using mobilization and alternative forms of communication in more authoritarian contexts. Within the gig economy of various national contexts, such forms of mobilizing and negotiation are common, where possible, as is the use of legal means and measures to force collective "agreements" (Smith 2021; Stewart & Standford 2017). The use of public spaces, social media and alternative forms of engagement to remit a demand and bargaining from a distance is common. Social media has many such uses in helping workers raise demands that may eventually be taken up by employers (see Geelan (2021) and Greene, Hogan & Grieco (2003) for a discussion on the role of the internet and social media in relation to union strategies).

Moreover, the social dimension of collective bargaining in terms of various actors is also important and central to the development of new forms of alliances at the local level (Holgate 2015). That is not say that more formal approaches to collective bargaining do not exist or are not relevant, but, rather, that these are increasingly complemented by new forms in terms of inter-actor relations and diverse approaches to the strategies deployed to alter established employer practices, whether formally agreed or not. This, thus, parallels the debate on organizing and union renewal that has pointed to viewing strategies of union action through a broader lens (Murray 2017). Much depends on how the spheres of the state, the labour market, the community/social and external societal factors, and the workplace are tied together through the strategies of actors and their agendas (Martínez Lucio 1988; Martínez Lucio & Mustchin 2019).

In some senses, while the debate in contexts such as Europe within more institutionalized circuits has revolved around sustaining coordinated forms of collective bargaining and – where possible – extending their thematic coverage in the face of neoliberal threats and policies of deregulation, the developments outlined in these final sections suggest that the strategic developments and forms of bargaining involve a wider set of actors and processes. To this extent, these new debates and approaches

15. One could, however, argue that there was never a golden age of collective bargaining, just as there was never a collective age of industrial relations, given the complexities and tensions identified earlier in the chapter and, if there was one, it was momentary, being present in the immediate post-Second World War period and specific to particular liberal democratic contexts and interests (Hyman 1989).

are becoming highly relevant to national cases where collective bargaining has not been as institutionally embedded or sustained. Issues of culture and inclusion through different forms of representation generating new forms of demands are important (Peró 2020). Forms of protest, mobilization, social alliance-making and broader patterns of representation allow for a greater postcolonial and decolonizing sensibility that moves away from American or Eurocentric approaches when studying such features of industrial relations and their formal dimensions (see also Alberti & Péro 2018). The decentred nature of joint regulation and the role of the state in certain contexts bring forth a need to think through patterns and forms of bargaining in new ways and to not assume a stable starting point in terms of institutional industrial relations.

CONCLUSION: RECLAIMING AND BROADENING THE UNDERSTANDING OF COLLECTIVE BARGAINING

The concept of collective bargaining has been a central feature of the way working conditions are determined. That the state and employers can set conditions unilaterally in some cases does not alter the reality that formal processes of bargaining and more discreet and strategic forms of bargaining "from a distance" can and do play an important role. As a concept it remains important, although, increasingly, we require a more flexible understanding of how negotiation works. Aspects of earlier forms of industrial relations thinking were not averse to this, and one could argue that, at the heart of industrial relations, there has existed an awareness of the complexities and ambiguities of representation at work.

Even before the increasing sensitivity to equality issues, among other matters, the discipline in some senses detected that representation at work was not a singular process but, rather, that it had a range of political tensions and differences aligned to it. The more radical traditions – especially within the Marxist schools of thought – were alert to this and viewed an awareness of instability as being key to any understanding of worker representation. Subsequent radical studies from a more sociological and equality-related perspective have developed this further and shown how the boundaries or collective bargaining and its processes are constructed through a range of power relations as outlined above. Hence, the chapter has tried to outline how debates regarding collective bargaining developed and identified this ambiguity. Over time, the need to locate the spaces of bargaining and its position became viewed from broader perspectives with regard to economic structure, social hierarchies and structures, and ideological perspectives. Understanding how systems of joint-regulation vary has become increasingly sensitized to a broader concern with context and the position of joint-regulation within this context.

Even the supposed crisis of collective bargaining raises some issues as to how we understand its role. The tendency to view its decline due to the role of anti-union and anti-collective state and employer strategies must be tempered by the fact that the content of collective bargaining has been expanding, and social sensitivities and broad issues of equality have been developing in and around it. This has led to key interventions noted earlier regarding the way collective bargaining is a much more contested space socially.

This "social dimension of collective bargaining" has implications not solely for issues of content, but also its form and, therefore, also implications for the processes and politics involved, which are clearly linked to broader sets of actors and new forms of connections that exist in relation to mobilization. This requires a more complex set of exchanges across different groups, voices and organizational actors, beyond management and even the state in some cases. In some cases this can facilitate collective bargaining strategies within the workforce and its representatives, but it can also generate problems as to how literally to close off agreements. Either way, we are seeing new forms emerging and they can reflect the changing nature of voice and social concerns within society.

To that extent, the chapter has concerned itself with the increasing interest in the context of collective bargaining, the manner in which changes are torn between complex economic and social contradictions, and the way in which bargaining processes are strategically approached through different union identities and politics. That is not say collective bargaining does not, in turn, have an effect on various social actors and their agendas, as the traditional view regarding the problems of institutionalization has outlined for some time – neither does it mean we should view collective bargaining as some form of dependent variable. However, there is a need to view collective bargaining as much more of a contended space contingent on various structural and strategic factors, especially if we are to include broader and more nuanced forms of bargaining and mobilization that allow for a move away from a Eurocentric approach.

Finally, there is a limit, as democratic "expansion" for this requires calls for broader representative structures and processes that work at more strategic levels of the economy and society. In this respect, the structural changes and expansion of collective bargaining and industrial relations processes is, perhaps, an indication of a deeper regulatory crisis that reflects the changing boundaries of work in relation to economy, society and the environment. This suggests that the relaunching of the debate on industrial democracy is the next suitable step for comprehending these sets of issues.

REFERENCES

Alberti, G. & D. Però 2018. "Migrating industrial relations: migrant workers' initiative within and outside trade unions". *British Journal of Industrial Relations* 56(4): 693–715.

Atzeni, M. 2010. *Workplace Conflict: Mobilization and Solidarity in Argentina*. Basingstoke: Palgrave Macmillan.

Baccaro, L. & C. Howell 2017. *Trajectories of Neo-liberal Transformation: European Industrial Relations since the 1970s*. Cambridge: Cambridge University Press.

Bacon, N. & J. Storey 1993. "Individualization of the employment relationship and the implications for trade unions". *Employee Relations* 15(1): 5–17.

Barranco, O. & O. Molina 2021. "Continuity and change in trade union frames: evidence from general strikes in Spain". *Economic and Industrial Democracy* 42(4): 1232–53.

Batstone, E. 2015. "Arms'-length bargaining: industrial relations in a French company". *Historical Studies in Industrial Relations* 36(1): 73–137.

Blackett, A. & C. Sheppard 2002. "The links between collective bargaining and equality". International Labour Office Working Paper 10.

Bosch, G., S. Lehndorff & J. Rubery 2009. "European employment models in flux: pressures for change and prospects for survival and revitalization". In *European Employment Models in Flux: A Comparison of Institutional Change in Nine European Countries*, edited by G. Bosch, S. Lehndorff & J. Rubery, 1–56. London: Routledge.

Bourguignon, R. & C. Coron 2021. "The micro-politics of collective bargaining: the case of gender equality". *Human Relations* 76(3): 395–419.

Burchill, F. 1999. "Walton and McKersie, a behavioral theory of labor negotiations 1965". *Historical Studies in Industrial Relations* 8: 137–68.

Clegg, H. 1976. *Trade Unionism under Collective Bargaining*. Oxford: Blackwell.

Connolly, H., S. Marino & M. Martínez Lucio 2019. *The Politics of Social Inclusion and Labor Representation: Immigrants and Trade Unions in the European Context*. Ithaca, NY: Cornell University Press.

Dahrendorf, R. 1959. *Class and Class Conflict in Industrial Society*. Stanford, CA: Stanford University Press.

Dobbins, T., E. Hughes & T. Dundon 2021. "'Zones of contention' in industrial relations: framing pluralism as praxis". *Journal of Industrial Relations* 63(2): 149–76.

Druker, J. 2016. "Blacklisting and its legacy in the UK construction industry: employment relations in the aftermath of exposure of the Consulting Association". *Industrial Relations Journal* 47(3): 220–37.

Dundon, T. & T. Dobbins 2015. "Militant partnership: a radical pluralist analysis of workforce dialectics". *Work, Employment and Society* 29(6): 912–31.

Dundon, T. & G. Gall 2013. "Anti-unionism: contextual and thematic issues". In *Global Anti-Unionism* edited by G. Gall & T. Dundon, 1–17. London: Palgrave Macmillan.

Edwards, G. 2009. "Public sector trade unionism in the UK: strategic challenges in the face of colonization". *Work, Employment and Society* 23(3): 442–59.

Edwards, P. 1986. *Conflict at Work*. Oxford: Blackwell.

Fernández González, C. & M. Martínez Lucio 2013. "Narratives, myths and prejudice in understanding employment systems: the case of rigidities, dismissals and flexibility in Spain". *Economic and Industrial Democracy* 34(2): 313–36.

Flanagan, R. 2008. "The changing structure of collective bargaining". In *The Sage Handbook of Industrial Relations*, edited by P. Blyton *et al.*, 406–19. London: Sage.

Geary, D. 2001. "The 'union of the power and the intellect': C. Wright Mills and the labor movement". *Labor History* 42(4): 327–45.

Geelan, T. 2021. "Introduction to the special issue: the internet, social media and trade union revitalization: still behind the digital curve or catching up?" *New Technology, Work and Employment* 36(2): 123–39.

Goldthorpe, J. 1974. "Industrial relations in Great Britain: a critique of reformism". *Politics & Society* 4(4): 419–52.

Goodman, J. 1984. *Employment Relations in Industrial Society*. London: Philip Allan.

Gorz, A. 1987. *Farewell to the Working Class: An Essay in Post-Industrial Socialism*. London: Pluto.

Greene, A.-M., J. Hogan & M. Grieco 2003. "E-collectivism and distributed discourse: new opportunities for trade union democracy". *Industrial Relations Journal* 34(4): 282–9.

Greene, A.-M., G. Kirton & J. Wrench 2005. "Trade union perspectives on diversity management: a comparison of the UK and Denmark". *European Journal of Industrial Relations* 11(2): 179–96.

Hall, P. & D. Soskice 2001. " Introduction". In *Varieties of Capitalism: The Institutional Foundations of Comparative Advantage*, edited by P. Hall & D. Soskice, 1–70. New York: Oxford University Press.

Hall, S. 1985. "Authoritarian populism: a reply to Jessop *et al.* ". *New Left Review* 151(1): 115–23.

Heery, E. 2016. *Framing Work: Unitary, Pluralist, and Critical Perspectives in the Twenty-First Century*. Oxford: Oxford University Press.

Holgate, J. 2015. "Community organising in the UK: a 'new' approach for trade unions?". *Economic and Industrial Democracy* 36(3): 431–55.

Holgate, J., M. Simms & M. Tapia 2018. "The limitations of the theory and practice of mobilization in trade union organizhing". *Economic and Industrial Democracy* 39(4): 599–616.

Hyman, R. 1975a. *Marxism and the Sociology of Trade Unionism*. London: Pluto.

Hyman, R. 1975b. *Industrial Relations: A Marxist Introduction*. London: Macmillan.

Hyman, R. 1989. *Political Economy of Industrial Relations: Theory and Practice in a Cold Climate*. Basingstoke: Macmillan.

Hyman, R. 2000. *Understanding European Industrial Relations*. London: Sage.

Ibarra-Colado, E. 2006. "Organization studies and epistemic coloniality in Latin America: thinking otherness from the margins". *Organization* 13(4): 463–88.

Inversi, C. 2019. Exploring the Concept of Regulatory Space: Employment and Working Time Regulation in the Platform-Economy. PhD thesis, Manchester University.

Jefferys, S. 2015. "Arms'-length or nose to nose? Eric Batstone and bargaining in 1970s France". *Historical Studies in Industrial Relations* 36(1): 59–72.

Joyce, S. 2016. Revisiting Shop Stewards and Workplace Bargaining: Opportunities, Resources and Dynamics in Two Case Studies. PhD thesis, University of Hertfordshire.

Kelly, J. 1998. *Rethinking Industrial Relations: Mobilisation, Collectivism and Long Waves*. London: Routledge.

Kirk, E. 2018. "The reorganisation of conflict at work: mobilisation, counter-mobilisation and the displacement of grievance expressions". *Economic and Industrial Democracy* 39(4): 639–60.

Kirton, G. & A.-M. Greene 2005. "Gender, equality and industrial relations in the 'New Europe': an introduction". *European Journal of Industrial Relations* 11(2): 141–9.

Machin, S. & S. Wood 2005. "Human resource management as a substitute for trade unions in British workplaces". *Industrial and Labour Relations Review* 58(2): 201–18.

MacKenzie, R. & M. Martínez Lucio 2005. "The realities of regulatory change: beyond the fetish of deregulation". *Sociology* 39(3): 499–517.

MacKenzie, R. & M. Martínez Lucio 2014. "The colonisation of employment regulation and industrial relations? Dynamics and developments over five decades of change". *Labor History* 55(2): 189–207.

Marginson, P. 2015. "Coordinated bargaining in Europe: from incremental corrosion to frontal assault?". *European Journal of Industrial Relations* 21(2): 97 114.

Martínez Lucio, M. 1988. Trade Unions in Post-Franco Spain. PhD thesis, University of Warwick.

Martínez Lucio, M. 2006. "Trade unionism and the realities of change". In *Employment Relations in a Changing Society*, edited by L. Alonso & M. Martínez Lucio, 200–14. Basingstoke: Palgrave Macmillan.

Martínez Lucio, M. 2010. "Labour process and Marxist perspectives on employee participation". In *The Oxford Handbook of Participation in Organizations*, edited by A. Wilkinson *et al.*, 105–30. Oxford: Oxford University Press.

Martínez Lucio, M. 2020. "Trade unions and stress at work: the evolving responses and politics of health and safety strategies in the case of the United Kingdom". In *The Handbook of Stress and Wellbeing in the Public Sector*, edited by R. Burke & S. Pignata, 15–32. Cheltenham: Edward Elgar.

Martínez Lucio, M., A. Koukiadaki & I. Távora 2019. *The Legacy of Thatcherism in European Labour Relations: The Impact of the Politics of Neo-Liberalism and Austerity on Collective Bargaining in a Fragmenting Europe*. Liverpool: Institute of Employment Rights.

Martínez Lucio, M. & S. Mustchin 2019. "The politics and diversity of worker representation: the increasing fluidity and challenge of representation". In *Handbook of the Politics of Labour, Work and Employment*, edited by G. Gall, 144–60. Cheltenham: Edward Elgar.

Martínez Lucio, M. & D. Simpson 1992. "Discontinuity and change in industrial relations: the struggles over its social dimensions and the rise of human resource management". *International Journal of Human Resource Management* 32(3): 173–90.

Martínez Lucio, M. *et al.* 2023. "Work, spatial changes and 'arms-length' bargaining in the platform economy". In *Missing Voice? Worker Voice and Social Dialogue in the Platform Economy*, edited by A. Wilkinson *et al.*, 130–53. Cheltenham: Edward Elgar.

McIlroy, J. 1991. *The Permanent Revolution? Conservative Law and the Trade Unions*. London: Spokesman Books.

Moore, S. 2010. *New Trade Union Activism: Class Consciousness or Social Identity?* London: Palgrave Macmillan.

Moore, S. & T. Wright 2012. "Shifting models of equality? Union equality reps in the public services". *Industrial Relations Journal* 43(5): 433–47.

Morgan, G. & M. Hauptmeier 2021. "The social organization of ideas in employment relations". *Industrial and Labor Relations Review* 74(3): 773–97.

Murray, G. 2017. "Union renewal: what can we learn from three decades of research?". *Transfer: European Review of Labour and* Research 23(1): 9–29.

Mustchin, S. 2019. "Right-wing pressure groups and the anti-union 'movement' in Britain: aims of industry, neo-liberalism, and industrial relations reform, 1942–1997". *Historical Studies in Industrial Relations* 40(1): 69–101.

Park, H. 2013. "Overdetermination: Althusser versus Resnick and Wolff". *Rethinking Marxism* 25(3): 325–40.

Però, D. 2020. "Indie unions, organizing and labour renewal: learning from precarious migrant workers". *Work, Employment and Society* 34(5): 900–18.

Pohler, D. 2018. "Collective bargaining". In *The Routledge Companion to Employment Relations*, edited by A. Wilkinson *et al.*, 235–50. Abingdon: Routledge.

Pohler, D. & C. Riddell 2019. "Multinationals' compliance with employment law: an empirical assessment using administrative data from Ontario, 2004 to 2015". *Industrial and Labour Relations Review* 72(3): 606–35.

Poole, M. 1981. *Theories of Trade Unionism: A Sociology of Industrial Relations*. London: Routledge.

Purcell, J. 1991. "The rediscovery of the management prerogative: the management of labour relations in the 1980s". *Oxford Review of Economic Policy* 7(1): 33–43.

Rubery, J. & C. Fagan 1995. "Comparative industrial relations research: towards reversing the gender bias". *British Journal of Industrial Relations* 33(2): 209–36.

Scott, R. 2001. *Institutions and Organizations*. Second edn. Thousand Oaks, CA: Sage

Slichter, S. 1941. *Union Policies and Industrial Management*. Washington, DC: Brookings Institution.

Smith, H. 2021. "The 'indie unions' and the UK labour movement: towards a community of practice". *Economic and Industrial Democracy* 43(3): 1369–90.

Smith, P. 2009. "New Labour and the commonsense of neo-liberalism: trade unionism, collective bargaining and workers' rights". *Industrial Relations Journal* 40(4): 337–55.

Stewart, A. & J. Stanford 2017. "Regulating work in the gig economy: what are the options?". *Economic and Labour Relations Review* 28(3): 420–37.

Tailby, S. & S. Moore 2014. "Collective bargaining: building solidarity through the fight against inequalities and discrimination". *Cuadernos de Relaciones Laborales* 32(2): 361–84.

Távora, I. 2012. "The southern European social model: familialism and the high rates of female employment in Portugal". *Journal of European Social Policy* 22(1): 63–76.

Távora, I. & J. Rubery 2013. "Female employment, labour market institutions and gender culture in Portugal". *European Journal of Industrial Relations* 19(3): 221–37.

Terry, M. 1977. "The inevitable growth of informality". *British Journal of Industrial Relations* 15(1): 76–90.

Terry, M. 1993. "Workplace unions and workplace industrial relations: the Italian experience". *Industrial Relations Journal* 24(2): 138–50.

Traxler, F. 1996. "Collective bargaining and industrial change: a case of disorganization? A comparative analysis of eighteen OECD countries". *European Sociological Review* 12(3): 271–87.

Waddington, J., T. Müller & K. Vandaele 2019. "Setting the scene: collective bargaining under neo-liberalism". In *Collective Bargaining in Europe: Towards an Endgame, Volumes I–IV*, edited by T. Müller, K. Vandaele & J. Waddington, 1–32. Brussels: European Trade Union Institute.

Walton, R. & R. McKersie 1965. *A Behavioral Theory of Labor Negotiations: An Analysis of a Social Interaction System*. New York: McGraw-Hill.

Wever, K. 1998. "International labor revitalization: enlarging the playing field". *Industrial Relations* 37(3): 388–407.

Whitley, R. 2007. *Business Systems and Organizational Capabilities: The Institutional Structuring of Competitive Competences*. Oxford: Oxford University Press.

Wright Mills, C. & H. Schneider 1948. *The New Men of Power: America's Labor Leaders*. Bloomington, IL: University of Illinois Press.

CHAPTER 13

FROM CONTESTING MANAGERIAL PREROGATIVE TO PRODUCING WORKERS' CONTROL

Alan Tuckman

ABSTRACT

Workers' control has been a slogan and an aspiration of many unions and their activists and leaders. If the older craft societies were founded with collective bargaining as the *raison d'être* of their trade unionism, unionization of the unskilled constituted a challenge to this by straddling the "political" and "economic", conventionally posed as the division between party and union within the social democratic tradition. But workers' control was submerged in the integration of organized labour within commodity relations through collective bargaining with grievances converted into cash payments. Yet from its earliest expression, industrial action has also been linked with alternative organization for subsistence and meeting social needs, broadening and deepening the horizons of what labour unionism could be conceived for and, ultimately, raising the prospect of workers' control defined as a system such as socialism.

Keywords: Workers' power; managerial prerogative; challenges for unions

INTRODUCTION

Workers' control has been a slogan and an aspiration of many unions. If the older craft societies had been founded with collective bargaining as their *raison d'être*, the unionization of the unskilled rallied a challenge to the very system of capitalism. Rather than negotiation and compromise with employers, the "new unionism" posed direct action as a means of overthrowing capitalism. Not only the strike, the withdrawal of labour, but tactics like sabotage and occupation entered workers' repertoire. The skilled could see workers' control as a defence of their autonomy over the labour process, a defence of their job control. The unskilled could see the building of their union, or their workplace organization, as mobilizing for the "general strike" and prefigurative of the administration of a socialist society. Workers' control was the aspiration for revolutionary syndicalism and of other revolutionary and democratic socialists. In a more moderate formulation, workers' control has been associated with the establishment of cooperatives and other self-managed enterprises in a more gradual transformation of capitalism from within (Ranis 2016). Further, workers' control has been presented as the shifting frontier of struggle between labour and capital: "Militant trade unions ... able to wrest some, or most, of the prerogatives of management from the unilateral disposition of managers" (Coates & Topham 1972: 61). Workers' control is, therefore, far from an uncontentious concept, with ideological and political connotations, linked to the formation and development of worker organization, the development of union militancy and the dynamic conflict between management and workers.

Any approach to workers' control will, therefore, be ideological and political. This is apparent in attempting to unpackage the theory and the practice. First, therefore, it is worth pointing to some exclusions. There is often a conflation of the concept of workers' control with systems of participation, industrial democracy or representation established by management to give some voice to employees. Clegg (1960), for instance, argues that the establishment of collective bargaining was itself industrial democracy, with unions constituting a permanent opposition to management. Scargill and Kahn (1980), in contrast, rejects workers' control, which he interprets as the incorporation of unions into management through participation schemes. Running through these rejections is the conception of rigid boundaries between the roles of union and of management; management's right to manage and the union as voice of labour.

The notion of workers' control straddles the "political" and "economic", often being appended after demands for "nationalization" to differentiate proposals from conventional structures established after state takeover. This rarely gives an indication of what form this "workers' control" might take. At one extreme it might notify some system of worker participation or consultation with state managers, while at the other a structure of workers' councils as the very basis of a socialist "state" and civil society. Some proponents on the libertarian left would see "workers' control" as the very basis of socialism, being the prefigurative form of a "bottom-up" socialism in workers' organization at the workplace and across the economy, and the precursor of workers' councils.

Others, covering communists and social democrats, have argued that workers' control was a stage in the transition of means of production, distribution and exchange (PDE) to the state as ultimate owner of enterprise (see, e.g. Ramelson 1968, reprinted in Ramelson *et al.* 1970), being inherent in the turbulence of the transfer of power but only temporary as top-down authority is then established. There also seems an ebb and flow of concern with workers' control both ideologically and within the practice of organized workers. Montgomery (1983: 389), in considering the American experience, argues that we might "begin with the world of practice, and then ... some of the ideological implications of that practice; workers' control as a 'bottom up' rather than a 'top down'". It has also been related to the economic cycles of capitalism, with the relative strength of capital and labour. Ramsay (1977) periodized "cycles of control", others recognizing periodic "waves of involvement" (Ackers *et al.* 1992; Marchington *et al.* 1993).

This chapter first traces some of the conflicting and divergent notions and approaches to workers' control. We can see the emergence of workers' control with the rise of organized labour, then waxing and waning with the fortunes of that class. This chapter is principally concerned with the UK but the fundamental forces and tension within capitalism and the clash of class agency can be found everywhere there is wage labour so that there is the commodification of labour power, mediated and shaped by the experience of local history and institutions. Workers' control was initially articulated in the 1830s with the very first working-class organization and general labour unionism, reappearing around the turn of the twentieth century, again after wartime production and as the labour movement recovered in the 1930s. It has some limited voice in the debate around nationalization in the UK in the 1940s, largely lost but to then reappear at the end of the postwar boom in the 1960s and 1970s. While these appear disparate, it is possible to map a growing sentiment of historical practice and ideas shaken up in newer turbulent times. We might see in the re-emergent turbulence the tension between different approaches towards unions and organized labour.

To an extent this follows Kelly (1998), where militancy, waves of strikes and occupations, associated with workers' control, are clearly of mobilization. But, we must ask, why is it that workers often do not mobilize, and what forces demobilized workers' resistance in the face of such injustice? In this, we identify a tension between the commodification and the decommodification of the conflicts around labour power, a tension between needs and the capacity to provide them. On the one hand, there is a move towards collective bargaining as a defining relationship that seeks to commodify conflicts and grievances, converting them to money equivalents and payments, subsuming them within the cash nexus. In contrast are the forces of workers' control, with workers creating their own spaces to challenge the strategy and rationale of the cash nexus and so becoming a force for decommodification. Hence, there is the prospect of workers finding potential alternatives for subsistence in the break with the sale of labour power, and in imagining a rational outcome of human activity as satisfying human need.

UNIONS, INDUSTRIAL ACTION AND WORKERS' CONTROL

The first proposal for industrial action to be the weapon to transform society came from Chartist, William Benbow (1832). He proposes a "grand national holiday" when workers would withdraw their labour for the "holy" or "sacred" month. This was not just a strike but also the expropriation of the rich who "have become possessed of a most monstrous power, namely, the power of turning to their own advantage all the good things of life, without themselves creating themselves the smallest particle of any one of those good things" (Benbow 1832: 3). This was to be prefigurative of an economy devoid of the rich. Workers would establish local committees for the distribution of food and other necessities, and these constitute the foundation for a commission to take over national governance. At the end of the "sacred month", workers would have ownership and control of the workshops as the basis of the whole economy. Lessons were being learned from early disputes. The Grand National Consolidated Trade Union (GNCTU), the first general union formed, in part as a response to a protracted lockout of silk workers in Derby, proposed the establishment of alternative production to support strikers (Whitehead 2001). "In all cases of strikes and turn-outs," they resolved:

> where it is practicable, that the men be employed in the making or producing of all such commodities as would be in demand among their brother unionists, or any other parties, and that to effect this each lodge should be provided with a work-room or shop, in which those commodities may be manufactured on account of such lodge, which shall make proper arrangements for the supply of the necessary materials.
>
> (GNCTU 1834, reprinted in Cole 1938)

The early labour movement recognized the limited ownership of the means of PDE and the need for subsistence during industrial disputes. Inherent in their opposition to capitalism was the alternative of worker cooperation and, perhaps, a return to the ethos of pre-capitalist "commons" (Azzellini 2016). While with no direct involvement of Robert Owen, the GNCTU was essentially Owenite, with a belief in cooperatives as the basis of this alternative, different from the cooperative colonies and communities with which Owen was associated.

Workers' cooperatives grew out of the disputes and reflected the autonomy of workers before the putting-out system and other forms of subcontracting and penetration merchant capital into production (see Littler 1980; Marglin 1974). Consumer cooperatives, which may have been established to avoid the debasement of the company store, also had a role in supporting workers during disputes. As late as 1918, workers locked out for two years in the boot and shoe factories of Eyam and Stoney Middleton established an alternative factory with assistance of the wider labour movement and the Cooperative Wholesale Society (Bond & Taylor 2019).

POLITICS AND INDUSTRIAL CONFLICTS

By 1918, the time of the boot and shoemaker dispute, broader concerns of workers' control seemed imminent. In Russia, the establishment of soviets and factory committees prefigured the February and October 1917 revolutions (see Anweiler 1975; Mandel 2011). In Britain, discontent with wartime regulation and control, including the dilution of traditional craft skills by the employment of unskilled women workers, saw the rise of the shop stewards movement and the establishment of workshop committees on Clydeside and then in South Yorkshire (see Hinton 1973; Pribicevic 1959). The early twentieth century also saw the emergence of Guild Socialism, a particularly English advocacy of workers' control rooted, initially, in a romanticization of the medieval guild system (Penty 1906). This approach was adapted by Cole (1920a, 1920b), and others such as Bertrand Russell (1918), who rejected the Bolshevik model unfolding in Russia. They proposed "encroaching" on capitalist control through the establishment of guild organization for each trade:

> A policy directed to wresting bit by bit from the hands of the possessing classes, the economic power which they now exercise, by a steady transference of functions and rights from their nominees to representatives of the working class. ... "encroaching control" aims at taking certain powers

> right out of the employers' hands and transferring them completely to the organized workers. (Cole 1921: 177)

Just before the 1918 armistice, the Labour Party, the political wing of the labour movement in the UK, discussed its new constitution. This was principally designed to allow individual membership in response to limited female enfranchisement rather than through unions or affiliated political organizations, principally the Independent Labour Party, the Socialist Party and the Fabian Society. Its conference, held in January and February, agreed the constitution drafted and circulated by Sidney and Beatrice Webb at the end of 1917, which included Clause 4, giving the Labour Party's objective as:

> To secure for the workers by hand or by brain the full fruits of their industry and the most equitable distribution thereof that may be possible upon the basis of the common ownership of the means of production, distribution and exchange, and the best obtainable system of popular administration and control of each industry or service. (Labour Party 1918)

At least to its author, this was not to be the workers' control coming from the workshop committees, but achieved for, and not by, workers through the parliamentary activity of their elected MPs. While the language of the clause was designed to appease the more radical in the party, the Webbs were avoiding the inference of direct action.

Reforms were not just to formalize the political status of the Labour Party, but also to depoliticize the unions and contain them to purely industrial issues. The 1918 Labour election manifesto, an election called immediately after the armistice, demanded "the immediate nationalisation and democratic control of vital public services, such as mines, railways, shipping, armaments, and electric power, the fullest recognition and utmost extension of trade unionism, both in private employment and in the public services" (Labour Party 1918).

An important advocacy for nationalization came from Sidney Webb and Robert Smillie, the Miners' Federation General Secretary and President of the Triple Alliance. Both were appointed to the Coal Industry (Sankey) Commission (see Coates 1974; Gleason 1920). They used this opportunity to highlight the profiteering of the mine owners in contrast to the sacrifices of the miners. Smillie argues: "The mine owners say, 'We invested our money in those mines, and they are ours.' I say we invest our lives in those mines"(cited in Gleason 1920: 67). This statement was to have an echo as a basic expression of the fundamental differences between commodified and decommodified relations and the challenge to private property.

While united in their criticism of the mine owners, Webb and Smillie came with divergent views on the role of unions. Smillie, from the tradition for direct action and workers' control, posed the mutual support of unions in industrial action as part of the transition to a socialist society. Webb, and the Fabians, drew sharp division between the "industrial" and "political", considering workers' control and direct action as symptoms of the immaturity of the labour movement. Responsible and mature unions should be limited to collective bargaining with employers, and any discussion of political change was for Labour, the political wing of the labour movement, and achieved through parliamentary action.

Arguments on the divisions between the recently established Labour and the unions over concern for political and industrial issues were highlighted by the wartime conditions. Was it legitimate for the unions to take action to oppose military intervention in Russia? Was it for them to take action against conscription regulations? Since these were political, should they be left to parliamentary action by the Labour MPs? At its annual conference in 1918, the first conference after the adoption of its constitution, this division between "political" and "industrial" was one of the main areas of concern. Smillie, supporting the political use of a general strike asked:

> Where do political questions end, and industrial questions begin? Politicians say that the nationalization of mines is political, but does the conference condemn the miners who made up their minds they would strike if they did not get nationalization of mines? To me nationalization is a great labour question. Starved and kicked and kept in miserable houses for generations, the miners have been building up fortunes for the privileged class. Are the organized miners not to use the power of their organization to improve their conditions by nationalization of mines? (cited in Gleason 1920: 89)

In 1919 Sidney Webb (*New Statesman* 29 March 1919) outlined his vision of a general strike by the Triple Alliance, led by Smillie, which not only echoed his critique of direct action in the wake of the Russian revolution but also seemed to anticipate future media scares by later journalists:

> We would have been nearer a social revolution than anyone had previously thought possible. … The whole country would have been, in a week or two, fireless, foodless, trainless, and wage less. The Government would necessarily have stuck at nothing to suppress what would have been – lawful as it was – essentially an act of civil war, within twenty-four hours the whole country would have been in military occupation. … The mining districts would have been strongly garrisoned with soldiers, and the Government had made precautionary preparation for other steps of which we prefer to say nothing. Never in the whole history of this country should we have seen such a display of force against a popular movement, itself absolutely unexampled in magnitude. (cited in Gleason 1920: 148–9)

There would, Webb argues, also be the cooperative movement in support of the strike, allowing subsistence providing the necessities of life, the farms and factories of the societies providing the food for the strikers. He continued, observing that other unions would act in solidarity:

> Up and down the kingdom the mining districts and the great railway centres are the special strongholds of the Co-operative Movement, of which an enormous proportion of the million and a half strikers would have been members. … If the Government had … used all its resources to put down what it would

> have regarded as civil war, and had, in some unforeseeable way, succeeded, it would probably have kindled such a flame of industrial rebellion, or at least set smouldering such a persistent resentment, as would have had political as well as industrial consequences that no man can measure. (*Ibid.*: 149)

Finally, Webb drew attention to an extreme tactic, which workers might resort to if all else failed. Their most revolutionary act, in appropriating the means of PDE, often against the advice of their own unions, was "that there might be such a thing as a 'stay in' strike, to which beaten men, smarting under a sense of injustice, are apt to resort, even against all the efforts of their Trade Unions" (Ibid: 149). Such events had occurred in Russia and elsewhere in Europe and were to bring Italy to near revolution a year later (Spriano 1975), so it was the role of political action, according to Webb, to avoid such rebellion.

When the General Strike did occur in the UK in 1926, Beatrice Webb's (1926) diary was to leave the most illuminating account of their attitude to the strike and the role of unions. At 4am on the morning the strike began (3 May 1926) she appeared sympathetic, commenting in her diary: "Unless the government ends the reign of the profit maker it will end the government – may indeed break up the country." The next day, 4 May 1926, however, she noted that:

> ... we personally are against the use of the general strike in order to compel the employers or a particular industry to yield to the men's demands however well justified these claims may be. Such methods cannot be tolerated by any government – even a Labour government would have to take up the challenge.

At the collapse of the General Strike, called off by the TUC, she was to conclude that:

> The failure of the general strike of 1926 will be one of the most significant landmarks in the history of the British working class. Future historians will, I think regard it as the death gasp of that pernicious doctrine of workers control of public affairs through the trade unions, and by the method of direct action.

This was not to be attributed to the aspirations or actions of the working class but brought to them by political and union leaders. She continued: "This absurd doctrine was introduced into British working-class life by Tom Mann and the Guild Socialists and preached insistently, before the war, by the Daily Herald under George Lansbury." It seemed that by 1926 the death of workers' control could be announced, especially since this appeared an international development in socialism. Beatrice Webb (1926) also observed in her diary:

> In Russia, it was quickly repudiated by Lenin and the Soviets, and the trade unions were reduced to complete subordination to the creed autocracy of the Communist Party. In Italy the attempt to put this doctrine into practise by seizing the factories led to the fascist revolution. In Great Britain this belated an emasculated edition of the doctrine of workers' control will probably lead to a mild attempt to hamper trade union activities.

Both the Webbs' brand of social democracy and the Russian Bolsheviks had abandoned workers' control for the ethos of professional management and traditional hierarchy in the workplace. What is more, the major attempt in Italy, with unions accepting concessions for workers leaving the factories, had led to reaction (Spriano 1975) and the rise of fascism.

FRONTIER OF CONTROL

Gleason (1915) was a war correspondent from the US, volunteering as an ambulance driver in Belgium. He became interested in the postwar developments in the UK's labour movement, attending the proceedings of the Sankey Commission as well as Labour and TUC conferences in 1918 and 1919. He reported a particularly British, "slow and measured" revolutionary movement to his readers in the US:

> What the Russians grabbed for too swiftly in soviets and workers' committees, England is attaining step by step stumblingly in the Shop Stewards' Movement and shop and pit committees. It is control of industry by the producers (including, of course, foremen, managers, draughtsmen, directors, technical advisors) (Gleason 1920: 27)

As well as reporting the progress of socialism in the UK, Gleason collected key writings of the guild socialists, the shop stewards movement and unions for the American readership in *What the Workers Want: A Study of British Labor* (1920). This book was picked up and reviewed enthusiastically by a young US economist, Carter Goodrich, who, with Gleason's advice, headed for the UK to study worker organization. In the UK, with assistance from R. H. Tawney (who provided a Foreword to the published account), as well as from G. D. H. Cole, at the time the most prominent Guild Socialist, Goodrich (1921) undertook his research into mining conditions, visiting mining communities to talk to miners and miners' leaders. Goodrich (1921: 264) was aware of the different forms of control exerted in the workplace. He made a clear distinction between what he called "old craft" or "customary control". The former, which we might call "job control", was based upon "small groups of skilled workers", traditional workers "clinging on" to established controls. In contrast, he identified the demands for control spreading among the newly organized, what he referred to as "contagious control". Thus:

> The powers won by the shop stewards are being used up and down the country as a text for vigorous propaganda. The shop stewards' control was decidedly contagious control [as] its actual extent may be easily underrated by an outsider. It was recorded in no formal agreements. It rested on the war shortage of labor and was abruptly checked in the period of unemployment that followed the end of the war.
>
> (Goodrich 1921: 10–11, emphasis in the original)

This was also a critique of the Webbs' approach to unions, looking not at the formal structures and agreements that the unions might become party to, nor even to seeing unions as a "continuous association of wage earners", but concerned with the very informality and "custom and practice" that becomes established outside of such agreements. With the immediate postwar movement already waning, Goodrich looked to more subtle dynamics within the workplace. He was concerned with "moving tendencies" (Goodrich 1921: 260), the "moving", or "fighting", "frontier of control" between workers and management. He identified struggles over the form and method of wage payment, over working patterns, staffing levels, work organization and other terms and conditions, as well as other practices within the workplace. In a metaphor strikingly like Gramsci's notions of "war of position" and "war of manoeuvre" (Gramsci 1971: 229ff), Goodrich, drawing from both from the American "frontier" experience and from the European war, ranged against "executive control", argues that:

> The employer "by his absolute knowledge and mere motion" provides capital, decides what to produce and how to produce it, provides any sort of place to work, hires whom he likes, pays his hands any wages by any system, works them any number of hours he likes, drives them by any method and with any degree of supervision, promotes, fines, or dismisses them for any cause, trains any hand for any job, dictates every process in the minutest detail – and does all this and more "subject to change without notice".
>
> (Goodrich 1921: 52)

Perhaps, the clearest illustration of Goodrich's approach can be found in his own later study of US miners, *The Miner's Freedom* (1925). This study in industrial sociology examined the changes experienced in the industry with the introduction of long-wall working and mechanized coal-cutting. Essentially, he examined the loss of a "miner's freedom" expressed in loss of control over their time and working conditions. In the piecework system of short-wall or pillar working, the miners had control over when and how they worked.

In some ways this reflects the "Saint Monday", described by Thompson (1967) of an earlier generation of workers. Goodrich's analysis of coal mining can also be seen like the management attack on the miners' "indulgency pattern" in gypsum mining (Gouldner 1955). The discussion of changes in mining, and the influence of Taylorism and Fordism, also anticipate later discussion of labour process (see Braverman 1974), or even Goodrich's miners' enquiry of "How are the cars running?" to Burawoy's (1979) account of "making out" by machine workers at "Allied". The "frontier of control" identified the "soldiering" or "restriction of output", so central to management theorists, but saw it as an expression of workers' control. This approach has more recently been important in relation to the experience of the daily workplace struggles, with considerable relevance to long-term employer campaigns against "overmanning", "restrictive practices", and to increase "flexibility", and has, for example, been explicitly used in examining call centres (Taylor & Bain 2001), railway workers (Hughes & Dobbins 2020; Hughes, Dobbins & Murphy 2018) as well as in the elimination of the tea break in workplaces (Upchurch 2020).

THE "SECOND SHOP STEWARD MOVEMENT"

In the 1920s there appeared a foreclosure on debate. Beatrice Webb seemed to be talking for both communist and social democratic strands of socialism in arguing the end of the "pernicious doctrine of workers' control of public affairs", with the management of such affairs being the role of professionals or politicians. During the late 1930s and after 1945, there was some debate within Labour over nationalization (see Coates & Topham 1970: 302–30; White 1949), but only limited mechanisms for worker voice incorporated beyond union recognition and, perhaps, limited participation schemes. Postwar economic growth and the Keynesian policy of full employment facilitated union growth and a mobilization of the shop-floor challenge to management through a growth of shop steward numbers. Topham (1964: 4), influenced by Cole and Goodrich, wrote in *New Left Review* that:

> The quantitate growth of the shop stewards' strength in industry, the causes and numbers of strikes (particularly local, spontaneous strikes) are significant factors, and that the whole area of conflict surrounding the role of the shop steward is likely to intensify in the near future. ... whilst the Left's main task should be to assist at the birth of articulate and explicit demands for control at shop-floor level, we must insist upon the need to generalize these outwards to embrace the whole framework of social, economic, and political decision-making.

This was fast becoming a moral panic. The so-called "British disease", of the relative decline of the economy compared with its competitors, was claimed to be related to, if not caused by, the strength of shop-floor unionism via the growing number of shop stewards. Topham (1964: 4) argues, in contrast, that the very capacity for fighting for their members in shop-floor grievances, amounted to an "implicit demand for workers' control". However, he saw this as potentially leading to a more explicit and conscious movement. There was, of course, a paradox here in that this development was conditional upon the extension of collective bargaining into the workplace with the shop stewards and the development of the "two systems" of the mid-1960s highlighted by the Donovan Commission (1968). National negotiations could only pick up the general issues affecting all the members in an industry, while the shop steward system developed through, and had to tackle, the immediate specific problems at each workplace. At plant level, problems were transformed into monetary payments; qualitative issues of work intensity, safety, working hours and a plethora of others entered the sphere of the cash nexus (see Marsh & Coker 1963; Terry 1983). Commodity fetishism found concrete manifestation in the complexity of rates and bonus payments for various and diverse jobs.

Topham's article was linked to a seminar to be held in April 1964, which he described as the "most representative of its kind on the issue of Workers' Control since the 1920s" (Topham 1964: 13). Part of its purpose was to pressure the Labour Party into adopting some forms of workers' control into any new nationalization plans. In the debate on this, Hugh Scanlon, newly-elected president of the Amalgamated Engineering Union, echoing Robert Smillie a generation before, articulated what he thought the job rights of workers should be:

> The traditional position of the Labour movement, concerning the rights of the workers, is that this right of property has been established by the workers themselves, workers whether by hand or by brain, who have built up this property by applying the Labour power and therefore have as much right to its direction as the shareholders. (Scanlon c.1970: 3)

The failure of the first Wilson government to adopt any policies that gave workers a voice in nationalized industries added to a mounting disillusion with the Labour government by the established Left. Disillusion with the policies of the Wilson government led to the establishment of the Institute for Workers Control (IWC) in 1968 to coordinate the programme of seminars, conferences and publications. The founders of the IWC saw its purpose in rescuing the earlier tradition (Coates & Topham 1970) and developing its contemporary relevance (Coates 1968). The IWC, picking up on the expansion of shop-floor organization, drawing on the older tradition, and particularly the work of G. D. H. Cole, argues that in extending its power, union organization was an encroachment of workers' control against managerial prerogative (Coates & Topham 1972).

Disillusion also permeated the Communist Party, with much of its strength in industrial branches among shop stewards. Revelations about Stalin, followed by the suppression of the 1956 Hungarian uprising, led to some re-evaluating the "top-down" model of socialism that had dominated since the 1920s. This "new Left" that emerged, and from which the IWC was established (Barratt Brown 2012), began to explore alternative models, principally that of Yugoslavia's system of self-management (see Barratt Brown 1960; Singleton & Topham 1963). The system of self-management, the workers' councils established in Hungary in 1956 (Anderson 1964; Lomax 1976, 1980), as well as the Algerian liberation movement (Ben Bella 1963; Clegg 1971), influenced a resurrection of workers' control ideas, particularly among this new Left. Even for the more moderate, the record of postwar growth in the West German economy was often attributed to their adoption of co-determination, with worker representation on works councils and supervisory boards (Spiro 1954).

By the late 1960s there were seen to be mounting problems in the UK economy – while growth was continuing, it was not as strong as its competitors. As part of its modernization recipe, the so-called "white heat of technological revolution", the 1964 Labour government promoted corporate mergers. Government established, through its Industrial Reorganisation Corporation, companies they hoped could compete internationally. Included in this policy was Upper Clyde Shipbuilders (UCS), formed from separate yards, as well as newly expanded conglomerates such as GEC/AEI/EE and Lucas, all expanded in a wave of mergers and takeovers.

THE BIG FLAME?

"Stay-down strikes", or "downers", had occurred in the mines in the 1930s (for a fictionalized account, see Slater 1936) and were used especially by workers in mass production (Fore nd; LCSST 1971; *The Times* 4 February 1961; Turner, Clack & Roberts 1967; Weller 1962). In 1968 the occupation of factories – the most extreme expression of workers' control under capitalism – became an important part of the events in France as militancy migrated from the students, engaged in university sit-ins, to industrial workers (Hoyles 1973). In a repeat of events of 1936, when an occupation movement erupted, workers across France took over their factories (Danos & Gibelin 1986). In 1969 a television play written by Jim Allen and directed by Ken Loach, *The Big Flame*, portrayed a fictional takeover by its workers of the Liverpool docks.

By the late 1960s some major worker actions against closures and rising unemployment were much anticipated, although it was recognized that a traditional strike, a "withdrawal of labour", would be counterproductive, but no realistic alternative had been tried. In 1969, shop stewards at GEC/AEI/EE, a company formed out of mergers facilitated by the Labour government, proposed an occupation to resist redundancies, a plan initially accepted at a mass meeting. Without immediate occupation of the factories, the employer was able to mobilize opposition, including recruiting support from some of the stewards. As doubts developed on the implication of a takeover, for safety and insurance while in occupation, and, perhaps most importantly, what impact resistance might have for unemployment benefits and for any redundancy package, the plan was voted down by a later mass meeting (Chadwick 1970; IWC 1969; Newens 1969).

The events at GEC/AEI/EE indicate the uncertainty concerning occupation. By taking over the workplace rather than leaving it, the occupation is fundamentally different from a strike. At least tacitly this becomes a challenge to property rights against the more conventional withdrawal from the sale of labour power. Even more than in strikes, the occupation needs organization. In a work-in, the workers exercise their control and organization of the means of production, which could be argued to be more radical than a sit-in. However, the adoption of a sit-in or work-in was a practical decision rather than an ideological one. While some argue that occupation is "a tactic of class struggle – not an experience in workers' control" (Sherry 2010: 126), it could be considered as prefigurative of workers' self-management, not only taking over but also using worker collective expertise in running the enterprise. Occupations can vary greatly across a spectrum, from short small-scale protests to the takeover of a whole establishment with the continuation of production, and perhaps organization of the supply chain and distribution, by the workforce. It can also lead to reflection on the very purpose of the work they are fighting for (Cooley 1980).

On 30 July 1971, Jimmy Reid, the convenor of shop stewards at UCS, announced that:

> The world is witnessing the first of a new tactic on behalf of workers. We're not going to strike. We're not even having a sit-in strike. We're taking over the yards because we refuse to accept that faceless men, or any group of men in Whitehall or anyone else, can take decisions that devastate our livelihoods

> with impunity ... The shop stewards representing the workers are in control of this yard. Nobody and nothing will come in and nothing will go out without our permission. (in Foster & Woolfson 1986: 199–200)

The receiver appointed by the government to sell off the yards remained in ultimate control, but, as workers received their redundancy, the shop stewards took over the paying of wages drawn from outside donations to the work-in. While they denied that this was any form of workers' control, and while they held expensive ships as "hostage", their actual control over the yards was limited. However, as the government was to admit: "shop stewards ... had secured a monopoly of publicity in the press and on television and would probably continue to oppose any solution which did not embrace all four of the yards concerned'(Cabinet Office 1971: 7). Despite this limited actual control exerted by shop stewards at UCS, and long before their success in keeping the yards open, it proved an inspiration for an extension of worker militancy into work-ins, sit-ins and other forms of occupation involving taking control of the workplace. These became most associated with resistance to redundancies and closure, but often these occurred in the general escalation of conflict in employment. Within a year, and before there was any idea that the UCS work-in would save any jobs, there were to be around a hundred other occupations in some ways claiming its inspiration. This form of militant action, unlike the strike, appeared to be seen as legitimate by the public. The industrial correspondent for *The Times* (22 May 1972) noted:

> Workers' occupations as a form of planned industrial action have acquired a certain respectability ... They are no less disruptive than an all-out strike ... [but it is] ... a curious fact that the public generally seem to take a more amicable view of men who turn up at their factory each day and then sit about doing nothing, than it does of those who just stay at home in bed. Going to work – as against doing any – is apparently seen as something of a virtue in its own right.

In the following decade in the UK, there were reports of over 260 instances, with perhaps a further 70 by 2018 (Tuckman 2021b).

It was one thing to occupy a plant but soon some important questions needed facing. What exactly was the purpose of the occupation, and what outcome was sought? A buyout by a new owner or takeover of the workplace themselves in some way – or improved redundancy terms? These questions often had to be addressed pragmatically. The first workers' cooperative established in this wave was almost as a pragmatic move by a small group of women workers to help support themselves during their occupation, a move reminiscent of the action of the boot and shoe workers locked out in Derbyshire more than half a century before. When Sexton, Son and Everard decided to rationalize their shoemaking business by closing a small plant in Fakenham, the women workers, abandoned by their union, decided to support themselves by manufacturing leather bags from the scrap materials available and selling them on local markets. They continued as a cooperative at a nearby factory (Wajcman 1983), which they sustained for five years (*The Guardian* 2 August 1977).

Occupations entered the mainstream of industrial relations, including a wave associated with engineering industry national bargaining. In 1971–72, the negotiations on a "long-term agreement" broke down. The unions were pushing for extended holidays, a shorter working week and "moves towards" equal pay for women, with the employers refusing to increase their offer of £1 per week and with no improvements on the other demands. In each case the national claim, often compounded by local grievances, led to employer threat of lockouts with prompt worker retaliation of an occupation. The occupations constituted a major escalation in conflict as employers attempted to shift the frontier of control, taking back control, which had been encroached by shop stewards, and returning to "orderly bargaining" with the formal officers of the unions.

Although this section has concentrated upon the UK, occupations, and cooperatives set up from them, were an international phenomenon (see, e.g. Burgmann, Jureidini & Burgmann 2012 on Australia, and Reid 2018 on France).

POSING AN ALTERNATIVE?

When Labour was elected to government in 1974, Tony Benn, the new Secretary of State for Industry and an IWC supporter, gave assistance to plans for cooperatives already at Norton Villiers Triumph's Meriden motorcycle plant, at IPD, the collapsed successor to Fisher Bendix where workers were again resisting closure, and at the *Scottish Daily News* in Glasgow after the *Scottish Daily Express* was closed by Beaverbrook newspapers. These "Benn cooperatives" were all short-lived, established with very limited funding, most of which went to repay previous owners for the rundown and obsolete plant that the new enterprises were supposed to be built on. While only attempted in a few cases, this wave of occupations soon became associated with the establishment of workers' cooperatives (Coates 1976).

The threat of redundancy could be divisive for workers. Lucas Industries had grown dramatically in the takeover and merger boom of the 1960s (CIS Anti-report 1975), and, like GEC-AEI and other similar companies, was moving into a phase of "rationalization". Shop stewards across its 13 plants established a combine committee during a dispute sparked by the standardization of the pension scheme across the merged company. They soon found other issues arising out of the company's strategy, which involved redundancies and plant closures.

Mobilization of support against redundancies or closure at one plant could be seen as undermining potential security at another. Benn was surprised to face booing at the NVT Small Heath when he announced his support for the plant at Meriden (*The Guardian* 17 July 1975).While the redundancies at Lucas Aerospace affected all the factories, one, at Willesden, had been earmarked for closure. The Lucas Aerospace Combine Committee was first fully mobilized around a strike and occupation at Burnley, part of the national engineering dispute in 1972 (Wainwright & Elliott 1982: 64–9). When the national claim was submitted, management offered a 50 pence a week increase. After five weeks of industrial action, occupation followed by strike, they received an increase of £1.50. The Combine Committee increasingly mobilized over corporate strategy, with action that attempted to avoid pitting plant against plant in such rationalizations.

While they could be reasonably effective in this, they also began to question some of the implications of employment in the aerospace industry integrated within armament production.

Some workers in occupation had already looked at what they were producing, mostly in terms of minor technical refinement rather than ethics of the product itself. NVT workers attempted to produce a new motorcycle, and at Imperial in Hull they designed a new portable typewriter. However, like much domestic research and development, these were revisions based on existing components rather than involving any innovation. The Lucas Combine, prompted by a meeting with Benn, countered redundancies by surveying plant capacity and products, as well as worker skills, which took shape as their "alternative plan" launched in 1976. In following through the rationale of resistance, it was clear, as Mike Cooley, the chair of the combine, put it: "There is no point occupying a factory if the products it makes are not in demand" (*The Guardian* 23 January 1976).

But this was not an appeal for a resurgence in the market or for greater state spending on armaments to avoid the proposed 13,500 redundancies. Instead, workers assessed their own skills as producers and the capacity of Lucas Aerospace for "socially useful" alternatives (Cooley 1980). As well as challenging the deskilling of labour through the dynamics of capitalist enterprise, their plan proposed a new range of products. These included kidney dialysis machines in short supply to the National Health Service (NHS). This range also included other products using their skills, some revolutionary for the 1970s, which they could develop. These included new sources of renewable energy, heat pumps and hybrid vehicles, as well as a road-rail vehicle, which they developed and that toured Britain (LACSSC 1978). Opposed by the company, considered unviable and marginalized and ignored by the government, as well as by their own unions as an "unofficial body" (LACSSC 1979, 1981), the combine's plans remained within an alternative undercurrent of workers' control, raised when arms production is challenged and plants are threatened with closure (Sprung 2020).

COMMODIFYING JOBS

Occupation only occurred in a small number of cases of major redundancy or closure (Tuckman 2021a). Even in the most highly organized of workplaces such resistance was rare with the strength of union organization acting to counter mobilization, with unions more attuned to bargaining redundancy terms. In many cases, especially in the motor industry, where union organization was considered strong and militant, there was little resistance even to major closures. This absence can be attributed to a strategy of demobilization and avoidance practices by managements (IPM 1976; IWC 1976; Metra 1972).

In 1963 the Contract of Employment Act gave the requirement for one month's notice for large-scale redundancies, then the Redundancy Payments Act of 1965 introduced payment based on length of service and age (Parker *et al.* 1971). Legislation was designed to enhance the mobility of labour, allowing workers in declining industry to move to areas of expansion. However, in large well-organized sites it principally focused union action on negotiation of a redundancy package rather than mobilization and resistance. Even where there was considerable support for resistance and direct action, this was often dissipated by an improved offer. Redundancy payments could be enhanced in negotiation. Also, it was often threatened to be withdrawn, or at risk, if there was any resistance. Such a management strategy contributed to the defusing of the occupation plan at GEC-AEI in 1969, and was refined elsewhere (IWC 1976). In 1975 the passing of the misnamed Employment Protection Act rather than protecting jobs went on to cement their commodification.

At British Leyland in Speke, Beynon (1978) noted the announcement of closure during a long-running strike, with a returning to work to organize resistance. With initial rejection of a redundancy agreement and talk of resistance at a union mass meeting, "as closure date drew near ... a new offer appeared":

> ... with all the warnings about the need for orderly rundown – any trouble and you could forget the cash. This offer shifted the balance. A very high proportion of the labour force at Speke had worked in the plant for less than two years. Employed in 1976 to produce large numbers of the TR7 they had no entitlement under the Redundancy Payments Act. In its final offer the Company shifted the balance of the funds in their direction offering an, across the board, increase of twelve weeks pay. Very cunning. Tear them apart. At a mass meeting a shop stewards' recommendation to reject the offer and fight the closure was turned over. It was an emotional meeting. Some men fought, others cried, one shop steward had a heart attack. But the vote was to accept. The plant was to close without a fight.
>
> (Beynon 1978: 32)

Similar stories can be found at other major redundancies and closures at, for example, the closure of Chrysler, Linwood or Dunlop on Merseyside (Marren 2016). The effect of the Redundancy Payments Act had been to commodify jobs, giving them a monetary value and allowing them to be bought out in negotiation (Fryer 1973). The period of notice allowed workers to search for new employment as well as dissipate resistance with an inevitable drift away during the period leading up to the final redundancies. The decision whether to take some resistance was left to this remaining rump, often alongside managements promise of short-term continued employment for selected workers in the tasks needed to close down the site.

One area that did begin to see resistance, and worker takeovers, was the NHS. In 1976, Labour not only abandoned the postwar policy of full employment (Callaghan 1976), but also initiated a programme of public sector cuts moving into reverse 30 years of expansion. At the same time, the NHS experienced its own rationalization, centralizing provision into large district hospitals, with the closure of specialist and cottage hospitals. In the initial campaign developed at the Elizabeth Garrett Anderson Hospital in central London, medical and nursing staff continued the care of patients after the formal closure by the health authority. Supporters from unions, community groups and the women's movement, as well as patients, linked with staff at the specialist women's hospital, named after the first woman doctor to qualify in the UK. The hospital was picketed but continued in-care of patients during a two-year work-in. This was followed by more than 20 further occupations to keep hospitals, or wards or departments, open (Anon 2020; COHSE nd; Past Tense 2013).

These hospital occupations united hospital workers and the local communities – the producers and consumers of health services – in joint campaigns to maintain a recognized social need. Staff remained at the hospital, and were paid by the authority, while patients were being treated. In some cases, where patients were removed by health authorities and staff transferred, the occupation was continued. At Hounslow Hospital, where staff initially staged a work-in, a dawn raid by the local health authority removed the remaining 21 in-patients, wrecking of the two wards (Heron 1978). The hospital remained occupied by supporters for a further year.

While occupations might have peaked in the early 1970s, they did not disappear (Mustchin 2021), with important cases continuing into the 1980s and later. Often the employer had not followed the strategy of commodification through redundancy, giving no notice of closure, and had not made any final payments to workers. At Meccano (Campbell 2021), a three-month occupation started when the toy-making workers were given 40 minutes' notice of closure (*Liverpool Echo* 30 November 2019). Sometimes with wages or other payments outstanding, or the failure to negotiate with unions over redundancies, the sense of injustice led to mobilization of workers (see Clark 2013, 2021; Gibbs 2021; Gibbs & Kerr 2020; Wright, Philips & Tomlinson 2021).

But, given the level of closure and redundancy, particularly in the 1980s with the consolidation of neoliberal policies under the Thatcher government, resistance went into decline. With ideological reinforcement of the market and self-interest, enhanced by a raft of union legislation that increasingly kettled the action of workplace activity, resistance was all but neutered (Tuckman 2018). The introduction of the requirement for industrial action ballots, with periods of notice for any action, further reinforced the period of notice for redundancy to stifle the prospects for direct action by workers. The most direct anti-Thatcher action occurred at Hotform Engineering in Preston where workers staged a brief sit-in, more a traditional "downer", which was long enough to bring about the cancellation of a visit by the prime minister to the factory (*The Times* 18 February 1983). In the changes associated with Thatcherism, and neoliberalism more globally, we can see a major management advance in managerial prerogative within Goodrich's "fighting frontier of control". This advance, characterized by appeals to shareholder values and entrepreneurship and undermining of collectivism, was often reinforced by tightened legislation and more authoritarian measures, often justified as "abolition of regulation", meaning an undermining of the protections and controls won by workers.

RETURN OF RESISTANCE?

Even with this sizeable shift in the frontier of control, some workers did mobilize resistance against declining employment conditions, especially when they involved redundancy and closure. The financial crash led to some escalation in the use of occupations. In March 2009, 12 workers at the Prisme packaging works in Dundee in Scotland were told of their immediate redundancy without any redundancy pay (Labour Research

August 2009). They staged a six-week occupation before establishing a cooperative (Whyte 2010). Workers at three plants owned by Visteon, sold off by Ford and in 2009 facing closure, occupied two of the plants as hostage against their claimed outstanding pension and redundancy pay from continuous service dating back to Ford ownership of Visteon. Thirty workers at the Vesta wind turbine plant occupied when the workforce of 600 were declared redundant and were supported by a climate camp of more than 200 supporters outside the plant.

Examining the eight occupations in the UK between 2007 and mid-2011, Gall (2011: 613–15) cited five stimuli to occupation: collectivized nature of redundancy, immediate and unforeseen nature of redundancy, loss of deferred wages and compensation, pre-existing collectivization, and positive demonstration effect. With the re-emergence of occupations in the UK, some commentators were seeing a return to the 1970s wave (*The Guardian* 15 June 2009; Labour Research August 2009).

While there were some incidents of occupation, these did not approach the scale of the 1970s, at least not in the UK. The main stimulus to occupations in twenty-first century UK, the cases covered by Gall (2011: 613), tended to be "loss of deferred wages and compensation". Other reported cases were typically very short occupations, more the "downers", and a return to spontaneous short disputes frustrated by post-1980s legislation and the realities of precarious working.

Typical was the short occupation at the bedmakers, Kozee Sleep, West Yorkshire. Workers were given no notice of closure. When the receiver later announced that it was uncertain when wages and other outstanding money could be paid to workers, they staged a short sit-in (*Huddersfield Daily Examiner* 22 May 2015). It was discovered that the company was employing slave labour (*Huddersfield Daily Examiner* 9 December 2015). At Gate Gourmet in 2005, flight catering staff based at Heathrow arrived at work to find agency workers in their place. Adjourning to the canteen to see whether shop stewards could resolve the issue, the workers were first detained in the canteen by management, then dismissed. The action, and sympathy action by other staff at the airport, were later considered unlawful by the courts (Anitha & Pearson 2018).

There was justification for the view of a wave of occupations if considered internationally (Democracy Now 2009). The occupation of Republican Doors and Windows in Chicago (Lydersen 2009) also resulted in the creation of a workers' cooperative from the closed factory (Rabble 2013). What became labelled as "recuperated", "reclaimed" or "recovered" factories were workplaces that were first closed and then taken over by their workers or local community, with the space used to provide work or other social purpose. Azzellini (2017, 2018) visited 24 such workplaces, mostly in Venezuela, but also in other parts of Latin America, as well as in Italy, France and Greece. Some managed to resume some level of production, often modified to suit social need. Voi.me in Greece changed the chemical plant to produce environmentally friendly household cleaners (Vio.me nd). Officine Zero, created from the closed rail maintenance and repair yard in the centre of Rome, was occupied by some of the sacked staff, who opened the space to other workers. It was:

> … founded as an eco-social factory. Officine Zero means literally "workshops zero": "zero bosses, zero exploitation, zero pollution", … In half a dozen workshops as carpentry, padding, metal works and general repairs the workers at Officine Zero dedicate themselves mainly to recycling domestic

> appliances, computers and furniture. The common project is to turn the former sleeping car repair facility into an industrial reuse and recycle center. OZ is administrated horizontally by all workers, from the workshops together with the precarious workers sharing an office floor in the former administration building. (Azzellini & Ressler 2015)

The reclaimed workplace movement of the twenty-first century, while experienced in many countries, has been most associated with Argentina. Ruggeri (2016) estimates that from around 20 recuperated workplaces in 2000, there were 160 there by 2016. Vieta (2021: 115) cites University of Buenos Aires research, which indicated 400 recuperated workplaces had existed between the early 1990s and 2016, with almost 90 per cent surviving as workers' cooperatives. Many of these started when workers took over after the site was abandoned by the owners, as "hostage" against owed wages. Sometimes, the takeover was of abandoned sites by local communities. In all cases the spaces were put to use, either in some adaption of previous use or new uses (Lavaca Collective 2007; see also Lewis & Klein 2004).

Among the worker cooperatives established, across a wide range of sectors of the economy, was the Hotel Bauen in central Buenos Aires, where hotel workers took over in 2003 and ran it as a hotel and social centre while fighting off attempts by the previous owners to repossess it. It closed due to the Covid-19 epidemic in 2020 (Sobering 2022). Covid-19 also influenced direct action in Paris, where McDonald's workers, facing temporary closure because of the pandemic, took over with local residents and used the fast-food establishment to distribute meals to those in need (Azzellini 2021: 430). The Kurdish struggle for independence in northern Syria, with their establishment of Rojava, has seen the initiation of popular assemblies of direct democracy within a libertarian municipalism (see, for example, Bookchin, Şahin & Sitrin 2019; Cemgil & Hoffman 2016)

WORKERS' CONTROL: PAST, PRESENT AND FUTURE

Aspiration towards workers' control has been rooted in unions and labour movements from their inception, perhaps harking back to traditional forms work organization and production for social need before penetration by commodity relations and the profit motive. Early expressions of direct action with the first propagation of industrial action were linked to maintaining means of PDE – including founding workers' cooperatives – as temporary or, in the case of the general strike or "grand national holiday", a permanent alternative to capitalism. In the UK this aspiration informed new unionism, the shop stewards' movement and guild socialism. Elsewhere the organization and mobilization of direct action, not just for better pay and conditions but as the very means of transforming society, were at the centre of all forms of syndicalism.

The emergence of social democratic parties led to demarcation disputes over who had domain over industrial and political spheres. Strike action was depoliticized; direct action became marginalized against the legitimation of parliamentary action; unions increasingly were drawn into collective bargaining, arguing over monetary reward; and differences between the interests of labour and capital could be bought and sold. Strong and even militant union organization did not necessarily mean fighting against the commodification of labour power, but could be an indication of workers being drawn more fully into the cash nexus. Unions have always been Janus-like, both increasingly integrated into capitalism through collective bargaining and also mobilizing direct action for the aspirations of workers for an alternative society. While increasingly marginalized, the aspiration for workers' control never disappeared, becoming associated with encroachment by workplace organization over management prerogative, and informal worker practices fighting management on a frontier of control, which we might see as the precondition or prefigurative for a more dramatic expression involving taking command of the means of PDE.

The aspiration for worker control also reappears in debates on any extension of state control of industry (as in the UK in the 1930s, 1940s and 1970s). The aspiration for workers' control can re-emerge in periods of crisis and withdrawals of capital through job losses and plant closure. As capital withdraws, workers take over, usually on a relatively small-scale but indicating a bigger possibility. While capital survives, experience has shown the continued capacity of groups of workers to take control and reshape production to meet social need.

REFERENCES

Ackers, P. *et al.* 1992. "The use of cycles? Explaining employee involvement in the 1990s". *Industrial Relations Journal* 23(4): 268–83.

Anderson, A. 1964. *Hungary 56*. London: Solidarity.

Anitha, S. & R. Pearson 2018. *Striking Women: Struggles & Strategies of South Asian Women Workers from Grunwick to Gate Gourmet*. London: Lawrence & Wishart.

Anon 2020. *South London Women's Hospital Occupation 1984–85*. London: Past Tense. http://libcom.org/library/south-london-womens-hospital-occupation-1984-85

Anweiler, O. 1975. *The Soviets: The Russian Workers, Peasants, and Soldiers Councils, 1905–1921*. New York: Pantheon.

Azzellini, D. 2016. "Labour as a Commons: the example of worker-recuperated companies". *Critical Sociology* 44(4/5): 763–76.

Azzellini, D. 2017. *Communes and Workers' Control in Venezuela: Building 21st-Century Socialism from Below*. Leiden: Brill.

Azzellini, D. 2021. "Class struggle from above and from below during the Covid-19 pandemic". *Journal of Labor and Society* 24(3): 418–39.

Azzellini, D. & O. Ressler 2015. "Occupy, resist, produce – officine zero". Workerscontrol.net. www.workerscontrol.net/authors/occupy-resist-produce-%E2%80%93-officine-zero

Barratt Brown, M. 1960. "Yugoslavia revisited". *New Left Review* 1: 39–43.

Barratt Brown, M. 2012. "The institute for workers control". *The Spokesman* 116: 47–53.

Ben Bella, A. 1963. *Ben Bella on Workers Control*. London: British Aid to Algeria Committee.

Benbow, W. 1832. *Grand National Holiday, & Congress of the Productive Classes*. London: Journeyman Press.

Beynon, H. 1978. *What Happened at Speke?* Liverpool: 6-612 Branch, TGWU.

Bond, S. & P. Taylor 2019. "The air of freedom: a story of hard times and striking boot and shoemakers". In Eyam & Stoney Middleton 1918–1920. Sheffield: Philip J. Taylor.

Bookchin, D., E. Şahin & M. Sitrin 2019. "Eyewitnesses to the Rojava revolution: women empowerment". ROAR Magazine. roarmag.org/essays/eyewitnesses-rojava-revolution-bookchin-sahin.html

Braverman, H. 1974. *Labor and Monopoly Capital: The Degradation of Work in the Twentieth Century*. New York: Monthly Review Press.

Burawoy, M. 1979. *Manufacturing Consent: Changes in the Capitalist Labor Process Under Capitalism*. Chicago, IL: University of Chicago Press.

Burgmann, V., R. Jureidini & M. Burgmann 2012. "Doing without the boss: workers control experiments in Australia in the 1970s". *Labour History* 103: 103–22.

Cabinet Office 1971. "CM(71) 47th Conclusion". https://discovery.nationalarchives.gov.uk/details/r/C8727509

Callaghan, J. 1976. "Leader's speech, Blackpool 1976". British Political Speeches. www.britishpoliticalspeech.org/speech-archive.htm?speech=174

Campbell, G. 2021. "Meccano factory occupation timeline". libcom.org. https://libcom.org/history/meccano-factory-occupation-timeline

Cemgil, C. & C. Hoffman 2016. "The 'Rojava revolution' in Syrian Kurdistan: a model of development for the Middle East?". *IDS Bulletin* 47(3).

Chadwick, G. 1970. "The big flame – an account of the events at the Liverpool factory of GEC-EE". In *Trade Union Register*, edited by K. Coates, T. Topham & M. Barratt Brown. London: Merlin.

CIS Anti-report 1975. *Where is Lucas Going?* London: Counter Information Services, Anti-Report No 12.

Clark, A. 2013. "'And the next thing, the chairs barricaded the door': the Lee Jeans factory occupation, trade unionism and gender in Scotland in the 1980s". *Scottish Labour History* 48: 116–34.

Clark, A. 2021. "Workplace occupations in British labour history: rise, fall, and historical legacies". *Labour History Review* 86(1): 1–7.

Clegg, H. 1960. *A New Approach to Industrial Democracy*. Oxford: Blackwell.

Clegg, I. 1971. *Worker's Self-Management in Algeria*. London: Allen Lane.

Coates, K. 1968. *Can the Workers Run Industry?* London: Sphere/Institute for Workers' Control.

Coates, K. (ed.) 1974. *Democracy in the Mines: Some Documents of the Controversy on Mines Nationalisation Up to the Time of the Sankey Commission*. Nottingham: Spokesman Books.

Coates, K. (ed.) 1976. *The New Worker Co-operatives*. Nottingham: Spokesman Books.

Coates, K. & T. Topham (eds) 1970. *Workers Control: A Book of Readings and Witnesses for Workers' Control*. London: Panther.

Coates, K. & T. Topham 1972. *The New Unionism: The Case for Workers' Control*. London: Peter Owen.

COHSE nd. "Hospital occupations in Britain". http://cohse-union.blogspot.co.uk/2010/10/hayes-cottage-hospital-occupation-25th.html

Cole, G. 1920a. *Guild Socialism*. London: Fabian Society.

Cole, G. 1920b. *Guild Socialism Re-Stated*. London: Leonard Parsons.

Cole, G. 1921. *Guild Socialism: A Plan for Economic Democracy*. New York: Frederick A. Stokes.

Cole, G. 1938. "A study in British trade union history: attempts at 'General Union' 1829–1834". *International Review for Social History* 4(1): 359–462.

Cooley, M. 1980. *Architect or Bee? The Human/Technology Relationship*. Slough: Langley Technical Services/Hand & Brain.

Danos, J. & M. Gibelin 1986. *June '36: Class Struggle and the Popular Front in France*. Translated by P. Marsden, P. Fysh & C. Bourry. London: Bookmarks.

Democracy Now 2009. "Fire the boss: Naomi Klein and Avi Lewis on 'The worker control solution from Buenos Aires to Chicago'". www.democracynow.org/2009/5/15/fire_the_boss_naomi_klein_avi

Donovan Commission 1968. Royal Commission on Trade Unions and Employers Associations. Report, Cmnd 3623.

Fore, M. nd. "Strategy for industrial struggle". Bromley: Solidarity Pamphlet No 37.

Foster, J. & C. Woolfson 1986. *The Politics of the UCS Work-in: Class Alliances and the Right to Work*. London: Lawrence & Wishart.

Fryer, R. 1973. "Redundancy, values and public policy". *Industrial Relations Journal* 4(2): 2–19.

Gall, G. 2011. "Contemporary workplace occupations in Britain". *Employee Relations* 33(6): 607–23.

Gibbs, E. 2021. "'It's not a lot of boring old gits sitting about remembering the good old days': the heritage and legacy of the 1987 Caterpillar factory occupation in Uddingston, Scotland". *Labour History Review* 86(1): 117–44.

Gibbs, E. & E. Kerr 2020. "Mobilizing solidarity in factory occupations: activist responses to multinational plant closures". *Economic and Industrial Democracy* 43(2): 612–33.

Gleason, A. 1915. *With the First War Ambulance in Belgium: Young Hilda at the War*. New York: A. L. Burt.

Gleason, A. 1920. *What the Workers Want: A Study of British Labor*. New York: Harcourt, Brace & Howe.

GNCTU 1834. "Resolutions of the grand national consolidated trades union: extracted from a report of the proceeding of a special meeting of trades union delegates, held in London on the 13th, 14th, 15th, 17th, 18th, and 19th of February, 1834". *The Pioneer*, or *Trades Union Magazine*, 8 March.

Goodrich, C. 1921. *The Frontier of Control: A Study in British Workshop Politics*. London: Harcourt, Brace. Reprinted in 1978 by Pluto Press, edited and introduced by R. Hyman.

Goodrich, C. 1977 [1925]. *The Miner's Freedom: A Study of the Working Life in a Changing Industry*. Boston, MA: Marshall Jones.

Gouldner, A. 1955. *Wildcat Strike: A Study of an Unofficial Strike*. London: Routledge & Kegan Paul.

Gramsci, A. 1971. *Selections from Prison Notebooks*. Edited and translated by Q. Hoare & G. Nowell Smith. London: Lawrence & Wishart.

Heron, L. 1978. "Hounslow: raid and rally". *Spare Rib* 65: 30–31.

Hinton, J. 1973. *The First Shop Stewards' Movement*. London: Allen & Unwin.

Hoyles, A. 1973. *Imagination in Power: The Occupation of Factories in France in 1968*. Nottingham: Spokesman Books.

Hughes, E. & T. Dobbins 2020. "Frontier of control struggles in British & Irish public transport". *European Journal of Industrial Relations* 33(1): 174–83.

Hughes, E., T. Dobbins & S. Murphy 2018. "'Going underground': a tube worker's experience of struggles over the frontier of control". *Work, Employment and Society* 33(1): 174–83.

IPM 1976. *Sit-ins and Work-ins*. London: Institute of Personnel Managers.

IWC 1969. *GEC-EE Workers Take-over*. IWC Pamphlet Series 17. Nottingham: Institute for Workers' Control.

IWC 1976. "Sit-ins: the employers counterattack". *Workers' Control Bulletin*, 14–16.

Kelly, J. 1998. *Rethinking Industrial Relations: Mobilisation, Collectivism and Long Waves*. London: Routledge.

Labour Party 1918. *1918 Labour Party Manifesto*.

LACSSC 1978. *Lucas: An Alternative Plan. Nottingham: Lucas Aerospace Shop Stewards' Combine Committee.* Nottingham: Institute for Workers' Control.

LACSSC 1979. *Democracy Versus the Circumlocution Office*. Nottingham: IWC Pamphlet No. 65.

LACSSC 1981. *Lucas: Diary of Betrayal*. London: CAITS.

Lavaca Collective 2007. *Sin Patrón: Stories from Argentina's Worker-Run Factories*. Chicago, IL: Haymarket Books.

LCSST 1971. *The Arrow Mill Sit-in*. Lancashire Campaign for Shop Stewards in Textiles (LCSST).

Lewis, A. & N. Klein 2004. *The Take*. London: Artefact Films.

Littler, C. 1980. "Internal contract and the transition to modern work systems: Britain and Japan". In *The International Yearbook of Organizational Studies*, edited by D. Dunkerley & G. Salaman, 186–216. London: Routledge & Kegan Paul.

Lomax, B. 1976. *Hungary 1956*. London: Allison & Busby.

Lomax, B. 1980. *Eyewitnesses in Hungary: The Soviet Invasion of 1956*. Nottingham: Spokesman.

Lydersen, K. 2009. *Revolt on Goose Island: The Chicago Factory Take-Over, and What it Says About the Economic Crisis*. New York: Melville House.

Mandel, D. 2011. "The factory committee movement in the Russian revolution". In *Ours to Master and to Own: Workers' Control from the Commune to the Present*, edited by I. Ness & D. Azzellini, 104–29. Chicago. IL: Haymarket Books.

Marchington, M. *et al.* 1993. "The influence of managerial relations on waves of employee involvement". *British Journal of Industrial Relations* 31(4): 553–76.

Marglin, S. 1974. "What do bosses do? The origins and functions of hierarchy in capitalist production". *Review of Radical Political Economics* 6(2): 33–60.

Marren, B. 2016. *We Shall Not Be Moved: How Liverpool's Working Class Fought Redundancies, Closures and Cuts in the Age of Thatcher*. Manchester: Manchester University Press.

Marsh, A. & E. Coker 1963. "Shop steward organization in the engineering industry". *British Journal of Industrial Relations* 1(3): 170–90.

Metra 1972. *An Analysis of Sit-ins*. London: Metra Consulting Group.

Montgomery, D. 1983. "The past and future of workers' control". In *Workers' Struggles, Past and Present: A "Radical America" Reader*, edited by J. Green, 389–405. Philadelphia, PA: Temple University Press.

Mustchin, S. 2021. "Job destruction and closures in deindustrializing Britain: the uses and decline of workplace occupations in the 1980s". *Labour History Review* 86(1): 91–116.

Newens, S. 1969. "The GEC/AEI take-over and the fight against redundancy at Harlow". *Trade Union Register* 171: 4.

Parker, S. *et al.* 1971. *Effects of the Redundancy Payments Act: A Survey Carried out in 1969 for the Department of Employment*. London: Office of Population Censuses & Surveys, Social Survey Division.

Past Tense 2013. *Occupational Hazards: Occupying Hospitals: Inspirations and Issues from our History*. London: Past Tense.

Penty, A. 1906. *The Restoration of the Guild System*. London: Swan Sonnenschein.

Pribicevic, B. 1959. *The Shop Stewards' Movement and Workers' Control, 1910–1922*. Oxford: Blackwell.

Rabble 2013. "Chicago workers open new cooperative factory five years after Republic Windows occupation". Video. http://rabble.ca/rabbletv/program-guide/2013/05/best-net/video-chicago-workers-open-new-cooperative-factory-five-year

Ramelson, B. 1968. "The possibilities and limitations of workers control". *Marxism Today*, February: 296–303.

Ramelson, B. *et al.* 1970. *The Debate on Workers' Control: A Symposium from Marxism Today*. Nottingham: Institute for Workers' Control.

Ramsey, H. 1977. "Cycles of control: worker participation in sociological and historical perspective". *Sociology* 11: 481–506.

Ranis, P. 2016. *Cooperatives Confront Capitalism: Challenging the Neoliberal Economy*. London: Zed.

Reid, D. 2018. *Opening the Gates: The Lip Affair, 1968–1981*. London: Verso.

Ruggeri, A. 2016. "The worker-recovered enterprises in Argentina: the political and socioeconomic challenges of self-management". www.workerscontrol.net/authors/worker-recovered-enterprises-argentina-political-and-socioeconomic-challenges-self-managemen

Russell, B. 1918. *Roads to Freedom: Socialism, Anarchism and Syndicalism*. London: Allen & Unwin.

Scanlon, H. c.1970. *Workers' Control and the Transnational Company*. IWC Pamphlet. Nottingham: Institute for Workers' Control.

Scargill, A. & P. Kahn 1980. "The myth of workers' control". Occasional Papers in Industrial Relations, University of Leeds and University of Nottingham.

Sherry, D. 2010. *Occupy! A Short History of Workers' Occupations*. London: Bookmarks.

Singleton, F. & T. Topham 1963. *Workers' Control in Yugoslavia*. London: Fabian Society.

Slater, M. 1936. *Stay Down Miner*. London: Martin Lawrence.

Sobering, K. 2022. *The People's Hotel: Working for Justice in Argentina*. Durham, NC: Duke University Press.

Spiro, H. 1954. "Co-determination in Germany". *American Political Science Review* 48(4): 1114–27.

Spriano, P. 1975. *The Occupation of Factories: Italy 1920*. London: Pluto.

Sprung, S. 2020. "The plan that came from the bottom up". theplandocumentary.com

Taylor, P. & P. Bain 2001. "Trade unions, workers' rights and the frontier of control in UK call centres". *Economic and Industrial Democracy* 22(1): 39–66.

Terry, M. 1983. "Shop stewards through expansion and recession". *Industrial Relations Journal* 14(3): 49–58.

Thompson, E. 1967. "Time, work-discipline and industrial capitalism". *Past & Present* 38: 56–97.

Topham, T. 1964. "Shop stewards and workers' control". *New Left Review* 1(25): 3–16.

Tuckman, A. 2018. *Kettling the Unions? A Guide to the 2016 Trade Union Act*. Nottingham: Spokesman Press.

Tuckman, A. 2021a. "After UCS: workplace occupation in Britain in the 1970s". *Labour History Review* 86(1): 7–35.

Tuckman, A. 2021b. "Workplace occupation in the United Kingdom, 1971–2019". Appendix to special issue of Labour History Review. *Labour History Review* 86 (1): 165–89.

Turner, H., G. Clack & G. Roberts 1967. *Labour Relations in the Motor Industry*. London: Allen & Unwin.

Upchurch, M. 2020. "Time, tea breaks, and the frontier of control in UK workplaces". *Historical Studies in Industrial Relations* 41(1): 37–64.

Vieta, M. 2021. *Workers' Self-Management in Argentina: Contesting Neo-Liberalism by Occupying Companies, Creating Cooperatives, and Recuperating Autogestión*. Chicago, IL: Haymarket Books.

Vio.me n.d. "Vio Me: Occupy, Resist, Produce!". www.viome.org/search/label/English

Wainwright, H. & D. Elliott 1982. *The Lucas Plan: A New Trade Unionism in the Making?* London: Allison & Busby.

Wajcman, J. 1983. *Women in Control: Dilemmas of a Workers' Co-operative*. Milton Keynes: Open University Press.

Webb, B. 1926. Manuscript Diary. LSE Digital Library. https://digital.library.lse.ac.uk/objects/lse:beg266zey

Weller, K. 1962. "Sit-in at BMC". *Solidarity: For Workers Power* 2(2): 3–6.

White, E. 1949. *Workers' Control: The Challenge of 1950*. London: Fabian Publications & Victor Gollancz.

Whitehead, B. 2001. "The Derby lock-out 1833–4 and the origins of the labour movement". Unison South Derbyshire Health Branch.

Whyte, D. 2010. "'Sin Patron' in Dundee?" *Red Pepper* 172: 28.

Wright, V., J. Philips & J. Tomlinson 2021. "Defending the right to work: the 1983 Timex workers' occupation in Dundee". *Labour History Review* 86(1): 63–91.

CHAPTER 14

FROM SECTIONALISM AND SECTIONALITY TO INTERSECTIONALITY

Jenny K. Rodriguez

ABSTRACT

This chapter interrogates sectionalism and sectionality, discussing intersectionality and its potential to rethink ideas about union collective identity and organizing concerning diversity and difference. The chapter contends that the traditional use of sectionality seems obsolete and there is clear potential for more inclusive and democratic organizing and mobilization through an intersectional approach that places closer attention to how socially constructed categories of difference interplay with the work and employment experiences of particular groups. The framing of single identity in struggle is insufficient to capture the needs of worker groups sitting at the intersection of multiple marginalities, so adopting intersectional organizing would enable developing collective identity and organizing activities that resonate with the lived experiences of diverse groups and promote intersectional solidarity to support coalitionist alliances that have stronger potential for social justice and change.

Keywords: Collective identity; intersectionality; challenges for unions

INTRODUCTION

In this chapter, I discuss intersectionality and its potential to rethink ideas about union collective identity and organizing.[1] One of the main challenges of advancing discussions about collective identity in unions is both the fixed understanding about union-related collective identity as well as how this collective identity develops and is – or should be – maintained. There are two issues here. First, the historical

1. I wish to thank Gregor Gall for his comments on this chapter. His critical insight and engagement with the ideas I discuss helped me to clarify points I make in this chapter.

prevalence of White, male-dominated logics that has shaped union activities and priorities has meant that the interests of specific groups remained largely ignored. Second, scholarship problematizing key ideas pertaining to unions, such as power relations and ideology, has been developed without consideration of the complex identities of those it seeks to theorize about. However, exceptions are found in some more recent work in the 2000s (see Colgan & Ledwith 2000; Dickens 2000; Greene & Kirton 2002; Holgate 2005; Kirton & Healy 2004; Ledwith & Colgan 2002; McBride 2001; Noon & Hoque 2001; Wajcman 2000). While these works set the foundation for the analysis of gender and race in unions, this remains at the margins of dominant discussions about unions.

Against this backdrop, narratives about solidarity and collectivism presented by unions need to pay closer attention to the role of difference(s) in the way workers understand the collective and how it mediates notions of collective identity and organizing. Efforts at union revitalization have engaged with diversity from a position of limited understanding of difference(s) and, as a result, have failed to successfully accomplish the key aim of broadening the membership base. Several questions are of importance here, not least the extent to which diverse workers (e.g. young workers, migrant workers, workers in the informal sector or in the gig economy, etc.) simply do not relate to the idea of collective identity as presented by unions. An important consideration is that while the experiences of workers have dramatically changed and employment issues, such as precariousness, have become more complex and nuanced, discussions about the impact of unions on employment remain scarce (Darvas, Gotti & Sekut 2023).

In this context, sectionalism and sectionality present an inherent tension, because, while looking to create collective categorizations to unite workers, they could also be seen to divide them because they overlook issues more pressing concerning the inequalities and disadvantages experienced in society and in the labour market, which go beyond the class struggle argument that underpins "worker" and "sectoral" categorizations. This tension raises the question about the extent to which the prevalent forms of sectionalism and sectionality support achieving the aim of labour unionism to unite and align as many workers as possible. We could, indeed, think about potential strategies to deal with difference, such as acceptance, denial or reconfiguration (e.g. forging a higher level of identity and consciousness).

Sense of community and belonging in struggle are essential to collective efforts. However, the increasing inequalities faced by individuals and groups, which go beyond the class-based struggle, require engagement with ideas about diversity and difference in a way that does not articulate a single identity in struggle but, rather, recognizes the importance of intersectional diversity and solidarity in struggle. Consequently, this chapter is organized into four sections. The first section provides and overview of sectionalism and sectionality, problematizing collective identity and class consciousness as underpinning concepts. The second section discusses intersectionality and organizing, focusing upon the potential offered by intersectionality as a more appropriate way to deal with worker organizing, and addressing some criticisms to intersectionality. The third section delineates ideas about intersectional organizing, and the last section concludes on the issues at hand.

SECTIONALISM AND SECTIONALITY

Sectionalism and sectionality have been central to understanding union collective identity, membership and participation in unions and social movements more generally. Gall (2008, 2012) defines sectionalism as competitive and highly differentiated workplace labour unionisms based upon trade. Trade is seen as fundamental to the notion of "collective" (identity, organizing, action and bargaining) because professional orientation is seen to facilitate collectivism (Apitzsch *et al.* 2022) and workers use demarcations around job boundaries to articulate their concerns (Roberts & Cullinane 2023).

As the strategies used to group workers, sectionalism and sectionality give us insight into what drives categorizations. Unions have articulated collective identity in relation to "trade", "section", "industry", "worker" and "class", historically using these to drive efforts to organize. The argument about the strength of collective identity in generating and sustaining commitment and cohesion has been used to explain why people mobilize and participate, as well as the tactical choices they make to achieve change (Flesher Fominaya 2010; Polletta & Jasper 2001). This is, perhaps, the first issue that needs problematization for several reasons. The centrality of class in the worker struggle does not consider the racialized and gendered logics of capitalism, which have historically shaped inequalities affecting minoritized groups. In this respect, the class struggle is fragmented in relation to intra-group disadvantage, which is not recognized in discussions about sectionalism and sectionality. Perhaps, more importantly, the lack of consideration of how workers' experiences unfold at multiple points of intersecting oppression and disadvantage could explain why workers may perceive that union agendas exclude their interests (see Jefferys & Oual 2007; Thomas 2016).

Collective identity is a very abstract, opaque, multifaceted, hybrid concept with a highly contradictory nature (Flesher Fominaya 2010; Jacobs 2013; see also Chapter 1). Conversely, generating a collective identity is crucial for a movement to emerge (Taylor & Whittier 1992). In effect, consideration of issues related to developing and maintaining a collective identity have been central to union efforts to organize. The idea of collective identity is predicated upon the assumption of collective interest which Kelly (1998: 27–33) argues "is always understood through a collective identity which is built and changeable by various factors: a sense of injustice, social attributions, social identification and perception of a chance to change the situation by collective agency".

Most discussions within studies of unions have primarily focused on working-class consciousness and activity as the key departure point for unionism and collective organizing. A problem with thinking in terms of working-class consciousness is that the understanding of class has a root in a particular form of capitalism that is not universal, and overlooks colonial histories of oppression. For instance, Sullivan (2021) notes that the history of the working class in Britain has focused solely on the struggles of the industrial working class in the British Isles, which reproduces a particular understanding of nation-state that is sustained by state institutions and notions of citizenship that overlook the histories of workers from across the Empire. Obscuring the stories, struggles and role of British overseas workers in the history of organized working class has meant that the narratives about the working-class struggle are presented in relation to the dominant histories of Empire and, in turn, "provide a basis for nationalist based narratives in current debates on immigration and asylum" (Sullivan 2021: 4). These narratives ultimately support specific ideas about what the class struggle is and the meaning of resistance in ways that hinder efforts to organize many racially minoritized groups.

Something similar could be argued in relation to the narrow understanding of collective identity that comes from sectionalism and sectionality. There are challenges in the way that unions define their agendas for action, which are ultimately driven by union leaderships rather than by union memberships. In this respect, sectionalism accommodates to, at some level, but also perpetuates categorizations that are artificial and problematic because framing individuals' identities in occupational terms does not necessarily then also engage with the complexities of the experiences of disadvantage that emerge from marked social identities, categories of difference, differentiation processes and systems of domination (see Dhamoon 2011). These complexities might be more important for workers than an occupational categorization, given their role in shaping experiences of inequality within an occupational collective. For instance, unions that represent professionals in highly skilled occupations appear detached from a common "reality" when making claims of worker exploitation similar to the ones that have traditionally informed collective organizing in sectors primarily characterized by low pay and poor work conditions. Sectional distinctions are reductive because they do not recognize the porosity of oppression and disadvantage that brings together groups based on how different forms of sectionality intersect for particular groups within a larger collective.

In a discussion of the challenges in organizing Polish workers by two unions – Unison (UK) and Unia (Switzerland), Rogalewski (2022: 386) notes that the involvement of Polish workers in these unions was "predominantly influenced by the need for social justice as well as employment protection". We could see the claim for social justice as more closely linked to how nationality (being Polish) and citizenship status (being a migrant) intersect to articulate social identities that create distinct disadvantage for this group beyond employment protection. In this respect, sectionality categorization obscures the complexity of their struggle and could be seen as a form of epistemic "Othering" (see Dhamoon 2011) because the sectoral commonality does not capture the fundamental granularity required to understand migrants' "specific social and workplace needs" (Alberti, Holgate & Tapia 2013: 4140).

The fragmentation of the collective could be seen in the UK in the responses to the 2021–22 disputes between the University and College Union and the Universities and College Employers' Association, which saw academics and academic-related staff go on strike in the middle of the pandemic over pay, equality, workload, casualization and pensions. These disputes created division within union members and criticism from the public given the perceived status of academics as a privileged group, where the national average salary of an academic was much higher than the median annual pay for full-time employees (ONS 2022). Similarly, a large majority of academics were able to work from home, which reinforced their perceived status as a privileged group during the pandemic. This was despite the realities of precariousness experienced by many academics that could put them on par with other workers in precarious jobs in other sections.

An important consideration is that the explanation of the reasons underlying the dispute were also contentious because they appeared to essentialize the struggles as similar in nature, overlooking the socio-historical patterns of subordination that are institutionally reproduced for some groups of workers. For instance, when mentioning intersectional pay inequality as part of the #ucuRISING campaign, the University and College Union (2022: 2) noted that:

> Our sector is blighted by inequality with a pay gap between Black and white staff of 17 per cent, a disability pay gap of 9 per cent and a mean gender pay gap of 16 per cent. Employers are failing to tackle this situation and that's why we are demanding national action to monitor and end pay inequality.

While there is merit in highlighting these categorical differences, this is not an intersectional overview of the problem because the mutually constitutive impacts are not clearly articulated. Groups are also essentialized within single categories (e.g. race) and excluded (e.g. women of colour are neither Black nor White). Against the backdrop of persistent disadvantage experienced by Black women and women of colour in academia resulting from structural racism and sexism, Blell, Liu & Verma (2023) note that the reliance on the equality narrative of policy implementation means that universities would be unlikely to assess the intersectional impact of the pandemic on minoritized members of staff, something that the dispute did not engage with either.

Ultimately, many issues coexist and intersect to shape collective identity and equally make it a fluid notion. Worker diversity and new work arrangements mean that sectionalities are increasingly more complex. Thinking about workers in occupational and sectoral terms overlooks how their experiences of inequality are marked by their social identities. For example, following the death of George Floyd in the US, police violence against Black people and the institutional racism that is said to characterize the police force have been called out by supporters of the Black Lives Matter movement. This has put police unions in the spotlight as they defend the collective rights of their members (which include Black members), who actively participate and are considered "a significant instrument of that racism" (Unger 2020: 28). More problematic is the argument made by some (see Thomas & Tufts 2020: 127) who note that "the construction of solidarity within police unions serves to undermine and criminalise movements that contest police violence against racialised communities and to more broadly counter working-class resistance to austerity and right-wing authoritarian populism".

Intersecting social identities might be more important to workers because they are part of a spectrum of histories and struggles of disadvantage and subordination, and are embedded in regulatory regimes, institutions, practices and attitudes that have created structural barriers throughout their lives, often many times in intergenerational ways (Yuval-Davis 2015). In addition, while it could be argued that the fragmentation of employment structures has led to more awareness of diversity and difference, these issues are discussed in a way that adopts a single-axis approach that looks at socially constructed categories of difference (e.g. gender, race, disability) in isolation. In some cases, essentializing categorizations are used, such as focusing on women to signify gender, or focusing on ethnic minorities to signify race. In other cases, the complexity of identity is not fully problematized. For instance, discussing trans-liberation and labour unionism in relation to the work of transgender activist Leslie Feinberg, Salah (2014) notes that despite almost 50 more years of scholarship on queer solidarity, there is a lack of academic legitimation of this discussion, which is considered newly established and emergent.

The previous points highlight that discussions are not just about the class struggle but also the tensions between dynamics that are simultaneously racialized, gendered and classed, among others, and that are played out in social and work life, with societal inequalities that emerge as a result. Reflecting on the construction of collective identity of Black members of police unions, there is an undeniable fragmentation within this group because the sectionality of union members conflicts with features salient to other forms of collective identity that some share, such as being a Black member in an institution with a history of violence against Black people. This fragmentation, however, is not new and perhaps has more visibility now as a result of the more open discussion about race and institutional racism and the calls for accountability in the public domain – for example, Marks (2000) reports on the organizing efforts of Black police officers in the South African Police Force in the 1980s and 1990s, which led to the formation of a union and a Black management network. However, these efforts continue to be isolated and somewhat underplayed in the grand narrative of discussions about union collective identity and union organizing.

The previous point highlights the need to understand the roots of inequality as multilayered, so while it is not just about class, it is also not just about race (and gender or other social categories). Rather, it is about the intersection of minoritized statuses (e.g. disabled migrant women) as well as the ways in which some privileged characteristics intersect with forms of oppression and disadvantage (e.g. White working-class women). In this chapter, my use of the term "privilege" draws upon discussions about intersectionality and social identity (see Carbado 2016) to refer to how individuals' social identities interplay with access to and allocation of power and control over opportunities, resources and outcomes, and create "multilayered and routinized forms of domination" (Crenshaw 1991: 1245). These dynamics are embedded in what Collins (1990) terms a matrix of domination, which sees privilege and oppression as intersecting along a spectrum of hierarchized social identities (see Anthias 2012).

This diversity and the privileges and inequalities related to it must also be situated in the context of societal changes (e.g. Brexit and the Covid-19 pandemic). In this respect, the metaphor of intersectionality is useful to shift the discussion in a way that both captures the complexity of worker identities and problematizes how these identities inform the forms of engagement with and participation in unions, alongside fractures and necessary splits in order to reframe the role of unions in a way that recognizes both the collective as well as group-specific differences within it that drive specific concerns. The latter is relevant for unions because "a notion of shared identities as a basis of solidarity tends to privilege the voice and preferences of dominant groups within movements" (Tormos 2017: 711).

Discussions about sectionality have traditionally focused on trade, occupations and industry. My argument is, therefore, that we need to frame the relationship between identity and union engagement as related to the shift both in the understanding of unionism and the salience of identity characteristics that may go beyond these affiliation and associations. For example, the idea of shopfloors as distinctly sectionalized from each other no longer applies to the realities of most workplaces given the significant changes to the structure of work and workplaces in terms of the structures of employment, participation and composition of the workforce, the organization of work, and working patterns and utilization of labour (Jansson 2022; Rodriguez, Johnstone & Procter 2017). These changes have had a significant impact on how union activity is structured, yet unions have struggled to reconcile these differences in the articulation of collective interests.

The sectionalized union structure has been historically debated. On the one hand, it is seen to promote a stronger sense of collective identity, which is linked with the aim and purpose of unions. On the other hand, it has been criticized for promoting a fragmented and decentralized structure that challenges that very aim and purpose. The challenges of organizing labour are related to finding common ground for a collective to address issues of importance and understanding commonalities in the politics of struggle. In this respect, the idea of sectionality could be counter-intuitive to the idea of creating and strengthening an overall class-based collective because it assumes that a generalized struggle is what brings people together. There is a valid argument that "being workers" is the overarching collective framework. However, the different forms of inequality, and how they simultaneously intersect and are multiplicative for particular groups, raises the issues of what a collective would constitute and whose interests it would represent.

At the same time, framing collective action in relation to things such as collective bargaining or resistance to workplace conditions or mistreatment makes sense if looked at from the perspective of how these issues develop across the board. There is unlikely to be a scenario where different industries experience similar conditions and, perhaps more importantly, where the labour landscape is strictly similar. Nevertheless, the separation between, for example, professionals and labourers in some industries overlooks disadvantages between members who share similar social identities, which would be relevant to address the structural inequalities that exist beyond industries for specific groups.

Two different stories of Indigenous women can help to illustrate this point. First, Horowitz's (2017) study of Kanak women's involvement with mining highlights the symbolic and cultural violence as well as the hegemonic ideologies of women's subordination deployed by executives at two Falconbridge refineries in New Caledonia to intersectionally oppress this group of women, who are young, rural and poor and, as Indigenous women, have been historically treated as inferior as part of an oppressive colonial history. Second, Sonia Maribel Sontay Herrera, an Indigenous woman and Organizational and Advocacy Coordinator of Majawil Q'ij Association, a Guatemalan organization that works to empower Indigenous women in local communities in Guatemala, who recounts the discrimination, racism and stereotyping she faced when looking for professional work after finishing her studies, where many employers stated that they only had domestic work to offer her – "They see us as domestic workers; when they see an Indigenous woman, they assume that's all we can do" (UN Women 2020). Both these accounts speak to the centrality of intersecting social identities (gender, ethnicity, age and class) in shaping the disadvantages experienced by women in ways that, for instance, would not be experienced by either White women or Indigenous men). In this respect, the question ultimately remains in terms of whether the "class character" is the appropriate notion around which to articulate a common struggle facing workers.

Melucci (1995) problematizes what underpins the motivation for individuals to organize as part of a collective and how they maintain this motivation over time. An important point here relates to "how the meaning of collective action derives from structural preconditions or from the sum of individual motives" (Melucci 1995: 42). Arguably, those individual motives are related to everyday experiences and relationships. In discussing this point, Hyman (1975: 42) questions the centrality of consciousness of class identity, which he sees as "far removed from the everyday processes of industrial relations". In effect, in between the Marxist distinction of a proletariat that sells its labour and a bourgeoisie employer that exploits it, lie nuances in relation to intra-group degrees of exploitation that have not been comprehensively unpacked in discussions about labour unionism.

While recognizing that the raising of intra-group degrees of exploitation risks undermining narratives of "collective struggle" that underpin labour unionism, the issue perhaps is the extent to which it has been easier for unions to, first, maintain a Marxist class-based narrative because the other option is to engage in the complex interrogation of the ways in which capitalist accumulation is not only classed but also racialized, gendered and able-ized, among others, and second, to continue to focus their efforts on the preferences of an increasingly narrow group of dominant union members. The argument here is not one that looks to promote a form of individualism but one that looks to propose that sectionalism and sectionality are, by their very nature, exclusive and othering. An important tension pertains to what is seen as important by academics theorizing and researching unions and by workers themselves.

Hyman (1979: 94–5) argues that "the contradictory relationship between sectionalism and broader class consciousness and organisation needs a more solid theoretical interrogation so that more effective strategies for anti-capitalist struggles are developed". Presenting the struggle from a perspective that assumes that collective interests are uniform overlooks the importance of recognizing difference within the broader argument for social justice. For instance, in their discussion about migrant labour organizing, Tapia and Alberti (2019) interrogate the idea of "migrant worker", noting that migrant intersectionalities is a more appropriate way to understand the inequalities experienced by these workers.

Collective organizing is linked to coming together through common interests aligned with fighting against exploitation, inequalities and oppression. However, it must be recognized that exploitation, inequalities and oppression are embedded in institutional structures and arrangements that operate in neither linear ways nor in compartmentalized ways – they emerge from overlapping structures of domination that are mutually constitutive (Cho, Crenshaw & McCall 2013). In this respect, variability in (collective) associational power resulting from diverse institutional and strategic constraints needs consideration and, in turn, a priority should be given to developing solidarity and cooperation mechanisms among unions. This, according to Caraway and Ford (2017), is an area for which more academic research is needed. Indeed, one of the key weaknesses of union revitalization strategies is that they have not engaged with these ideas with critical intentionality. For example, despite looking to attract more diverse groups (e.g. young workers), unions' main internal struggle with diversity remains largely unaddressed. Similarly, the andro-centric environment in unions continues to be an unresolved issue, especially in leadership, which poses challenges to gender equality given the prioritization of male interests (see Dickens 1997, 2000; Kirton & Healy 2013; Ledwith & Munakamwe 2015).

INTERSECTIONALITY AND ORGANIZING

This section discusses the potential offered by intersectionality as a more appropriate way to deal with worker organizing. While intersectionality is not seen as a corrective to sectionality, intersectionality does provide much needed nuance to understand the relationship between identities and inequalities, in a way that is useful and more generative to explore how diversity is integrated in discussions about collective identity.

Intersectionality is a complex term because there is no clear definition of it, but rather it is a metaphor used to illustrate the argument that the intersection of diverse social categories of difference (e.g. gender, race, ethnicity, age, disability, nationality, religion, etc.) intersect to create forms of privilege and oppression. The idea behind intersectionality is that these categories are inseparable and operate across individual, institutional, cultural and societal domains (Rodriguez *et al.* 2016). That inseparability is essential to understand inequalities and the systems of power they maintain and perpetuate. In particular, single categories (e.g. gender or race) overlook that subordination is multidimensional. For example, even as we recognize that there is gender inequality, this experience differs for Black women and women of colour, given the contested nature of Blackness as a visible marker of identity.

Collins and Bilge (2016: 25–30) note that there are four ideas that underpin intersectionality: first, systems of oppression (e.g. racism, sexism, class exploitation, ableism) are interconnected and mutually constitutive; second, social inequality configurations take form within intersecting oppressions; third, the way social actors understand and perceive social problems shows their historically and socially rooted location within power relations; and fourth, individuals' and groups' distinctive standpoints on social phenomena are explained by the different ways in which they are located within intersecting oppressions.

It is important to position intersectionality as a theory about structural forms of power and not as a theory about identity politics (Bryson 2021). In this sense, the importance of intersectionality for union organizing lies both in how it recognizes identity complexity and brings it to the fore to interrogate institutionally based inequality. Ferree (2009) stresses that intersectionality must be understood as an integrated framework, rather than analysed at a single (individual or institutional) level of analysis, and agrees with other intersectional scholars, such as Collins (2001), in arguing that neither structural locations nor axes of oppression are static but play out in "organizational fields in which multidimensional forms of inequality are experienced, contested and reproduced in historically changing forms" Ferree (2009: 87). As a result, intersectionality can more fully and intentionally capture multiplicity of identities in a way that allows to identify more clearly how these identities interact with ideologies, issues, frames, collective action repertoires and organizational forms.

The idea of intersectionality has a root in social justice and posits that social categories of difference intersect to shape the experiences of oppression and privilege of individuals and groups. According to Collins (1990), the practices that sustain oppression and privilege constitute a matrix of domination that operates as a heuristic device that perpetuates them. In many ways, at the core of intersectionality is the interest to explain privilege and oppression from a perspective that recognizes that these are multilayered and cannot be explained by a single point. For instance, in their work looking at migrant workers, Alberti, Holgate & Tapia (2013) argue that the universalistic approach that underpins looking at migrants simply as workers overlooks the centrality of their identities as migrant workers – this is to suggest that the ways in which the experiences of this group unfold is never disconnected from the category of migrant. This approach, they argue, is particularistic and they note it should be integrated with the universalistic approach to understand union organizing.

Thus, a fundamental question is: to what extent is it possible to create commonality of interest and cohesion from diversity? This has been an important question in the context of the current decline in union density and often ineffectual efforts at union revitalization. The drivers of union revitalization speak to a recognition of the importance of diversity but not necessarily of what multidimensional diversity is and how it shapes the lived experiences of oppression and disadvantage for particular groups. Perhaps a rethinking of this question is also important. The aim of achieving commonality of interest and cohesion in the development of class consciousness cannot just be articulated in relation to one dimension of the work experience and not consider that the work experience is linked to identity. This recognition seems explicit in how discussions about organizing are being framed around intersectional identities of oppressed workers. For example, discussing LGBTQ+ organizing, O'Brien (2021: 825) found:

> Organisers and worker-leaders consistently reported queer and trans workers were more likely than their straight and cis counterparts to join new organising campaigns quickly, more likely to be motivated to organise and protest, and more likely to play leadership roles in their organising than their straight or cis counterparts.

This explains their propensity to collective action as related to their experiences of discrimination and alienation, as well as their prior social movement activity.

Some criticisms

Any discussion about intersectionality must recognize its contentious relationship with those who advocate for the centrality of class – this despite the fact that that experiences of oppression have always had intersectional potential because the very discussion of class difference is implicitly gendered and racialized. Nevertheless, there is tension between Marxist feminists and intersectionality scholars in relation to the place of class. One of the key criticisms of intersectionality discussions is that class was ignored in the initial discussions about intersectionality, which focused on gender and race (see Crenshaw 1989).

In her Marxist critique of intersectionality, Foley (2019: 10) notes:

> … "race" does not cause racism, gender does not cause sexism. But the ways in which "race" and gender – as modes of oppression – have historically been shaped by the division of labor can and should be understood within the explanatory framework supplied by class analysis, which foregrounds the issue of exploitation.

As interesting as this critique is, it resembles the reification of class that perpetuates the universalizing of the interest of White, middle-class, Western feminism and continues to "other" non-White racialized and minoritized groups, especially women. As Bannerji (2005: 145) notes: "A trade union cannot properly be said to be an organisation for class struggle if it only thinks of class in economic terms, without broadening the concept of class to include "race" and gender in its intrinsic formative definition."

Another criticism of intersectionality, and perhaps one that speaks directly to discussions about union organizing and collective action, is that intersectionality is empirically impractical, and this has been highlighted in reports about how equality representatives struggle to deal with cases of discrimination on multiple grounds (see Moore, Wright & Conley 2012). This criticism reflects how national context interplays with equality policies and legal frameworks. For instance, while the European Union has been seen to encourage member states to merge single equality bodies into integrated ones that cover multiple inequality strands, legislation at national level complicates this (Verloo *et al.* 2012).

Despite recognizing that we need to analyse how intersecting oppressions embedded in institutional structures and arrangements underpin systems of domination (Collins 2019), beyond presenting itself as a sensibility (see Crenshaw 1989; Healy, Bradley & Forson 2011), intersectionality is seen as unable to provide "an adequate explanatory framework for addressing the root causes of social inequality in the capitalist socio-economic system" (Foley 2019: 11). Furthermore, some like Mojab and Carpenter (2019: 277) raise questions about the material potential of intersectionality to achieve social justice: "What is to become of us after our gender, race, class, sexuality, disability, and more intersect? What is being socially constituted by these identity lines crisscrossing and intersecting each other?"

In this respect, while intersectionality and Marxism share a similar ideological foundation around which notions of domination and oppression are built, intersectionality is seen as more limited in relation to how it can be applied to achieve social justice. This has hindered engagement with intersectionality in discussions about concrete actions that unions can take to develop intersectional agendas of social justice at work, being an issue that reflects the view that the only way to equality is through the solidarities of sameness, while seeing dynamics of difference as vexed and too complex to engage with (see Cho, Crenshaw & McCall 2013). The key strength of intersectionality is its usefulness as a theoretical framework to raise awareness about the multilayered complexity of identity and its relationship with multiple oppressions in ways that can inform intersectionally conscious forms of solidarity (see Tormos 2017).

INTERSECTIONAL ORGANIZING

An important avenue of exploration for union activism is the consideration of how union policies and practices embed structural and political intersectionality in the ways it develops agendas for collective, progressive social change. This should start with what Tormos (2017: 712) terms the recognition of the "intersectional contour of oppression", which should drive a reassessment of practices so that decision-making structure and leadership are inclusive, distinct social groups are supported to organize autonomously within the collective and advocacy is directed to social policies to address multiple forms of oppression.

Going back to the issue of union revitalization, an important question pertains to the possibility to create commonality of interest and cohesion from diversity. There are concerns about heterogeneity, diversity and inclusiveness in collective identity (Flesher Fominaya 2010), and while thinking about this intersectionally will complicate these terms, what intersectionality offers is the possibility to develop a more integrated understanding of difference that can support solidaristic activism. Intersectional solidarity is the foundation of intersectional organizing and refers to an inclusive approach that focuses upon building connections and coalitions across difference to achieve social change through the recognition of difference, the examination of power and the understanding of how intersecting social structures shape experience (Ellison & Langhout 2020; Hancock 2011).

So far there has been a focus on diversity in (re)framing solidarity and collective identity, as well as on forms of organizing. Similarly, when thinking about groups for whom specific social categories of difference are seen as defining to their lived experiences, it is important to recognize that diverse sectionality has shaped many discussions, with works expanding from looking at gender or race to exploring other categories such as disability. For instance, Bacon and Hoque (2015) argue that unions have increasingly paid more attention to the needs of disabled people, noting, for example, Trades Union Congress evidence of unions initiating claims under the Disability Discrimination Act 1995, representing workers with mental health problems and tackling disability discrimination in sickness absence procedure. Similarly, in a study of the Transport Salaried Staffs' Association Neurodiversity Project, a UK Union Learning Fund project designed to represent neurological-impaired employees in the UK rail/transport industry, Richards and Sang (2016) found that the project generated identified solidarity and reported successes both in individual case work as well as in shaping trade union-employer bargaining agendas.

Broadly, some works have reported on self-organized groups within unions (Bairstow 2007; Briskin 1999; Humphrey 2000; Parker & Douglas 2010; Parker & Foley 2010; Towle 2011), some of which explicitly allude to intersectional concerns. For instance, in their discussion about political organizing and social formations during the Covid-19 pandemic, Schirmer and Tarlau (2022: 18) recount how the Movement of Rank and File Educators, a social justice caucus within the United Federation of Teachers in New York

City in the US, successfully brought together teachers, parents and students to support its "outright rejection of both the union's and school leadership's response to Covid". The actions of these members, Schirmer and Tarlau (2022: 22) note, show how they "exercise power, not just petition for it".

Other forms of unionism have also emerged that speak to alternative ways of organizing, such as community unionism (e.g. Martínez Lucio & Perrett 2009) and the emergence of "indie" unions (Però 2020; Smith 2021).[2] All these different forms of organizing break from hegemonic unions and generate solidaristic movements for specific groups of workers who remain marginalized within the union movement. For instance, Martínez Lucio and Perrett (2009) explain that labour market exclusion and segregation experienced by Black and minority ethnic workers means that community and territorial structures of union organization may be more appealing to these groups.

A potential avenue of exploration is not only the preferences of these groups but also more explicitly whether and how they have attempted to transform unions and make them more inclusive, and what those experiences have been like. Some groups have chosen to act within the union structure to look to change it. In their study of the experiences of Black women trade unionists, Kennedy-Macfoy, Gausi & King (2021: 515) note "the intersection of gender, race, social class, migration status, and age exposes Black women workers to the specific harms of racist, capitalist patriarchy". This goes back to the idea of moving away from a class-centred approach and identifying how the union narrative is framed in relation to the disadvantages experienced by particular groups, which start in many cases with histories of coloniality, extraction and exploitation.

In some respects, one could argue that the survival or "true" renewal of unions relies upon challenging the insider-outsider model and looking within to understand the oppressive dynamics that the class-based discussion perpetuates. It is fundamental to articulate strategies and initiatives for collective action that consider the socio-historical location and experiences of different groups. While there is value in drawing from existing evidence and developing grassroots insight into the potential ways in which organizing could be made more inclusive, the idea of renewal by diversity needs to be further developed through more understanding of intersectional appreciation of dynamics of oppression and privilege. The potential here is to develop coalitionist alliances rather than remain within traditional sectional commonalities; as noted by Smith and Smith (2015: 124), "the strongest politics are coalition politics that cover a broad range of issues. There is no way that one oppressed group is going to topple a system by itself. Forming principled coalitions around specific issues is very important."

CONCLUSION

Speaking to colleagues about my interest in intersectionality and how it can be used in the theory and praxis of unions, some contend that focusing upon difference undermines the idea of the collective. As a migrant woman of colour, inhabiting complex intersectional

2. In the UK, the United Voices of the World union is an example of the global phenomenon of unions of and for migrant workers.

differences that are constantly invoked by institutional structures, and wishing to be part of the union collective, this argument appears reductionist and rooted in logics of disconnected exclusivity, which could be called "White privilege". As previously noted, discussions about intersectionality have been challenged by Marxist feminists. In reality, the shared interest in fighting exploitation and achieving social justice makes both more attuned with each than some scholars are perhaps willing to recognize.

Labour politics within the neoliberal capitalist system cannot overlook the ways in which the intersections of social categories of difference create multiple oppressions for particular groups and that each configuration of inequality has its own set of social barriers. The Marxist position could be said to recognize that no form of oppression can be understood in isolation and those ideas, brought together, could be the way forward for a new generative dialogue about collective identity that brings to the fore mobilizing underpinned by solidarity in difference. More importantly, the recognition of these differences has more potential for connecting different movements and enhancing solidarity and collective activism. Allan and Robinson (2022: 602) note that in order for unions to rebuild power resources, they must understand internal differences in collective identities, noting that "future union success and solidarity amongst workers might be dependent on the ability of unions to recognize and negotiate multiple collective identities."

The union and labour movements seem to have understood difference as having to identify one encompassing category around which to build collective identity, perhaps forgetting that collective identity is "simultaneously a characteristic of collectivities and people" (Van Stekelenburg 2013: 219), so the cause that drives the collective identity is important insofar as individuals see it as relevant to their personal identity and sense of self. In particular, the focus on shared work grievances seems limiting to grievances experienced by particular groups that systematically undermine them in the workplace. A way forward in the twenty-first century is to develop a Marxist analysis of capitalism that engages with the idea of capitalism as intersectional, which means treating class, race, gender and sexuality as well as other social categories of difference as always intersecting and shaping dynamics of capitalist accumulation (see Bohrer 2018: 46). The importance of intersectionality as part of strategies of resistance and to achieve social justice, relies upon thinking more intentionally about bringing diversity and difference into action in ways that move beyond numerical representation (for example, how diverse union membership is). Instead, it is about a combination that looks to reformulate the types of questions asked about social justice, to think more specifically about how structures as well as mechanisms of voice operate intersectionally within unions.

An important structural challenge that speaks to conceptually problematizing the challenges unions are facing given the increasing presence of interest groups or caucuses, where particular groups of workers (e.g. racially minoritized, women, migrants, disabled, etc.) come together around their interests within union structures. While these groups do not challenge unions' overall principles of organizing and action, they do engage in targeted action that addresses specific concerns of the members of these groups (see, e.g. Parker & Douglas 2010; Parker & Foley 2010; Towle 2011). There is much potential to conceptually expand on what this means for union organizing and collective identity. While these groups create fragmentation within unions because they are normally formed "in opposition to a given union leadership" (Schirmer & Tarlau 2022: 7), given a more specific focus, they can bring workers together around specific social and political goals and devote their energy to developing political power to address their social justice vision that resonates strongly with group members. In this context, while it does not appear to be that sectionality is abandoned, intersectionality is brought in.

Finally, the scrutiny of unions as organizations is also important to understand how they adapt and evolve, especially how they engage with revitalization efforts (see Bamber, Jerrard & Clark 2022). We also need more understanding of what explains the way unions operate beyond the assumption that it is always about the collective struggle – for example, despite the popularity of agendas of gender equality, there is evidence of gender pay gap in union pay structures (see Aleks, Saksida & Kolahgar 2021; Cooper 2012). In this context, the traditional use of sectionality seems obsolete and there is clear potential for more inclusive and democratic organizing and mobilization through an intersectional approach that centres social categorical differences. Reflecting on the future of unions, Hyman (2002) notes the importance of union agendas reflecting the diverse ways in which work connects to life, or how workers themselves would like them to connect, for unions to appeal to a broader constituency. This can be accomplished, or at least started, by developing intersectional organizing.

REFERENCES

Alberti, G., J. Holgate & M. Tapia 2013. "Organising migrants as workers or as migrant workers? Intersectionality, trade unions and precarious work". *International Journal of Human Resource Management* 24(22): 4132–48.

Aleks, R., T. Saksida & S. Kolahgar 2021. "Practice what you preach: the gender pay gap in labor union compensation". *Industrial Relations* 60(4): 403–35.

Allan, K. & J. Robinson 2022. "Working towards a green job? Autoworkers, climate change and the role of collective identity in union renewal". *Journal of Industrial Relations* 64(4): 585–607.

Anthias, F. 2012. "Intersectional what? Social division, intersectionality, and levels of analysis". *Ethnicities* 13(1): 3–19.

Apitzsch, B. *et al.* 2022. "Labour market collectivism: new solidarities of highly skilled freelance workers in medicine, IT and the film industry". *Economic and Industrial Democracy* 44(4). https://doi.org/10.1177/0143831X221120534

Bacon, N. & K. Hoque 2015. "The influence of trade union disability champions on employer disability policy and practice". *Human Resource Management Journal* 25(2): 233–49.

Bairstow, S. 2007. "'There isn't supposed to be a speaker against!' Investigating tensions of 'safe space' and intra-group diversity for trade union lesbian and gay organization". *Gender, Work and Organization* 14(5): 393–408.

Bamber, G., M. Jerrard & P. Clark 2022. "How do trade unions manage themselves? A study of Australian unions' administrative practices". *Journal of Industrial Relations* 64(5): 623–44.

Bannerji, H. 2005. "Building from Marx: reflections on class and race". *Social Justice* 32(4): 144–60.

Blell, M., S. Liu & A. Verma 2023. "Working in unprecedented times: intersectionality and women of color in UK higher education in and beyond the pandemic". *Gender, Work and Organization* 30(2): 353–72.

Bohrer, A. 2018. "Intersectionality and Marxism: a critical historiography". *Historical Materialism* 26(2): 46–74.

Briskin, L. 1999. "Autonomy, diversity, and integration: union women's separate organizing in North America and Western Europe in the context of restructuring and globalization". *Women's Studies International Forum* 22(5): 543–54.

Bryson, V. 2021. *The Futures of Feminism*. Manchester: Manchester University Press.

Caraway, T. & M. Ford 2017. "Institutions and collective action in divided labour movements: evidence from Indonesia". *Journal of Industrial Relations* 59(4): 444–64.

Carbado, D. 2016. "Privilege". In *Everyday Women's and Gender Studies*, edited by A. Braithwaite & C. Orr, 141–6. New York: Routledge.

Cho, S., K. Crenshaw & L. McCall 2013. "Toward a field of intersectionality studies: theory, applications, and praxis". *Signs* 38(4): 785–810.

Colgan, F. & S. Ledwith 2000. "Diversity, identities and strategies of women trade union activists". *Gender, Work and Organization* 7(4): 242–57.

Collins, P. 1990. *Black Feminist Thought: Knowledge, Consciousness, and the Politics of Empowerment*. New York: Routledge.

Collins, P. 2001. "Like one of the family: race, ethnicity and the paradox of US national identity". *Ethnic & Racial Studies* 24: 3–28.

Collins, P. 2019. *Intersectionality as Critical Social Theory*. Durham, NC: Duke University Press.

Collins, P. & S. Bilge 2016. *Intersectionality*. Cambridge: Polity.

Cooper, R. 2012. "The gender gap in union leadership in Australia: a qualitative study". *Journal of Industrial Relations* 54(2): 131–46.

Crenshaw, K. 1989. "Demarginalizing the intersection of race and sex: a black feminist critique of antidiscrimination doctrine, feminist theory and antiracist politics". *University of Chicago Legal Forum* 1(8): 139–67.

Crenshaw, K. 1991. "Mapping the margins: intersectionality, identity politics, and violence against women of color". *Stanford Law Review* 43(6): 1241–99.

Darvas, S., G. Gotti & K. Sekut 2023. "Trade unions, collective bargaining, and income inequality: a transatlantic comparative analysis". In *The Future of Work: A Transatlantic Perspective on Challenges and Opportunities*, edited by Transatlantic Expert Group on the Future of Work, 159–87. Brussels: Bruegel AISBL, The German Marshall Fund of the United States and the European Union.

Dhamoon, R. 2011. "Considerations on mainstreaming intersectionality". *Political Research Quarterly* 64(1): 230–43.

Dickens, L. 1997. "Gender, race and employment equality in Britain: inadequate strategies and the role of industrial relations actors". *Industrial Relations Journal* 28(4): 282–91.

Dickens, L. 2000. "Collective bargaining and the promotion of gender equality at work: opportunities and challenges for trade unions". *Transfer: European Review of Labour and Research* 6(2): 193–208.

Ellison, E. & R. Langhout 2020. "Embodied relational praxis in intersectional organizing: developing intersectional solidarity". *Journal of Social Issues* 76(4): 949–70.

Ferree, M. 2009. "Inequality, intersectionality and the politics of discourse". In *The Discursive Politics of Gender Equality*, edited by E. Lombardo, P. Meier & M. Verloo, 86–104. London: Routledge.

Flesher Fominaya, C. 2010. "Collective identity in social movements: central concepts and debates". *Sociology Compass* 4(6): 393–404.

Foley, B. 2019. "Intersectionality: a Marxist critique". *New Labor Forum* 28(3): 10–13.

Gall, G. 2008. "Multi-unionism and the representation of sectional interests in British workplaces". *Industrielle Besiehungen/German Journal of Industrial Relations* 15(4): 356–75.

Gall, G. 2012. "The engineering construction strikes in Britain, 2009". *Capital & Class* 36(3): 411–43.

Greene, A.-M. & G. Kirton 2002. "Advancing gender equality: the role of women-only trade union education". *Gender, Work and Organization* 9(1): 39–59.

Hancock, A.-M. 2011. *Solidarity Politics for Millennials: A Guide to Ending the Oppression Olympics*. Basingstoke: Palgrave Macmillan.

Healy, G., H. Bradley & C. Forson 2011. "Intersectional sensibilities in analysing inequality regimes in public sector organizations". *Gender, Work and Organization* 18(5): 467–87.

Holgate, J. 2005. "Organizing migrant workers: a case study of working conditions and unionization at a sandwich factory in London". *Work, Employment and Society* 19(3): 463–80.

Horowitz, L. 2017. "'It shocks me, the place of women': intersectionality and mining companies retrogradation of indigenous women in New Caledonia". *Gender, Place and Culture* 24(10): 1419–40.

Humphrey, J. 2000. "Self-organization and trade union democracy". *Sociological Review* 48(2): 262–82.

Hyman, R. 1975. *Industrial Relations: A Marxist Introduction*. London: Palgrave Macmillan.

Hyman, R. 1979. "British trade unionism in the 70s". *Studies in Political Economy* 1(1): 93–112.

Hyman, R. 2002. "The future of unions". *Just Labour* 1: 7–15.

Jacobs, R. 2013. "The narrative integration of personal and collective identity in social movements". In *Narrative Impact: Social and Cognitive Foundations*, edited by M. Green, J. Strange & T. Brock, 205–28. Hove: Psychology Press.

Jansson, J. 2022. "Re-inventing the self: implications of trade union revitalization". *Economic and Industrial Democracy* 43(1): 450–68.

Jefferys, S. & N. Ouali 2007. "Trade unions and racism in London, Brussels and Paris public transport". *Industrial Relations Journal* 38(5): 406–22.

Kelly, J. 1998. *Rethinking Industrial Relations: Mobilisation, Collectivism and Long Waves*. New York: Routledge.

Kennedy-Macfoy, M., T. Gausi & C. King 2021. "When a movement moves within a movement: Black women's feminist activism within trade unions". *Gender and Development* 29(2/3): 513–28.

Kirton, G. & G. Healy 2004. "Shaping union and gender identities: a case study of women-only trade union courses". *British Journal of Industrial Relations* 42(2): 303–23.

Kirton, G. & G. Healy 2013. "Commitment and collective identity of long-term union participation: the case of women union leaders in the UK and USA". *Work, Employment and Society* 27(2): 195–212.

Ledwith, S. & F. Colgan 2002. "Tackling gender, diversity and trade union democracy: a worldwide project?" In *Gender, Diversity and Trade Unions: International perspectives*, edited by F. Colgan & S. Ledwith, 1–27. London: Routledge.

Ledwith, S. & J. Munakamwe 2015. "Gender, union leadership and collective bargaining: Brazil and South Africa". *Economic and Labour Relations Review* 26(3): 411–29.

Marks, M. 2000. "Transforming police organizations from within". *British Journal of Criminology* 40(4): 557–73.

Martínez Lucio, M. & R. Perrett 2009. "Meanings and dilemmas in community unionism: trade union community initiatives and black and minority ethnic groups in the UK". *Work, Employment and Society* 23(4): 693–710.

McBride, A. 2001. *Gender Democracy and Trade Unions*. Aldershot: Ashgate.

Melucci, A. 1995. "The process of collective identity". In *Social Movements and Culture*, edited by H. Johnston & B. Klandermans, 41–63. London: Routledge.

Mojab, S. & S. Carpenter 2019. "Marxism, feminism, and 'intersectionality'". *Journal of Labor and Society* 22(2): 275–82.

Moore, S., T. Wright & H. Conley 2012. *Addressing Discrimination in the Workplace on Multiple Grounds: The Experience of Trade Union Equality Reps*. London: ACAS.

Noon, M. & K. Hoque 2001. "Ethnic minorities and equal treatment: the impact of gender, equal opportunities policies and trade unions". *National Institute Economic Review* 176: 105–16.

O'Brien, M. 2021. "Why queer workers make good organisers". *Work, Employment and Society* 35(5): 819–36.

Office for National Statistics 2022. Employee earnings in the UK: 2022. Measures of employee earnings, using data from the Annual Survey for Hours and Earnings (ASHE). www.ons.gov.uk/employmentandlabourmarket/peopleinwork/earningsandworkinghours/bulletins/annualsurveyofhoursandearnings/2022

Parker, J. & J. Douglas 2010. "The role of women's groups in New Zealand, UK and Canadian trade unions in addressing intersectional interests". *International Journal of Comparative Labour Law and Industrial Relations* 26(3): 295–319.

Parker, J. & J. Foley 2010. "Progress on women's equality within UK and Canadian trade unions: do women's structures make a difference?" *Relations Industrielles/Industrial Relations* 65(2): 281–303.

Però, D. 2020. "Indie unions, organizing and labour renewal: learning from precarious migrant workers". *Work, Employment and Society* 34(5): 900–18.

Polletta, F. & J. Jasper 2001. "Collective identity and social movements". *Annual Review of Sociology* 27(1): 283–305.

Richards, J. & K. Sang 2016. "Trade unions as employment facilitators for disabled employees". *International Journal of Human Resource Management* 27(14): 1642–61.

Roberts, R. & N. Cullinane 2023. "Skilled maintenance trades under lean manufacturing: evidence from the car industry". *New Technology, Work and Employment* 38(1): 103–24.

Rodriguez, J., S. Johnstone & S. Procter 2017. "Regulation of work and employment: advances, tensions and future directions in research in international and comparative HRM". *International Journal of Human Resource Management* 28(21): 2957–82.

Rodriguez, J. *et al.* 2016. "The theory and praxis of intersectionality in work and organisations: where do we go from here?". *Gender, Work and Organization* 23(3): 201–22.

Rogalewski, A. 2022. "Trade unions challenges in organising Polish workers: a comparative case study of British and Swiss trade union strategies". *European Journal of Industrial Relations* 28(4): 385–404.

Salah, T. 2014. "Gender struggles: reflections on trans liberation, trade unionism, and the limits of solidarity". In *Trans Activism in Canada: A Reader*, edited by D. Irving & R. Raj, 149–68. Toronto: Canadian Scholars' Press.

Schirmer, E. & R. Tarlau 2022. "Never let a good crisis go to waste: labor organizing during COVID-19", *Journal of Labor and Society* 1(aop), 1–34.

Smith, B. & B. Smith 2015. "Across the kitchen table: a sister-to-sister dialogue". In *This Bridge Called My Back: Writings by Radical Women of Color*, edited by C. Moraga & G. Ansaldua, 111–26. Albany, NY: SUNY Press.

Smith, H. 2021. "The 'indie unions' and the UK labour movement: towards a community of practice". *Economic and Industrial Democracy* 43(3): 1369–90.

Sullivan, W. 2021. "Race, colonialism, resistance and denial". *International Union Rights* 28(3): 3–4.

Tapia, M. & G. Alberti 2019. "Unpacking the category of migrant workers in trade union research: a multi-level approach to migrant intersectionalities". *Work, Employment and Society* 33(2): 314–25.

Taylor, V. & N. Whittier 1992. "Collective identity in social movement communities: lesbian feminist mobilization". In *Frontiers in Social Movement Theory*, edited by A. Morris & C. Mueller, 349–65. New Haven, CT: Yale University Press.

Thomas, A. 2016. "Degrees of inclusion: free movement of labour and the unionization of migrant workers in the European Union". *Journal of Common Market Studies* 54(2): 408–25.

Thomas, A. 2020. "Cross-border labour markets and the role of trade unions in representing migrant workers' interests". *Journal of Industrial Relations* 62(2): 235–55.

Thomas, M. & S. Tufts 2020. "Blue solidarity: police unions, race and authoritarian populism in North America". *Work, Employment and Society* 34(1): 126–44.

Tormos, F. 2017. "Intersectional solidarity". *Politics, Groups, and Identities* 5(4): 707–20.

Towle, C. 2011. "Highlighting the B in LGBT: the experiences of one UK trade union". *Journal of Bisexuality* 11(2/3): 317–19.

Unger, D. 2020. "Which side are we on? Can labor support #BlackLivesMatter and police unions?". *New Labor Forum* 29(3): 28–37.

UCU 2022. *Postgraduate Researchers: Your Guide to the UCU Strikes*. London: University and College Union.

UN Women 2020. *Intersectional Feminism: What it Means and Why it Matters Right Now*. New York: United Nations Women.

Van Stekelenburg, J. 2013. "Collective identity". In *The Wiley-Blackwell Encyclopedia of Social and Political Movements*, edited by D. Snow *et al.*, 219–25. Malden, MA: Wiley-Blackwell.

Verloo, M. *et al.* 2012. "Putting intersectionality into practice in different configurations of equality architecture: Belgium and the Netherlands". *Social Politics* 19(4): 513–38.

Wajcman, J. 2000. "Feminism facing industrial relations in Britain". *British Journal of Industrial Relations* 38(2): 183–201.

Yuval-Davis, N. 2015. "Situated intersectionality and social inequality". *Raisons Politiques* 2: 91–100.

CHAPTER 15

THE RATIONALITY AND LIMITATIONS OF LABOUR UNION BUREAUCRACY

David Camfield

ABSTRACT

The idea that unions are bureaucratic organizations is an old one. However, there is no consensus on what bureaucracy is, why unions are so often bureaucratic, what the interests of full-time officials are and whether the union officialdom is inevitably a conservative influence within unions. This chapter provides an overview of efforts to answer these questions and attempts to advance thinking about the union officialdom. It considers the ideas of both academic writers and political thinkers committed to working-class movements. It begins by considering why workers create, maintain and join formal organizations and what generates bureaucracy in them. It then examines efforts to explain what bureaucracy is and why workers' organizations are so often bureaucratic, before turning to the much-debated question of whether the interests of full-time union officials are different from those of the workers they represent and whether the officialdom is always a conservative force within unions.

Keywords: Union organization; union democracy; union bureaucracy; challenges for unions

INTRODUCTION

Union activists and academic analysts of unions have long thought about unions as bureaucratic organizations. However, there is no consensus on what exactly bureaucracy is, why unions are so often bureaucratic, what the interests of full-time union officials – as the personnel of the bureaucracy – are and whether the union officialdom is inevitably a conservative influence within unions. This chapter provides an overview of efforts to answer these questions and seeks to advance thinking about the union officialdom. It considers answers proposed by both a range of academic writers and by political thinkers committed to working-class movements. It begins by considering why workers create, maintain and join formal organizations and where bureaucracy arises from in these circumstances. It then examines efforts to explain what bureaucracy is and why workers' organizations are so often bureaucratic before turning to the much-debated question of whether the interests of full-time union officials are different from those of the workers they represent and whether the officialdom is always a conservative force within unions.

FORMALITY AND ORGANIZATION

Bureaucracy is a feature of formal organizations (Beetham 1987). While workers have long created and sustained unions and other formal organizations of one kind or another, it is worth pausing to examine why this is the case before considering the phenomenon of bureaucracy itself. Clarity about why workers create, maintain and join formal organizations is helpful for the study of bureaucracy.

"Yet what force on earth is weaker than the feeble strength of one?" asks a line in venerable labour song, "Solidarity Forever". This rhetorical question of composer Ralph Chaplin conveys the idea – the most elementary insight of the working-class movement – that collective action creates possibilities for exercising power that are not available to workers who try to negotiate with their employers on a one-to-one basis. Workers create organizations in order to act collectively. These organizations can be informal, without codified practices and rules. The history of wage earners acting collectively to improve their pay and working conditions or influence the state precedes the establishment of even rudimentary unions. For example, in England:

> The various skilled trades involved in the shipbuilding industry provide an excellent example of effective industrial action in the absence of permanent trade union organisation. Situated in an expanding industry, although one subject to severe fluctuations, the shipwrights, caulkers, ropemakers, ships' carpenters and other skilled crafts had a tradition of strike activity stretching back to the end of the seventeenth century. (Stevenson 1992: 158)

Workers continue to organize informally in the twenty-first century. A noteworthy case is the situation in contemporary China. There, underground informal organizing is a pragmatic necessity for worker activists because of the character of the official unions of the All-China Federation of Trade Unions (ACFTU) – "an arm of the party-state rather than an organ that workers can use to fight for and defend their rights and interests" (Yu & Ruixue 2014: 53) – and the high level of state repression faced by people who organize for workers' rights independently of the ACFTU. However, informal organizing by workers has manifest limitations. First, it is difficult to sustain over time given the demands of paid and unpaid work on working-class people; when a participant ceases to participate in informal organizing, finding a replacement may be difficult. Second, informal organizing means that people who take on responsibilities or carry out actions are not subject to clear mechanisms of accountability to other workers.[1] Third, employers may be especially reluctant to negotiate with representatives of informally

1. For a classic feminist discussion of this issue, see Freeman (nd).

organized workers, who lack any legal standing and whose representativeness they may have reason to doubt. Fourth, by their nature, informal organizations are unable to exercise whatever rights the state confers on formal organizations of workers, so it is easier for states to control formal organizations that they have officially recognized than informal organizations, which tend to be more volatile and unpredictable and, thus, it is to the advantage of states to recognize workers' organizations and grant them at least some rights.

These limits have long created very strong pressures for workers concerned with defending and improving their pay and working conditions to create and maintain formal organizations of some kind, even if their structures are minimal and they are forced to operate in a clandestine manner. For example, in the UK, formal unions date back to the 1700s. Thompson (1968: 174n) notes: "Although the Combination Act was not passed until 1799, this only strengthened *existing* legislation against trade unions." Records of the earliest unions themselves are scarce but laws designed to criminalize unions testify to workers' efforts to create them. Once established, formal workers' organizations may coexist with informal organization – the Facebook groups that teachers and other education workers in the US have used successfully to organize strike action outside official union structures are one example of this dynamic (Weiner 2018), while underground networks of labour organizers in China are another. Across working-class history, though, formal organizations have tended to eclipse informal ones due to their greater visibility, stability, legitimacy and effectiveness. It is the existence of formal workers' organizations that sets the stage for considering the phenomenon of union bureaucracy.

WHAT IS BUREAUCRACY, AND WHY ARE WORKERS' ORGANIZATIONS USUALLY BUREAUCRATIC?

It is often taken for granted that unions and other workers' organizations are and must be bureaucratic, often without much clarity about what is meant by bureaucracy. This means that the fundamental question of why unions are bureaucratic is never posed. This matters for two reasons. First, to avoid sloppy thinking, it is important to clarify precisely what bureaucracy is. Second, our answers have consequences for how we assess contemporary unions and possibilities for progressive change. These questions have been discussed, with more or less clarity, for over a century. This section will review some noteworthy attempts to answer them.

Mainstream social science

In the late nineteenth century, academic social science began to provide answers to these questions. The contributions of Max Weber have played an influential role in the study of bureaucracy. For Weber, bureaucracy is a matter of full-time office-holders in hierarchical organizations who are paid in money to administer rules.

> Bureaucratic administration means fundamentally the exercise of control on the basis of knowledge. This is the feature of it which makes it specifically rational. This consists on the one hand in technical knowledge which, by itself, is sufficient to ensure it a position of extraordinary power. But in addition to this, bureaucratic organizations, or the holders of power who make use of them, have the tendency to increase their power still further by the knowledge growing out of experience in the service.
>
> (Weber 1952a: 26)

Bureaucracy's key features are hierarchy, continuity, impersonality and expertise (Beetham 1987: 12). In Weber's (1952a: 24) view, bureaucracy is, from a technical perspective, the most efficient form of organization:

> It is superior to any other form in precision, in stability, in the stringency of its discipline, and in its reliability, [so much so that] it would be sheer illusion to think for a moment that continuous administrative work can be carried out in any field except by means of officials working in offices ... bureaucratic administration is, other things being equal, always, from a formal, technical point of view, the most rational type.

The growth of administrative tasks, both quantitatively but above all in terms of a qualitative intensification and expansion of tasks, is what drives the bureaucratization of organizations (Weber 1952b: 66). The only way to weaken bureaucracy is through retrogressing to "small-scale organization" (Weber 1952a: 25). Although Weber paid very little attention to unions (Strasseri 2015), his theory of bureaucracy has often informed how unions are understood as bureaucratic organizations.

British academics and social reformers, Beatrice and Sidney Webb, were contemporaries of Weber. Unlike him, however, they showed a great deal of interest in unions and wrote on their history and contemporary organization at length. Much less given to theorizing than Weber, the Webbs treat union bureaucracy in a similar manner. They dub the earliest unions "free alike from permanently differentiated officials, executive council or representative assembly", as examples of "primitive democracy" (Webb & Webb 1965: 8). They focus on the rise of full-time officials. They suggest that it was the demands of conflict with employers and state repression more than the volume of administrative tasks that initially "led to a departure from this simple ideal" (Webb & Webb 1965: 9). But it is "administrative efficiency" and the need for expert specialists that drives the development of bureaucracy (Webb & Webb 1965: 59). The creation of the first full-time elected administrative positions "laid the foundation of a separate governing class" (Webb & Webb 1965: 15) in unions. As they carry out their duties, such officials develop administrative expertise. At the same time, they become well known among their unions' members, which bolsters their status and makes it more likely that they will be re-elected (Webb & Webb 1965: 16). The history of unions bears witness to "the long and inarticulate struggle of unlettered men to solve the problem of how to combine administrative efficiency with popular control" (Webb & Webb 1965: 58). In their view, the combination of permanent full-time staff with full-time elected representatives who oversee them is a reasonable solution (Webb & Webb 1965: 59–60).

German-Italian sociologist, Robert Michels, for a time a collaborator of Weber, shares a similar basic understanding of what bureaucracy is. His classic 1911 book *Political Parties* is also concerned with unions. Perusek (1995: 61) effectively summarizes its central argument:

> Stated briefly, Michels' thesis in *Political Parties* is this, Democracy is the self-rule of the mass of workers. This must be based on small, face-to-face groups. But modern society requires that workers combine in large organizations. Self-rule is impossible in these large organizations, for reasons technical, psychological, and material.

In Michels' (1959: 348) words, "The organic structure of the trade unions is based upon the same foundation as that of the political party of the workers, namely, the representation of the interests of the rank and file by individuals specially elected for that purpose." Even radical syndicalist unions are "unable to escape the oligarchical tendencies which arise in every organization" (Michels 1959: 355). The basis for these tendencies – Michels' so-called "iron law of oligarchy" – is that durable organization beyond a small scale requires "technical specialization" and thus "expert leadership". Thus "organization implies the tendency to oligarchy", the division between a minority of leaders and a mass of followers (Michels 1959: 31, 32). This oligarchy, while not identical to bureaucracy in Weber's sense, tends to be bureaucratic because of the role of technical knowledge in large organizations. On a deeper level, Michels (1959: 86–7) argues:

> The incompetence of the masses is almost universal throughout the domains of political life, and this constitutes the most solid foundation of the power of the leaders. The incompetence furnishes the leaders with a practical and to some extent with a moral justification. Since the rank and file are incapable of looking after their own interests, it is necessary that they should have experts to attend to their affairs.

As Barker (2001) suggests, this leads to explaining bureaucracy as a consequence of human nature. In fact, for Michels, according to Barker (2001: 25), "there is a twin inevitability, organisation engenders both oligarchy, the emergence of leaders who can prevent challenges to their rule, and conservatism, the diversion of democratic movements from their original goals".

Several major themes of Weber and Michels were embedded in the dominant intellectual tradition within Western mainstream social science during the years of the long boom that followed the end of the Second World War, of which the US structural-functionalist sociologist Talcott Parsons was a leading exponent. Here, though, the inevitability of bureaucracy was often linked to technological developments (Gouldner 1955). These themes are visible in influential academic work on unions published during that era. A notable example is *Union Democracy* by US social researchers Seymour Martin Lipset, Martin Trow and James Coleman (1956), who argue:

> Unions, like all other large-scale organizations, tend to develop a bureaucratic structure, that is, a system of rational (predictable) organization which is hierarchically organized. Bureaucracy is inherent in the sheer problem of administration, in the requirement that unions be "responsible" in their dealings with management (and responsible for their subordinate units), in the need to parallel the structures of business and government, in the desire of workers to eliminate management arbitrariness and caprice, and in the desire of the leaders of unions to reduce the hazards to their permanent tenure in office. The price of increased union bureaucracy is increased power at the top, decreased power among the ordinary members. (Lipset, Trow & Coleman 1956: 8)

Here the influence of Weber and Michels is easy to see. The foremost contemporary critic within US social science to the approach found in *Union Democracy* was Alvin Gouldner. In "Metaphysical pathos and the theory of bureaucracy", Gouldner (1955) argues that Weber fails to consider historical examples of large-scale endeavours that have been organized without substantial bureaucracy, citing the construction of the Egyptian pyramids. He also notes, "Weber never considers the possibility that it is not 'large size' as such that disposes to bureaucracy; large size may be important only because it generates other social forces which, in their turn, generate bureaucratic patterns" (Gouldner 1955: 500). Gouldner accuses Michels of ignoring the tendency of people to push for democracy, to which the tendency of organizations to become oligarchic is a response (Gouldner 1955: 506). Although not discussed with reference to unions, these considerations are relevant to the study of union bureaucracy.

Early radical answers

Outside universities and within the working-class movement itself, there emerged a very different tradition of analysis of union bureaucracy that was strongly critical of it. An early English-language expression of an emerging perspective can be found in *The Miners' Next Step*, a pamphlet issued in 1912 by a group of syndicalist-influenced Welsh socialist miners. It notes that there are leaders who are "'trade unionists by trade' and their profession demands certain privileges. The greatest of all these are plenary [decision-making] powers. Now every inroad the rank and file make on this privilege lessens the power and prestige of the leader" (Unofficial Reform Committee 1912). Somewhat similar ideas can be found in the writings of some other contemporary socialists, including Rosa Luxemburg (McIlroy 2014: 504–5).

Several Marxist thinkers developed explanations for why unions had become bureaucratic, going beyond observing symptoms and politically criticizing union officials and bureaucratic union practices. However, these generally did not focus on clarifying precisely what bureaucracy was. In Italy, Antonio Gramsci, analysing unions as organizations through which workers sell their labour power to capital and emphasizing the market character of this activity, observes that these efforts lead to bureaucracy:

> [Workers] have created this enormous apparatus of concentration of flesh and graft ... They have assumed from outside or they have generated from within a trusted administrative personnel, expert in this kind of speculation, up to the job of dominating the conditions of the market, capable of stipulating contracts, of assessing commercial vagaries, of initiating economically useful operations. (Gramsci 1919)

In the draft of an article left unfinished at the time of his assassination by an agent of the Stalinist rulers of the USSR in 1940, the Russian revolutionary, Leon Trotsky (1990), stressed the influence on unions of the capitalist state. He emphasizes that unions are "drawing close to and growing together with the state power" because of "social conditions common for all unions" (Trotsky 1990: 39).[2] He points to how, in an era of monopoly capitalism, unions are faced with centralized firms closely linked with state power, and claims that this compels unions to converge with states.

Belgian Marxist, Ernest Mandel, later proposed a very different explanation that linked bureaucracy to the ebbs and flows of class struggle in capitalist societies. Some of the minority of workers who keep union activity going through periods when the level of struggle is low, "together with middle-class intellectuals who have access to cultural skills from which the bulk of the working class is excluded, must take on responsibility for administering the unions ... created by periodic upsurges of mass activity" (in Post 1999: 123). Later, Mandel (1992: 60) argued that this creates the "risk that working-class organisations will themselves become divided between layers exercising different functions. Specialisation can result in a growing monopoly of knowledge, of centralised information. Knowledge is power, and a monopoly of it leads to power over people."

In the late 1950s, Greek-French theorist, Cornelius Castoriadis, writing as he moved from Marxism towards a social theory of his own but while he was still a revolutionary socialist, addressed the issue from a different angle. Castoriadis (1988: 201) asserts that working-class organizations have degenerated, "becom[ing] integrated into the system of exploitation". Part and parcel of this is "their bureaucratization ... the formation within them of a stratum of irremovable and uncontrollable leaders" (Castoriadis 1988: 201). This reproduces within the workers' movement what Castoriadis (1988: 201) views as "the fundamental social relationship of modern capitalism, the relationship between directors and executants". Thus, he links the bureaucratic character of unions to this broader trend in society. Within unions, the officials direct and worker activists execute decisions made on their behalf. The relationship between unions and the working class also reproduces the same deeper pattern he identifies, with workers usually following unions led by the stratum of officials (Castoriadis 1988: 201). At the same time, however, Castoriadis vigorously rejects the belief that mass organizations like unions must inevitably be bureaucratic. This idea "boils down to the assertion that the problem of

2. For brief critical observations about Trotsky's view of the condition of capitalism at the end of his life, see Molyneux (1981: 176–9).

centralization can be solved only by bureaucracy", which he claims reflects "the most deeply rooted of bourgeois prejudices" (Castoriadis 1988: 207). How to handle centralization is an inescapable challenge for workers' organizations. It can be addressed either with the "bourgeois-bureaucratic solution" of bureaucracy, assigning responsibility to a layer of full-time officials, or with the proletarian solution of "direct democracy and the election of recallable delegates" (Castoriadis 1988: 207).

Later radical answers

In the late 1970s, British socialist academic Richard Hyman started to rethink the understanding of union bureaucracy inherited from the Trotskyist group of which he had been a member, the International Socialists (later the Socialist Workers' Party). Hyman (1989: 158) argues:

> There is an important sense in which the problem of "bureaucracy" denotes not so much a distinct *stratum of personnel* as a *relationship* which permeates the whole practice of trade unionism. "Bureaucracy" is in large measure a question of the differential distribution of expertise and activism, of the *dependence* of the mass of union membership upon the initiative and strategic experience of a relatively small cadre of leadership – both "official" and "unofficial".

In another formulation, Hyman characterizes bureaucracy as a "corrosive pattern of *internal social relations*" in unions that may be encouraged by leaders but "can readily develop in the absence of such manipulative strategies" (Hyman 1989: 246). This tendency flows from the central activity of unionism: "collective bargaining undertaken by 'specialist' negotiators *on behalf* of the broader membership consolidates a representative hierarchy functionally oriented towards accommodation and compromise with capital and its agents" (Hyman 1989: 181). This conception reflects the influence of Gramsci's previously mentioned insight, while shifting the focus away from full-time union officials (bureaucrats) to bureaucracy understood as a relationship, but in a different way than Castoriadis understands it as a relationship.

A quarter of a century later, Sheila Cohen (2006: 162), a British socialist academic and activist with a background similar to Hyman, drew from her analysis of workplace struggles that:

> ... the beginnings of bureaucratisation are rooted not in explicit "social democratic" ideology or organisational structures as such, but in the central contradiction between spontaneous "sectional" militancy and the need to organise effectively around workplace issues by uniting the membership and thus, on many occasions, suppressing the specific interest of sectional groups.

However, the development of bureaucracy out of this contradiction in workplace union activity is not inevitable, for "a process of conscious recognition" can allow activists who understand that the tension is real to "maintain a (precarious) balance between effective organisation and direct democracy" (Cohen 2006: 163).

Dissatisfied with the lack of theoretical clarity and other weaknesses in previous radical discussions of union bureaucracy, I have attempted to develop a historical materialist understanding that starts with the foundational concept of bureaucracy itself, the meaning of which has often been murky. Bureaucracy, I argue, is "best understood as a mode of existence of social relations in which people's activity (labour) is organized through formal rules that limit their ability to determine its character and goals, and which they themselves are not able to alter with ease" (Camfield 2013: 138). Although bureaucracy existed in pre-capitalist societies, in capitalist societies it is generated by the separation of conception from execution in human activity at and beyond the paid workplace, as well as by the monopolization of intellectual labour by groups powerful enough to organize the life activity of other people with formal rules, including scientific and religious elites (Camfield 2013: 138–41).[3]

In unions, bureaucracy arises out of negotiating the sale of workers' labour power to employers. Workers tend to lock in the gains collectively bargained with employers in formal contracts; situations in which workers' bargaining power is so strong that the times they prefer not to do so are rare and short-lived. For their part, employers try to embed restrictions on strikes and their control over the labour process into contracts. When they succeed, these formal rules are bureaucratic. Bureaucracy also develops in unions in ways that reproduce the tendency to separate conception (planning) from execution (doing) that is pervasive under capitalism after its earliest phases, although this is challenged when workers' self-activity, in which workers take matters into their own hands in directly democratic ways, generates an anti-bureaucratic counter-tendency.

Another cause of union bureaucracy is the way in which states use laws and regulations to force union activity into channels moulded by their constricting formal rules. Full-time union officials also tend to bureaucratize union activity, seeking to shape it in ways that reflect their interests as a distinct social layer whose existence as such depends on the stability of unions as institutions. Thus, they tend to favour their own control over union activity (Camfield 2013: 142–9). This approach is an effort to clarify the meaning of bureaucracy itself and to explain why workers' organizations are so often bureaucratic without relying upon assumptions, like those of Weber and Michels, which assume the inevitability and superiority of bureaucracy. It distinguishes bureaucracy as a way of organizing people's activity from the union officialdom as a social layer. This facilitates the analysis of contemporary unions as bureaucratic organizations headed by a bureaucratic officialdom. It also represents an alternative to Hyman's approach, which draws attention to unions as bureaucratic organizations at the expense of also considering the role of the officialdom, and to approaches that focus on full-time officials and have little or nothing to say about bureaucracy as a pervasive feature of unionism at all levels.

3. Bray (2019: 58) argues that, as capitalism develops, the separation of conception from execution where, "the division between mental and manual labour ... becomes increasingly central, polarized, and pervasive ... increasingly divides the entire social field, as the majority of workers have the conception and control of their labour processes stripped from them, while a growing but still-minority fraction of labourers monopolizes those powers, performing the functions of capital".

WORKERS AND BUREAUCRACY: IN WHOSE INTERESTS?

In the study of unions, most discussion of bureaucracy has not been about what union bureaucracy is as a social phenomenon or about its causes. Rather, it has been about the politics of union bureaucrats – usually meaning full-time union officials. The central questions here have been whether the interests of full-time officers and staff differ from those union members they purport to represent as a whole and whether the officialdom is a conservative influence on unions and workers' struggles.[4] This is a live issue among union activists and in labour studies research, and has been so since the late nineteenth century.

Mainstream social science

A classical pessimistic perspective from sociology is provided by Michels. Since he believes that unions' "organic structure ... is based upon the same foundation as that of the political party of the workers" (Michels 1959: 348), his beliefs that "power is always conservative" and "[a]s the organization increases in size, the struggle for great principles becomes impossible" (Michels 1959: 366) clearly apply to unions. Unions as bureaucratic organizations run by full-time officials concerned to preserve their positions and the funds required to pay their salaries are threatened by "bold and enterprising tactic[s] ... For these reasons the idea of such a tactic becomes more and more distasteful" (Michels 1959: 373). Union officials can be expected, in a self-interested way, to oppose anything that jeopardizes the future of the organization upon which they depend in order to continue as officials. "Thus, from a means, organization becomes an end ... Henceforward the sole preoccupation is to avoid anything which may clog the machinery" (Michels 1959: 373).

For their part, the Webbs, having spent a great deal of attention studying the history of early British unions, conclude, as Perusek (1995: 59) puts it in a discussion of their views, that full-time officials tend to be conservative because their "*social role*" as the unions' "guardians of growing financial affairs" affect "their vision of strategy and tactics".

Early radical answers

Writing a few years before Michels, Polish-German revolutionary socialist, Rosa Luxemburg, worried about the "bureaucratism and a certain narrowness of outlook" among German union officials, which she saw as the result of their specialized union activity as well as a period of capitalism in which the level of class struggle was low. She comments on what Michels would soon write about, an "over-valuation of the organisation, which from a means has gradually been changed into an end in itself, a precious

4. Throughout this chapter I avoid using the term "bureaucrat" when discussing union officials (elected or appointed, officers or staff) for the sake of encouraging reflection on bureaucracy as a social phenomenon that should not be conflated with them. I refer to union officials collectively as the officialdom rather than "the bureaucracy".

thing, to which the interests of the struggles should be subordinated" (Luxemburg 1986: 87).[5] Luxemburg also notes officials' aversion to actions seen as risky for union stability (Luxemburg 1986: 87). This amounts to the identification of certain conservative tendencies that lead officials to act in ways that are not, in Luxemburg's view, in the interests of the working class. However, it is not yet a developed theory of the union officialdom.

In his analysis of imperialism prompted by the First World War and support for their respective states in the war by most of the leaders of the European socialist movement, Russian revolutionary socialist, Vladimir Lenin (1965: 10), proposed that imperialism had created "a stratum of bourgeoisified workers, or the 'labour aristocracy'". These he calls "the labour lieutenants of the capitalist class" (Lenin 1965: 10). This layer is created by the capitalist class, he argues, using super-profits extracted from colonies. With these profits, "it is *possible to bribe* the labour leaders and the upper stratum of the labour aristocracy" (Lenin 1965: 9). Lenin has no doubt that the officialdom is inherently a counter-revolutionary force. This sweeping criticism encompasses full-time union officials without clearly distinguishing them from high-paid workers in imperialist countries. It does not analyse any specific activities or characteristics of the union officialdom.[6]

For his part, Trotsky's political assessment of the union officialdom was as uncompromising as Lenin's. His analysis of it is similar, being almost as imprecise, although he pays somewhat more attention to the officialdom as distinct from the labour aristocracy. Writing shortly before his assassination, he mentions that a labour aristocracy and an officialdom exist in colonial and semi-colonial countries too. In keeping with his analysis of how unions are converging with states in an age of monopoly capitalism, he attributes the existence of a union officialdom in such countries to government sponsorship (Trotsky 1990: 40).

Later radical answers

Analyses by radicals of the character of union officialdom proliferated during the years of the long economic boom that followed the end of the Second World War. In this era unions enjoyed unprecedented growth and institutional stability in advanced capitalist countries, stimulating radicals to pay more attention to trying to understand labour union officialdom. In the US, an influential contribution early in this period was made by C. Wright Mills. Mills was educated and employed as a sociologist but was, at the time he wrote *The New Men of Power, America's Labor Leaders* (1948), also an unaffiliated radical influenced by both Weber and anti-Stalinist Marxism. Mills argues that union leaders mobilize workers against corporations and the state during union

5. In the first decade of the twentieth century Michels was for several years an active member of the Socialist Party of Italy and the Social Democratic Party (SDP) in Germany, as a radical supporter of syndicalism. In 1907–08, he abandoned socialism and embraced elite theory, eventually becoming a fascist (Beetham 1977). During the years when Michels was involved with the SDP, Luxemburg was a leader of the party's radical wing.

6. For an excellent critique of theories of the labour aristocracy, see Post (2010).

organizing campaigns. "Yet even as the labor leader rebels, he holds back rebellion. He organizes discontent and then he sits on it, exploiting it in order to maintain a continuous organization; the labor leader is a manager of discontent. He makes regular what might otherwise be disruptive" (Mills 2001: 9). Mills's explanation for this evocative description flows from his perception of "the partial integration of company and union bureaucracies" (Mills 2001: 224) at the level of the workplace and the firm. Union officials – the stability of whose positions, it should be noted, had been reinforced by the union rights won in the US in the 1930s – strove to cooperate with managers "in making and administering company rules" (Mills 2001: 224). In exchange for this role:

> ... something must be given. The integration of union with plant means that the union takes over much of the company's personnel work, becoming the disciplining agent of the rank and file within the rules set up by the joint committee. The union bureaucracy stands between the company bureaucracy and the rank and file of the workers, operating as a shock absorber for both ... Responsibility is held for the contract signed by the company to uphold this contract, the union must often exert pressure upon the workers. (Mills 2001: 224)

Mills extends his assessment of union officials as a brake on workers' struggles at the point of production beyond full-time union officials to include the lowest level of those holding some kind of union position, namely shop stewards: "The rank-and-file leaders of the union, the shop stewards, operating as whips within the plant, become rank-and-file bureaucrats of the labour leadership ... primarily answerable to the labor union hierarchy, rather than to the rank and file who elect them" (Mills 2001: 224).

The late 1960s and the 1970s saw high levels of workers' collective action, including unauthorized strikes and other forms of rebellion by rank-and-file workers that brought them into conflict with union officials. This fed the growth of socialist organizations that saw workplace struggles as politically important, and led to efforts to develop Marxist analyses of the union officialdom more sophisticated than those inherited from earlier classical Marxism. The early writings of Hyman, some of whose later rethinking was discussed previously, was an important contribution in the UK. In the course of a Marxist study of industrial relations, Hyman (1975: 91) suggests that the reason that officials are committed to the defence of "existing bargaining arrangements and the terms of existing collective agreements" against workers' rebellion as well as against managerial violations is that their function as a group is "the negotiation and renegotiation of order within constraints set by a capitalist economy and a capitalist state". While managing discontent, with Hyman (1975: 195) citing Mills on this issue, officials tend "to allow their role as advocate of the members' own express interests to yield precedence to their role as guardians of organisational interests (that is, as a means of transmitting the pressures of external power on to their own members)".

In the US, historian Robert Brenner offered a Marxist analysis of the officialdom, drawing on discussions among socialist activists in the 1970s. Brenner (1985: 44) argues that union officials (along with social democratic politicians and middle-class leaders of communities of oppressed people) are "*a distinct social layer* with distinctive interests quite different from those of the mass of the working class". The basis for this differentiation is that, as full-time officials, their conditions of existence are not the same as those of the workers they represent. The officialdom is "not *directly affected*" (Brenner 1985: 45) by employer attacks on workers' wages and working conditions. To preserve their existence as officials, they do not have to organize collectively in the way that workers do. Instead,

> The organization – and the bureaucratic group which founds itself upon it – not only provides the officials with their means of support, thereby freeing them from the drudgery of manual labour and the shop floor. It constitutes for them a whole way of life – their day-to-day function, formative social relationships with peers and superiors on the organizational ladder, a potential career, and, on many occasions, a social meaning, a raison d'etre. To maintain themselves as they are, the whole layer of officials must, first and above all, maintain their organizations. It is thus easy to understand how an irresistible tendency emerges ... to treat their organizations as ends in themselves, rather than as the means to defend their memberships.
>
> (Brenner 1985: 45)

The politics that correspond to this social layer's interests are, according to Brenner, reformist. They seek to win reforms within capitalist society, avoiding militant, highly disruptive forms of collective action in favour of routine collective bargaining and electoral politics. Officials act in this way to avoid threatening the union institutions on whose existence they depend (Brenner 1985: 45–6); they seek to be accepted by capital and "to free themselves from dependence on their working-class base" (Brenner 1985: 44).

By the end of the 1970s, Hyman's perspective had changed in a significant way. The way in which he proposed the different understanding of what union bureaucracy is, discussed earlier – bureaucracy as a certain kind of relationship of dependence of the led upon leaders that existed across unions – is accompanied by a criticism of "the notion of 'trade union bureaucracy'" as a "descriptive category or derogatory slogan rather than an analytical concept adequately embedded in a serious theory of trade unionism" (Hyman 1989: 149). This can be used to "present trade union officialdom as scapegoats for contradictions inherent in trade unionism as such" (Hyman 1989: 149). At the same time, Hyman maintains that people who function as union representatives on an ongoing basis are important mediators between workers, employers and the state and tend to reproduce "an accommodative and subaltern relationship" to the latter two forces, which try to reduce workers' disruption of capital accumulation (Hyman 1989: 149–50). Drawing attention to the spread of shop stewards engaged in full-time or almost full-time union work in the 1970s, he questions a paradigm that depicts unions in terms of a clash between "the union bureaucracy" and "the rank and file".

A much further-reaching criticism from within the radical left of ideas about the bureaucratic conservatism of union officials was presented a decade later by John Kelly, another British industrial relations researcher. After a survey of socialist arguments about union officials, Kelly argues that:

> ... *in general*, there is no convincing evidence that union officials, on the whole, are more "conservative" than their memberships ... Clearly there are "conservative" officials and militant workers, but there are also militant officials and conservative workers, and the precise balance between these groups is likely to vary with circumstances. (Kelly 1988: 182)

Kelly (1988: 183) does acknowledge that collective bargaining in capitalist society has a conservative aspect, "reproducing the legitimacy of two parties in industry and their more or less peaceful co-existence", and that union full-timers are preoccupied with this activity.

More recently, other British socialist academics have weighed in to these discussions. Darlington and Upchurch (2012: 79) question Hyman's rethinking, suggesting that in response to a crude account of the relationship between unionized workers and the union officialdom he "arguably 'threw the baby out with the bathwater'". Darlington and Upchurch (2012: 80) maintain that it is the combination of full-time officials' activity in bargaining and "other important *sociological* and *political* factors ... such as their specific social role as intermediary and mediator between capital and labour, their substantial material benefits, and their political attachment to social democracy" that makes them "a distinct social stratum with interests different from, and sometimes in antagonism to, their rank-and-file members".

In contrast, McIlroy, influenced by Hyman's rethinking of bureaucracy (McIlroy 2014: 500), argues from a Marxist perspective, "Distrust of leaders is a healthy instinct. But trade unionism is the overriding, animating issue, not any distinctive interests of its staff." At the same time, he recognizes that the officialdom's "insulation from members, tendencies to autonomy and oligarchy, their power and salaries, demand reform and democratisation" (McIlroy 2014: 518). Full-time officials are "agents with appreciable if limited leeway", with particular "structural location, functions and interests" (McIlroy 2014: 501) in bureaucratic union organizations in capitalist societies.

Union officialdom, different interests and a conservative force?

The questions of whether the interests of union officials as a group are different from those of the members they represent as a whole, and whether they exert a conservative influence on workers' struggles, are more controversial, more politically charged than the questions about bureaucracy explored in the previous section of this chapter. As these are distinct questions, they deserve to be discussed separately. My answers aim to build upon the insights of previous contributions to an important debate in the study of unions that has largely died out or become stale.

In order to address the question of the officialdom's interests, it is first necessary to touch on the issue of needs. A strong case can be made that human beings have objective needs or life requirements, which must be distinguished from mere preferences. Life requirements are things that, when humans are deprived of them, "results in harm to our organism or our humanity" (Noonan 2012: 47). Canadian philosopher, Jeff Noonan (2012), argues that humans have physical, socio-cultural and temporal needs; the ways of realizing those needs are many and varied. Similarly, humans flourish when we realize our capacities to feel, think and act creatively, which as social beings we can only do in society.

That humans in general have these needs does not, however, mean that all groups of people have identical interests. Much social science treats interests as subjective – people's interests are just whatever it is that they prefer, as revealed by their actions. Nevertheless, there are also compelling arguments that objective interests exist. One such account is offered by British Marxist social theorist, Alex Callinicos, building on the work of the sociologist Anthony Giddens. This proposes that objective interests are connected to people's ability to achieve what they want, which, Callinicos (1987: 132) argues, depends on where they are located within their society's class relations. "The class struggle," he argues, "is, in a sense, the process through which agents discover their interests by exploring the extent of their powers." Revising this from a reconstructed historical materialist perspective, people's interests can be seen as contingent upon where they find themselves within the interlocking matrix of social relations (not limited to class) of which their society is composed.[7] These social relations shape whether or how individuals are able to attain their life requirements.

If we accept this or a similar theory of interests, we can develop an account of the common interests of persons who belong to the working class in capitalist societies. Within the framework of a historical materialist perspective of how capitalism operates, such persons' interests lie in countering the "separation, division and atomisation" (Lebowitz 2003: 99) that it is in capital's interest to increase among and between them. As Canadian economist, Michael Lebowitz, contends (2003: 99), "Only by reducing the degree of separation ... only through combination and unity can wage-labourers capture the fruits of cooperation *for themselves* and realise their 'own need for development.'" This involves collective struggle against capital in the labour market, in the labour process and as the owner of what workers produce (Lebowitz 2003: 89–99). Solidarity – workers acting to support each other in ways that reduce the separation of which Lebowitz writes, at whatever scale – is the key at all times and in all places. The significance of unions, therefore, is as a means (but not the only one) for members of the working class to conduct their struggles.

Are the interests of full-time union officials the same as those of workers? The first part of an answer to this question must be to appreciate that the union officialdom does constitute a social layer distinct from the working class as a whole. Although most officials sell their ability to work in exchange for wages and do not have significant managerial authority – this is reserved for the higher-level officials to whom they report – their position within class relations is unique. The officialdom as a whole, from the presidents of central union bodies down to full-time union representatives and specialists such as researchers and lawyers, exists as a mediator between workers on the one hand and employers and the state on the other. It negotiates the terms and conditions of the relationship between wage workers and their employers, whether private or public sector. As such, its members' interests cannot be identical to those of

7. On reconstructed historical materialism and societies as interlocking matrices of social relations, see Camfield (2016) or, for a more accessible presentation of this approach, see Camfield (2017).

the working class out of which it arose historically and from whose ranks it is chiefly replenished.

As to what the officialdom's interests are, a point of Brenner's is crucial: union officials can only continue to subsist as officials if union institutions can continue to operate. Unlike unionized workers, full-time union officials can only function as such if unions can pay their salaries. Consequently, it is in their interests to oppose the development of a situation in which the ability of the union to do so is jeopardized. If this is true in general, the constraints that shape how particular groups of union officials carry out their task of mediating workers' relations with employers and state power vary widely. These constraints influence the interests of a given group of officials in important ways. Crucial here is how a state's bureaucratic legal and regulatory framework affects what officials are allowed to do and what penalties are imposed for transgressions of that pattern of political administration.[8]

For example, in Canada, the right to strike is subject to very tight statutory restrictions, only unionized workers may strike, some are denied the right to strike at all, and those who do have the right are permitted to withdraw their labour only after their collective agreement has expired and in the course of negotiating a new one (which is not dissimilar to Germany). All other strikes are illegal. If a labour board finds that a strike is illegal, unions and union officials may be fined. Courts can find union officials in civil or criminal contempt for non-compliance with labour board rulings or court orders, with imprisonment a possible penalty (Fudge & Glasbeek 1992). Governments in Canada also have the authority to end strikes with back-to-work legislation, which generally contains steep penalties for non-compliance, and have not been shy to exercise it (Smith 2019). In the UK, there is no right to strike as such, merely the legal privilege of unions not to be sued for loss of business by an employer directly affected by a strike and subject to various restrictions. The key restriction is that strikes are lawful only if they are in contemplation of the furtherance of a trade dispute; that is, connected to a limited range of employment-related issues. Employers can seek court injunctions against unions for unlawful strikes. Failure to comply with an injunction is contempt of court, and courts can levy fines and sequester union assets as punishment (Gall 2017).

The other side of the pressures operating upon the officialdom is the extent to which union members are able to direct or at least influence their official representatives. The officialdom's everyday social conditions are different from those of union members. However, if the number of workers paying union dues falls enough that it seriously affects the union as an institution, this threatens the officialdom. Worker dissatisfaction with their union is one possible cause of a decline in dues-payers. To the extent that worker dissatisfaction will affect union finances, it is in the interests of officials to mitigate this sentiment (where labour laws require dues payment from all workers represented by a union, members or not, the officialdom is insulated from this to some extent – this remains the case, for example, in all Canadian jurisdictions). There are many ways officials can try to do this, from leading militant action to urging workers to accept that the union is earnestly attempting to represent them but that forces beyond its control mean that the results of union efforts can only be meagre. Individual officials

8. On the concept of political administration, see Neocleous (1996).

can certainly act in ways that contravene the collective interests of the officialdom as a social layer; for example, by encouraging workers to strike even though the state is likely to impose harsh fines on the union as a consequence. However, for a union's officials to act together in ways that put the union's institutional survival at risk requires an extremely high level of ideological commitment to that course of action and/or a lot of pressure from the workers they represent. The refusal of Arthur Scargill and the leadership of the National Union of Mineworkers (NUM) in the UK to back down in the face of a court ruling that declared NUM strike action unlawful during the union's epic 1984–85 struggle, even though this resolve led to the sequestration of the NUM's assets (Milne 2014), stands out as one of the rare examples of officials persevering with militancy in spite of the severe consequences for their organization.

The preceding discussion has analysed the officialdom in terms of people who work full-time on union activity, whether as elected or appointed officers or as union staff who are hired (or, occasionally, elected) to work for a union. The larger the union, the greater the amount of activity is required to represent the membership. When unions are large bureaucratic organizations, they are also more likely to employ staff who are specialists in one area of activity or another (for example, in health and safety issues, law or communications). We must also recognize that there are many workers who carry out union activity, hold union office as shop stewards or branch/local executive members and are allowed a certain amount of time away from their regular jobs to conduct union business (what in UK union nomenclature is called "facility time"). Managers are paid to manage on a full-time basis, while the shop stewards and other union activists who deal with them have to do their regular jobs as well as union work. For this reason, when union activity requires lots of meetings with managers to represent workers and other time-consuming tasks, union officials will often negotiate arrangements that allow worker activists to take time away from their jobs for union business. The more time such workers spend away from their jobs on union activity, the more their everyday lives become similar to those of full-time officials and they become subject to the pressures that affect full-timers; conversely, the more time they spend in their regular jobs, the more their conditions are similar to those of their co-workers and they are under rank-and-file influence.

Do its distinctive interests make the officialdom an inevitably conservative force, even though few actually make this claim today? Here the meaning of "conservative" calls out for clarification. If it is understood to mean being opposed to militancy, the claim is clearly wrong; there is plenty of historical evidence of union officials encouraging members to act collectively and lead militant collective action. For example, in the UK, Sharon Graham, General Secretary of the Unite union, is promoting sector-wide combines of shop stewards, workplace organizing and coordinated strike action to strengthen workers in relation to employers and government. Across the Atlantic, the executive of the Canadian Union of Postal Workers has adopted a plan to train many members in workplace organizing, with the aim of building the power needed to affect conditions on the job and deter government interference in collective bargaining with Canada Post (Canadian Union of Postal Workers 2022).

In both these cases, as in many similar ones, the efforts of certain top officials to encourage militancy have been welcomed by many activists but have been much less well received by other officials of the union in question. If it means being opposed to law-defying militant action, the claim that the officialdom is inevitably conservative has more credibility, but examples of union officials engaging in such action are still not difficult to find even if they are uncommon. In addition to the Scargill-led NUM, one can cite, to give Canadian examples, defiance of back-to-work legislation by the Hospital Employees' Union in 2004, a 2005 British Columbia Teachers' Federation strike deemed illegal from the start, and some sympathy strike action to support these strikes that was, in reality, led by Left officials (Camfield 2006, 2009), as well as a law-defying strike in 2022 by education workers in Ontario (Lukacs & Paling 2022). If conservative refers to a political commitment to capitalism, there are also many examples of unions whose leaders have been pledged to anti-capitalist politics.

The weight of historical evidence does, though, support a softer claim. Because of their interests as a distinct social layer, union officials, regardless of their politics, have a strong incentive to oppose anything that threatens their ability to continue to function as union officials. They will tend to oppose an escalation of workers' struggle that puts the union as an institution, upon which they depend for their ability to operate as union officials, at serious risk of being unable to function (for example, because of lack of funds or the refusal of employers infuriated by union-sanctioned workers' militancy to deal with officials) or even of collapse. In other words, they tend towards institutional conservatism. However, the extent to which this is true in the contemporary world does vary. It is affected by how institutionally secure a union is, how bureaucratic it is, the character of state regulation of union activity, how intense class struggle is in a given context, and by union ideology. Moreover, institutional conservatism can coexist with some kinds of militancy and political radicalism (Camfield 2013).

CONCLUSION

The principal aim of this chapter has been to provide an overview of major contributions to debates about two interconnected aspects of unions as workers' formal organizations: first, the nature of union bureaucracy as a social phenomenon and, second, the politics of the labour officialdom. We have seen that the dominant approach within academic social science considers, in the tradition of Weber, that bureaucratic ways of organizing are seen to be superior by virtue of their superior administrative efficiency and that this technical superiority explains why unions became bureaucratic as they grew in size. To this, Michels adds an emphasis upon the inherent inability of workers to run their own affairs. Parallel to this approach have been numerous radical perspectives on union bureaucracy. Early radical analysts point to the dynamics of negotiating the sale of labour power, the influence of capitalist states on unions, the rhythms of class struggle, the lack of access of most workers to cultural skills, and how capitalism organizes societies in ways that separate conception and execution as sources of union bureaucracy. Later radicals have drawn on such ideas to develop more sophisticated explanations, paying more attention to clarifying what exactly union bureaucracy is.

In the analysis of what is often called "the bureaucracy" – the union officialdom – Michels proposes an early influential account of their self-interested conservatism. Outside the academy, Lenin and Trotsky offer crude and imprecise radical theories and Luxemburg a few specific observations. Radical analysis of union officialdom developed further in the decades after the Second World War, giving attention to officials' role in collective bargaining under pressure from employers and state and to the dependence of the officialdom on the institutional preservation of union institutions. While some radicals have argued that a focus on officials as a distinct group is an unfortunate distraction from analysing the contradictions of unionism itself in capitalist society and/or that the officialdom is not inherently conservative, others maintain that these concerns remain intellectually and politically significant even if a theory of the officialdom must be refined. As long as the level of workplace collective action by unionized workers remains low and it is rare to see conflict between militant workers and union officials of the kind that was much more common in many countries in the 1960s and 1970s, we are unlikely to see new answers to questions about the interests of the union officialdom and bureaucratic conservatism that will substantially advance understanding of these issues.

REFERENCES

Barker, C. 2001. "Robert Michels and the 'cruel game'". In *Leadership and Social Movements*, edited by C. Barker, A. Johnson & M. Lavalette, 24–43. Manchester: Manchester University Press.

Beetham, D. 1977. "From socialism to fascism, the relation between theory and practice in the work of Robert Michels: from Marxist revolutionary to political sociologist". *Political Studies* 25(1): 3–24.

Beetham, D. 1987. *Bureaucracy*. Minneapolis, MN: University of Minnesota Press.

Bray, M. 2019. *Powers of the Mind, Mental and Manual Labour in the Contemporary Political Crisis*. Bielefeld: Transcript.

Brenner, R. 1985. "The paradox of social democracy: the American case". In *The Year Left: An American Socialist Yearbook*, edited by M. Davis, F. Pfeil & M. Sprinker, 32–86. London: Verso.

Callinicos, A. 1987. *Making History, Agency, Structure and Change in Social Theory*. Cambridge: Polity.

Camfield, D. 2006. "Neoliberalism and working-class resistance in British Columbia, the hospital employees' union struggle, 2002–2004". *Labour/Le Travail* 57: 9–41.

Camfield, D. 2009. "Sympathy for the teacher, labour law and transgressive workers' collective action in British Columbia, 2005". *Capital & Class* 99: 81–107.

Camfield, D. 2013. "What is trade union bureaucracy? A theoretical account". *Alternate Routes* 24: 133–56.

Camfield, D. 2016. "Theoretical foundations of an anti-racist queer feminist historical materialism". *Critical Sociology* 42(2): 289–306.

Camfield, D. 2017. *We Can Do Better: Ideas for Changing Society*. Winnipeg: Fernwood.

Canadian Union of Postal Workers 2022. "Empowering CUPW members to fight forward". www.cupw.ca/en/empowering-cupw-members-fight-forward

Castoriadis, C. 1988. "Proletarian and organization, I". In *From the Workers' Struggle Against Bureaucracy to Revolution in the Age of Modern Capitalism*, edited by D. Curtis, 193–222. Vol. 2 of *Cornelius Castoriadis Political and Social Writings*. Minneapolis, MN: University of Minnesota Press.

Cohen, S. 2006. *Ramparts of Resistance: Why Workers Lost Their Power and How to Get it Back*. London: Pluto.

Darlington, R. & M. Upchurch 2012. "A reappraisal of the rank-and-file versus bureaucracy debate". *Capital & Class* 36(1): 77–95.

Freeman, J. n.d. "The tyranny of structurelessness". https://j0ofreeman.com/joreen/tyranny.htm

Fudge, J. & H. Glasbeek 1992. "Alberta nurses v. a contemptuous Supreme Court of Canada". *Constitutional Forum* 4(1): 1–5.

Gall, G. 2017. "Injunctions as a legal weapon in collective industrial disputes in Britain, 2005–2014". *British Journal of Industrial Relations* 55(1): 187–214.

Gouldner, A. 1955. "Metaphysical pathos and the theory of bureaucracy". *American Political Science Review* 49(2): 496–507.

Gramsci, A. 1919. "Unions and councils". www.marxists.org/archive/gramsci/1919/10/unions-councils.htm

Hyman, R. 1975. *Industrial Relations: A Marxist Introduction*. London: Macmillan.

Hyman, R. 1989. *The Political Economy of Industrial Relations: Theory and Practice in a Cold Climate*. London: Macmillan.

Kelly, J. 1988. *Trade Unions and Socialist Politics*. London: Verso.

Lebowitz, M. 2003. *Beyond Capital, Marx's Political Economy of the Working Class*. Second edn. Basingstoke: Palgrave Macmillan.

Lenin, V. 1965. *Imperialism, the Highest Stage of Capitalism*. Peking: Foreign Languages Press.

Lipset, S., M. Trow & J. Coleman 1956. *Union Democracy: The Internal Politics of the International Typographical Union*. New York: Anchor.

Lukacs, M. & E. Paling 2022. "The inside story of how education workers beat back Doug Ford". The Breach. https://breachmedia.ca/the-inside-story-of-how-education-workers-beat-back-doug-ford/

Luxemburg, R. 1986. *The Mass Strike*. London: Bookmarks.

Mandel, E. 1992. *Power and Money: A Marxist Theory of Bureaucracy*. London: Verso.

McIlroy, J. 2014. "Marxism and the trade unions: the bureaucracy versus the rank and file debate revisited". *Critique* 42(4): 497–526.

Michels, R. 1959. *Political Parties: A Sociological Study of the Oligarchical Tendencies of Modern Democracy*. Translated by E. & C. Paul. New York: Dover Publications.

Mills, C. 2001. *The New Men of Power: America's Labour Leaders*. Chicago, IL: University of Chicago Press.

Milne, S. 2014. *The Enemy Within: The Secret War Against the Miners*. London: Verso.

Molyneux, J. 1981. *Leon Trotsky's Theory of Revolution*. Brighton: Harvester.

Neocleous, M. 1996. *Administering Civil Society: Towards a Theory of State Power*. Basingstoke: Macmillan.

Noonan, J. 2012. *Materialist Ethics and Life-Value*. Montreal, ON: McGill-Queen's University Press.

Perusek, G. 1995. "Classical political sociology and union behavior". In *Trade Union Politics, American Unions and Economic Change, 1960s–1990s*, edited by G. Perusek & K. Worcester, 57–76. Atlantic Highlands, NJ: Humanities Press.

Post, C. 1999. "Ernest Mandel and the Marxian theory of bureaucracy". In *The Legacy of Ernest Mandel*, edited by G. Achcar, 119–51. London: Verso.

Post, C. 2010. "Exploring working-class consciousness, a critique of the theory of the 'labour-aristocracy'". *Historical Materialism* 18: 3–38.

Smith, C. 2019. "Class struggle from above, the Canadian state, industrial legality, and (the never-ending usage of) back-to-work legislation". *Labour/Le Travail* 86: 109–22.

Stevenson, J. 1992. *Popular Disturbances in England, 1700–1832*. Second edn. Harlow: Longman.

Strasseri, V. 2015. "Weber and the 'labour question': an initial appraisal". *Max Weber Studies* 15(1): 69–100.

Thompson, E. 1968. *The Making of the English Working Class*. Harmondsworth: Penguin.

Trotsky, L. 1990. "Trade unions in the epoch of imperialist decay". In *Trade Unions in the Epoch of Imperialist Decay*, edited by J. Riddell. New York: Pathfinder.

Unofficial Reform Committee 1912. *The Miners' Next Step*. https://libcom.org/library/miners-next-step-swmf-1912

Webb, S. & B. Webb 1965. *Industrial Democracy*. Second edn. New York: Augustus M. Kelley.

Weber, M. 1952a. "The essentials of bureaucratic organization, an ideal-type construction". In *Reader in Bureaucracy*, edited by R. Merton, B. Hockey & H. Selvin, 18–27. Glencoe, IL: The Free Press.

Weber, M. 1952b. "The presuppositions and causes of bureaucracy". In *Reader in Bureaucracy*, edited by R. Merton, B. Hockey & H. Selvin, 60–68. Glencoe, IL: The Free Press.

Weiner, L. 2018. "Walkouts teach US labor a new grammar of struggle". *New Politics* 17(1): 3–10.

Yu, A. & B. Ruixue 2014. "Autonomous workers' struggles in contemporary China". In *New Forms of Worker Organization: The Syndicalist and Autonomist Restoration of Class-Struggle Unionism*, edited by I. Ness, 39–61. Oakland, CA: PM Press.

CHAPTER 16

EXPLAINING UNION PARTICIPATION AND CITIZENSHIP: IMPLICATIONS FOR UNION STRATEGY

Ed Snape

ABSTRACT

This chapter considers the factors associated with members' commitment to, and participation in, their unions, with the latter conceptualized broadly to encompass formal and informal activities. Key findings from the research literature are highlighted and the policy implications for unions discussed. A notable finding is that instrumentality does not fully explain why members commit to and participate in unions, with pro-union attitudes and ideological motivations also significant. Research findings suggest that unions need to go beyond narrow instrumentalism and should instead pursue a broader social justice agenda, perhaps, encompassing notions such as "community unionism". Findings also suggest that this should be associated with a local union leadership style that emphasizes member participation and aims to build members' sense of shared mission and self-worth as a union member. Finally, the chapter reviews future research needs, focusing on how researchers can help inform more effective union strategy.

Keywords: Union commitment; union participation; union effectiveness; challenges for unions

INTRODUCTION

Unions in the UK, the US and elsewhere have been in crisis in terms of absolute and relative membership levels and in economic and political influence. "Union renewal" has been an ongoing project for some time, linked to an attempt to focus more strongly on union organizing, rather than simply servicing existing members (e.g. Heery *et al.* 2000). A key concern is with the extent to which unions can wield the necessary resources to maintain their existing organization as well as to organize new sectors, especially given the limited numbers of employed union officials and their dependence upon lay activists to shoulder much of the burden of member servicing and recruitment (Gall &

Fiorito 2012b). Although some have suggested that member activism is not necessarily the key to union renewal in every situation, with central leadership strategy and the role of full-time officer involvement more significant in some campaigns, member activism emerges as a key factor in many cases (Hickey, Kuruvilla & Lakhani 2010). Furthermore, in recent years, lay activism itself has been in decline as members are more reluctant to take on active roles (Gall & Fiorito 2012b). Hence, in focusing upon members' union participation, this chapter deals with an issue of key importance to the future of unions.

The chapter considers the factors associated with members' commitment to and participation in their unions. Participation is conceptualized broadly to encompass formal and informal activities, including what may be considered as pro-union "citizenship", in the sense of behaviours likely to benefit the union that are voluntary on the part of the individual member. Key findings from the research literature are highlighted and the policy implications for unions are discussed, addressing Gall and Fiorito's (2012b) suggestion that the union commitment and renewal literatures need to be better integrated if they are to provide effective guidance to unions. The aim is not to provide a meta-analysis of the studies, nor even a comprehensive review. Rather, the chapter provides a selective but reasonably representative review of the research, examining the nature and antecedents of union participation, including factors such as demographics, socialization, leadership and workplace climate, and then discussing the analysis of commitment profiles as an alternative research approach. Next, the chapter draws out the implications for union strategy, organizational practices and leadership, before discussing future research needs, outlining what needs to be done if the literature is to provide more specific guidance to unions.

LITERATURE ON UNION PARTICIPATION

This section discusses the largely social psychological literature on union participation, examining underlying theoretical frameworks, and the specific antecedents, mediators and moderators identified in the empirical studies. First, it discusses what is meant by union participation.

What is union participation?

Union participation is defined as "the behavioural involvement of members in the operation and activities of the union, including participation in administrative and democratic structures, day-to-day discussion of union affairs with fellow members, reading union literature, and taking part in union campaigns and industrial action" (Redman & Snape 2004: 847). Participation may, thus, be seen as involving some form of more or less active involvement, going beyond simply joining the union and remaining as a member. Clearly, there are potentially a broad range of different activities involved.

A few studies have argued for a unidimensional approach (Kuruvilla *et al.* 1990), and Kelloway and Barling (1993) argue that participation follows a cumulative pattern, from relatively easy to more demanding activities. More generally, the literature has discussed participation as a multidimensional construct, although there is a lack of consensus on the dimensions involved. For example, McShane (1986a) defines three dimensions: "meeting participation" (attending union meetings), "voting participation" (voting in union elections) and "administrative participation" (holding office or sitting on a committee). In contrast, Kelly and Kelly (1994) suggest two dimensions, distinguished in terms of the amount of effort involved, with "easy" participation typical of many rank-and-file members, and "difficult" participation tending to be the reserve of more committed activists. This lack of consistency may be in part artefactual, reflecting the sampling of different behaviours and the use of different items across studies (Kelloway, Catano & Carroll 1995). It should also be recognized that various union, occupational, industrial relations and legal contexts may be associated with somewhat different opportunities and forms of participation. In spite of all this, many studies echo Kelly and Kelly (1994) in distinguishing between dimensions relating to rank-and-file and activist participation (e.g. Smith, Duxbury & Halinski 2019).

According to Tan and Aryee (2002), the lack of consensus on dimensionality sparked an interest in conceptualizing union participation as union citizenship behaviour (UCB). The suggestion is that union participation shares many of the features of what has been referred to as "organizational citizenship behaviour" (see, e.g. Organ, Podsakoff & MacKenzie 2006), in that it is usually voluntary, benefits the organization concerned (in this case, the union), and cannot be made contractual or subject to penalties for non-performance (Fullagar *et al.* 1995; Martínez-Íñigo, Zacharewicz & Kelloway 2020).

One of the first studies to adopt the UCB approach identified four dimensions: activist behaviour, compliance or sportsmanship, voice within the union, and grievance-related or helping behaviours. Following Skarlicki and Latham (1996), Tan and Aryee (2002) suggest a more parsimonious model based on two dimensions, UCB-organization, including behaviours such as standing for union office, which are likely to benefit the union as an organization, and UCB-individual, based on behaviours such as interpersonal helping, likely to benefit fellow members individually. Often following this two-dimensional approach, research on union citizenship has entered the mainstream, with many studies adopting UCB as an indicator of participation (Chan *et al.* 2006; Martínez-Íñigo, Zacharewicz & Kelloway 2020; Snape & Redman 2004).

Modelling union participation

A primary focus has been evaluating the antecedents of participation, with union commitment typically seen as a mediator between antecedent factors such as job satisfaction, organizational commitment, union instrumentality and pro-union values on the one hand, and union participation on the other (Barling, Fullager & Kelloway 1992; Iverson & Kuruvilla 1995; Newton & McFarlane Shore 1992). The level of analysis has been the individual member and relationships with the union. Bamberger, Kluger & Suchard (1999) provide a meta-analysis of the early work, finding support for an "integrative model" of union commitment and participation (as shown in

Figure 16.1). Union instrumentality refers to the degree to which the union is perceived as addressing members' needs in the employment relationship, such as pay, benefits and other rewards, reflecting an economic exchange perspective, with participation seen as reciprocation for benefits received (Fullagar & Barling 1989). Pro-union attitudes refer to an individual's perceived desirability of unions in general, reflecting a broader social perspective and a values-based motivation for participation (McShane 1986b). Job satisfaction and organizational commitment reflect the influence of organizational factors on union commitment and participation (Barling, Fullager & Kelloway 1992: 80).

The meta-analytic findings suggest that, while both instrumentality and pro-union attitudes were positively associated with union commitment and participation, "prounion attitudes appeared to have a consistently stronger impact on union commitment [and participation] than ... union instrumentality" (Bamberger, Kluger & Suchard 1999: 314). This suggests a narrow instrumentality view of member-union relations, as emphasized in traditional US-business union models, for example, which may be insufficient to effectively motivate widespread member participation, a point

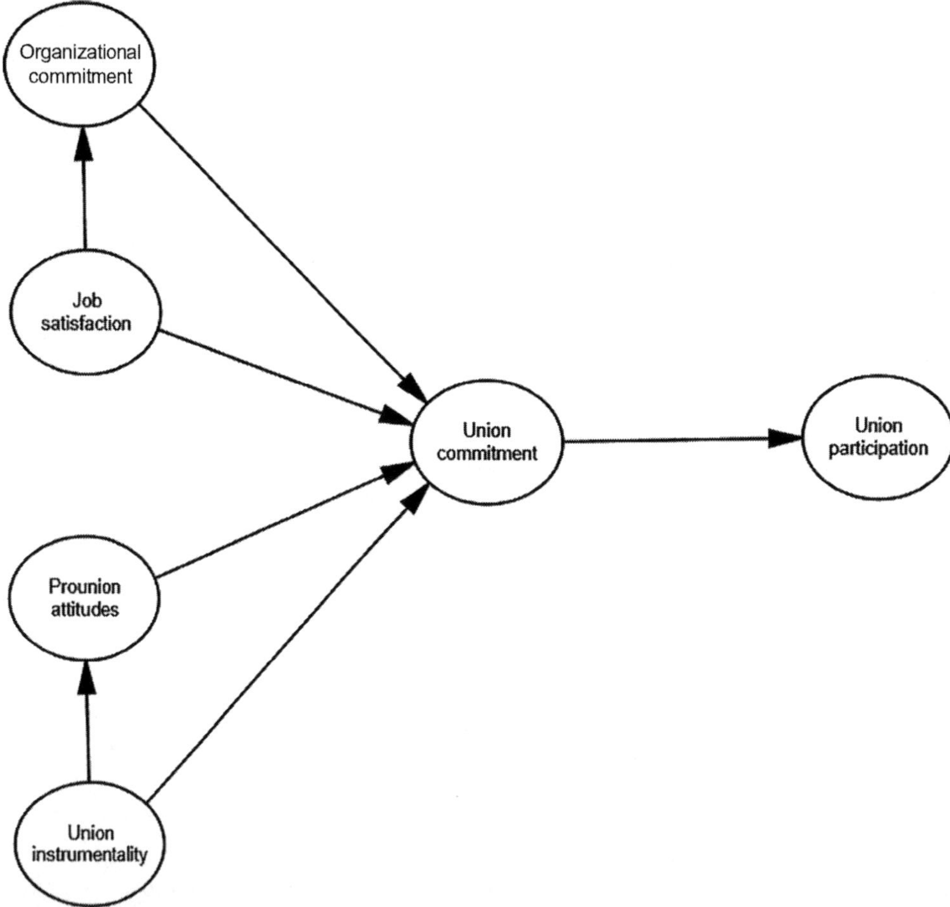

Figure 16.1 Bamberger, Kluger and Suchard's "integrative model"
Source: Bamberger, Kluger & Suchard (1999).

returned to later. Organizational commitment was also positively associated with union commitment and participation, perhaps suggesting "dual loyalty" as individuals commit to both employer and union, with the roles of loyal employee and union member seen as non-conflictual and even complementary (e.g. Reed, Young & McHugh 1994). Finally, findings on the association between job satisfaction and union commitment and participation were less clear cut, with only a weak negative association with union commitment and participation, and with better model fit once this construct was omitted (Bamberger, Kluger & Suchard 1999: 314).

Subsequent studies have examined the generalizability of these findings, with some using UCB measures of participation (e.g. Tan & Aryee 2002). The analysis has been replicated using primary data from union members in a non-Western context, providing considerable support for the generalizability of the integrative model (Chan *et al.* 2006; Tan & Aryee 2002). Echoing the Bamberger, Luger & Suchard (1999) findings, subsequent studies have produced weak and mixed results on the role of job satisfaction (Chan *et al.* 2006; Redman & Snape 2014; Tan & Aryee 2002), suggesting the possibility that this may be moderated by contextual factors (Fuller & Hester 1998).

There have been attempts to evaluate the extent to which the findings "may vary with the composition and nature of the workforces examined as well as with environmental characteristics, such as the industrial relations context" (Bamberger, Kluger & Suchard 1999: 315). For example, Redman and Snape (2014) tested the model in two teachers' unions in the UK, the moderate, professionally oriented Voice, and the more traditionally "unionate" and militant National Union of Teachers (NUT). Findings based on separate samples of members of each union provided support for the basic Bamberger, Kluger and Suchard findings overall, but in the NUT, pro-union attitudes were more potent as an antecedent of union commitment and participation, and the link between union commitment and participation was stronger. Such findings were as anticipated and provide support for the notion that the relative importance of factors influencing union commitment and participation may vary according to context, in this case with values-based motivation stronger and the motivators of active participation more potent in the more traditionally "militant" union. Regarding the importance of the workplace context, there is a developing literature on the role of industrial relations climate (see below). Overall, such studies provide an indication that union participation models are indeed subject to workforce compositional and contextual influences.

The research generally assumes that the causal ordering is from attitudes and commitment to participation, consistent with an attitudes-to-behaviour paradigm. There is empirical evidence to support this. For example, Fullagar and Barling's (1989) longitudinal study measured both self-rated union loyalty and participation at eight-month intervals, with cross-lagged regression indicating a unidirectional link from union loyalty to union participation, with no evidence of reverse causation. Fullagar *et al.* (2004) performed a similar analysis with a ten-year interval between measures, again with cross-lagged regressions suggesting a causal order from union commitment to participation. It is also worth noting that the former study looked at more formal participation (for example, participation in meetings, and elections and serving as a union officer), while the latter looked at informal activities as envisaged in the UCB measures, suggesting that the causal ordering applies to both formal and informal union activities.

Gender and demographic factors

Demographic factors have tended not to be the central focus of union participation research, nor have they generally been theorized in the studies. Nevertheless, there are some findings, not least as demographic characteristics have often been included as control variables. In studies of union commitment and participation, gender often emerges as nonsignificant. However, a few studies suggest that women are more likely to show higher levels of attitudinal commitment, but that men have a higher willingness to actively work for the union (Snape, Redman & Chan 2000). This may reflect the traditional domestic responsibilities of women limiting the time available for active participation, and also the relative lack of female role models in union leadership (Gallagher & Clark 1989). However, it is fair to say that gender has not been a primary focus of the research and this should be a priority for future studies. Marital status has not usually featured in the studies, something that might be addressed in future, perhaps interacting with gender.

It appears that ethnic minority workers are more likely to join a union, out of both instrumental and solidaristic motivations. Evidence suggests they are more likely to join because unions offer some protection against discrimination (Rosenfeld 2014). In addition, a US study suggests that African-Americans may be more likely than White workers to vote "yes" in a union certification election, and that this effect is partially mediated by their pro-social orientation (Gumber & Padavic 2020). The findings on age are mixed, with some studies finding that older workers show higher levels of union commitment, but with others finding no such effect (Snape, Redman & Chan 2000). Occupational and job tenure, while positively associated with organizational commitment, perhaps surprisingly has shown no significant association with union commitment. Educational level has been included in some studies, but the findings are mixed, with most finding no significant relationship with union commitment and just a few finding a negative relationship (Snape, Redman & Chan 2000).

A recent study examined the impact of generational cohort on the antecedents of union participation (Smith, Duxbury & Halinski 2019), finding that an individual's favourable perception of their relationship with their union was positively associated with union participation for all age cohorts, but that this association was stronger for "Baby Boomers" than for "Generation X" and "Millennials", consistent with the idea that these later cohorts are less readily motivated to participate actively. An intriguing finding is that the association between "passive" forms of participation (for example, discussing the union with co-workers and promoting the union) and hours actually spent participating was stronger for "Generation X" and "Millennials" than for "Baby Boomers". This may suggest that younger generations have different patterns of activism, emphasizing face-to-face interaction with members and co-workers, rather than the formal committees and leadership positions more typical of "Baby Boomer" activists. Of course, it may be that this simply reflects relative seniority in their activist career, with older activists more likely to be engaged in the formal union hierarchy, but the possibility that new patterns of activism among younger workers are being seen is worthy of future research (see below).

Overall, given the mixed findings on many of the demographic variables, it is likely that the occupational and industrial relations context moderate such relationships, and the findings of individual studies may reflect the occupational mix of the samples. The conclusion of Barling, Fullager & Kelloway (1992: 102) is probably still appropriate:

> Research tends to indicate that demographic-related variation in union participation is attributable to factors such as industrial and occupational distributions, exposure to or experiences with unions, different levels of satisfaction with particular job facets, differences in union attitudes, instrumentality and efficacy beliefs, and other factors that may vary with demographic characteristics.

Union socialization

Several studies have demonstrated that the way in which members are socialized into the union has an effect on their union commitment and participation (e.g. Fullagar, McCoy & Shull 1992; Fullagar *et al.* 1995; Gordon *et al.* 1980; Tan & Aryee 2002). This has usually been assessed over the first year of membership, examining both formal orientation sessions and more informal events such as being personally introduced to the union representative. The findings suggest that the extent to which a member receives orientation and socialization into the union early in their membership has a positive impact on their union commitment and participation, with evidence that this is mediated in part by a positive effect on their pro-union attitudes (Fullagar, McCoy & Shull 1992; Tan & Aryee 2002).

Beyond the union itself, there is a possibility that union participation is subject to family socialization. Early research on union activism points to the importance of family background and pro-union attitudes prior to joining the union (Barling, Fullager & Kelloway 1992: 133). A US study demonstrated that students' perceptions of their parents' union participation and union attitudes were positively associated with students' own attitudes to unions and their willingness to join (Barling, Kelloway & Bremermann 1991). In a study utilizing the British Household Panel Survey, Bryson and Davies (2019) found that the probability of union membership among younger workers was higher for those whose parents were members, with stronger effects for daughters and for those with both parents as members. Furthermore, these effects were stronger in areas with higher union density, pointing to a role for community influences in underpinning family socialization.

Union leadership and member participation

Studies have also examined the impact of union leadership on members' union commitment and participation. There is a strong tradition of qualitative research in this field, pointing to the importance of participatory union leadership at workplace level. For example, based on a study of five workplaces in the UK, Fosh (1993) suggested the level of union commitment and participation may ebb and flow in response to industrial relations developments in the workplace, but that participation surges in response to such developments may be enhanced with a local union leadership that encourages membership involvement, rather than adopting a "leave it to me" approach to union representation. Similarly, Greene, Black & Ackers (2000) drew upon case study evidence to identify local leaders with a participatory style, emphasizing the importance of communication and members' involvement in decision-making, and a collectivist outlook that sees issues as reflecting shared experience rather than just a collection of individual grievances. Such leaders, they argue, can more effectively build on surges in members' involvement and participation.

There have also been quantitative studies, many again focusing on workplace union leadership. In an early US study, Kelloway and Barling (1993) found that local shop stewards' leadership, measured using items from the Bass MLQ measure of transformational leadership, was positively associated with members' participation in the union. Transformational leadership behaviours involve going beyond a rewards-based transactional approach, to include leadership behaviours that inspire a sense of shared mission and pride in the organization, providing followers with individual encouragement and intellectual stimulation (Bass & Riggio 2006). This points again to the importance of values-based motivations for union participation.

Consistent with this, members' relationship with their union may go beyond a sense of exchange, with participation in the union also due to shared values and arising out of a concern for the union and for fellow workers. This may be seen as a "covenantal" relationship rather than one based purely on economic or social exchange (Snape & Redman 2004). Drawing upon this perspective, Twigg, Fuller & Hester (2008) examined the process by which the transformational leadership of local shop stewards may impact on members' commitment and UCB. They suggest that transformational leaders establish a sense of a covenantal relationship between member and union, whereby members feel that the union is highly supportive, fostering a feeling of obligation, trust and a sense of self-worth as a member of the union.

Cregan, Bartram & Stanton (2009) examined a nurses' union in the process of an organizing campaign aimed at mobilizing member support through building identification with the union, providing leadership training for local union representatives and attempting to win public sympathy for the union's cause. They found that local representatives' transformational leadership was positively associated with members' union loyalty and willingness to work for the union, mediated by members' social identification with the union, again suggesting that local union leaders may develop members' loyalty in a way that transcends a purely exchange relationship.

Fortin-Bergeron, Doucet & Hennebert (2017) suggest a preoccupation with transformational leadership risks providing an incomplete view of the behaviours displayed by local union leaders. They examined the relative significance of rival theories in a study of the transformational and authentic leadership behaviours of local union representatives. Authentic leadership refers to the extent to which leaders "know their strengths and weaknesses, are transparent, rely on rigorous ethical principles to make decisions and act in keeping with their values and principles" (Fortin-Bergeron, Doucet & Hennebert 2017: 795). Such qualities may be particularly relevant to local union leaders concerned about building members' trust and participation, and this is consistent with earlier research suggesting that "honesty, transparency, and active listening" are critical in making a good union representative (Fortin-Bergeron, Doucet & Hennebert 2017: 795; see also Kahn & Tannenbaum 1957; Nicholson 1976). Findings suggest that both transformational and authentic leadership styles are positively associated with members' UCB, with the effects of authentic leadership mediated by member-leader value congruence and those of transformation leadership mediated by value congruence and also by members' collectivist orientation.

These studies have used standard leadership models imported from organizational research. However, the qualitative research cited above points to some very specific patterns of local leader behaviour, so that it is worth considering the specific form of leader behaviour more carefully. Metochi's (2002) study of Cypriot public sector workers used a measure of "service-oriented leadership" grounded in the literature on steward behaviour, focusing on daily interactions between stewards and members, and drawing in particular on Nicholson, Ursell and Lubbock's (1981) notion of leader "accessibility". Metochi (2002) found evidence for both a direct effect of steward leadership on members' willingness to participate in the union, and some evidence of mediation by members' union loyalty and collectivist orientation.

Regardless of the leadership theory or operationalization of leader behaviours used, there remains the question of the relative importance of leadership and members' own independent orientations in motivating participation. The organizational leadership literature includes a discussion of "substitutes for leadership" theory, dealing with the question of the importance of leadership relative to other factors (Kerr & Jermier 1978). This question is relatively neglected in the context of unions, but in a sample of shop-floor workers in Korea, Kwon (2013) found a positive association between the transformational leadership of local union leaders and members' willingness to participate in the union, but negative interactions between leadership and both "them and us" attitude and union instrumentality. This pattern of interactions suggests that members' own attitudes may, indeed, be acting as substitutes for leadership, in the sense that where such attitudes are present, union leadership becomes less significant as an independent influence on member participation. The implication is that union leadership may be especially significant in motivating union participation in contexts where workers' underlying pro-union attitudes and sense of union solidarity are limited, in effect requiring leadership to provide motivation.

Finally, it is notable that the studies examined here have tended to focus on the role of union leadership as an independent motivator of members' union participation, with evidence that this is significant. However, more neglected is the possibility that the role of local union leaders is also to help translate industrial relations experiences and union commitment into active participation. This is what authors such as Fosh (1993) imply. Stated more formally, there is a need for studies not only of the main effects of local leadership but also of how such leadership may moderate the relationship between union commitment and active participation, with a more participative union leadership presumably strengthening this link as attitudinal commitment is more readily translated into active participation.

The role of workplace climate

Most of the research on union participation has focused upon the significance of individual attitudes, with little or no consideration of contextual factors. An exception to this is the industrial relations climate of the workplace, usually defined as the degree to which union–management relations are seen by organizational members as either cooperative or conflictual (Hammer, Currall & Stern 1991). Fuller and Hester (1998) suggest that climate will impact work-related attitudes, since, according to social information processing theory, the social environment influences individuals' perceptions and attitudinal judgements concerning the job and work context.

Several studies have measured perceived workplace industrial relations climate using member survey data, usually assessing the main effects on union commitment and participation. However, findings have been mixed with regard to whether a more positive and "cooperative" industrial relations climate is positively or negatively associated with union commitment and participation. It has been suggested that a perception of the management–union relationship as cooperative, trusting and respectful will be reflected in stronger union commitment and participation, since there is then greater cognitive consistency between the role of employee and union member as the two are seen as non-conflictual.

Angle and Perry (1986) and Magenau, Martin and Peterson (1988) provided empirical support for this in North America, as did Deery, Erwin and Iverson's (1999) study of an Australian automotive manufacturer that had recently attempted to develop more cooperative working with the union. However, in a sample of Australian public sector workers, Deery, Iverson and Erwin (1994) found that industrial relations climate was *negatively* related to union commitment. They suggest that this may reflect a strongly adversarial industrial relations tradition in this research site, so that more conflictual perceptions of the industrial relations climate are associated with a perceived need for union protection and, hence, greater union commitment and participation. These differing cognitive consistency and union protection arguments are worthy of future research.

Most of the industrial relations climate literature has measured climate using employee survey ratings in a single organization or workplace, then assessing the association between individuals' perceptions of climate and their attitudes and behaviours. This exclusively individual-level analysis examines perceptions of climate and how they vary across individuals within a given workplace. As such, these studies are measuring what can be referred to as "psychological" climate rather than workplace climate, or perceptual differences concerning a single workplace rather than differences between workplaces (Dastmalchian 2008).

Recognizing this limitation, Snape and Redman (2012) examined workplace climate across multiple workplaces, providing an assessment of climate in each workplace by aggregating individual employee assessments of climate by workplace. They assessed inter-rater agreement within workplaces and the between workplace variance, thereby providing evidence that a consensual assessment of climate is, indeed, meaningfully differentiated across workplaces. Using a cross-level approach, these authors then found that a more cooperative workplace industrial relations climate had a significant negative association with individuals' union commitment, while a psychological industrial relations climate had no significant association with commitment. In other words, the evidence suggested that employees in a workplace with a more positive or cooperative workplace climate had lower levels of union commitment. On the face of it, this provides support for the union protection argument referred to above, although it would be useful to see additional studies to provide corroboration and to reflect different industrial relations contexts.

The implications of the role of industrial relations climate are practically significant, in that a positive industrial relations climate-union participation link as envisaged in the cognitive consistency view suggests that unions may prosper by cultivating labour-management cooperation, with members then feeling less conflict between their roles as loyal employee and union member. However, the reverse union protection finding implies that members may ally more strongly with their unions where they feel themselves to need protection from their employer, while a strong emphasis on union-management partnership risks undermining such support for the union. Given the significance of this issue and the limited nature of the existing research, this is clearly an area where further research is needed, particularly studies that operationalize climate at the workplace or work-unit level.

There is also a suggestion that industrial relations climate may moderate the relationships in the union participation model. Using an indirect assessment of climate (the correlation between union commitment and organizational commitment – a positive correlation regarded as a cooperative industrial relations climate) in a meta-analysis of prior studies, Fuller and Hester (1998) found the positive associations between union instrumentality and socialization, on the one hand, and union commitment, on the other, were significantly stronger where the industrial relations climate was adversarial, while the job satisfaction-union commitment link was negative in adversarial climates and positive in cooperative climates. The association between union commitment and participation did not vary with climate. These findings, while using an indirect measure, suggest that climate is significant as a moderator of the relationships in the union participation process. Again, this issue deserves further research with direct measures of workplace climate.

The exploration of union commitment profiles: a "person-based" approach

Commitment research has, as has been seen, traditionally involved a construct-based approach, whereby the associations between pro-union attitudes, instrumentality, organizational commitment, job satisfaction, union commitment and union participation are examined using techniques such as multiple regression or structural equation modelling. The aim is to test hypotheses about the relationships between variables. This is the dominant approach in much of quantitative social science. In contrast, a "configurational" approach focuses upon individuals, identifying groups or clusters of individuals based on their distinctive "profiles" across the study variables. This approach has become more common in the broader organizational and multiple commitment research (Meyer & Morin 2016), but it is still rare in the study of union commitment.

An early example using this approached is provided by Heschiser and Lund (1997), who categorized members based on their "instrumental" and "normative" commitments, seeing the former as reward-based and the latter as ideologically based attachments to the union. Looking just at those committed members with better or worse than neutral attitudes on each construct, they generated a four-group classification of union members, labelled as "identifiers" (highly positive on both), "positive free agents" (low but positive on both), "expressives" (high on normative commitment only), and "instrumental" (high on instrumental commitment only). They found, as expected, that the identifier and expressive groups had higher extra-role union participation than the other two groups.

A more sophisticated approach to the analysis of profiles involves the use of statistical cluster analysis to generate groups based on the actual distribution of the sample, typically designed to maximize similarity on the variables within clusters and dissimilarity across clusters. This has the advantage of identifying actually occurring clusters of individuals, rather than imposing groups based on arbitrarily cut-off points, so helping to ensure that the groups are meaningful in the population under analysis. Akoto (2020) performed such an analysis of union members in the US and Ghana based on measures of "instrumental" and "ideological" commitment, deriving a very similar four-group solution in each sample, labelled "invested" (strong instrumental and moderate ideological commitment), "free agents" (moderate instrumental, weak ideological), "complacent" (weak instrumental, moderate ideological), and "allied" (moderate on both). Group membership was associated with UCB, broadly as anticipated. In the US sample, free agents had a significantly lower level of UCB than the three other groups, which did not differ significantly. In the Ghanaian sample, the invested and allied groups had significantly higher levels of UCB than both the complacent and free agent groups for UCB-Organizational (UCB-O) and the invested group had higher UCB-Individual (UCB-I) than the complacent group.

Stable union commitment profiles have been demonstrated in a longitudinal study over a ten-year period using Kelloway's 13-item union commitment scale (Kelloway, Catano & Southwell 1992), with four groups emerging based on the three dimensions of union loyalty, responsibility to the union and willingness to work for the union (Morin *et al.* 2021). Four groups emerged, highly committed, moderately committed, moderately uncommitted and highly uncommitted, and, as anticipated, members of the high commitment group had significantly higher levels of UCB-O and UCB-I than members of the low-commitment group. However, the four groups simply reflected a uniform distribution of high to low scores on each dimension, with the three dimensions covarying, suggesting that the three dimensions used in this study may be redundant and best considered as a unidimensional construct. Having said that, while the study included what may be seen as affective and normative dimensions (loyalty and responsibility), along with willingness to participate, no instrumental bond was included, so that the results are not necessarily comparable with the studies reviewed earlier. Finally, the study again confirms the importance of ideological beliefs about unions in general, with such beliefs a significant antecedent of commitment group membership, while satisfaction with the specific union was not.

Overall, the findings from these studies show that profile membership meaningfully differentiates UCB, suggesting that the profile-based approach has potential. Union research using the profile approach is still at an early stage, and there is a need to develop this work further, perhaps incorporating a broader range of constructs to provide a more complete characterization of member types and an analysis of their implications across a range of unions and sectors. A key advantage of the profile approach is that it analyses the significance of specific combinations or patterns of factors in motivating participation, in contrast to the largely factor-by-factor approach of traditional regression-based analysis (Meyer & Morin 2016; Morin *et al.* 2021). It may, therefore, have practical utility for unions in providing insights for recruitment targeting of specific groups.

IMPLICATIONS FOR UNIONS

While not all cases of successful union organizing rely on lay activists (Hickey, Kuruvilla & Lakhani 2010), it is likely that in general unions will need the active participation of members if they are to successfully renew their organizations. Unions have long lacked a sufficient number of paid officers to sustain their organizing and servicing needs, being historically dependent on lay activists in "self-organized" workplaces (Snape 1994). In recent years, the situation has deteriorated as paid union officials find themselves drawn in to do more of the workplace organizing and servicing as there are insufficient numbers of committed lay activists (Gall 2009). The number of lay activists in unions is generally very low compared to other kinds of voluntary organizations, and appears to be lower than 5 per cent of members in many unions (Gall & Fiorito 2012b). Hence, addressing the factors associated with lay member participation is a key issue for unions. The literature surveyed earlier provides us with important insights that have practical implications for unions.

Social justice and community unionism

As stated earlier, a notable finding from the literature is that narrow instrumentality considerations may not be the most important explanation of why members commit to and participate in their union. In particular, the extent to which the individual member holds pro-union attitudes appears to be more significant (e.g. Bamberger, Kluger & Suchard 1999), and there is a suggestion that the member-union relationships are often "covenantal" rather than purely exchange-based (Snape & Redman 2004). In practical terms, this suggests that a largely instrumentality-based "business union" or "service union" model will underperform in attracting member support and participation, and that a broader conceptualization of what a union is and does seems more appropriate.

Even in traditional unionized job territories, an appeal to members that emphasizes the broader social justice mission of unions seems justified. If unions are to broaden their appeal, there may be a need to embrace notions such as "social movementism" or "community unionism" (Holgate 2015). This may involve trying to build solidarity through community-wide campaigns, focusing, for example, on low pay and the "living wage", migrant workers or disadvantaged occupational groups such as contract cleaners, often working in collaboration with community groups (e.g. Wills 2008, 2009). This echoes back to the earlier history of labour unionism, where unions were often more closely linked into their communities, forming an integral part of a wider social movement (Holgate 2015). The idea is that while unions act as a "vested interest", pursuing members' immediate instrumental needs, they may go beyond this to address wider notions of social justice, serving also as a "sword of justice" (Flanders 1975: 15).

As has been seen, there have been mixed findings on the association between demographic variables such as gender, ethnicity, educational level, age and tenure on the one hand, and union commitment and participation on the other. It may be that occupational factors and experience with unions are at least as significant as demographics (see Barling, Fullager & Kelloway 1992). Importantly, there is little or nothing in the research to indicate that such demographic characteristics necessarily preclude the development of strong union commitment and active participation. The implication is that unions should seek to address the specific concerns of groups in need of union protection, afford opportunities for participation and promote role models for those who are currently underrepresented among activists. Echoing arguments above, this may form part of a community-based approach to union campaigns and should be framed in terms of social justice, encompassing sectors and occupations that have typically included large numbers of female workers, ethnic minorities, migrants, and younger workers.

To the extent that such community unionism draws participating unions and their members into broader social campaigns with a social justice agenda, unions may serve as training grounds for "democratic citizenship", and the community unionism initiatives of unions, such the Unite union, in the UK have tried to link "union consciousness" and "community consciousness" (Holgate 2015). However, the non-union community organizations involved in such campaigns as the living wage have their own agendas and may be seen as independent "new actors" in industrial relations (Heery, Abbott & Williams 2012). Unions have in some cases experienced difficulties working with such organizations, not least because their single issue and relatively opportunistic nature contrasts with the bureaucratic nature of unions, so that the coalitions formed can be difficult to sustain (Holgate 2013).

Tapia (2013) argues there are differences in culture and member motivations, as unions develop a "service-driven culture", emphasizing "instrumental commitment" among members, while community organizations develop a "relational culture", focusing on "social commitment". While there is probably something in this comparison, as has been seen, the research findings on union membership, commitment and participation suggest that values-based motivations are generally significant for union members (e.g. Bamberger, Kluger & Suchard 1999), and that socialization, including pre-employment family and community socialization (Bryson & Davies 2019), is important in shaping attitudes and behaviour towards unions. This suggests union initiatives based upon the community unionism model are not necessarily at odds with the traditional member-union relationship, although it seems likely that they will work to re-emphasize values-based motivations (Tapia 2013: 684).

There is a tentative suggestion in the findings on younger-age cohorts that they may have a different pattern of activism than "Baby Boomers", participating in their unions through face-to-face interaction with members and co-workers, rather more than through the formal committee membership and leadership positions typical of "Baby Boomer" activists (Smith, Duxbury & Halinski 2019). If this is confirmed in other studies, it may suggest that a new pattern of people- and relationship-centred activism among "Millennials" and "Generation X" is being seen, and one that may fit comfortably with the community unionism approach. At least one study of community-based organizations has suggested that they seek to develop a strong "relational culture", based on one-on-one or small group meetings and having members share their experiences at meetings to build communal bonds and solidarity (Tapia 2019).

Beyond the workplace?

Community unionism has been characterized as a union strategy that involves appeals to members that go beyond the individual workplace, which is important given the decline of large workplaces and the shift away from strong workplace organization (Cobble 1991; Milkman 2013). Community unionism may encompass work-based and occupational communities, but there is also a focus on what might be called "place-based" communities, encompassing wider social networks and community associations such as religious groups, political and cultural organizations (Holgate 2015). So, community unionism offers a possible way forward for unions that promises to reconnect with the community and tap into sources of influence beyond the workplace.

A second approach to transcending the workplace is for the union to directly provide individual services to members; for example, education, training and advisory services, addressing members' needs, without having to engage with the employer (James & Karmowska 2016). The research on union commitment and participation provides at least some support for this as a supplementary strategy, since instrumentality-based appeals have at least some traction in motivating union commitment and participation and perhaps in helping shape attitudes to unions (e.g. Bamberger, Kluger & Suchard 1999). However, the findings suggest that the values-based motivations are more potent, indicating a wider community-based approach may be more effective than a purely services-based approach (Snape & Redman 2004). Furthermore, case study evidence on the individual services approach, while providing evidence of initial membership gains, suggests that there are often difficulties in membership retention and in building on the initial success in promoting union participation (James & Karmowska 2016; Milkman 2013), although the same may also be said of some community-unionism campaigns, at least in the UK (Mustchin 2012). Of course, individual services may be providing within a broader community unionism approach and, indeed, the two approaches may be mutually reinforcing.

Organizational practices and leadership

The literature suggests that there is a great deal that unions can do in terms of organizational practices and leadership. First, research suggests the extent to which members receive union orientation and socialization into the union early on is positively associated with their union commitment and participation. Orientation programmes need to be developed and maintained, including both formal programmes and informal practices such as meeting union representatives and other members. Second, leadership behaviour of union representatives is positively associated with members' union commitment and participation. There are implications for how union representatives are selected and trained, with the literature suggesting that a transactional, "leave it to me" leadership approach should be discouraged, with union leaders needing to emphasize accessibility to members (Metochi 2002), and to focus upon more positive leader behaviours, perhaps including those suggested by "transformational" and "authentic" leadership theory (Cregan, Bartram & Stanton 2009; Fortin-Bergeron, Doucet & Hennebert 2017), even if these modus operandi are more resource and time intensive. The aim is to help build members' sense of shared mission, obligation, trust and a sense of self-worth as a member of the union (Cregan, Bartram & Stanton 2009).

The profile analysis research is as yet at an early stage, but the identification of member groups – and, perhaps, also non-member groups – based on their attitude profiles may provide useful insights for the targeting of membership recruitment and retention campaigns, and in the identification of groups who may be more likely to participate in union activities. Additional research may provide useful insights in the design and development of specific recruitment and other campaigns.

Management–union partnerships

There is a question as to whether unions are better able to win member support by cultivating a positive relationship with employers, as seen in positive "management–union partnerships" (Marchington 1998), or alternatively whether members are more supportive of unions that are more independent and, perhaps, adversarial in their dealings with employers. The research on industrial relations climate provides insights on this, although findings have been mixed. Studies that compare individuals' perceptions of climate have often found a positive association between industrial relations climate and union commitment and participation, apparently supporting the cooperative management–union partnership perspective. Studies have suggested that cooperative management–union partnerships can deliver benefits to the employer, union and employees alike (e.g. Deery, Erwin & Iverson 1999).

Based upon available evidence, the possibility cannot be ruled out that a positive management–union partnership may, at least, in some cases be associated with higher union commitment and participation. However, a study comparing workplace industrial relations climates found a negative association with union commitment (Snape & Redman 2012), suggesting that unions may more readily win support where industrial relations are adversarial. This is corroborated by meta-analytic findings suggesting that union instrumentality and socialization are more strongly associated with union commitment with a more adversarial industrial relations climate (Fuller & Hester 1998). On balance, it seems more likely that an adversarial industrial relations climate will be associated with stronger union commitment and participation, perhaps as in such situations members are more likely to value union protection. If this is confirmed by future research, then it seems that collaborative union–management partnerships are not necessarily the only, or even the best, way forward for unions. Of course, in most instances such an option is not available.

FUTURE RESEARCH NEEDS

The literature provides some useful insights for practice. However, there are also some limitations and, looking forward, there are outstanding research needs. This penultimate section provides some suggestions for this, focusing especially on how the research can better meet the needs of unions.

First, recognition is needed of the importance of the industrial relations context in shaping research findings. At the most basic level, more studies in a range of occupations and sectors, and also different types of union (Redman & Snape 2014) are needed, so that the question of generalizability can be better addressed. Recognition is also needed that some of the literature is now quite dated and there is an urgent need for studies examining union participation in the contemporary economic and social climate (Gall & Fiorito 2012a). It is also necessary to include certain aspects of context within individual studies; for example, with study designs encompassing multiple workplaces and assessing the impact of workplace-level factors on participation models (e.g. Snape & Redman 2012). More generally, researchers should be strongly encouraged to provide a detailed consideration of their study context, and to consider this when discussing their findings and their implications.

Second, there is a need to further consider the conceptualization and operationalization of participation (Gall & Fiorito 2012a). There has been a commonly used two-dimensional measure of UCB that differentiates between activities that may benefit the union as organization (UCB-O) and fellow union members (UCB-I) (Skarlicki & Latham 1996), thus trying to tap into a breadth of different types of activity. There are also studies that differentiate between "easy" participation, typical of rank-and-file members, and "hard" activities by definition typical of hard-core activists (e.g. Kelly & Kelly 1994), but other dimensions are also used (e.g. Redman & Snape 2004). The orthodox social-psychological approach is generally to try to establish common operationalization of constructs to underpin generalization, but in the case of context-specific behaviours such as union participation and, for that matter, organizational citizenship behaviours (Organ, Podsakoff & MacKenzie 2006), the lack of consistency should come as no surprise. Indeed, there is a case for studies to take a more context-specific approach to defining and measuring union participation, grounded in the experience of the occupation, the union and the workplace concerned.

Third, there is a need for alternative methodological approaches. Much can be achieved using the traditional socio-psychological approaches, particularly if multi-level and longitudinal designs are included, but there is a need to broaden the research. Traditional industrial relations case study work has a role (see, e.g. Fosh 1993; Greene, Black & Ackers 2000), particularly in providing a context-rich and in-depth understanding of the processes involved in the emergence of union activity. Also, as suggested earlier, it will be useful to have more work on the profile approach, which will allow us to learn more about which combination of antecedent factors are likely to be associated with participation, rather than simply studying the impact of each factor individually. Such studies may also be helpful in identifying groups for recruitment targeting and internal union campaigning.

However, new approaches are also called for, and one in particular will be discussed here. It has been suggested that the existing literature has more to say on relatively common forms of union participation and citizenship behaviour, such as might be undertaken by a large number of rank-and-file members, but that it says rather less on the motivation and emergence of more activist forms of participation (Gall & Fiorito 2012a), which is typically the reserve of a very small number of activists (Gall & Fiorito 2012b). This is due in part to the methodology used, with analyses based on general surveys of members likely to reflect common forms more strongly.

An alternative approach would be to develop a "process theory" of the individuals' path to union activism, based perhaps on examples such as Lee and Mitchell's (1994) "unfolding model" of employee turnover. Such a model could be developed based on a moderate number of detailed interviews with activists, revealing their paths to activism, systematically incorporating common precipitating events or shocks, plans of action, image violations, underlying attitudes and solicitations to become an activist. The aim would be to systematically analyse individuals' experiences and derive a generalizable model, perhaps with specific hypotheses for further testing (Lee *et al.* 1999). There are already biographical studies of workers and union activists, based on biographical interviews (e.g. Mrozowicki & Trappmann 2020; Mrozowicki & Trawinska 2013). This kind of study of union activists in different occupational and union contexts would reveal more about how and why individuals become union activists and would significantly extend our knowledge of participation for this key group.

Finally, it is worth noting that the research on union commitment and participation originated in a period when the traditional industries and occupations were much larger, and union strongholds were dominated by male workers. Clearly, there is a need for future research to reflect the labour market changes of recent years, in particular the larger numbers of female workers and atypical workers, and the general diversity of the workforce, including migrants and members of ethnic minorities. New sectors, occupations and organizational forms should also be reflected in future research, as should the implications of globalization for labour and product markets.

CONCLUSION

This chapter has considered the factors associated with members' commitment to, and participation in, their unions, with the latter conceptualized broadly to encompass formal and informal activities, including what may be considered as pro-union "citizenship" behaviour. Key findings from the literature have been highlighted and the implications for unions discussed. Finally, areas for future research have been considered.

While much of the literature reviewed is social-psychological, with a methodology based upon responses from individual employees and members, an important conclusion to emerge is that members look beyond individual instrumentalism in evaluating their unions. Union-member relations have been shown to go beyond instrumental exchange, to encompass values-based motivations for union commitment and participation, with member-union relationships more accurately labelled as "covenantal" rather than entirely transactional. The findings reviewed are consistent with the argument that unions need to go beyond narrow instrumentalism and should instead pursue a broader social justice agenda. This should be associated with a local union leadership style that emphasizes member participation and aims to build members' sense of shared mission and self-worth as a union member.

The literature has not paid a great deal of attention to directly investigating the association between union participation and wider community citizenship. However, some insights can be derived into the notion of unions as stimulating interest and involvement in broader campaigns for social justice. First, as has been seen, the literature provides support for the notion of "community unionism", suggesting that a focus on narrow economic instrumentalism is likely to be a less effective way forward for unions. To the extent that they are implemented, such approaches draw participating unions into broader social campaigns, involving the championing of a social justice agenda in cooperation with community organizations. Second, there is emerging evidence that members' union commitment and participation may be stronger where the industrial relations climate is seen to be more conflictual. To the extent that this is true, for unions to remain viable they must be essentially independent of management, and not merely passive, cooperative partners in creating harmonious relations. Taking these points together, the vision of unions acting in pursuit of social justice as part of a wider community movement rings true.

REFERENCES

Akoto, E. 2020. "Profiles of ideological and instrumental union bonds: a test in the US and Ghana contexts". *International Journal of Human Resource Management* 31(7): 835–58.

Angle, H. & J. Perry 1986. "Dual commitment and labor-management relationship climates". *Academy of Management Journal* 29(1): 31–50.

Bamberger, P., A. Kluger & R. Suchard 1999. "The antecedents and consequences of union commitment: a meta-analysis". *Academy of Management Journal* 42(3): 304–18.

Barling, J., C. Fullager & E. Kelloway 1992. *The Union and its Members: A Psychological Approach*. New York: Oxford University Press.

Barling, J., E. Kelloway & E. Bremermann 1991. "Preemployment predictors of union attitudes: the role of family socialization and work beliefs". *Journal of Applied Psychology* 76(5): 725–31.

Bass, B. & R. Riggio 2006. *Transformational Leadership*. Mahwah, NJ: Lawrence Erbaum.

Bryson, A. & R. Davies 2019. "Family, place and the intergenerational transmission of union membership". *British Journal of Industrial Relations* 57(3): 624–50.

Chan, A. *et al.* 2006. "Union commitment and participation in the Chinese context". *Industrial Relations* 45(3): 485–90.

Cobble, S. 1991. "Organizing the post-industrial workforce: lessons from the history of waitress unionism". *Industrial and Labor Relations Review* 44(3): 419–36.

Cregan, C., T. Bartram & P. Stanton 2009. "Union organizing as a mobilizing strategy: the impact of social identity and transformational leadership on the collectivism of members". *British Journal of Industrial Relations* 47(4): 701–22.

Dastmalchian, A. 2008. "Industrial relations climate". In *The Sage Handbook of Industrial Relations*, edited by P. Blyton *et al.*, 548–71. Los Angeles, CA: Sage.

Deery, S., P. Erwin & R. Iverson 1999. "Industrial relations climate, attendance behaviour and the role of trade unions". *British Journal of Industrial Relations* 37(4): 533–58.

Deery, S., R. Iverson & P. Erwin 1994. "Predicting organizational and union commitment: the effect of industrial relations climate". *British Journal of Industrial Relations* 32(4): 581–97.

Flanders, A. 1975. *Management and Unions: The Theory and Reform of Industrial Relations*. London: Faber.

Fortin-Bergeron, C., O. Doucet & M.-A. Hennebert 2017. "Relative influence of authentic and transformational leadership of local union representatives on the adoption of union citizenship behaviors". *Leadership and Organization Development Journal* 38(6): 794–811.

Fosh, P. 1993. "Membership participation in work-place unionism: the possibility of union renewal". *British Journal of Industrial Relations* 31(4): 577–92.

Fullagar, C. & J. Barling 1989. "A longitudinal test of a model of the antecedents and consequences of union loyalty". *Journal of Applied Psychology* 74(2): 213–27.

Fullagar, C., D. McCoy & C. Shull 1992. "The socialization of union loyalty". *Journal of Organizational Behavior* 13(1): 13–26.

Fullagar, C. *et al.* 1995. "Impact of early socialization on union commitment and participation: a longitudinal study". *Journal of Applied Psychology* 80(1): 147–57.

Fullagar, C. *et al.* 1995. "Organizational citizenship and union participation: measuring discretionary membership behaviors". In *Changing Employment Relations: Behavioral and Social Perspectives*, edited by L. Tetrick & J. Barling, 311–31. Washington, DC: American Psychological Association.

Fullagar, C. *et al.* 2004. "Union commitment and participation: A 10-year longitudinal study". *Journal of Applied Psychology* 89(4): 730–37.

Fuller, J. & K. Hester 1998. "The effect of labor relations climate on the union participation process". *Journal of Labor Research* 19(1): 173–87.

Gall, G. (ed.) 2009. *Union Revitalisation in Advanced Economies: Assessing the Contribution of Union Organising*. Basingstoke: Palgrave Macmillan.

Gall, G. & J. Fiorito 2012a. "Toward better theory on the relationship between commitment, participation and leadership in unions". *Leadership and Organization Development Journal* 33(8): 715–31.

Gall, G. & J. Fiorito 2012b. "Union commitment and activism in Britain and the United States: searching for synthesis and synergy for renewal". *British Journal of Industrial Relations* 50(2): 189–213.

Gallagher, D. & P. Clark 1989. "Research on union commitment: implications for labor". *Labor Studies Journal* 14(1): 52–71.

Gordon, M. *et al.* 1980. "Commitment to the union: development of a measure and an examination of its correlates". *Journal of Applied Psychology* 65(4): 479–99.

Greene, A.-M., J. Black & P. Ackers 2000. "The union makes us strong? A study of the dynamics of workplace union leadership at two UK manufacturing plants". *British Journal of Industrial Relations* 38(1): 75–93.

Gumber, C. & I. Padavic 2020. "Race differences in motivations for joining unions: the role of prosocial beliefs". *Social Science Quarterly* 101(2): 490–502.

Hammer, T., S. Currall & R. Stern 1991. "Worker representation on boards of directors: a study of competing roles". *Industrial and Labor Relations Review* 44(4): 661–80.

Heery, E., B. Abbott & S. Williams 2012. "The involvement of civil society organizations in British industrial relations: extent, origins and significance". *British Journal of Industrial Relations* 50(1): 47–72.

Heery, E. *et al.* 2000. "Organizing unionism comes to the UK". *Employee Relations* 22(1): 38–57.

Heshiser, B. & J. Lund 1997. "Union commitment types and union activist involvement: lessons for union organizers and labor educators?". *Labor Studies Journal* 22(2): 241–66.

Hickey, R., S. Kuruvilla & T. Lakhani 2010. "No panacea for success: member activism, organizing and union renewal". *British Journal of Industrial Relations* 48(1): 53–83.

Holgate, J. 2013. *Trade Union Involvement in Broad-Based Community Organising: A Comparative Study of London, Sydney and Seattle*. Leeds: Centre of Employment Relations, Innovation and Change.

Holgate, J. 2015. "Community organising in the UK: a 'new' approach for trade unions". *Economic and Industrial Democracy* 36(3): 431–55.

Iverson, R. & S. Kuruvilla 1995. "Antecedents of union loyalty: the influence of individual dispositions and organizational context". *Journal of Organizational Behavior* 16(6): 557–82.

James, P. & J. Karmowska 2016. "British union renewal: does salvation really lie beyond the workplace?". *Industrial Relations Journal* 47(2): 102–16.

Kahn, R. & A. Tannenbaum 1957. "Union leadership and member participation". *Personnel Psychology* 10(3): 277–92.

Kelloway, E. & J. Barling 1993. "Members' participation in local union activities: measurement, prediction and replication". *Journal of Applied Psychology* 78(2): 262–79.

Kelloway, E., V. Catano & A. Carroll 1995. "The nature of member participation in local union activities". In *Changing Employment Relations: Behavioral and Social Perspectives*, edited by L. Tetrick & J. Barling, 333–47. Washington, DC: American Psychological Association.

Kelloway, K., V. Catano & R. Southwell 1992. "The construct validity of union commitment: development and dimensionality of a shorter scale". *Journal of Occupational and Organizational Psychology* 65(3): 197–211.

Kelly, C. & J. Kelly 1994. "Who gets involved in collective action? Social-psychological determinants of individual participation in trade unions". *Human Relations* 47(1): 63–88.

Kerr, S. & J. Jermier 1978. "Substitutes for leadership: their meaning and measurement". *Organizational Behavior and Human Performance* 22(2): 375–403.

Kuruvilla, S. *et al.* 1990. "Union participation in Japan: do western theories apply?". *Industrial and Labor Relations Review* 43(3): 374–98.

Kwon, S. 2013. "The impact of transformational leadership on members' willingness to participate in union activities in Korea: exploration of universality and substitutability for antagonism". *International Journal of Human Resource Management* 24(2): 265–84.

Lee, T. & T. Mitchell 1994. "An alternative approach: the unfolding model of voluntary labor turnover". *Academy of Management Review* 19(1): 51–89.

Lee, T. *et al.* 1999. "The unfolding model of voluntary turnover: a replication and extension". *Academy of Management Journal* 42(4): 450–62.

McShane, S. 1986a. "The multidimensionality of union participation". *Journal of Occupational Psychology* 59(1): 177–87.

McShane, S. 1986b. "General union attitude: a construct validation". *Journal of Labor Research* 7(4): 403–17.

Magenau, J., J. Martin & M. Peterson 1988. "Dual and unilateral commitment among stewards and rank-and-file members". *Academy of Management Journal* 31(2): 359–76.

Marchington, M. 1998. "Partnership in context: towards a European model?". In *Human Resource Management: The New Agenda*, edited by P. Sparrow & M. Marchington, 208–25. London: Financial Times Pitman.

Martínez-Íñigo, D., T. Zacharewicz & E. Kelloway 2020. "The mediating role of shop stewards' union citizenship behavior in the relationship between shop stewards' and union members' loyalty: a multilevel analysis". *European Journal of Work and Organizational Psychology* 29(6): 880–8.

Metochi, M. 2002. "The influence of leadership and member attitudes in understanding the nature of union participation". *British Journal of Industrial Relations* 40(1): 87–111.

Meyer, J. & A. Morin 2016. "A person-centered approach to commitment research: theory, research and methodology". *Journal of Organizational Behavior* 37(4): 584–612.

Milkman, R. 2013. "Back to the future? US labor in the new gilded age". *British Journal of Industrial Relations* 51(4): 645–65.

Morin, A. *et al.* 2021. "Investigating the dimensionality and stability of union commitment profiles over a 10-year period: a latent transition analysis". *Industrial and Labor Relations Review* 74(1): 224–54.

Mrozowicki, A. & V. Trappmann 2020. "Precarity as a biographical problem? Young workers living with precarity in Germany and Poland". *Work, Employment and Society* 35(5): 221–38.

Mrozowicki, A. & M. Trawinska 2013. "Women's union activism and trade union revitalization: the Polish experience". *Economic and Industrial Democracy* 34(2): 269–89.

Mustchin, S. 2012. "Unions, learning, migrant workers and union revitalization in Britain". *Work, Employment and Society* 26(6): 951–67.

Newton, L. & L. McFarlane Shore 1992. "A model of union membership: instrumentality, commitment and opposition". *Academy of Management Review* 17(2): 275–98.

Nicholson, N. 1976. "The role of the shop steward: an empirical case study". *Industrial Relations Journal* 7(1): 15–26.

Nicholson, N., G. Ursell & J. Lubbock 1981. "Membership participation in a white-collar union". *Industrial Relations* 20(2): 162–78.

Organ, D., P. Podsakoff & S. MacKenzie 2006. *Organizational Citizenship Behavior: Its Nature, Antecedents and Consequences*. Thousand Oaks, CA: Sage.

Reed, C., W. Young & P. McHugh 1994. "A comparative look at dual commitment: an international study". *Human Relations* 47(6): 1269–93.

Redman, T. & E. Snape 2004. "Kindling activism? Union commitment and participation in the UK fire service". *Human Relations* 57(7): 845–69.

Redman, T. & E. Snape 2014. "The antecedents of union commitment and participation: evaluating moderation effects across unions". *Industrial Relations Journal* 45(6): 486–506.

Rosenfeld, J. 2014. *What Unions No Longer Do*. Cambridge, MA: Harvard University Press.

Skarlicki, D. & G. Latham 1996. "Increasing citizenship behavior within a labor union: a test of organizational justice theory". *Journal of Applied Psychology* 81(2): 161–9.

Smith, C., L. Duxbury & M. Halinski 2019. "It's not just about paying your dues: impact of generational cohort on active and passive union participation". *Human Resource Management Journal* 29(2): 371–94.

Snape, E. 1994. "Union organizing in Britain: the views of local full-time officials". *Employee Relations* 16(8): 48–62.

Snape, E. & T. Redman 2004. "Exchange or covenant? The nature of the member-union relationship". *Industrial Relations* 43(4): 855–73.

Snape, E. & T. Redman 2012. "Industrial relations climate and union commitment: an evaluation of workplace-level effects". *Industrial Relations* 51(1): 11–28.

Snape, E., T. Redman & A. Chan 2000. "Commitment to the union: a survey of research and the implications for industrial relations and trade union". *International Journal of Management Reviews* 2(3): 205–30.

Tan, H. & S. Aryee 2002. "Antecedents and outcomes of union loyalty: a constructive replication and an extension". *Journal of Applied Psychology* 87(4): 715–22.

Tapia, M. 2013. "Marching to different tunes: commitment and culture as mobilizing mechanisms of trade unions and community organizations". *British Journal of Industrial Relations* 51(4): 666–88.

Tapia, M. 2019. "'Not fissures but moments of crises that can be overcome': building a relational organizing culture in community organizations and trade unions". *Industrial Relations* 58(2): 229–50.

Twigg, N., J. Fuller & K. Hester 2008. "Transformational leadership in labor organizations: the effects on union citizenship behaviors". *Journal of Labor Research* 29: 27–41.

Wills, J. 2008. "Making class politics possible: organizing contract cleaners in London". *International Journal of Urban and Regional Research* 32(2): 305–23.

Wills, J. 2009. "Subcontracted employment and its challenge to labor". *Labor Studies Journal* 34(2): 441–60.

CHAPTER 17

COMMITMENT TO, AND ACTIVISM WITHIN, LABOUR UNIONISM

Jack Fiorito, Andrew Keyes, Zachary A. Russell and Pauline de Becdelièvre

ABSTRACT

This chapter addresses commitment and activism in labour unions. We first define our topic and the relationship between these two key concepts and their relationships with other "unionism terms". Next, we review selected theoretical perspectives and selected "classics" in the scholarly literature on commitment and activism. Following that, we review findings on possible antecedents, emphasizing meta-analytic and recent reports. We then turn attention to a few critical issues for the future, among them younger workers, new technology and measurement of commitment. We conclude with a brief comment on the importance of commitment and activism for union vitality and renewal.

Keywords: Union commitment; union activism; union activists; challenges for unions

INTRODUCTION

In this chapter we focus upon the relationship between workers and their unions, and the role that relationship plays in worker activism on behalf of unions. Various terms have been used to refer to the relationship between workers and unions, including "attachment". A more common term in the scholarly literature is "commitment", which is often recognized as multidimensional, and consequently differentiated into sub-dimensions such as belief in unionism, loyalty, responsibility and willingness-to-work on behalf of the union. Additional closely related terms include union citizenship behaviours (see Chapter 16) and allegiance. All these terms focus heavily on the affective, attitudinal and intentional orientation of the members towards their unions. Many other union-related concepts, such as union satisfaction (satisfaction with the union) or perceived union support (of workers), may be related to union commitment, but are seen as distinct concepts.

There are other relationships between workers and their unions that are not usually considered part of the commitment construct. Notably, workers have quasi-contractual relationships with unions wherein they agree or at least accept that the union will be their exclusive bargaining agent, their advocate for public policy issues, a provider of direct benefits and (often) that the union will be compensated for its services, typically through dues payments. To this point we have deliberately avoided the term "union member" to emphasize that the relationships of interest can also involve non-members represented by the union but who do not pay dues, also known as "free riders". Most often, however, discussions of union commitment refer to relationships between union members and their unions, with little or no reference to non-members (see Conlon & Gallagher 1987).

Activism here refers to voluntary behaviours by workers intended to support their unions. An alternative term, "participation", is a close synonym, as is "union citizenship behaviour", but the activism term is sometimes preferred to emphasize that the more traditional "participation" term can refer to rather passive actions, such as reading a union newsletter, attending a union meeting, voting in a union election or a union contract ratification ballot, or any number of newer low-demand acts, such as posting or reading material on X, WhatsApp, Snapchat or Facebook. Sometimes participation solely in these low-demand activities is even disparaged as "slacktivism". More demanding participation or activism forms, such as chairing or serving on a union committee, serving as a steward or officer, recruiting new members, walking a picket line, carrying a protest sign, striking, arguing for a particular agenda or mobilizing other members, might be more properly termed "activism". Reading a union newsletter and similarly low-demand actions are, nonetheless, forms of activism or participation. That they are relatively easy actions and far less difficult for union leaders to elicit from workers makes this no less true. That lower-demand acts may be prerequisites or "gateways" to more extensive and more demanding activism also cautions against their neglect. As with commitment, activism discussions are usually in reference to union members only, although clearly non-members can engage in many of the acts just described.

Why do these concepts, commitment and activism, matter? At least two main reasons come readily to mind. First, unions are by tradition and to some extent by law expected or required to be democratic organizations responsive to those they represent, and in particular their members. Participation in unions is one important indicator of union democracy in the sense of reflecting a "representational imperative" (Child, Loveridge & Warner 1973; also see Chapter 16). Second, an "administrative imperative", referring to how unions function as organizations, their effectiveness, in particular, requires that unions focus resources upon achieving union goals. For most unions, member volunteers are a critical resource. Although union staff and some officers are paid to serve members, members generally seem unwilling to "tax" themselves sufficiently through dues to provide an acceptable level of service. Unions, therefore, rely heavily upon volunteer efforts, namely union participation by members, to supplement limited paid staff and officer resources. "[M]any members see dues as a 'payment' for a union to provide representation services in the workplace. Unions are, however, in the peculiar situation of then having to turn around and ask those same members to volunteer to help them deliver those services" (Clark 2009: 15). As noted elsewhere and with a nod to Abraham Lincoln, unions are *of*, *by* and *for* the workers. Most discussions of "union renewal" – the revitalization of what in many instances seem to be moribund unions – cite a critical role for activism (see, e.g. Fosh 1993; Gall & Fiorito 2012b; Hickey, Kuruvilla & Lakhani 2010; Jódar, Vidal & Alós 2011).

Although our response to the "why it matters" question thus far has focused upon participation or activism, the role of commitment is clearly evident. Perhaps most obviously, the widely recognized union commitment dimension of "willingness-to-work", that is, volunteer for the union, is a prerequisite for activism. Indeed, prior studies have documented a strong link from commitment to activism, and both theory and evidence support a causal attribution running from commitment to activism (Bamberger, Kluger & Suchard 1999; Barling, Fullagar & Kelloway 1992: 109; Fullagar *et al.* 2004; Gall & Fiorito 2012b). Both union commitment and union activism have been justly referenced as important to "the very fabric of unions" (Barling, Fullagar & Kelloway 1992: 95; Gordon *et al.* 1980: 480). And importantly, of all the myriad factors that influence union success or failure, member attitudes, beliefs and intentions regarding unions are among the critical factors over which unions have greatest control (Clark 2009: 18–19; Gallagher & Clark 1989).

Our goal in this chapter is to offer a brief review of the general literature on union commitment and activism. Following that we identify a limited number of critical contemporary issues, summarizing and assessing the state of knowledge on those, providing suggestions for future research and implications for unions and public policy.

THEORY AND LITERATURE ON UNION COMMITMENT AND ACTIVISM

We cannot do justice to extant theorizing and research on commitment and activism within one chapter (see also Chapter 16). Instead, we identify selected contributions that have previously summarized knowledge on these topics or that represent particularly notable contributions, most of these dating from the previous century. We will then focus on selected contributions in the current century and some of the issues that are most salient today.

Research on union commitment and activism dates back to the 1950s and earlier. Some of the earliest work focused on the phenomenon of "dual allegiance" (e.g. Purcell 1954; Stagner 1954). Interest in this topic was in part based on the observation that workers, despite the sentiment of the labour ballad, "Which Side Are You On?" and Marxist and pluralist tenets on the inevitable conflict of interests between workers and capitalists (Budd 2005), often expressed loyalty or allegiance towards both their union and their employer. Research on activism began to appear at roughly the same time, with an initial emphasis on identifying activist types.

A resurgence in union commitment research occurred in the early 1980s. This integral stage of union commitment literature developed from advancements in the more general organizational commitment literature (Barling, Fullagar & Kelloway 1992; Morin *et al.* 2021). Indeed, Gordon *et al.*'s (1980) seminal work that initiated the resurgence of union commitment research had built upon this literature (e.g. Porter, Crampon &

Smith 1976; Steers 1977), and many of the items were sourced directly from existing organizational commitment measures. Since Gordon and colleagues' work, thousands of articles and book chapters have investigated the antecedents, consequences and contextual factors related to union commitment.

Theoretical perspectives on commitment and activism

Klandermans (1986: 189–91) identifies three main theories that are used in research on union activism: frustration-aggression theory, rational choice theory and interactionist theory. In the first, union activity is a reaction to "frustration, dissatisfaction, or alienation in the work situation" (Klandermans 1986: 190). Rational choice theory emphasizes the individual's perceived "costs and benefits" of activism (Klandermans 1986: 190). Finally, interactionist theory emphasizes "the networks and groups" inside and/or outside the workplace, and that activism "is inextricably bound up in group culture" (Klandermans 1986: 190). Klandermans subsequently discusses motives for union joining, which he considers a form of union activism or participation. The essential ideas clearly apply more broadly to various forms of labour unionism. These motives relate to varying degrees to the aforementioned theories, and are described as:

> 1. instrumental motives: people participate because they think they stand to gain by it ...;
> 2. ideal-collective motives: people participate because of the general societal functions of the union movement as a movement of change [and];
> 3. social pressure: people participate because of social pressure from colleagues, family members, etc.
>
> (Klandermans 1986: 192)

Over time, researchers have tended to rely increasingly on a general theoretical framework for behavioural intentions and behaviour advanced by Ajzen (1991), called the Theory of Planned Behaviour (TPB), and its forerunners including the Theory of Reasoned Action (e.g. Fishbein & Ajzen 1975). In its later variations of TPB, the concept of perceived control or self-efficacy was added, although this influence was recognized earlier and alluded to multiple times in Klandermans (1986: 193); for example, in citing a 1981 finding that "participation in union decision-making strengthened commitment to the union". The Ajzen TPB framework (1991) can be summarized thus: Behavioural intentions and subsequent behaviour are a function of:

1. Attitudes towards the behaviour;
2. Subjective norms (salient others' attitudes towards the behaviour);
3. Perceived behavioural control (i.e. self-efficacy regarding the behaviour).

Attitudes and affect (feelings) are based upon beliefs, and most pertinent attitudes here could include perceptions of union instrumentality, perceptions of unions' broader social roles, and other beliefs and feelings such as perceived union support and union satisfaction. Salient others include colleagues or co-workers (and supervisors), family and friends. Beliefs about union democracy can be considered another basis for attitudes towards the behaviour. They might also be seen as an indicator of perceived control, not in the narrower sense of whether the behaviour is volitional as emphasized in Ajzen's 1991 explication, but in the broader sense that workers perceive their efforts will make a difference (Cregan, Bartram & Stanton 2009: 705; Fiorito, Padavic &DeOrtentiis 2015). Ajzen also noted that for any specific behaviour or behavioural intention there may be situational "background factors". Job satisfaction comes to mind as a potential background factor in considering union commitment and activism. Job satisfaction is not a belief or attitude concerning the behaviour or behavioural intention per se (say, willingness-to-work for the union), but job dissatisfaction may be a motivating force for seeking change, with union activism seen as a possible mechanism for bringing about positive change per frustration-aggression theory.

Selected classics

Barling, Fullagar and Kelloway (1992) provide the most thorough review of union commitment and activism up to that time, providing in-depth chapter-length reviews on each topic. Gallagher and Strauss (1991) offer a slightly broader focus but with considerable attention to commitment and activism in a more concise presentation. Snape, Redman and Chan's (2000) review article is specifically focused on union commitment, but also offers some effort to link commitment findings to other "downstream" phenomena such as membership, activism and renewal. Clark (2009) does not expressly focus on commitment and activism, but gives considerable attention to the issue of new-member socialization as a means to build commitment and activism. Whereas Barling, Fullagar and Kelloway (1992), Gallagher and Strauss (1991) and Snape, Redman and Chan (2000) tell us what the literature says in considerable detail, Clark (2009) offers less detail on that and more of a "how to" approach for building commitment and activism based on the literature.

Gordon *et al.*'s (1980) detailing of the development and testing of an instrument for measuring union commitment also deserves mention among the classics. They offered what was arguably the first psychometrically rigorous effort to measure union commitment. Their influential work became a stepping-off point for much subsequent research that tested, validated, challenged and refined theorizing about union commitment. Bamberger, Kluger and Suchard's work (1999) also merits mention. Most notable about it is its use of meta-analysis to quantitatively summarize evidence on relations between commitment, participation and other variables, and the use of that evidence to evaluate alternative theoretical models via structural equation modelling. None of the models considered produced an adequate fit, but an "integrative model" based on melding rival models produced the best fit.

Although there are other contributions that merit placement on our list of "classics", just one more work will be mentioned, namely Klanderman (1986), as an example of an impressive body of work. He is notable for succinctly summarizing three main theories and reviewing evidence from both North America and Europe. Also noteworthy is his conceptualization of various union member actions as manifestations of "participation", although conceptualizing these and other union-related acts and cognitive states (beliefs, feelings, attitudes, intentions, etc.) as forms of "unionism" might also be useful (see Fiorito, Padavic & Russell 2018).

What the classics tell us

One of the persistent findings in studies of union commitment and activism is their multidimensionality. Gordon *et al.* (1980) identify distinct commitment dimensions of belief in unionism, responsibility, loyalty and willingness-to-work. Subsequent studies sometimes challenge this schema, with some arguing that belief in unionism is more an antecedent to commitment rather than a facet of commitment. Others call for a simplification along lines of a distinction between feelings (i.e. loyalty) and behavioural intentions (responsibility and willingness). Some argue for a different but related schema specifying affective commitment and continuance (of membership) commitment. Taking a person-centred approach, Morin *et al.* (2021) challenge the utility of union commitment facet distinctions, finding distinct "profiles" or types of commitment reflecting different levels, but not qualitative differences; that is, a low-commitment member is typically low on loyalty, responsibility and willingness-to-work for the union, and a high-commitment member is typically high on all three facets.

Similarly, studies of activism draw various distinctions. Passive (e.g. reading a union communication) and active participation (e.g. serving on a committee) are differentiated, and a parallel distinction is drawn between low-demand and high-demand forms of activism (Kelly & Kelly 1994). McShane (1986a, 1986b) identifies "administrative participation", a construct defined to include relatively high-demand activities such as running for or holding elected office, serving on a committee or editing a newsletter, but also certain less-demanding activities such as proposing a motion, seconding a motion, or offering views at a union meeting, after finding these latter agenda-setting activities related more closely to the more active forms than to the more clearly passive act of attending meetings. McShane (1986b) and more recent studies (e.g. Fiorito, Padavic & DeOrtentiis 2015; Gall & Fiorito 2012a, 2012b; Smith, Duxbury & Halinski 2019) tend to favour the distinction between more passive and more active forms of activism, arguing that determinants of each may differ and that more active forms are in greatest need to advance union renewal.

Distinctions have also been drawn between formal and informal participation, with formal participation (e.g. holding elected office) generally a more active form. Gall and Fiorito (2012a) also expressly note a need for "an activism of conception rather than just execution", recognizing value in creative efforts such as agenda-setting. McKay *et al.* (2020) advocate the person-centred approach, emphasizing that there are various stable types of activists, but that individuals sometimes transition among types. In McKay *et al.*'s (2020) schema, many unions would like to transition some of their "silent supporters" to "high participators" or "leadership participators".

Before turning to a review of findings on commitment and activism determinants, we register our agnosticism about levels of aggregation for commitment and activism.

"Researchers choose levels of abstraction consistent with their research interests" (Schwab 2005: 28). Those who prefer using specific facets such as commitment's "willingness-to-work" or activism's "administrative participation" often express a concern that particular facets may have different determinants or that determinants will not be uniformly important for all facets. There is some evidence in support of these views. But there is also evidence that union commitment facets are strongly positively correlated, suggesting there may be little gained in differentiating among them when discussing commitment (Morin *et al.* 2021). The broader concepts of unidimensional commitment and activism are meaningful, often used in research and have proved useful in advancing understanding of commitment and activism, their causes and consequences.

Findings on union commitment and activism

In our summary of findings for union commitment and activism, we favour meta-analyses, post-2000 studies and reviews, and attempt to include studies from several nations over and above the UK and the US. Apart from the meta-analyses, it is notable that most studies have used fairly homogenous samples from a single worksite or local union. This leads to a relative neglect of situational factors that are constant within samples but may vary considerably across samples (see Fuller & Hester 1998 on labour relations climate; and Chapter 16).

Demographic characteristics

Several studies examine various demographics such as gender, age, ethnic or racial groupings and union membership tenure. To some extent these are measures of convenience or they are used to address the question of *who* is active, but, in many instances, there are substantive attributions that involve notions such as shared experiences of particular groups. Results for gender measures reflect varied interpretations or influences. Women have been found to express greater loyalty to unions, perhaps stemming from union pressures against discrimination, but lesser willingness-to-work for the union and lower participation (e.g. Snape & Redman 2004; Snape, Redman & Chan 2000). The contrast is attributed to societal gender-related role demands. This is suggested in one study where being female had no apparent effect on participation, but being a mother was associated with lower participation (Kuruvilla *et al.* 1990). Most studies find no gender effect per se.

Age variables also show mixed results, with some indication of greater commitment but no greater participation or even lower levels of participation for older workers. Variables for Hispanic and Black ethnic or racial identification yield effects generally ranging from none to slightly positive influences on commitment and activism (Fiorito, Gall & Martinez 2010). Interpretations suggest shared group experiences of discrimination foster greater appreciation for collective action through unions. Age shows little relation to participation, although a negative effect is sometimes found. Union tenure shows a positive or nonsignificant relation to participation (Fiorito, Padavic & DeOrtentiis 2015; Snape & Redman 2004), as do more liberal political leanings (Fiorito, Gall & Martinez 2010).

What is perhaps most striking about demographic influences from our selective review is the general lack of attention to demographics in certain contexts, and in those studies that do examine demographics, the abundance of mixed and nonsignificant results. Regarding the first point, such omissions may be particularly notable among studies using structural equation modelling, such as those by Bamberger, Kluger and Suchard (1999), Fullagar *et al.* (2004) and Smith, Duxbury and Halinski (2019). Whether this possible neglect stems from desires for parsimony, previous mixed and nonsignificant findings, or structural equation modelling limitations is unclear. We would like to think it is the first two factors, consistent with our second point, echoing Snape, Redman and Chan's (2000: 210–11) review of findings on "personal characteristics", which reports few if any persistent or strong effects, and Klandermans' (1986: 194) earlier conclusion that "d"emographic variables seem to be of little influence, especially in multivariate analysis'. Morin *et al.* (2021) provide one of the few instances where authors initially consider demographic influences and then explicitly discard such measures based on their initial findings of little or no influence.

Cognitive influences on activism

Researchers have found more fertile ground in the exploration of cognitive influences on union commitment and activism. We refer here to influences such as beliefs, affect or feelings, and attitudes. To some extent, common method variance in single-administration surveys may be at play, although such concerns can be overblown (Lance *et al.* 2010). Construct overlap may also be a factor cautioning against making too much of some findings. Still, sound theoretical bases (e.g. Ajzen 1991) and repeated findings in various settings suggest most findings in this area do, in fact, represent causal relations among the constructs of interest.

The most persistent finding on union commitment and activism is the positive and substantial relationship between the two, in multiple meta-analyses (e.g. Bamberger, Kluger & Suchard 1999; Chen 2020; Fuller & Hester 1998; Monnot, Wagner & Beehr 2011), qualitative reviews (e.g. Barling, Fullagar & Kelloway 1992; Clark 2009; Gallagher & Strauss 1991; Snape, Redman & Chan 2000), and original empirical studies (e.g. Dawkins 2016; McKay *et al.* 2020; Morin *et al.* 2021; Tetrick *et al.* 2007), using either commitment facets or unidimensional commitment measures. Special note among the empirical studies is due to Fullagar *et al.*'s (2004) study focusing on the causal direction of the commitment-activism relationship, in which they provide persuasive evidence that the dominant influence runs from commitment to activism rather than the reverse. Using cross-lagged regression, they show that union commitment predicted union activism ten years later, but activism did not predict commitment ten years on. These strong, persistent and theoretically important findings naturally draw our attention to commitment and its antecedents in the causal chain that promotes activism, but first we should note some other findings regarding determinants of activism.

Results are less convincing on other activism influences, but some are strong enough and persistent enough to warrant mentioning. General union attitudes are often posited as a commitment antecedent, but direct positive effects on activism have also been noted in meta-analyses and numerous original studies (e.g. Chen 2020; Fiorito, Gall & Martinez 2010, 2015; Kuruvilla *et al.* 1990). In addition, union instrumentality has shown direct effects (e.g. Kuruvilla *et al.* 1990) as well as indirect effects mediated via general union attitude (e.g. Snape & Redman 2004). Measures reflecting subjective norms – the views of important others (Ajzen 1991) – have also been linked to activism (e.g. Sverke & Kuruvilla 1995). An additional influence on activism grounded in Ajzen's (1991) 'TPB', perceived control or self-efficacy, has shown a strong influence (e.g. Fiorito, Padavic & DeOrtentiis 2015; Peetz, Webb & Jones 2002) and was recognized as an influence much earlier, as were others' intentions. Gallagher and Strauss (1991: 157 emphasis added) noted that Klandermans' (1984) resource mobilization theory suggested that "members participate in union activities when they are convinced the goal is important, that *their own participation will make a difference, that others will participate, and that together they will be successful*".

Cognitive influences on union commitment

As Chen (2020) concludes from a recent meta-analysis, union commitment and particularly its willingness-to-work-for-the-union facet, have done "yeoman's work" in predicting union activism. Accordingly, considerable attention has focused on determinants of union commitment. Persistent findings have been noted for several variables.

General union attitude and similar measures (e.g. belief in unionism, feelings towards unions, pro-union attitude) show persistent, positive impacts (e.g. Monnot, Wagner & Beehr 2011). In Gordon *et al.* (1980), belief in unionism is considered a facet of union commitment, raising some questions about whether these findings are partly due to construct overlap. Over time, however, the literature has moved towards conceiving general union attitude or belief in unionism – sometimes referenced in terms such as "ideological" or "identification" – as antecedent to union commitment both in unidimensional form and in terms of the loyalty, responsibility and willingness-to-work facets. This makes sense in that beliefs are a key basis for feelings and attitudes.

At least one recent study, namely Fiorito and Padavic (2022), encourages researchers to move away from reliance upon general union attitude to instead focus on the underlying beliefs that give rise to such attitudes that over time become the "emotional residue" of previous and now-detached beliefs (Montgomery 1989: 266–7). Morin *et al.* (2021), like Fiorito and Padavic (2022), note that general union attitude measures tend to focus more on the role of unions in the broader society, or pro-social attributes, and that the role of these pro-social aspects relative to union instrumentality, which focuses more on self-interest or more immediate gains in one's workplace, may be understated. Cregan, Bartram and Stanton (2009: 703, 705–6) similarly argue that collectivism, operationalized as union commitment's loyalty and willingness-to-work facets, is a "social pattern" of willingness to give "priority to group goals over ... personal goals", with "social identification" playing a key mediating role between transformational leadership and collectivist attitudes. As Flanders (1970: 15) argued long ago: "[I]t is as a sword of justice rather than a vested interest that [the union movement] generates loyalties and induces sacrifices among its own members, and these are important foundations of its strength and vitality."

Union instrumentality or effectiveness perceptions have long been thought to motivate union commitment, as suggested in the quote summarizing Klandermans' (1984) theory on participation above. Multiple meta-analyses and original studies specify union instrumentality as a determinant of both general union attitude and union commitment, with general union attitude as a partial or full mediator of instrumentality's influence on commitment (e.g. Monnot, Wagner & Beehr 2011). In other words, union instrumentality perceptions may bolster pro-union attitudes but also directly increase union commitment, at least unidimensional commitment and certain facets. Original studies have repeatedly failed to find a relationship between union instrumentality and willingness-to-work for the union, attributing such motivation to more ideological or pro-social rather than pragmatic self-interest motivations.

Perceived union support has been found to enhance commitment in multiple original studies since 2000, namely those by Fuller and Hester (2001), Snape and Redman (2004) and Tetrick *et al.* (2007). Its distinctness from general union attitude, union instrumentality and union satisfaction is somewhat unclear, however, and findings for its influence in the absence of such predictors leave some doubt as to whether omitted variables bias is inflating findings for perceived union support. At least one study, namely Snape and Redman (2004), found effects on commitment for both perceived union support and for general union attitude when both are specified as predictors. The conceptual basis for a "perceived union support" effect is strong, yet doubt remains as to distinctness among it, general union attitude and perhaps satisfaction with union representation.

Job satisfaction has been proposed as an inhibitor of union commitment, based on the plausible argument that dissatisfied workers see union commitment and participation as means to address and remedy sources of job dissatisfaction. Findings for this hypothesis have been mixed and, interestingly, Fuller and Hester (1998) find evidence for a moderating effect of industrial relations climate. In positive climate settings, job satisfaction shows a positive effect on commitment, whereas in negative climate settings, job satisfaction shows a negative effect on union commitment. Where relations are good, unions appear to share the credit, but where relations are bad, better terms and conditions are apparently credited to the employer and reduce union commitment.

As noted above, organizational (employer) and union commitment were linked long ago in findings of "dual allegiance". Studies continue to find positive relationships between the two commitment foci (e.g. Bamberger, Kluger & Suchard 1999; Chen 2020; Monnot, Wagner & Beehr 2011). Whether there is a causal link is less clear, as this positive relationship may also be attributable to individual predispositions or correlated causes. Given that job satisfaction has been linked strongly to organizational commitment (e.g. Bamberger, Kluger & Suchard 1999; Monnot, Wagner & Beehr 2011), it is surprising that union satisfaction has rarely been considered as a predictor of union commitment in recent studies, and it is absent from meta-analyses by Bamberger, Kluger and Suchard (1999), Chen (2020), Fuller and Hester (1998), and Monnat, Wagner & Beehr (2011).

Morin *et al.* (2021: 243) report moderately strong correlations between union satisfaction and union commitment measures, but in their models predicting commitment profile membership, models including predictors for general union attitude and union instrumentality, "the effects of union satisfaction were limited".

Finally, a smattering of findings has revealed other antecedents of commitment, most notably two influences that are "social" in nature. These include subjective norms (e.g. Vandaele 2020) and early socialization (Clark 2021; Fullagar *et al.* 1995; Fuller & Hester 1998). Subjective norms refer to perceptions of salient others' views of the activity or intentions in question (Ajzen 1991). Those others may include friends, relatives and co-workers. Parental or spousal union membership is sometimes a proxy in this vein. In any case, the central concept is that people feel a desire to conform to expectations of those whose views they value (Bendix 1956: 314; Mayo 1933). Although more direct measures are rarely examined, there is some support for the concept in findings.

The second and loosely related influence in this vein refers to early socialization-to-the-union experiences, both in terms of formal "orientation" and informal socialization via co-workers. An interesting finding in this realm is that the informal socialization experiences appear more influential than formal programmes (Fullagar *et al.* 1995). This suggests that unions should encourage their members to welcome and orient new members, but later results also support union efforts to develop effective formal new member orientation (NMO) programmes: "Data indicate that members who had a positive NMO experience had higher levels of overall union commitment, and higher levels for the three dimensions of commitment than members who did not have a positive experience" (Clark 2021: 166–7). As Clark and others (e.g., Clark 2009; Fullagar *et al.* 1995) have long emphasized, union socialization or orientation programmes are one of the few union commitment influences over which unions have relatively unfettered control; that is, largely free from employer influence.

CRITICAL ISSUES IN COMMITMENT AND ACTIVISM FOR THE FUTURE

In this section, we take up selected commitment and activism issues that are critical for future research and practice. Some of these have been touched upon above, but they deserve further attention. These include younger workers, new technology, synergy between research on organizational and union commitment, and issues at the intersection of union commitment, activism and renewal.

Younger workers

In an effort to stem declining membership in countries all over the world, scholars and unions have turned their attention to young workers, often viewed as those under the age of 30 (e.g. Haynes, Vowles & Boxall 2005; Tailby & Pollert 2011) or under the age of 35 (e.g. Tapia & Turner 2018; Vandaele 2020). In many countries, including Canada, Western European nations, the US, Australia and New Zealand, the young worker unionization rate is much lower than in the overall workforce (Bailey *et al.* 2010; BLS 2023;

Budd 2010; Haynes, Vowles & Boxall 2005; Serrano Pascual & Waddington 2000, cited in Waddington & Kerr 2002; Tailby & Pollert 2011), and the average age of union members has increased since the historical peak of unionization rates (Schmitt & Warner 2010). The increase in the average age of union members and the low unionization rate among young people have raised questions about the causes of these changes and whether they are related to young workers' attachment or commitment to labour unions.

Specifically, do the prior predictors of union commitment and activism apply to contemporary young workers? Some scholars posit there are generational differences that may account for the decline among unionized young workers – with these generational changes reflected in youth's attitudes about unions. Younger people are viewed as more individualistic and instrumental than prior generations. Consequently, younger workers identify with unions due to more instrumental reasons (Smith & Duxbury 2019). It is thought these differences have caused a misalignment between young workers and traditional union values (Freeman & Diamond 2003; Waddington & Kerr 2002). Because of this, fewer young workers may be committed or active.

Smith, Duxbury and Halinski (2019) report that "Millennials" (born between 1980 and 2000) reported lower activism than "Generation Xers" (born 1965–79), who in turn reported lower activism than "Baby Boomers" (born 1945–64). These statistically significant differences held for both passive and active forms of union participation. A key independent variable, "relationship with union", measuring how satisfied members are with how the union connects and interacts with them, enhanced participation for all three cohorts, but with some differences in how influential it was across cohorts, and evidence that age moderates the relationship between it and both passive and active participation.

Alternatively, others attribute the decline in young workers in unions to changes in the broader labour market and the political economy (Hodder 2015; Hodder & Kretsos 2015). Since the height of unionization, employment growth has occurred in sectors with lower unionization rates (Waddington & Kerr 2002), such as the service sector; new forms of work have increased part-time and temporary employment, which has made it more difficult for unions to organize these workers (Fiorito *et al.* 2021; Hodder 2015); and employers have increased their resistance to unions in the workplace (Waddington & Kerr 2002). Together, it is suggested these labour market conditions have reduced the ability of unions to organize young workers. In addition to labour market conditions, government reforms are also proposed to have influenced unionization rates. These include government actions, including the outlawing of the closed shop in the UK (Tailby & Pollert 2011), the Macron Ordonnances in France (Fulton 2020; Rehfeldt & Vincent 2018), and limits to public sector bargaining in the US (Milkman 2013). In summary, one assumed cause of lower rates of unionization among young workers is the structure of modern labour markets and current government regulation.

Finally, the actions of unions themselves have also been held responsible for the decline of young workers participating in unions. Because union socialization is believed to influence union commitment (Clark 2021; Gall & Fiorito 2012b), it is important for unions to socialize new members, but sometimes unions do an insufficient job, and in extreme cases some young workers report they were never asked to join the union (Waddington & Kerr 2002). As a result, reasons cited for the decline of unionized young workers are that unions have been "inefficient" in recruiting them (Hodder 2015; Hodder & Kretsos 2015) and unions were slow to adopt strategies and programmes targeted at young workers. In addition to socialization, union leadership, union organization and union policies are also thought to influence commitment towards unions (Clark 2009; Gall & Fiorito 2012b; Smith, Duxbury & Halinski 2019). Thus, socialization, leadership and structures of unions all guide young worker commitment to unions.

Of the reasons cited for decline among union members and union commitment, attitudes towards labour unions are especially important because they are thought to be an antecedent of union commitment (Bamberger, Kluger & Suchard 1999; Gall & Fiorito 2012b). Despite the various explanations proposed by scholars for the decline of youth unionization, several studies report that younger generations do not have more negative attitudes towards labour unions (e.g. Bryson, Gomez & Willman 2010; Freeman & Diamond 2003; Haynes, Vowles & Boxall 2005). In fact, using time-lag data on high school seniors, Aleks, Saksida and Wolf (2021) conclude that attitudes towards unions have remained largely consistent since the 1970s, but the share of youth having no opinion about unions has more than doubled. Haynes, Vowles and Boxall (2005) conclude that individualism among younger workers is not the cause of the decline of young union members. These findings suggest that other factors have contributed to the decline. Instead of differences in attitudes, changes in the labour market, employer opposition to unions, and government regulation hostile to unions are proposed as contributors to the decline in young worker unionization (Bryson, Gomez & Willman 2010; Haynes, Vowles & Boxall 2005; Waddington & Kerr 2002).

A few studies have examined additional predictors of union commitment specifically among young workers. A study using a Swedish sample proposed that young workers were more instrumentally oriented than their older counterparts (Alivin & Sverke 2000), and this was also true in a sample of young workers from New Zealand (Haynes, Vowles & Boxall 2005). Using a sample of Canadian young workers, union instrumentality was found to influence union commitment, and, as occurred with prior studies that included all workers, general attitudes towards unions were also predictive of support for unions (Hennebert, Fortin-Bergeron & Doucet 2021). The study also observed that alignment with the union's values and leadership style increased commitment. Contrary to prior findings of general union commitment, job satisfaction and organizational commitment were not predictive of union commitment (Hennebert, Fortin-Bergeron & Doucet 2021). Thus, several, but not all, of the predictors of union commitment among all workers were also predictive of union commitment among young workers.

Outside of the individual characteristics of young workers, limited scholarship has also considered the role of technology, especially social media. Although it is an overstatement to say that today's young workers view printed newspapers, paper flyers, email and other traditional union communication tools as part of a bygone era along with the telegraph and "snailmail", it is, nonetheless, the case that younger workers are far more reliant upon text messaging and social media platforms (e.g. Facebook, Tik Tok, X, WhatsApp) for communication.

Views of technology's role have been mixed. Some scholars believe it is a promising tool to help reach and influence the attitudes of an already tech-savvy group of young workers and that the lack of its embrace by unions has contributed to the decline of young workers as a share of union members (Bryson, Gomez & Willman 2010). Geelan (2015: 6) asserts that "the Internet has clearly become the most important form of communication for young people", but others are sceptical of the ability of the internet to reach young people and caution against strategies that "pigeon-hole or patronise young workers" (Hodder & Houghton 2020: 41). Scholars of this tradition often emphasize the view that young people are also workers, so union strategies to increase commitment should focus more broadly on all workers. In a related vein, Fiorito *et al.* (2021) suggest that more pro-union attitudes for youth reported in some polls may reflect the fact that younger workers are more likely to be employed in precarious work situations, not from their youth per se.

Scholars originally proposed that basic websites associated with Web 1.0 would help unions to reduce barriers to activism and mobilize workers (Saundry, Stuart & Antcliff 2007). Then, with the introduction of social media associated with Web 2.0 came the promise of more interaction among activists and unions. In reality, the use of social media platforms, such as X, varies by union with some using it purely to disseminate information, as opposed to interacting with online users, but those social media accounts run by union sections of young workers are more likely to actively engage with users (Hodder & Houghton 2020).

Despite the promise of technology to engage young workers, technology and social media are not an absolute solution to several challenges faced by the labour movement. With algorithms associated with many social media sites and the rise of "echo chambers" on social media (Blanc 2022), unions face a challenge to understand whether the engagement on its social media sites is representative of its membership. The structure and algorithms associated with social media also present a challenge for unions to attract young workers because they must either know about unions and seek them out on social media or "like" other similar accounts, which could then subsequently cause the algorithms to recommend labour movement-related social media pages.

Thus, without knowing about unions, it can be challenging for unions to reach young workers online. Hodder and Houghton (2020: 44) also suggest that "assumptions that young people will automatically like or follow unions on social media are misguided". Unions must give young workers a reason to want to follow them online, and even if young workers do follow unions online, social media only supplements, but does not replace, traditional organizing (Blanc 2022; Simms, Holgate & Roper 2019). In summary, although technology will help unions reach workers, there remain several challenges in doing so.

Union leaders are aware of the "union gap" in membership, commitment and activism. While expressing optimism about polls showing young worker interest, they also recognize the need for change. Richard Trumka, then AFL-CIO president, said:

> These men and women need a strong voice. But when they look at unions, they don't see themselves – only a grainy, faded picture from another time … We have to change our approach to organizing and representation to better meet their needs. And we will! (Shuler & Marschall 2014: 6)

Shuler and Marschall (2014: 6) wrote that, for several years, the AFL-CIO "made a concerted effort to take on this challenge … engaging and preparing young trade unionists nationwide for leadership roles". In the UK, Trades Union Congress General Secretary, Paul Nowak, notes: "Critically, unions have to do more to engage with young workers … using social media and new technology effectively" and explore the potential of new membership forms that provide a "gateway" membership for young workers (Nowak 2014: 690).

Effect of new technologies on activism and commitment

The evolution of new technologies may significantly affect activism and commitment. In terms of activism, both active and passive forms are impacted. According to literature (e.g. Fiorito, Padavic & DeOrtensiis 2015; Gall & Fiorito 2012a, 2012b; Smith, Duxbury & Halinski 2019), passive forms of activism, such as reading union communication materials, have undergone changes. This includes accessing information through websites, downloading mobile applications or watching videos created by union leaders. Alongside the transformation of passive activism, it is important to consider the changes in active forms of activism as well (Kelly & Kelly 1994; Kerr & Waddington 2014).

First, the emergence of new technologies has reshaped the leadership dynamics among activists. While some activists traditionally demonstrated leadership skills during meetings, others have gained recognition for their ability to engage through social networks, create video content or respond to comments. Consequently, activism has evolved to incorporate this important dimension of leadership.

Second, the literature emphasizes the influence of these new communication forms on individuals' social networks. The adoption of digital platforms, particularly social media, has significantly broadened the reach and impact of activists. It allows them to connect with a larger and more diverse audience, engaging in conversations that transcend traditional boundaries. These new communication channels possess the potential to facilitate the creation of extensive networks and the widespread dissemination of activist messages to a much wider audience. As a result, social media has become a powerful tool for activists to amplify their voices, raise awareness and mobilize support for their causes (Li *et al.* 2015).

Third, activists appears to be influenced by these digital tools on a personal level (Kerr & Waddington 2014). Electronic forms of communication reintegrate domestic and professional activities as the spatiotemporal dimension of communication is reconfigured. A blend of private and union activities is emerging. This mix can create personal tensions that can be difficult to manage. More broadly, this can lead to more links between activists who mix their private and union lives.

Fourth, the latest evolution in activism is the approach to negotiation. Activists engage in constant negotiation. However, digital tools have been extensively utilized by unions and management during the Covid-19 crisis, altering the way negotiations take place and how activists operate. The traditional negotiation frameworks outlined in the literature have become outdated. New forms of communication are emerging and warrant further examination. Numerous studies have explored unions in the context of a crisis (Hyman

2010; Roche, Teague & Coughlan 2015), but few works have started to analyse negotiation in this new context (Ajzen & Taskin 2021; de Becdelièvre 2021). The latest studies have concluded that interactions within the realm of social dialogue have become less dramatic, and behind-the-scenes relationships have been weakened. The shift away from negotiations appears to reinforce pre-existing dynamics: trust is strengthened in the context of integrative dialogue, while mistrust is heightened in distributive interactions. Alongside this initial observation, a second observation arises, indicating that this shift could further amplify the power asymmetry in favour of the employers' side.

However, this evolution of activism concerning digital technology should not overshadow the transformation of commitment. New technologies have altered the relationship between unions and their members. They now communicate with them much more rapidly (Lévesque & Murray 2010). As soon as a reply is requested by one of the members, the union has a duty to respond, whether by messages on WhatsApp or Facebook (Panagiotopoulos & Barnett 2015). Previously, information flow was predominantly top-down, but the advent of videos and instant messaging through platforms such as Facebook or TikTok has facilitated much greater interaction. Commitment now goes beyond simply reading leaflets; it involves active engagement with union leaders. This increased level of interaction allows individuals to quickly assess the instrumentality or effectiveness of the union, which in turn serves as a motivating factor for union commitment (Klandermans 1984).

Workers' commitment to unions is also now expressed in different ways. While membership used to represent a key part of commitment, this is now demonstrated through various means. For instance, it can be measured by the number of "likes" on videos, the number of views a video receives or the quantity and content of comments left. These digital interactions have become important indicators of engagement and support for the union (Kerr & Waddington 2014; de Becdelièvre 2021).

Commitment has also evolved through the inclusion of different types of workers who have joined unions. Initially, commitment was primarily analysed from the perspective of employees, but, later, self-employed workers were also taken into account. This expanded understanding of commitment recognizes that individuals across various employment statuses can demonstrate dedication and involvement in union activities (Deng, Joshi & Gelliers 2016). The development of digital technology has seen the emergence of new platform workers with self-employed status (Peticca-Harris, De Gama & Ravishankar 2018). Initially, union leaders did not think they were morally obliged to represent these workers and opposed these new jobs. However, these workers are part of the labour force, so it is recognized that they should be represented by unions (Polkowska 2021). Finally, this transformation of activism and commitment due to these new forms of communication alters unions as organizations (Markus 2011). It helps unions to be more democratic, as activists can easily express their opinions and receive feedback (Kerr & Waddington 2014).

Construct conceptualization, definition and measurement

As scholars continue to seek to advance our understanding of union commitment, it makes sense to consider periodically deliberating upon how the construct is conceptualized and measured. As mentioned earlier, Gordon *et al.* (1980) initially developed a measure of union commitment comprised of union loyalty, responsibility to the union, willingness-to-work for the union and belief in unionism. Although many early studies used the measure, researchers soon began using measures consisting of various sub-parts. To help address the inconsistency across studies, which makes it difficult to interpret and compare study results, Kelloway, Catano and Southwell (1992) developed a measure capturing three of the four dimensions (their measure did not capture belief in unionism). There is a degree of consensus regarding the validity of their measure, and it is probably the most commonly used since its development. Still, many scholars note high correlation between the dimensions and argue for the use of a unidimensional measure (e.g. Tan & Aryee 2002). Recently, Morin *et al.* (2021: 250) conducted an in-depth analysis of the measure and found "there may be limited value to the adoption of a multidimensional approach to the assessment of union commitment". The authors further called for the development of a new, simpler measure.

Prior to considering the development of a new measure, it is prudent to first reconsider our conceptualization of what union commitment is and what it is not. As mentioned earlier, much of the union activism research employs TPB developed by Ajzen (e.g. 1991), emphasizing the role of attitudes in predicting behaviour (activism). Relatedly, some union commitment research also conceptualizes union commitment as an attitude (e.g. Clark 2021; Fullagar *et al.* 2004). This notion of commitment as an attitude differs from much of the commitment-related research. Although commitment scholars historically considered commitment an attitude, beginning around 2000 many researchers began describing commitment as a psychological state, a "mindset" (e.g. Meyer & Herscovitch 2001).

To aid in the discussion regarding the nature of union commitment, it may be useful to consider advancements in fields related to unions such as employing organizations (Meyer & Allen 1991) and leadership (Becker 1992), and how they complement recent union-focused conceptualizations (e.g. Fiorito, Padavic & Russell 2018; Morin *et al.* 2021). Although many findings in related fields corroborate those found for union commitment, others provide opportunity for debate and application in the union context. Meyer and Herscovitch (2001) note that the definition of employment-related commitment is mostly consistent across target entities, with the exception of unions. This is important, as how a construct is defined relates directly to its dimensionality and, in turn, how it is measured. Although not frequently addressed in comparison with other fields of research, within the union commitment literature there is debate regarding its definition and dimensionality, and a reconsideration of its nature and definition is worth discussing. Researchers should consider answering recent calls for a new measure (e.g. Morin *et al.* 2021) and, in the process, reassess the nature and definition of the union commitment construct.

Other issues for the future

There are many other issues that we cannot address adequately in this chapter but seek to highlight some of them in the hope of encouraging researchers to explore and develop these concepts more fully.

Developmental perspectives on commitment and activism

New union members are unlikely to be highly committed and active upon joining. Many activism roles may seem foreign and daunting. Low-demand activism can form a "gateway" to more demanding activism roles as new members become familiar with various activist roles, their importance and the camaraderie that collective action entails. Even modest accomplishments with easy tasks can help to build self-efficacy and collective efficacy perceptions that can support further activism. The notion of starting with a small "ask" or task that is often associated with "organizing model" wisdom seems to work. Both informal and formal new member socialization are critical to begin the process of developing willingness to take on the wide range of activism – from attending meetings to holding committee chair and other leadership roles. Recent work on person-centred approaches to commitment (e.g. Morin *et al.* 2021) can be extended to activism, and may provide valuable insights into how the union can influence transition probabilities to move more members from "easy" activism forms to more impactful forms.

Commitment is not just about conventional activism

Most definitions of activism involve some specific contribution of one's time to an action, but what about political contributions and endorsements? Lobbying by union members would more clearly fit the conventional activism definition, but monetary contributions and endorsements can also help the union achieve its public policy goals, which may vary in importance across unions but are clearly critical for many. The commitment-political support link was recognized long ago (e.g. Fields, Masters & Thacker 1987) but still deserves more attention.

Union beliefs, general union attitude and union social roles

Measures reflecting concepts such as general attitude to unions have been influential in studies of union commitment and activism. However, such measures are a bit of a "black box" that should be opened and examined in better light. Beliefs provide at least a major foundation for attitudes, and often statements used in union belief and general union attitude measures involve unions' social roles. Unions' workplace roles in pursuit of their members' self-interest are important, to be sure, but as Flanders (1970) argues, union social roles may be the more critical consideration for members and the general public. Future research should strive to reduce reliance on somewhat vague concepts such as "general union attitude" and focus more attention on the underlying beliefs that condition attitudes and how those beliefs are formed.

CONCLUDING COMMENTS

At the outset, we noted that a major reason for studying union commitment and union activism is the belief that these are essential for union democracy and union renewal. Although union democracy and renewal can have many meanings (Strauss

1991; Fiorito 2004, respectively), it is difficult to conceive of any meanings that do not require union member commitment and activism. Even in a study that focused on organizing success as its renewal meaning (Hickey, Kuruvilla & Lakhani 2010), the authors found that member activism and mobilization played a primary causal role in most cases of successful organizing (18 of 32), and a contributing role in a majority of the remaining cases (8 of 14). The importance of commitment and activism for renewal have been argued in depth elsewhere (e.g. Gall & Fiorito 2012a, 2012b; McAlevey 2015). The question for researchers and unions now is how to build and enhance commitment and activism.

REFERENCES

Ajzen, I. 1991. "The theory of planned behaviour". *Organizational Behaviour and Human Decision Processes* 50: 179–211.

Ajzen, M. & L. Taskin 2021. "La concertation sociale au temps du coronavirus". *Revue Francaise de Gestion* 298(5): 79–95.

Aleks, R., T. Saksida & A. Wolf 2021. "Hero or villain? A cohort and generational analysis of how youth attitudes towards unions have changed over time". *British Journal of Industrial Relations* 59(2): 532–67.

Alivin, M. & M. Sverke 2000. "Do new generations imply the end of solidarity? Swedish unionism in the era of individualization". *Economic and Industrial Democracy* 21(1): 71–95.

Bailey, J. *et al.* 2010. "Daggy shirts, daggy slogans? Marketing unions to young people". *Journal of Industrial Relations* 52(1): 43–60.

Bamberger, P., A. Kluger & R. Suchard 1999. "The antecedents and consequences of union commitment: a meta-analysis". *Academy of Management Journal* 42(3): 304–18.

Barling, J., C. Fullagar & K. Kelloway 1992. *The Union and Its Members: A Psychological Approach.* Oxford: Oxford University Press.

Becker, T. 1992. "Foci and bases of commitment: are they distinctions worth making?". *Academy of Management Journal* 35(1): 232–44.

Bendix, R. 1956. *Work and Authority in Industry.* New York: Wiley.

Blanc, E. 2022. "How digitized strategy impacts movement outcomes: social media, mobilizing, and organizing in the 2018 teachers' strikes". *Politics and Society* 50(3): 485–518.

BLS 2023. "Union members summary – 2022". Washington, DC: Bureau of Labor Statistics. www.bls.gov/news.release/archives/union2_01192023.htm

Bryson, A., R. Gomez & P. Willman 2010. "Online social networking and trade union membership: what the Facebook phenomenon truly means for labor organizers". *Labor History* 51(1): 41–53.

Budd, J. 2005. *Employment with a Human Face.* Ithaca, NY: Cornell University Press.

Budd, J. 2010. "When do US workers first experience unionization? Implications for revitalizing the labor movement". *Industrial Relations* 49(2): 209–25.

Chen, L. 2020. Union Commitment Facets (Loyalty, Responsibility, and Willingness): A Meta-analysis. Undergraduate honors thesis, Department of Psychology, University of Illinois, Chicago.

Child, J., R. Loveridge & M. Warner 1973. "Towards an organizational study of trade unions". *Sociology* 7(1): 71–91.

Clark, P. 2009. *Building More Effective Unions (2nd edition).* Ithaca, NY: Cornell University Press.

Clark, P. 2021. "The impact of new member orientation programs on union member commitment: evidence from a national study in a post-Janus setting". *Labor Studies Journal* 46(2): 158–81.

Conlon, E. & D. Gallagher 1987. "Commitment to employer and union: effects of membership status". *Academy of Management Journal* 30(1): 151–62.

Cregan, C., T. Bartram & P. Stanton 2009. "Union organizing as a mobilizing strategy: the impact of social identity and transformational leadership on the collectivism of members". *British Journal of Industrial Relations* 47(4): 701–22.

Dawkins, C. 2016. "A test of labor union social responsibility: effects on union member attachment". *Business and Society* 55(2): 214–45.

de Becdelièvre, P. 2021. "La militance à distance: une transformation du militantisme". In *Dialogue social. L'avènement d'un modèle?* edited by F. Géa & A. Stévenot, 551–67. Brussels: Bruylant.

Deng, X., K. Joshi & R. Gelliers 2016. "The duality of empowerment and marginalization in microtask crowdsourcing: giving voice to the less powerful through value sensitive design". *MIS Quarterly* 40(2): 279–302.

Fields, M., M. Masters & J. Thacker 1987. "Union commitment and membership support for political action: an exploratory analysis". *Journal of Labor Research* 8(2): 143–57.

Fiorito, J. 2004. "Union renewal and the organizing model in the United Kingdom". *Labor Studies Journal* 29(2): 21–53.

Fiorito, J., G. Gall & A. Martinez 2010. "Activism and willingness to help in union organizing: who are the activists?". *Journal of Labor Research* 31: 263–84.

Fiorito, J. & I. Padavic 2022. "What do workers and the public want? Unions' social value". *Industrial and Labor Relations Review* 75(2): 295–320.

Fiorito, J., I. Padavic & P. DeOrtentiis 2015. "Reconsidering union activism and its meaning". *British Journal of Industrial Relations* 53(3): 556–79.

Fiorito, J., I. Padavic & Z. Russell 2018. "Pro-social and self-interest motivations for unionism and implications for unions as institutions". *Advances in Industrial and Labor Relations* 24: 185–211.

Fiorito, J. *et al.* 2021. "Precarious work, young workers, and union-related attitudes: distrust of employers, workplace collective efficacy, and union efficacy". *Labor Studies Journal* 46(1): 5–32.

Fishbein, M. & I. Ajzen 1975. *Belief, Attitude, Intention, and Behaviour*. Reading, MA: Addison-Wesley.

Flanders, A. 1970. *Management and Unions: The Theory and Reform of Industrial Relations.* London: Faber.

Fosh, P. 1993. "Membership participation in workplace unionism: the possibility of union renewal". *British Journal of Industrial Relations* 31(4): 577–92.

Freeman, R. & W. Diamond 2003. "Young workers and trade unions". In *Representing Workers: Union Recognition and Membership in Britain*, edited by H. Gospel & S. Wood, 29–50. London: Routledge.

Fullagar, C. *et al.* 1995. "Impact of early socialization on union commitment and participation: a longitudinal study". *Journal of Applied Psychology* 80(1): 147–57.

Fullagar, C. *et al.* 2004. "Union commitment and participation: a 10-year longitudinal study". *Journal of Applied Psychology* 89(4): 730–7.

Fuller, J. & K. Hester 1998. "The effect of labor relations climate on the union participation process". *Journal of Labor Research* 14(1): 171–87.

Fuller, J. & K. Hester 2001. "A closer look at the relationship between justice perceptions and union participation". *Journal of Applied Psychology* 86(6): 1096–105.

Fulton, L. 2020. *National Industrial Relations: An Update.* London/Brussels: Labour Research Department/European Trade Union Institute (ETUI).

Gall, G. & J. Fiorito 2012a. "Toward better theory on the relationship between commitment, participation, and leadership in unions". *Leadership & Organization Development Journal* 33(8): 715–31.

Gall, G. & J. Fiorito 2012b. "Union commitment and activism in Britain and the United States: searching for synthesis and synergy for renewal". *British Journal of Industrial Relations* 50(2): 189–213.

Gallagher, D. & P. Clark 1989. "Research on union commitment: implications for labor". *Labor Studies Journal* 14(1): 52–71.

Gallagher, D. & G. Strauss 1991. "Union attitudes and participation". In *The State of the Unions*, edited by G. Strauss, D. Gallagher & J. Fiorito, 139–74. Madison, WI: Industrial Relations Research Association.

Geelan, T. 2015. "Danish trade unions and young people: using media in the battle for hearts and minds". In *Young People and Trade Unions: A Global View*, edited by A. Hodder & L. Kretsos, 71–89. London: Palgrave Macmillan.

Gordon, M. *et al.* 1980. "Commitment to the union: development of a measure and an examination of its correlates". *Journal of Applied Psychology* 65: 479–99.

Haynes, P., J. Vowles & P. Boxall 2005. "Explaining the younger–older worker union density gap: evidence from New Zealand". *British Journal of Industrial Relations* 43(1): 93–116.

Hennebert, M.-A., C. Fortin-Bergeron & O. Doucet 2021. "Understanding union commitment among young workers: a cross-theoretical perspective". *Relations Industrielles/Industrial Relations* 76(2): 265–90.

Hickey, R., S. Kuruvilla & T. Lakhani 2010. "No panacea for success: member activism, organizing, and union renewal". *British Journal of Industrial Relations* 48(1): 53–83.

Hodder, A. 2015. "Young and unionised in the UK? Insights from the public sector". *Employee Relations* 37(3): 314–28.

Hodder, A. & D. Houghton 2020. "Unions, social media and young workers – evidence from the UK". *New Technology, Work and Employment* 35(1): 40–59.

Hodder, A. & L. Kretsos 2015. "Young workers and unions: context and overview". In *Young People and Trade Unions: A Global View*, edited by A. Hodder & L. Kretsos, 1–15. London: Palgrave Macmillan.

Hyman, R. 2010. "Social dialogue and industrial relations during the economic crisis: innovative practices or business as usual?" ILO Working Paper. Geneva: International Labour Office.

Jódar, P., S. Vidal & R. Alós 2011. "Union activism in an inclusive system of industrial relations: evidence from a Spanish case study". *British Journal of Industrial Relations* 49(S1): s158–80.

Kelloway, K., V. Catano & R. Southwell 1992. "The construct validity of union commitment: development and dimensionality of a shorter scale". *Journal of Occupational and Organizational Psychology* 65(3): 197–211.

Kelly, C. & J. Kelly 1994. "Who gets involved in collective action? Social psychological determinants of individual participation in trade unions". *Human Relations* 47(1): 63–88.

Kerr, A. & J. Waddington 2014. "E-communications: an aspect of union renewal or merely doing things electronically?" *British Journal of Industrial Relations* 52(4): 658–81.

Klandermans, B. 1984. "Mobilization and participation in trade union action: an expectancy-value approach". *Journal of Occupational Psychology* 57(2): 107–20.

Klandermans, B. 1986. "Psychology and trade union participation: joining, acting, quitting". *Journal of Occupational Psychology* 59(3): 189–204.

Kuruvilla, S. *et al.* 1990. "Union participation in Japan: do western theories apply?" *Industrial and Labor Relations Review* 43(4): 374–89.

Lance, C. *et al.* 2010. "Method effects, measurement error, and substantive conclusions". *Organizational Research Methods* 13(3): 435–55.

Lévesque, C. & G. Murray 2010. "Understanding union power: resources and capabilities for renewing union capacity". *Transfer: European Review of Labour and Research* 16(3): 333–50.

Li, N. *et al.* 2015. "Reading behaviour on intra-organizational blogging systems: a group-level analysis through the lens of social capital theory". *Information and Management* 52(7): 870–81.

Markus, M. 2011. "On the usage of information technology: the history of IT and organization design in large US enterprises". *Entreprises et Histoire* 60: 17–28.

Mayo, E. 1933. *The Human Problems of an Industrial Civilization.* Boston, MA: Harvard Business School.

McAlevey, J. 2015. "The crisis of New Labor and Alinksy's legacy: revisiting the role of the organic grassroots leaders in building powerful organizations and movements". *Politics & Society* 43(3): 415–41.

McKay, A. *et al.* 2020. "Types of union participators over time: toward a person-centered and dynamic model of participation". *Personnel Psychology* 73: 271–304.

McShane, S. 1986a. "The multidimensionality of union participation". *Journal of Occupational Psychology* 59: 177–87.

McShane, S. 1986b. "A path analysis of participation in union administration". *Industrial Relations* 25(1): 72–9.

Meyer, J. & N. Allen 1991. "A three-component conceptualization of organizational commitment". *Human Resource Management Review* 1(1): 61–89.

Meyer, J., N. Allen & C. Smith 1993. "Commitment to organizations and occupations: extension and test of a three-component conceptualization". *Journal of Applied Psychology* 78(4): 538–51.

Meyer, J. & L. Herscovitch 2001. "Commitment in the workplace: toward a general model". *Human Resource Management Review* 11(3): 299–326.

Milkman, R. 2013. "Back to the future? US labour in the new gilded age". *British Journal of Industrial Relations* 51(4): 645–65.

Monnot, M., S. Wagner & T. Beehr 2011. "A contingency model of union commitment and participation: meta-analysis of the antecedents of militant and non-militant activities". *Journal of Organizational Behaviour* 32(8): 1127–46.

Montgomery, R. 1989. "The influence of attitudes and normative pressures on voting decisions in a union certification election". *Industrial and Labor Relations Review* 42(2): 262–79.

Morin, A. *et al.* 2021. "Investigating the dimensionality and stability of union commitment profiles over a 10-year period: latent transition analysis". *Industrial and Labor Relations Review* 74(1): 224–54.

Nowak, P. 2014. "The past and future of trade unionism". *Employee Relations* 37(6): 683–91.

Panagiotopoulos, P. & J. Barnett 2015. "Social media in union communications: an international study with UNI global union affiliates". *British Journal of Industrial Relations* 53(3): 508–32.

Peetz, D., C. Webb & M. Jones 2002. "Activism amongst workplace union delegates". *International Journal of Employment Studies* 10(2): 83–107.

Peticca-Harris, A., N. De Gama & M. Ravishankar 2018. "Postcapitalist precarious work and those in the drivers' seat: exploring the motivations and lived experiences of Uber drivers in Canada". *Organization* 27(1): 1–24.

Polkowska, D. 2021. "Unionisation and mobilisation within platform work: towards precarisation: a case of Uber drivers in Poland". *Industrial Relations Journal* 52(1): 25–39.

Porter, L., W. Crampon & F. Smith 1976. "Organizational commitment and managerial turnover: a longitudinal study". *Organizational Behavior and Human Performance* 15(1): 87–98.

Purcell, T. 1954. *The Worker Speaks His Mind on Company and Union.* Cambridge, MA: Harvard University Press.

Rehfeldt, U. & C. Vincent 2018. "The decentralisation of collective bargaining in France: an escalating process". In *Multi-Employer Bargaining Under Pressure: Decentralisation Trends in Five European Countries* edited by S. Leonardi & R. Pedersini, 151–84. Brussels: European Trade Union Institute.

Roche, W., P. Teague & A. Coughlan 2015. "Employers, trade unions and concession bargaining in the Irish recession". *Economic and Industrial Democracy* 36(4): 653–76.

Saundry, R., M. Stuart & V. Antcliff 2007. "Broadcasting discontent: freelancers, trade unions and the internet". *New Technology, Work and Employment* 22(2): 178–91.

Schmitt, J. & K. Warner 2010. "The changing face of U.S. labor 1983–2008". *Working USA* 13(2): 263–79.

Schwab, D. 2005. *Research Methods for Organizational Studies*. Mahwah, NJ: Lawrence Erlbaum.

Serrano Pascual, A. & J. Waddington 2000. *Young People: The Labour Market and Trade Unions: A Report Prepared for the Youth Committee of the European Trade Union Confederation*. Brussels: European Trade Union Confederation Publications.

Shuler, E. & D. Marschall 2014. "AFL-CIO mobilizes young workers". *Perspectives on Work* 18: 6–10.

Simms, M., J. Holgate & C. Roper 2019. "The Trades Union Congress 150 years on: a review of the organising challenges and responses to the changing nature of work". *Employee Relations* 41(2): 331–43.

Smith, C. & L. Duxbury 2019. "Attitudes towards unions through a generational cohort lens". *Journal of Social Psychology* 159(2): 190–209.

Smith, C., L. Duxbury & M. Halinski 2019. "It is not just paying your dues: impact of generational cohort on active and passive union participation". *Human Resource Management Journal* 29: 371–94.

Snape, E. & T. Redman 2004. "Exchange or covenant? The nature of the member-union relationship". *Industrial Relations* 43(4): 855–73.

Snape, E., T. Redman & A. Chan 2000. "Commitment to the union: a survey of research and the implication for industrial relations and trade unions". *International Journal of Management Reviews* 2(3): 205–30.

Stagner, R. 1954. "Dual allegiance as a problem in modern society". *Personnel Psychology* 7: 41–7.

Steers, R. 1977. "Antecedents and outcomes of organizational commitment". *Administrative Science Quarterly* 22(1): 46–56.

Strauss, G. 1991. "Union democracy". In *The State of the Unions*, edited by G. Strauss, D. Gallagher & J. Fiorito, 175–201. Madison, WI: Industrial Relations Research Association.

Sverke, M. & S. Kuruvilla 1995. "A new conceptualization of union commitment: development and test of an integrated theory". *Journal of Organizational Behavior* 16: 505–32.

Tailby, S. & A. Pollert 2011. "Non-unionized young workers and organizing the unorganized". *Economic and Industrial Democracy* 32(3): 499–522.

Tan, H. & S. Aryee 2002. "Antecedents and outcomes of union loyalty: a constructive replication and an extension". *Journal of Applied Psychology* 87(4): 715–22.

Tapia, M. & L. Turner 2018. "Renewed activism for the labor movement: the urgency of young worker engagement". *Work and Occupations* 45(4): 391–419.

Tetrick, L. *et al.* 2007. "A model of union participation: the impact of perceived union support, union instrumentality, and union loyalty". *Journal of Applied Psychology* 92(3): 820–28.

Vandaele, K. 2020. "Newcomers as potential drivers of union revitalization: survey evidence from Belgium". *Relations Industrielles/Industrial Relations* 75(2): 351–75.

Waddington, J. & A. Kerr 2002. "Unions fit for young workers?" *Industrial Relations Journal* 33(4): 298–315.

CHAPTER 18

UNIONS WORKING WITH AND LEARNING FROM OTHER SOCIAL MOVEMENTS

Heather Connolly

ABSTRACT

There are strong arguments for unions to work with other social movements, especially in the context of neoliberalism and the growth of informal, contingent and precarious work. Social movements provide new ideas, new resources, new strategies and tactics and new sites of struggle as well as the potential to speak to younger, precarious and immigrant workers. There are also strong arguments for social movements to work with unions, who tend to be better resourced, with established consensus-based structures and representation in the workplace. This chapter considers the relations between unions and social movements in theory and practice, showing what unions and social movements can learn from each other, asking whether the relations between union and social movements are likely to be characterized by conflict, cooperation or indifference. More fundamentally, the question asked is whether other social movements are replacing organized labour as the main dynamic force for advancing workers' interests.

Keywords: Social movements; extra-workplace action; challenges for unions

INTRODUCTION

There is now a well-established set of studies in industrial and labour relations on unions adopting forms and practices of social movement unionism, and the links between unions and community groups (see Engeman 2015; Heery, Williams & Abbott; Holgate 2015; Tattersall 2011; see also Chapter 16). This chapter adds to the debate by considering the relations between unions and social movements in theory and practice, drawing on contemporary examples to show what unions and social movements can learn from each other and asking whether the relations between union and social movements are likely to be characterized by conflict, cooperation or indifference (Heery, Williams & Abbott 2012). More fundamentally, the question asked, using the example of the

"Yellow Vests" movement in France is whether other social movements are replacing organized labour as the main dynamic force for advancing workers' interests.

There are strong arguments for unions working with other social movements, especially in the context of globalization and neoliberalism, deindustrialization and the growth of informal, contingent and precarious work. Social movements provide new ideas, new resources, new strategies and tactics, new sites of struggle and the potential to speak to younger, precarious and immigrant workers. Voss and Sherman (2000) argue that if unions reject a linear, deterministic model, they could radicalize their goals and tactics, and mobilize workers through campaigns and break out of "bureaucratic conservatism". That said, without a minimum of leadership, some form of consensus-based structures and processes, movements risk dissipating in the long-term, as was the case with the Occupy! movement. There are also strong arguments for social movements to work with unions, who tend to be better resourced with established structures and representation in the workplace.

The chapter starts by revisiting familiar tensions and contractions in unions in order to clearly understand the limits and possibilities of unions working together with other social movements. Next, the chapter draws upon discussion and ideas in industrial relations and social movement studies, which provide sets of concepts to bring together the study of social movements and unions. To explore the practice of unions working with other movements, the chapter then looks at the relational dynamics between unions and social movements by drawing on contemporary examples. These examples substantiate some of the opportunities and challenges of union and other movements working together, and for unions adopting social movement unionism. The examples also support the argument that the primary – and potentially most sustainable – way in which labour and new social movements come together is through a process of "absorption", where the labour movement provides an institutional field upon which other movements can organize and campaign. Through a slow, cumulative process, union agendas can be radicalized and become more inclusive (Heery 2018). While the cases show the potential for rivalry and challenges to working together, this tends to be a case of institutional barriers and unions being caught defending a defensive position in the face of strong counter-mobilization, rather than fundamental differences of interests and methods.

UNDERSTANDING WHAT UNIONS DO AND THE TENSIONS THEREIN

Unions became a social movement – the labour movement – during the period of industrialization, when they challenged the "overall system of social control of economic resources", fighting against the exploitation of the working class and the domination of the elite (Touraine 1986: 154). With "industrial legality" (Gramsci 1971) and the organizational and material advantages for unions that accompany this, unions became more limited in scope and acted less like a social movement – weakening the organic, ideational resonance with workers whose interests unions sought to represent (Hyman & Gumbrell-McCormick 2017). Since the early 2000s, even in European countries where unions maintain high levels of associational and institutional power resources, unions have become increasingly marginalized (Baccaro & Howell 2017). Processes of co-optation and compromise further contributed to unions becoming a much weaker and less active movement (see Grote & Wagemann 2018). Consequently, from the 1980s onwards, industrial relations scholars' focus shifted to look at how unions could revitalize and maintain or recover their presence and influence. A focus of the union renewal and revitalization literature has been for unions to adopt social movement forms of unionism (Engeman 2015) and/or to maintain their relevance through working with and in the community and other social movement organizations (Heery, Williams & Abbott 2012; Holgate 2021).

From the 2010s, in response to the impact of the financial crisis of 2008, anti-establishment, anti-austerity, pro-democracy movements such as Occupy!, the Spanish *indignados* and, more recently, in 2018, the Yellow Vests (*gilet jaunes*) in France have emerged, with common sets of interests with unions. Coupled with new forms of work and organizing in the platform economy (Tassinari & Maccarrone 2020), which has accelerated since the financial crisis, and where we have seen the growth of independent unions, the future of unions appears increasingly dependent upon their ability to engage with a wider set of social and community-based interests and organizations.[1]

In order to make sense of how unions can work with and learn from social movements, it is important to revisit the contradictions and dualisms in union identity and organization. Hyman (1975, 2001), drawing upon writers such as Herberg (1947), explored the contradictory nature of union identities as both movements and organizations. On the one hand, unions are organizations enshrined in formal, official and often bureaucratic "representative" structures that prioritize collective bargaining and institutional survival related to bricks and mortar and financial assets (see also Chapter 15). On the other hand, unions comprise a movement, an organizational form that prioritizes workplace resistance, direct democracy, membership mobilization and radical economic and political aspirations (see also Chapter 15). To quote Herberg (1947: 6):

> This irreducible duality of nature necessarily results in a fundamental conflict of purpose and orientation. As a business-like service organization, the union requires efficient bureaucratic administration, very much like a bank or insurance company. The members of the union are merely clients who are entitled to the best service for their money but who should certainly not presume to interfere in matters of administration, since such matters are properly the function of trained and experienced officials specially selected for this purpose. But as a phase of democratic self-liberation movement of modern times, the union is an idealistic, quasi-religious collectivity; it is a crusading reform movement of which the members, the masses, and their democratic self-expression are the very essence. The union, as an institution, is thus in the grip of a very real contradiction.

The balance between these contradictory identities may, of course, differ markedly between unions, and may shift substantially over time, but the dualism itself is universal (Hyman 2001). In the trajectory of unions' development, the argument is that over

1. Also worth mentioning are land-based movements in the Global South (see Antunes 2001).

time the organization side of dualism becomes dominant. As they develop, unions are often seen as exemplifying the entrenched leadership and conservative transformation associated with Michels' (1915) "iron law of oligarchy". The objectives of members and union leaders diverge and this is often linked to the ways in which many workplace unions become bureaucratized, routinized and centralized. Internal union democracy can be achieved "only against external resistance and considerable odds" (Hyman 1975: 68).

At the same time, unions are also subject to democratic pressures from the rank and file, so rather than solely viewing the institutionalization of unionism as a constraint, leadership – and the institutional framework within which it functions – is essential for sustaining collective action. Any assessment of union action should analyse the ongoing and contested tensions between bureaucratized and democratic forms of organization (Hyman 1994; see also Fairbrother 2005). The organization side of the dualism is often neglected in studies on union renewal and revitalization strategies, with a lack of exploration of the ways in which unions manage tensions in their role and functioning. Radicalizing the union agenda long-term requires processes of deliberate and participative democracy to be strong in the union (Connolly 2020). Conversely, for social movements that fail to achieve longevity from a rejection of any (or at least very little) form of formalization, for example, Occupy! and the Yellow Vests, the strength of unions is found in the organization structures and democratic functioning.

The significance of the dominance of organization over movement takes on a greater or lesser importance depending on the identity of the movement in question. Hyman identifies three ideal types of union identity: first, as labour market regulators (market); second, as vehicles for raising workers' status (society); and third, as "schools or war" in a struggle between labour and capital (class). European unions have "tended to incline towards an often contradictory mixture of two of the three ideal types" (Hyman 2001: 4). In union movements where collective bargaining is the central focus of union practice, the goal of internal union democracy is less significant as the union's purpose is more narrowly defined and unions are accepted as being embedded in a range of institutions in order to engage in the main task of collective bargaining. In union movements with a radical conception of unionism, the goals of union democracy and independence become central features of unionism as the union acts "as a basis for collective struggle against as well as within capitalism, as an agency which ultimately can be effective only as a means of collective mobilisation of the working class" (Hyman 1979: 63). Understanding this dualism helps us to map different configurations of unions and the barriers but also the potential opportunities for working with and learning from social movements.

As unions became part of the institutional fabric of many advanced capitalist societies during the 1960s and 1970s, social movement scholars focused on the dynamics of "new social movements" and post-material activism, whereas industrial relations focused on worker representation and the specific bargains struck between workers and businesses. Whereas social movement scholars focused more broadly on the networks that compose movements – and the strengths and weaknesses of formalization, along with varieties of formalization – industrial relations scholars emphasized strategic choice and bargaining processes as critical elements that can enhance or prevent (worker) mobilization (Tapia, Elfström & Roca-Servat 2018).

Kelly's (1998) work has been important for (re)building bridges between social movement studies and industrial and labour relations disciplines. By refocusing the study of industrial relations away from institutions to the social processes underpinning collective action and identity, particularly in a time of declining union influence, Kelly helped push forward conceptual tools for understanding the possibilities for collectivism. The key question posed in Kelly's work is how individuals are transformed into collective actors willing and able to construct and sustain collective interests and engage in collective action against their employers. The process of moving from having common interests to understanding them as common and acting on them is filled with contingency, and worker solidarity is not an *a priori* fact, but grows out of the interactive processes among the workers in their confrontation with management. Mobilization theories emphasize the socially constructed nature of collective interests and mobilization and explain the presence or absence of collective organization and action.

Kelly's (1998: 126–9) approach has three advantages over extant approaches in industrial relations. First, instead of starting from the employer's need for cooperation and to secure work performance, it starts from injustice and exploitation. Second, it does not depend upon a simple distinction between individualism and collectivism, but distinguishes interest definition, organization, mobilization, and so on, treating "as problematic what previous industrial relations researchers often took for granted, namely the awareness by workers of a set of common interests opposed to those of the employer". Third, it helps address key issues such as how employees define interests in particular ways. In spite of the critiques of the approach and the lack of take-up in the field (see Gall & Holgate 2018), mobilization theory has revived the theoretical potential and wider application of industrial relations as a field, with the shift of emphasis from institutions to the micro-social processes of worker mobilization. While this way of thinking takes us away from the institutional bases of labour unionism, it also helpfully breaks down theoretical and practical boundaries between the study and the practice of building worker-interest based and wider social movements.

BUILDING BRIDGES IN SOCIAL MOVEMENT AND INDUSTRIAL RELATIONS THEORY

There are theoretical bridges between social movement and industrial relations theory in the concepts of internal mobilizing structures, external mobilizing structures and mobilizing cultures (Tapia, Elfström & Roca-Servat 2018). An awareness of the internal and external structures alongside mobilizing cultures helps us to understand and explain the extent to which unions may seek out broader alliances with social movements and the form that these alliances might take.

First, internal mobilizing structures concern organizational form, leadership and member resources. Different organizational structures affect the mobilization capacity or the mobilizing "repertoire" of a movement. The question then becomes less whether networks are formalized and more how they are formalized, and with what different effects. Referring back to the discussion on the movement/organization dialectic, here we see how unions, as institutions, have tended to formalize in such a way as to make them narrowly focused, particularly on workers, companies and/or sectors, moving away from broader representations of worker interests.

Second, political opportunity theorists highlight the importance of external mobilizing structures, political opportunities or constraints in explaining the emergence of collective action. There can be shifts within the political structure that create opportunities – or constraints – for social movements to emerge or thrive in. Social movement scholars tend to emphasize the shift or change in the political structure that gives way to opportunities or constraints, while industrial relations scholars tend to analyse the overall historical and institutional contexts in order to understand how these shape the behaviour of various actors, such as workers, unions or employers.

Third, creating mobilizing cultures is also an element for understanding the approaches in unions and social movements. Social movement studies have shown how movements use symbols, identity, emotions or other cultural dimensions, which in turn help produce solidarity, motivate participants and, thus, are a spur to collective action. Social movement scholars have studied the importance of framing or building class-based solidarity or working-class ideology as a way to motivate collective action. Kelly's (1998) work reinvigorated a focus upon the social processes underlying collectivism and the significance of feeling a sense of injustice. For Morgan and Pulignano (2020), the articulation and organization of solidarity involves the interaction of moral frameworks, political calculation and coalitions, and an understanding of the rituals, symbols and narratives of particular contexts. The more moral, political and performative elements of solidarity overlap and reinforce each other, the more potentially powerful movements can become.

Social movements and unions are concerned with collective action, where individuals share resources in pursuit of collective goals. For those studying social movements, the discussion on the relationship between movements and unions is hampered by not fully satisfactory conceptualizations of a social movement in analytic terms, namely, of the specific manner in which collective action is promoted within social movements (Diani 2018). Diani provides a useful way of exploring the different ways in which labour issues can be acted upon and explores the dilemma between "business unionism", focused on the representations of specific trades, and "social movement unionism", oriented to the building of broader platforms. Social movement unionism has been defined in different ways, but is generally conceived of as a union strategy that moves beyond member representation and collective bargaining to broader social movement mobilization, which requires alliances between unions and other community organizations with shared interests. Such a union-community alliance is a common feature of social movement unionism across multiple studies (see Engeman 2015).

Diani (2018) argues that unions and social movements are often approached as different types of organizations and the tendency has been to look for differences in their membership, in their organizational properties, in their repertoires of action and in their relation to the polity. Unions are seen to fall under the bureaucratic model and social movements under the informal, decentralized and participatory model. Instead, Diani suggests looking at the relationship between unions and other social and political organizations in the context of broader collective action fields. He suggests doing this by introducing different ways of promoting collective action, which he terms "modes of coordination", to show how unions can actually be embedded in different relational patterns to other actors. Using this approach enables us to take more explicitly into account the inherent complexity of union collective action. Diani looked at this at three analytical levels: individuals (in particular, union members/activists and individuals sympathizing with labour causes), organizations (unions and other social and political groups and associations) and events (collective gatherings focusing on labour as well as other issues). The structure of fields consists of the relational patterns linking individuals, organizations and events. This enables us to move our focus from the discussion of "unions as movements" (asking whether unions are social movement organizations or not) to "unions within movements" (exploring the role of unions within social movement processes).

Thus, Diani (2018) identifies two ideal types of relational patterns that unions can operate: a social movement mode of coordination or an organizational mode of coordination. The organizational mode of coordination broadly fits the experience of "business unionism", focusing on the representation of their specific membership constituency on a pretty pragmatic and instrumental basis, without aspirations to identify with or to get involved in the pursuit of a broader project of social change. The social movement mode of coordination precisely reflects the aspiration to link the struggles of specific sectors of workers to broader challenges through both the construction of extended alliances and the involvement of union members in numerous other social and political milieus.

This is the context in which we can situate and understand that unions have often struggled adequately to address the needs of the unorganized, disenfranchized and precarious workers. While institutional mechanisms, such as collective bargaining, were built to give power to the unions, at the same time they blocked broader mobilization efforts or the movement dimension of unions. In other words, while unions might be social movements per se – as Fairbrother (2008) argues – in practice, the movement or transformative dimension of the union has often been superseded by its more limited bargaining role. Therefore, unions continuously face tensions between being a potentially transformative social movement and an "ossified form of organisation" "that rests on stable relations, with limited goals" (Fairbrother 2008: 218).

MAPPING VARIETIES OF UNION AND SOCIAL MOVEMENT "FUSIONS"

Social movements can be defined as networks of informal interactions between a plurality of individuals, groups and/or organizations, engaged in political or cultural conflicts, on the basis of shared collective identities (Diani 1992). According to Grote and Wageman (2018), social movements contain the following core features: they are collectivities with a degree of temporal continuity; they are change-oriented and have some degree of organization; they have at least one target; they use extra-institutional tactics (for example, protests or demonstrations); and they have some form of leadership.

When talking about social movements, there are several terms and often a conflation between them. "New" social movements were originally described as "new" to distinguish them from modernist political parties and from the traditional labour movements of advanced capitalism – these new movements included women, peace campaigners and environmentalists. There are civil society organizations and "new actors" in employment relations, which can include community-based, campaigning, single-issue groups and, of course, advocacy organizations, such as Citizens Advice in the UK (Heery, Williams & Abbott 2012). However, distinctions between "new" and "old", and "modern" and "postmodern" are often misleading, as some of these movements have developed alongside labour movements and are, therefore, not their successors.

In many European countries the increase in insecurity and imposition of austerity has seen the emergence of new "outsider" movements, particularly in the worst affected countries, such as Greece and Spain. In 2018, in France, the Yellow Vests movement emerged, which was initially described as a populist movement, but since has been, often cautiously, associated with varying identities and interests. The unions in France were initially not supportive of the movement but then joined the call for action a month into the protests in December 2018. The Yellow Vests movement, discussed further below, is a good example of the dilemmas faced by unions in supporting movements that are seemingly populist and have right-wing support, but also reflect the interests of some of their main constituents.

Della Porta and Diani (2020) use the term "new, new" social movements to describe many of the post-financial crisis movements that emerged (see also Feixa, Pereira & Juris 2009). They describe these movements as straddling the frontier of physical and virtual space at the turn of the new millennium. They highlight the transformations and social conflicts associated with the consolidation of informational capitalism. The social base of these movements crosses generations, genders, ethnicities and territories. Their spatial base is no longer local or national, but is situated in globally networked space, like the neoliberal system these movements oppose. However, their decentralization constitutes a localized internationalism (glocality). Feixa, Pereira & Juris (2009: 437–8) describe them as anti-corporate globalization movements: "If new social movements were conceived as identity-based movements, 'new, new' social movements combine cultural and material demands, as well as local and global scales of action. 'New, new' social movements are also based on an infrastructural web of technical tools and new technologies." "New, new social movements" have been more reluctant to adopt formal structures and leadership, but experience the same pressures as other movements towards formalization and bureaucratization.

The variety of methods used by social movement organizations often overlap with methods of more partnership-oriented unions, including, for example, organizations that have entered in to discussions and/or relationships with employers – such as Citizens UK in the living wage campaign, where they appealed to employers' sense of social responsibility and make the business case for higher wages. Their use of methods that seek to draw employers into joint work has clear parallels with the labour movement. The shift from more agitational methods to more pro-business methods might be a question of maturity or, again, reflects Tarrow's (1989) "cycles of protest". Seeking a relationship with employers could lead to joint work with unions but also again lead to tension and rivalry as both unions and social movements present themselves as potential partners in regulating the employment relationship.

The anti-austerity movements in Greece and Spain, for example, have also shown some of the tensions apparent in unions working with and taking on the causes of social movements. Hyman and Gumbrell-McCormick (2017) identify key features of these movements, which overlap with the features identified with "new, new social movements" described above. First, a complex interconnection between national and supranational dynamics with mutual learning across frontiers – for example, the Spanish *indignados* or M15 movement, which was emulated across Europe. Second, a central role of social media in protest and resistance – for discussing grievances, formulating demands and coordinating action. Third, a reclaiming of public spaces as arenas for discussion and debate as well as demonstration. Finally, a "defence of the commons" – namely, resistance to privatization – particularly when imposed by the Troika of the European Central Bank, European Commission and International Monetary Fund, in the case of Greece.

While the unions have not necessarily initially worked with these movements and have been viewed as part of the problem in many cases, there have been examples of these movements acting together and the potential for forms of fusion developing. For example, the initial discourse of the *indignados* movement was very critical about the unions, for their conservatism and for opting for social dialogue over more direct action. Relations between the M15 movement and the Spanish unions moved from mistrust to convergence, especially with the radical unions, such as where tactics and interests were most similar (Roca & Días-Parra 2017).

Civil society certainly contains a variety of groups, movements and institutions, serving constituencies defined in a variety of ways and pursuing multiple objectives. The differences in these movements and the potential for fusions with unions should form part of the discussion and debate, but there is plentiful scope for joint working with the labour movement, and for unions to develop more social movement modes of coordination. Heery (2018) puts forward four different forms of cross-movement collaboration: union-social movement coalition; affiliation of unions to social movement organizations such as Amnesty, for example; union imitation of social movement practices – see Holgate's (2015) work on community unionism in London, Seattle and Sydney as an example of this; and union absorption of new social movements such as internal coalition building in unions, with the creation of women's committees and Black members' groups in unions in the UK, for example. Expression of new social movements through unions has been of major significance for the labour movement – and Heery (2018) argues that through this slow, cumulative and contested process of change within unions, the fusion thesis is most apparent.

UNIONS WORKING WITH AND LEARNING FROM SOCIAL MOVEMENTS IN PRACTICE

There is clear evidence of unions using different forms of cross-movement collaboration but often the conclusion is that there are strong barriers to change and "institutional sclerosis" often imposes a drag upon innovation (Holgate 2018). This section considers the development of relational dynamics between unions and social movements, drawing on a set of examples. The first example explores the links between Filipino workers in unions in the UK and Filipino community groups. The second example explores the links between undocumented migrant collectives and the General Confederation of Labour

(CGT) in France. Both of these examples demonstrate a form of fusion between union and social movement tending towards that of "semi-absorption" of social movements within the union. The unions provided a space for action and campaigning and for furthering workers' interests, but the social movements maintained a space outside the unions as well, in order to forward any interests that were not prioritized or were blocked by unions.

The third example explores the Yellow Vests movement in France, where, like the *indignados* in Spain, the movement explicitly rejected any links or joint actions with unions, at least initially and in a formal sense. The case makes visible the challenges for unions, as institutions embedded and reinforcing the current configuration of capitalism, to represent broader sets of interests. It also makes visible the challenges for social movement maintaining collective action and solidarity without the leadership and organization familiar in union organizations, and the meta-collective action frame of shared working-class interests.

Community-union working: Filipino migrant workers and community activists in the UK

Migrant workers make up a large and growing part of the public sector workforce in the UK, particularly in the health service and many outsourced areas of work, such as catering and cleaning. Many unions have recognized the need to recruit and represent these workers and adapt their practices to include greater links with community groups and local social movement and advocacy organizations.

A Labour government launched the Union Modernisation Fund (UMF) in 2005. Funding was initially provided to 35 innovative projects designed to speed unions' adaptation to changing labour market conditions. The projects were of variable duration and size, running from January 2006 to June 2009 and focused on numerous priority themes. Unison's Migrant Workers Participation Project sought to encourage greater levels of migrant worker participation in the democratic processes of the union, increase the numbers of migrant worker activists at all levels of the union, reduce economic and social exclusion of migrant workers in the workplace, and ensure that Unison's services better met the needs of migrant workers. The Filipino Activist Network (FAN), an informal network of Filipino activists, emerged out of the Migrant Workers Participation Project.

Much has been written about migrant worker organizing in Unison, including an evaluation report of the project, and it has been used as a case study example to show the links between unions and community organizations. The Filipino network has been seen as a success story of migrant worker self-organizing and, as a result of the strength of organizing among these workers, there has been a level of inclusion of Filipino workers within the power structures of Unison (Connolly, Marino & Martínez Lucio 2019; Connolly & Sellers 2017). This participation of migrant workers, particularly Filipino migrants, was evidenced in the growing number of Filipino activists who took up positions in local branches, for example.

There has been a strong level of community organizing in the Filipino community in the UK. Several community groups and advocacy organizations exist to help represent and campaign for the interests of Filipino migrants working in the UK. One under-explored area in the success of social movement/community-union links is the key role of leadership in driving forward links between community and union.

Arguably, the key to the development of joint working and campaigning was thanks the strategic leadership of a Filipino union official who spent 15 years as National Development Officer in Unison's Strategic Organising Unit, with lead responsibility in organizing migrant and precarious workers. She was also one of the founders of the Kanlungan Filipino Consortium and the Southeast and East Asian Centre, which are charitable organizations that provide community services among Filipinos and East and Southeast Asian Communities. Kanlungan Filipino Consortium is a London-based consortium of grassroots organizations advocating for Filipino migrants' rights. Kanlungan is a registered charity consisting of several Filipino community organizations working closely together for the welfare and interests of the Filipino and other migrant communities in the UK. The Filipino official was also one of the founders of the Status Now For all Network, which campaigns for regularization and granting of indefinite leave to remain for precarious and undocumented migrants and asylum seekers in the UK.

Having strategic leaders in the union who were also embedded and active within community groups, and where there is a crossover of activists from both social movements and union, helps shape a social movement mode of coordination in the union (see Diani 2018). There was recognition that there were structural barriers within the union that the Filipino activists had to fight to overcome. In cases where the union was slow or reluctant to support particular campaigns and/or issues, the Filipino activists looked to the community to mobilize and respond, organizing under the banner of the Filipino community. The majority of Filipino members were also members of community and/or religious groups, which facilitated the shifting between the two organizations to provide the necessary support and resources adapted to the particular campaign and/or issue.

Unison funded leadership training for activists in the FAN, with the aim of empowering these activists to take up positions within branches. The ambition of the leadership was to have a FAN activist on the executive committee of the union. The challenge of this type of strategy is that the success of inclusion and representation of migrant workers often relies on key activists and specific conditions (such as the UMF funding) to develop a strategic approach to union-community links. This is limited, because if the leader activists or officials leave the organization, momentum and support can be challenging to maintain, hinting at the significance of developing relational patterns between individual, organization and events to maintain a social movement "mode of coordination" (Diani 2018).

From civil protest to labour protest: undocumented migrants in France

The "immigrant worker" was a familiar figure of France's 1970s protest movements. However, during the 1980s and 1990s pro-regularization campaigns, when *sans papiers* – without papers – became an increasingly used term, there was little emphasis on employment and on identity of the migrants as workers (Barron *et al.* 2016). In the 1990s, the vulnerability frame was dominant in the *sans papiers* protests, which emphasized humanitarian representations of migrants and formulated their demands in general human rights terms. The union supported the *sans papiers* from the outside, and most political and labour organizations commonly referred to the need to help the powerless and to "common humanity" rather than to migrants' economic inclusion and the possible use of economic inclusion as leverage (Barron *et al.* 2016).

From the mid-1990s until the early 2000s, faced with the hardening of migration policy, migrant community solidarities crystallized in the form of a citizens' movement. Following mainly community-based action, there was a shift from 2006 when undocumented workers turned to the unions for support in the workplace. The decision taken by the CGT union to dedicate sustainable support to undocumented migrant workers opened up an institutional space within the union from around 2014, from when undocumented immigrant workers became a more integral part of French union action (Connolly & Contrepois 2018).

A key turning point in the collective identity of undocumented migrants being perceived and seeing themselves as "workers" occurred in September 2006, when a group of 22 undocumented workers in a small industrial laundry, Modeluxe, situated in Chilly-Mazarin went on strike when they were threatened with dismissal. The undocumented migrant workers reached out to a union, the CGT, and managed to gain support as workers. They leveraged employment legislation by going on strike, which meant their employment contracts were suspended, also suspending the redundancy process. Around 50 CGT militants camped outside, protecting the site from police entering the building. To take back the workers, the employer put pressure on the state (local prefecture) to be able to regularize the situation of the workers. The union sought public support for the strike and organized demonstrations and petitions.

Barron *et al.* (2016: 633) argue that the undocumented migrant workers' strike highlighted inclusion rather than dispossession by providing the key ideational condition for unions to embrace undocumented migrants as workers who are worth organizing and of making their regularization as a union fight. The emphasis on the awful working conditions of the undocumented workers – blatant violation of workers' rights – helped to gather support inside and outside of the company (Kahmann 2015). The union sought support from local politicians and the public through petitions and demonstrations. Three days after the beginning of the strike and occupation, regular co-workers joined the strike and brought the company to a standstill. All the workers obtained permits in January 2007 as an outcome of the action.

In 2007–08, the category of *travailleurs sans papiers* – workers without papers – appeared and achieved widespread use, as a result of intense social and institutional activity. The success of the strikes led to it being a "pole of attraction" for other undocumented workers, regardless of employment status. In the CGT, the boundaries of who is a *travailleurs sans papiers* has been the subject of framing disputes. Inclusion has been based on the undocumented migrants having to be "workers" in some form.

The campaigning and mobilizations organized by the CGT with undocumented migrant workers have led to concessions from various governments in the form of clarifying and formalizing the process for regularization of undocumented migrant workers. Undocumented migrant workers' experiences with the CGT have had their limits, however, as inclusion within the union was based upon the undocumented migrants having to be "workers". In order to support their rights as workers, many of the undocumented migrants were told to first try to find "fake papers" and get a job before bringing their cases to the CGT, which for the union was the best way to represent those considered "illegal" (Connolly & Contrepois 2018).

Divisions between the union and the undocumented migrant workers appeared during the actions for the *travailleurs sans papiers*, as the unions sought to limit the grouping to undocumented workers in order to be able to leverage the hard-won legislation for regularizing undocumented workers. Internal debates continue in the CGT around whether to broaden the campaigning to represent all undocumented migrants, regardless of employment status. Undocumented migrant movements staged protests outside the CGT headquarters to contest the approach adopted by the union. Action to support undocumented migrant workers has strengthened in the CGT, but the focus is on the legal routes afforded to the unions to further the interests of undocumented migrants who are working. The way in which the formalization of the "internal mobilizing culture" around representing undocumented migrant workers took place limited the shift towards a social movement mode of coordination.

A challenge to maintaining an institutional space for undocumented workers has been the transitory nature of the group – as in those who are part of the group are, in fact, seeking to change their situation in order to leave it (Barron *et al.* 2016). Yet their identification as "workers" opened up space for undocumented migrants to develop a sense of belonging to part of a broader grouping of workers.

The semi-absorption of migrant workers within the traditional structures of the CGT introduced shifts in the capacity of undocumented workers to exercise agency. Several of them took responsibilities within their company and sector union, on the basis of the experience they accumulated during the undocumented workers' strikes. When hundreds of undocumented migrants went on strike again in 2018, it was former or existing undocumented migrants who were able to take the lead in the movement.

Rejecting unions: the French Yellow Vests movement

The Yellow Vests movement began in 2018 as protest against a planned fuel tax rise, which President Emmanuel Macron insisted would aid the country's transition to green energy. At the time, polls suggested that the price of fuel was becoming a key concern for French citizens. The emergence of the movement was initially facilitated by the focus on the fuel tax rise, which resonated with a diverse group of French people online and led to collective actions and protests. On 17 November 2018, nearly 300,000 people dressed in "high-vis jackets" occupied roundabouts, tolls booths and roadblocks across France. After the third week of Saturday protests, the prime minister announced the suspension of the fuel tax rise for six months and a freeze of gas and electricity prices.

The movement was interpreted as a vehicle for a wider set of grievances, sometimes contradictory, and anger was directed towards the state, which led to recurring protests. What began as a fuel tax protest shifted into a wider anti-government movement, with an emphasis upon "justice", the redistribution of wealth and "ending social inequalities"

(Kouvelakis 2019: 77, 88). The Yellow Vests claims are pragmatic but diverse as they moved from the price of petrol to the break-up of large banks and to direct democracy. Among other demands, the movement has protested for higher minimum wages, fair taxes, the right to protest (and against police violence and repression), defence of public services, and was opposed to the increase in student registration fees and school reform. The main groupings of their demands are around fiscal injustice, social injustice and political representation. The movement has been described as a movement of citizens, consumers, taxpayers, workers – and it has not been about identifying as a member of a union or protester as part of a recognized group, but as a member of society as a taxpayer, or as someone doing a particular job (Maillard 2019: 29).

President Macron responded to the concerns of the movement by opening up a national debate for the protestors to voice their issues. The movement struggled to maintain the same numbers seen in the first few weeks and, after the government backed down on the fuel tax rise, participation in the movement declined gradually. By June 2019, after six months of protests, participation declined to around 7,000 protestors. Protests are still ongoing but there has been a rejection of any formalization of the movement, or of being co-opted by political parties and unions. There have been divisions into multiple collectives of Yellow Vests in different regions in the longer term.

The emergence of the Yellow Vests movement caught unions (and others) off guard (Maillard 2019: 29) as, unlike typical French protest movements, it sprang up online through petitions and posting videos on social media, without a set leader, union or political party behind it. The movement has left a permanent mark on French history, with the creation of a new category of social movement. While the numbers behind the movement have shrunk, the symbolism of the grouping and what they stand for has remained in public and media discourse, and has destabilized French politics and institutions, including unions, deemed no longer to represent the citizens in the movement. The discussion below draws out the varying features of the Yellow Vests movement, to understand the mobilizing culture and dynamics, the potential for joint working with unions and the challenges.

Understanding interests and identity

The longevity of the Yellow Vests appears to be related to the movement's ability to incorporate different and often competing "poles", from radicalism and pacifism, deliberation and action and occupation and outreach (de Raymond *et al.* 2023). The movement challenged many of the traditional sorts of collective identity, being framed around the right to "dignity" and "respect" and to live in a dignified manner – not traditional class, religious, national identity but "the people" wanting to be able to participate in society in a dignified way, and consume in society like those who have the means to can. Arguably, what united the movement in the early stages appeared to be a sense of shared experience. The experience of not being able to make ends meet when they felt that they were doing everything to avoid being dependent upon others, including the state and its welfare benefits, which they saw as humiliating handouts. They shared their experience online and bonded on an individual basis with millions of people without having to be a member of a party or trade union.

We find similar objectives in the unions and the Yellow Vests – the defence of rights as well as material and moral interests – but in the case of the Yellow Vests this was not conceptualized as collective but as a "community of individuals". The movement's recognition of inequality did not translate into a language of exploitation and emancipation but a "*mépris social*" or social contempt and lack of respect (Maillard 2019: 65).

If we understand collective identity as a form of identification that depends on a definition of "us" distinct from "them" – where an "us" can be an occupational group, a factory workforce, a religious or national identity, distinguished from "them", who is the "us" in the Yellow Vests? In surveys and studies about who the Yellow Vests are, we see a key feature has been the diversity of the movement. We see a dominant profile of what some have called "back-office" or "invisible workers" – couriers, building workers, and care workers mainly working in the private sector. As a collective, they have in common that they are a population that tends to live not from profits or rent, but from wages and social security. The movement was dismissed initially as right-wing and populist. However, in a survey by Knaebel and Hayns (2018), just over 40 per cent identified as left wing to some extent, and three times as many identified as far-left (14.9 per cent) as far-right (4.7 per cent), but by far the largest single group considered themselves as "neither left nor right" (33.1 per cent). A feature of the movement, as with other pro-democracy movements, is that they identify as "the people". The unanimity of the Yellow Vests is of a different order from other movements as it arises from a dimension of their identity; that is, not identifying as a movement, but as "the people" (Kouvelakis 2019).

A key element of the movement is that its prime identity has been formed around consumption rather than production. Yet production and consumption are intimately linked by the employee's wage or salary, and even the most consumption-oriented employee is concerned with his/her level of earnings (Kelly 1998). This complementarity of interests can provide the basis of joint campaigning between labour and these types of social movements. The Yellow Vests were not focused upon the issue of work directly, but the injustice felt by the movement participants emerged from their material conditions, intrinsically linked to their income and purchasing power (*pouvoir d'achat*). The initial demand of the movement to cancel a fuel tax rise relates to the ability for citizens' consumption rather than directly relating to incomes. Underlying this is the shift in capitalist employment relations, which have made work more precarious and incomes more unequal.

In the formation of collective interests and identity, who is "them"? The prime targets for blame have been President Emmanuel Macron, the elites and the government, with the framing of "us and them" directly in "le peuple et Macron".[2] Other institutions have also been targeted, primarily being viewed as not representing the interests of "the people" (for example, unions), who are seen as "social managers". The mainstream media became a target for the movement but also contributed to the way it was constructed. Representations of inequality – luxury shops/cars (images of a burning Porsche and shops along the Champs-Elysées) have also been the target. Yet the employer is not the target; the luxury goods are viewed as a reflection of inequality.

2. *J'veux du soleil* (2019) film by François Ruffin and Gilles Perret.

The Yellow Vests movement revealed the weaknesses of existing representative structures, not only of political channels but also of unions – especially in terms of the links with the grassroots, and consideration of individual demands and interests from those who feel left behind and forgotten in the current configuration of French capitalism. The Yellow Vests rejection of links with unions and the unions' initial rejection of the movement – despite commonalities of interests – exposed limitations in the model of union representation in France. French unions continue to have relatively high levels of institutional power and influence over public discourse, but are often characterized as representing the narrow interests of already protected workers, and the work of unions is increasingly undertaken by a professional cadre of "technocrats", with limited input and involvement from rank-and-file members. The movement exposed the possibilities for widespread collective action and resistance to state policies affecting working-class people, and was evidence of a wider fertile territory for solidarity that remains a dormant potential until activated by "moments of collectivism", sparked by a strong sense of injustice and/or crisis (Atzeni 2009).

A tentative convergence around the Yellow Vests demonstrations in the year following the emergence of the Yellow Vests, in 2019, culminated in the 5 December strikes, and showed how opportunities for cooperation exist between unions and this newer movement. The unions were initially against any links with the movement as a right-wing populist movement. The increasing joint action between the Yellow Vests and the unions has been interpreted as a shift to more left-wing politics in the movement (Bergem 2022). At an individual and local level there has been more joint-movement action and support between the unions and the Yellow Vests. With the Yellow Vests suspicion of unions, many union activists concealed their union membership status when participating in Yellow Vest demonstrations (Bergem 2022).

The high levels of participation in the protests and forcing the government to back down need to be set in context of the multiple failures of union-led collective action against major welfare and labour reforms since 2006, even where there have been high levels of participation and disruption and strong public support. In 2018, in the same year as the Yellow Vests movement emerged, unions on the railways, the bastion of French labour unionism, experienced an historic defeat over reforms to their status. Colfer and Basin (2022) directly compare the Yellow Vests movement with the actions of the CGT union in the SNCF railway company in 2018. The clearest difference between the two movements, they argue, is in the highly structured, hierarchical and organized nature of the CGT, and the decentralized and non-hierarchical street movement that makes up the Yellow Vests. As the Yellow Vests' rejection of the unions softened over time, the movement benefited from some of the permanency and structure of the unions. For instance, unions have been able to provide resources to the movement, such as offering union offices for the Yellow Vests general assemblies.

Unlike unions, the Yellow Vests have not organized around or in front of workplaces, companies or universities. They have occupied public spaces – symbolically the "roundabout". Maillard (2019) argues that this is a reflection of their interests and demands as being centred on consumption rather than production, namely, they are not blocking production but blocking the normal flow of consumption. They have also not focused on organized demos along grand boulevards of big cities – not the traditional "*saucisses – mergues*" demos, making reference to the mobile barbecues and sausages that were often part of traditional demonstrations (Maillard 2019: 25). The demonstrations have often seen riots and violence. But the movement learned quickly that violence worked, as the government made concessions fairly early by cancelling the planned fuel price increase.

By organizing through the internet and social media, a plurality of interests did not hamper aggregation, and individuality did not hamper coordination (Lianos 2019). The Yellow Vests do not "belong" to Facebook or X, as they would to a political party or a union. As they stated, they are "all or no one" and they do not wish to have representatives and leaders if they can avoid it, despite the pressure that institutions put upon them to enter into the established system of political exchange. This is a challenge for unions, working within formalized structures and processes, and with a dominant organization mode of coordination more than movement mode, but it shows again the opportunity for the unions to "absorb" the interests of these type of movements, building on individual cross-movement relationships and dynamics to develop social movement modes of coordination.

CONCLUSION

Unions emerged as forms of collective resistance challenging key elements of the existing capitalist social and economic order. With time, unions in many countries become increasingly dependent for their survival on institutional internal routines and formalized external relationships with employers and governments (Hyman & Gumbrell-McCormick 2017). Yet, many contemporary analyses of social movements mistake characteristics of newly created movements for essential features of the movements themselves. Part of what is seen to be lacking in unions is often a phase of development. If we look at the unions organizing workers in the "gig" economy, for example, we can see the contemporary relevance of this argument, with these unions adopting what some would see as tactics and methods from social movements, such as direct action and a focus on activist-led organization. Newly created social movements are likely to experience the same pressures towards formalization and bureaucratization. As Tarrow (1989) argues, the reliance upon direct action is common to both old and new movements during "cycles of protest".

Unions are organizations for which constructing solidarities has always been a goal, never an accomplishment (Hyman 1999). If we draw upon social movement theory to look at the creation of collective identity, Melucci (1989) argues that the creation of collective identity is a process of negotiation over time that involves fostering relationships among participants and stimulating an emotional dynamic among those involved. This is what Tapia (2013) describes as a relational culture, one based upon fostering relationships and creating trust between the members and the organization. Such processes may have been lost over time or underdeveloped in some unions but have been vital for many "new" social movements, which can make them seem and feel like they are more relevant and effective as forces for progressive change (Hyman & Gumbrell-McCormick 2017). An area not covered in the chapter, but which is clearly the next social movement space for unions to question their capacity for absorption, is climate change and green movements (see Kenfack 2020).

There are opportunities and potential for more social movement modes of coordination and mobilizing cultures in unions, recognizing what McAlevey (2016) calls the "whole worker". The "whole worker" organizing approach merges workplace and non-workplace interests and issues into a tight blend and, rather than building union-community coalitions, it is arguably more focused upon unions adopting the latter two forms of Heery's types of fusion: union imitation of social movement practices and union absorption of new social movements/community groups. When we look at the lessons from the Yellow Vests movement, the significance of recognizing and representing the "whole worker" and their surrounding networks becomes evident. As Hyman and Gumbrell-McCormick (2017) argue, developing synergies between the organizational capacity of the "old" and the imaginative spontaneity of the "new", drawing upon the strengths of each, is an important means to build effective resistance. The longevity of the union movement and the often-fleeting solidarity or action of some new movements confirms the strength, and/or potential strength of unions to contest varying conjunctures of neoliberal capitalisms.

REFERENCES

Antunes, R 2001. "Global economic restructuring and the world of labor in Brazil: the challenges to trade unions and social movements". *Geoforum* 32(4): 449–58.

Atzeni, M. 2009. "Searching for injustice and finding solidarity? A contribution to the mobilisation theory debate". *Industrial Relations Journal* 40(1): 5–16.

Baccaro, L. & C. Howell 2017. *Trajectories of Neo-liberal Transformation: European Industrial Relations Since the 1970s*. Cambridge: Cambridge University Press.

Barron, P. *et al.* 2016. "State categories and labour protest: migrant workers and the fight for legal status in France". *Work, Employment and Society* 30(4): 631–48.

Bergem, I. 2022. "When the revolution did not look like you thought it would: the yellow vest movement through radical-left activists' imaginaries". *Modern & Contemporary France* 30(3): 295–312.

Colfer, B. & Y. Basin 2022. "Organised labour and fluid organisations: insights from the gilets jaunes movement". In *European Trade Unions in the 21st Century: The Future of Solidarity and Workplace Democracy*, edited by B. Colfer, 149–72. London: Palgrave Macmillan.

Connolly H. 2020. "'We just get a bit set in our ways': renewing democracy and solidarity in UK trade unions". *Transfer: European Review of Labour and Research* 26(2): 207–22.

Connolly, H. & S. Contrepois 2018. "From social movement to labour protests". In *Austerity and Working-Class Resistance: Survival, Disruption and Creation in Hard Times*, edited by A. Fishwick & H. Connolly, 109–28. London: Rowman & Littlefield.

Connolly, H., S. Marino & M. Martínez Lucio 2019. *The Politics of Social Inclusion and Labor Representation: Immigrants and Trade Unions in the European Context*. Ithaca, NY: Cornell University Press.

Connolly, H. & B. Sellers 2017. "Trade unions and immigration in the UK: innovative responses in a cold climate". In *Trade Unions and Migrant Workers: New Contexts and Challenges in Europe*, edited by S. Marino, R. Penninx & R. Roosblad, 224–43. Cheltenham: Edward Elgar.

de Raymond, A. *et al.* 2023. "Les gilets jaunes): une révolte sans fin?". *Geneses* 130(1): 80–111.

Della Porta, D. & M. Diani 2020. *Social Movements: An Introduction*. Chichester: Wiley.

Diani, M. 1992. "The concept of social movement". *Sociological Review* 40(1): 1–25.

Diani, M. 2018. "Unions as social movements or unions in social movements?" In *Social Movements and Organized Labour*, edited by J. Grote & C. Wagemann, 43–65. London: Routledge.

Engeman, C. 2015. "Social movement unionism in practice: organizational dimensions of union mobilization in the Los Angeles immigrant rights marches". *Work, Employment and Society* 29(3): 444–61.

Fairbrother P. 2005. "Rediscovering union democracy: processes of union revitalization and renewal". *Labor History* 46(3): 368–76.

Fairbrother, P. 2008. "Social movement unionism or trade unions as social movements". *Employee Responsibilities and Rights Journal* 20(3): 213–20.

Feixa, C., I. Pereira & J. Juris 2009. "Global citizenship and the 'new, new' social movements: Iberian connections". *YOUNG* 17(4): 421–42.

Gall, G. & J. Holgate 2018. "Rethinking industrial relations: appraisal, application and augmentation". *Economic and Industrial Democracy* 39(4): 561–76.

Gramsci, A. 1971. *Selections from the Prison Notebooks of Antonio Gramsci*, edited by G. Nowell-Smith & Q. Hoare. London: International Publishers.

Grote, J. & C. Wagemann 2018. *Social Movements and Organized Labour: Passions and Interests*. Abingdon: Routledge.

Heery, E. 2018. "Fusion or replacement? Labour and the 'new' social movements". *Economic and Industrial Democracy* 39(4): 661–80.

Heery, E., S. Williams & B. Abbott 2012. "Civil society organizations and trade unions: cooperation, conflict, indifference". *Work, Employment and Society* 26(1): 145–60.

Herberg W. 1947. *Bureaucracy and Democracy in Labor Unions*. New York: Great Island Conference.

Holgate, J. 2015. "An international study of trade union involvement in community organizing: same model, different outcomes". *British Journal of Industrial Relations* 53(3): 460–83.

Holgate, J 2018. "The Sydney alliance: a broad-based community organising potential for trade union transformation?". *Economic and Industrial Democracy* 39(2): 312–31.

Holgate, J. 2021. "Trade unions in the community: building broad spaces of solidarity". *Economic and Industrial Democracy* 42(2): 226–47.

Hyman, R. 1975. *Industrial Relations: A Marxist Introduction*. London: Macmillan.

Hyman, R. 1979. "The politics of workplace trade unionism: recent tendencies and some problems for theory". *Capital & Class* 8: 54–67.

Hyman, R. 1994. "Theory and industrial relations". *British Journal of Industrial Relations* 32(2): 165–80.

Hyman, R. 1999. "Imagined solidarities: can trade unions resist globalization?". In *Globalization and Labour Relations*, edited by P. Leisink, 94–115. Cheltenham: Edward Elgar.

Hyman, R. 2001. *Understanding European Trade Unionism: Between Market, Class and Society*. London: Sage.

Hyman, R. & R. Gumbrell-McCormick 2017. "Resisting labour market insecurity: old and new actors, rivals or allies?". *Journal of Industrial Relations* 59(4): 538–61.

Kahmann, M. 2015. "When the strike encounters the sans papiers movement: the discovery of a workers' repertoire of actions for irregular migrant protest in France". *Transfer: European Review of Labour and Research* 21(4): 413–28.

Kelly, J. 1998. *Rethinking Industrial Relations: Mobilisation, Collectivism, and Long Waves*. London: Routledge.

Kenfack, C. 2020. "Labor environmentalism as a paradigm of social movement unionism: participation of Portuguese trade unions in the national climate jobs campaign". *Journal of Labor and Society* 23(2): 181–204.

Knaebel, R. & J. Hayns 2018. "The gilets jaunes and the unions: a convergence over what?" *Notes from Below*. https://notesfrombelow.org/article/gilets-jaunes-and-unions-convergence-over-what

Kouvelakis, S. 2019. "The French insurgency: political economy of the gilets jaunes". *New Left Review* 116/117: 75–98.

Lianos, M. 2019. "Yellow vests and European democracy". *European Societies* 21(1): 1–3.

Maillard, D. 2019. *Une colère française*. Paris: L'Observatoire.

McAlevey, J. 2016. *No Shortcuts: Organizing for Power in the New Gilded Age*. New York: Oxford University Press.

Melucci, A. 1989. *Nomads of the Present: Social Movements and Individual Needs in Contemporary Society*. Philadelphia, PA: Temple University Press.

Michels, R. 1915. *Political Parties: A Sociological Study of the Oligarchical Tendencies of Modern Democracy*. New York: The Free Press.

Morgan, G. & V. Pulignano 2020. "Solidarity at work: concepts, levels and challenges". *Work, Employment and Society* 34(1): 18–34.

Roca, B. & I. Días-Parra 2017. "Blurring the borders between old and new social movements: the M15 movement and the radical unions in Spain". *Mediterránean Politics* 22(2): 218–37.

Tapia, M. 2013. "Marching to different tunes: commitment and culture as mobilizing mechanisms of trade unions and community organizations". *British Journal of Industrial Relations* 51(4): 666–88.

Tapia, M., M. Elfström & D. Roca-Servat 2018. "Bridging social movement and industrial relations theory: an analysis of worker organizing campaigns in the United States and China". In *Social Movements, Stakeholders and Non-Market Strategy Research in the Sociology of Organizations*, edited by F. Briscoe, B. King & J. Leitsinger, 173–206. Bingley: Emerald.

Tarrow, S. 1989. *Democracy and Disorder: Protest and Politics in Italy, 1965–1975*. New York: Oxford University Press.

Tarrow, S. 1998. *Power in Movement: Social Movements and Contentious Politics*. Cambridge: Cambridge University Press.

Tassinari, A. & V. Maccarrone 2020. "Riders on the storm: workplace solidarity among gig economy couriers in Italy and the UK". *Work, Employment and Society* 34(1): 35–54.

Tattersall, A. 2011. "Labor-community coalitions, global union alliances, and the potential of SEIU's global partnerships". In *Global Unions: Challenging Transnational Capital through Cross-Border Campaigns*, edited by K. Bronfenbrenner, 155–73. Ithaca, NY: Cornell University Press.

Touraine, A. 1986. "Unionism as a social movement". In *Unions in Transition: Entering the Second Century*, edited by S. Lipset, 151–72. San Francisco, CA: ICS Press.

Voss, K. & R. Sherman 2000. "Breaking the iron law of oligarchy: union revitalization in the American labor movement". *American Journal of Sociology* 106(2): 303–49.

CHAPTER 19

CONCENTRIC CIRCLES OF CLASS STRUGGLE: FROM THE WORKPLACE TO THE WORLD

Marissa Brookes

ABSTRACT

Workers' organizing efforts have expanded over time, moving beyond the boundaries of individual workplaces to encompass larger circles of influence within corporate structures, entire industries and even whole global supply chains. This chapter explores the reasons behind this outward expansion and the factors that have facilitated or hindered it. Why have workers seen a need to organize more expansively? What is the relationship between local-level organizing and worker organization on the international scale? This chapter considers these questions through the lens of concentric circles of class struggle, which suggests that strong organization at the workplace level plays a vital role in helping workers achieve higher levels of organization. In other words, this smaller circle of worker organization provides the foundation for more encompassing circles of organization, including transnational labour alliances and other modes of collaboration involving a multitude of workers around the world.

Keywords: Union organization; scale of organizing; challenges for unions

INTRODUCTION

The first two decades of the twenty-first century saw a surge in strategic cooperation among workers and unions from different countries in response to myriad challenges and opportunities associated with economic globalization. Nevertheless, international organizing is neither new nor inevitable. Well before a wave of workers' transnationalism swept the 2000s and 2010s, labour unions in the nineteenth century were widening their circles to encompass their brothers and sisters in other countries. Indeed, the very idea of "the labour movement" was originally conceived of as international (Hyman 2005). Nevertheless, as the twentieth century commenced, unions found themselves caught between the enticements of internationalism and the relentless rise of the nation-state.

For the most part, the latter won. By the close of the twentieth century, however, disruptive forces were once more dramatically altering the nature of the terrain of work and employment with economies growing evermore global, and the relationship of workers and unions to the state – and to each other – started to shift yet again (see also Chapter 7). This led labour union leaders, who sought new strategies for sustainable and effective organizing, to expand their fields of vision and those of their unions ever more widely. In other words, outwards to the concentric circles beyond themselves.

Why have workers seen a need to organize in ever larger circles expanding outward from the workplace: within corporate structures, throughout entire industries and across whole global supply chains? What factors have inhibited or facilitated these outward expansions? What is the relationship between local-level organizing and labour organization on the international scale? This chapter considers these questions in light of concentric circles of class struggle. In particular, it analyses the relationship between the globalization of trade and finance, on one hand, and the scale of union organizing and labour mobilizations, on the other, while paying close attention to the persistent indispensability of the local scale for all forms of union organization.

The chapter begins by providing a general historical overview of union organizing in terms of both its expansion into more encompassing spaces and its contraction back into national and subnational circles. In doing so, it considers not only the difficulties posed by economic globalization but also how globalization has unexpectedly improved prospects for unions from different countries to connect, collaborate and take action. At the same time, however, the analysis also shows how unions' abilities to organize more or less widely can never be fully understood without taking into account the innermost circle of class struggle: the workplace.

WHY WIDEN THE CIRCLE?

Economic liberalism and the first great globalization

From the mid-nineteenth century through the onset of the First World War, labour mobilization in the Global North grew in response to the pressures and pain brought on by the Second Industrial Revolution and the rapid internationalization of trade and finance (see also Chapter 8). In this era, later labelled the "First Great Globalization", political elites oversaw the ascendency of economic liberalism – an ideology committed to free trade, the free flow of capital, limited state intervention and the privileging of market relations above all else – as imperialism forced open new markets overseas and the Gold Standard ensured the monetary stability needed for international trade and finance to thrive (Rodrik 2012). International migration also hit an all-time high, as nearly 10 per cent of the global population left their home countries for economic and political reasons.

Economic liberalism (see Chapter 8) inverted all previous social and political orders by disembedding the economy from society, thus subordinating social relations to the primacy of markets (Munck 2002; Polanyi 2001 [1944]). Abysmally low wages, long hours and horrendous working conditions in unsafe factories dominated the lives of individuals forced to abandon family- and village-based support systems for filthy, overcrowded urban centres across Europe and North America. By the late nineteenth century, factory workers were being subjected to overbearing supervision and dehumanizing levels of stress due to new production techniques informed by the emergence of scientific management, while massive corporate mergers in coal, lumber, meat packing, railroads and steel further consolidated the power of capital and pushed inequality to new extremes. Still, new ideas about democracy, individual rights and social justice arose in this era as well, inspiring workers in the industrializing countries to expand their organizing efforts to ever-wider circles, including internationally.

Early labour internationalism initially entailed gatherings of political parties, unions and intellectuals, who debated the creation of a world socialist community and promoted explicitly anti-capitalist, cross-border solidarity actions (Munck 2002: 135). Founded in 1864, the International Working Men's Association, also known as the First International, is considered the first organizational incarnation of labour internationalism. Born out of the revolutionary excitement of mid-nineteenth-century Europe and infused with intense intellectual debate, including the personal influence of Karl Marx, the First International promoted shorter working hours, wage protection and defence against lockouts and strike breaking by police and military forces (Munck 2002: 137). In the 1890s, national unions from various countries founded the International Trade Secretariats (ITSs), the economically oriented, sector-based international umbrella organizations that would, decades later, evolve into the Global Union Federations (GUFs) (McCallum 2013: 21).

The "First Great Globalization" was a time "when confidence in the regulatory effectiveness of purely national action was limited and when the interconnectedness of politics and economics was largely taken for granted" (Hyman 2005: 147). As nation states began to consolidate, however, workers and unions developed an overriding concern with securing labour rights through the then-expanding power of the state. Although the Second International, founded in 1889, launched the campaign for the eight-hour work day, helped workers win rights to strike and organize, and promoted May Day as an international holiday, it ultimately failed to ward off workers' new-found national loyalties in the face of world war and the imperialistic ambitions of their national governments (Munck 2002: 138; Waterman 2001: 19). Hence, at the peak of the "First Great Globalization", the scope of workers' struggles would collapse back into smaller circles and remain rooted in the national for some time.

In fact, many national labour movements supported only their own so-called "national" interests, adopting racist and xenophobic attitudes against other national labour movements as well as against fellow workers in their own countries. This phenomenon was influenced by various factors, including economic competition, social tensions and the political ideologies of the time. One of the key reasons for the alignment of labour movements with nationalistic sentiments was the intensifying economic competition brought about by globalization. As countries became more interconnected through trade and investment, workers started to perceive the presence of foreign labour as a threat to their own jobs and wages. This perception led to the development of protectionist attitudes, as workers sought to safeguard their economic interests by supporting policies that favoured domestic industries and workers over foreign competition.

Furthermore, social tensions arising from rapid industrialization and urbanization added fuel to the nationalist sentiments within labour movements. Workers faced harsh conditions, long hours and low wages, which often led to labour unrest and social discontent. In this context, workers began to associate their struggles with the interests of their nation, seeing themselves as defenders of their national identity and social order against perceived threats from other nations. Political ideologies of the time also played a significant role in shaping the relationship between labour movements and nationalism. Socialist and communist movements, which were gaining traction during this period, often emphasized the primacy of class struggle and the overthrow of capitalist systems. However, different labour movements interpreted and adapted these ideologies to suit their national contexts. In many cases, unions aligned themselves with nationalist agendas, viewing their struggle as a national one against foreign exploitation rather than as an international class struggle.

Unfortunately, the rise of nationalistic labour movements against other labour movements exacerbated the problems of racism and xenophobia. In an attempt to differentiate themselves from foreign workers, native-born workers and their organizations sometimes resorted to scapegoating and discriminatory rhetoric. This divisive environment fuelled animosity between different nationalities and led to the marginalization and mistreatment of immigrant workers. Racist ideologies and discriminatory practices gained ground as labour movements sought to protect the interests of their own members at the expense of others.

It is important to note, however, that not all labour movements succumbed to nationalist sentiments or embraced racism and xenophobia. Some labour organizations recognized the shared struggles of workers across national boundaries and sought international solidarity. They understood that the issues faced by workers were not limited to specific nationalities but were rooted in the exploitative nature of the capitalist system. These internationalist labour movements advocated for cooperation and unity among workers of different nations, aiming to build a global labour movement capable of challenging the inequalities and injustices of the time.

World wars and the "Great Depression"

All aspects of economic globalization – trade, finance and migration – contracted dramatically with the onset of the First World War. While globalization was grinding to a halt, labour's turn to nationalism had already been well underway, even as the Third International formed in 1919. Also known as the Communist International or Comintern, this last old wave of labour internationalism was mainly an organization of Moscow-aligned political parties used as a tool of Soviet foreign policy by Stalin after Lenin's death. Labour internationalism declined further throughout the Great Depression and the Second World War (McCallum 2013: 21–2).

On the national scale, there was much more movement. Through mass strikes and other forms of labour unrest, workers fought for wages sufficient to sustain their families. The Great Depression brought with it mass layoffs in agriculture, construction, retail and manufacturing, among other problems. Union organizing exploded in the second half the 1930s within the US with the passing of the National Labor Relations Act 1935 – known as the Wagner Act and which allowed workers to gain statutory certification for bargaining with employers – and the Roosevelt administration's efforts to tie the concept of industrial democracy to patriotic nationalism, which led workers to feel more entitled than ever to make demands on the state and their employers (Lichtenstein 2003). Organized labour across the industrialized countries used its new-found political power to fight for more worker protections.

By the mid-twentieth century, labour movements had linked their fortunes to those of their respective nations, and unions' transnational activities failed to seriously challenge either capital or state interests (McCallum 2013: 22–3; Wills 2002). Although this period was a high time for strikes and industrial actions in the automobile industry, textile industry and other prominent industries of the time (Silver 2003), union activity remained primarily national or subnational. Unions "were becoming an integral element of the nation-state", and "the practice of the unions was not only nationalist but also statist" (Munck 2002: 140).

The "Golden Age of Capitalism" and social democracy

The postwar era from 1945 to the early 1970s witnessed remarkable levels of prosperity and economic growth in the advanced capitalist democracies (see Chapter 9). Although devasted by the Second World War, Europe and Japan recovered and grew rapidly in this era thanks to the Marshall Plan and other well-orchestrated economic interventions. While some developing countries also grew rapidly, such as the oil-producing countries of the Middle East, much of the developing world lagged economically, with some areas still under colonial rule at the war's end. In the absence of serious foreign competition for manufactured goods, the US became a manufacturing powerhouse. Consumption in the industrialized countries shot up to record heights.

Unions played a key role in managing their national economies by cooperating with management to keep wage increases in line with productivity. With an expanding economic pie and a manufacturing-based economy, both unions and employers could easily ensure that higher wages would not cause inflation. The share of national income that went into wages stayed constant in this period, as did the share of national income going to profit. In West Germany, Sweden and Denmark, unions used highly coordinated and centralized collective bargaining to similarly restrain wage increases and keep inflation low. Throughout most of Western Europe, postwar welfare states blossomed with the creation or expansion of social safety nets such as unemployment insurance, healthcare, social security, disability pensions, veterans' benefits and agricultural price supports, all of which had a decommodifying effect on labour. With so much government intervention in the economy and employer acceptance of organized labour, thanks in part to Keynesian commitments to full employment and Cold War fears of communist alternatives, unions had little incentive to turn to internationalism.

What was left of labour internationalism came to be defined by the Cold War divide (McCallum 2013: 23). Just four years after the 1945 founding of the World Federation of Trade Unions (WFTU), which attempted to comprise all national union federations under one global umbrella, distaste for the organization's communist-dominated leadership and arguments over support for the Marshall Plan led several federations, including the US-based Congress of Industrial Organizations, the UK's Trades Union Congress and the French Confédération Générale du Travail – Force Ouvrière, or simply Force Ouvrière, FO, to secede and form the International Confederation of Free Trade Unions (ICFTU) to counter the pro-Soviet WFTU (Harrod & O'Brien 2002: 6; Munck 2002: 141–2). These rival organizations fought over recruitment of members in Asia, Africa and Latin America. In time, labour internationalism came to be associated with unions from the industrialized countries aiding their governments against "the communist threat" through strategic assistance to workers in the Global South. The American Federation of Labor and Congress of Industrial Organizations' (AFL-CIO) funding of anti-communist, business-friendly unions in Latin America is by now well known (Battista 2002; Garver 1989; Thompson & Larson 1978).

Meanwhile, the AFL-CIO pushed its affiliate unions to join the ITSs in an attempt to suppress the influence of European social democrats and spread the style of business unionism increasingly characteristic of industrial relations in the US. Despite the Americans' efforts, European dominance and social democratic principles pervaded both the ITSs and the ICFTU, which the AFL-CIO abandoned in 1969 (Hyman 2005: 141). Labour internationalism, thus, became split between the pro-Soviet WFTU, the increasingly European-dominated and bureaucratic ITSs and the ICFTU, the government-backed, anti-communist activities of the AFL-CIO, and newly emerging labour organizations in the developing countries, which took an ambivalent approach to offerings of foreign aid from all of the above (McCallum 2013: 22–3).

By the early 1970s, transnational corporations (TNCs) had begun to expand their power and reach. ITSs such as the International Metalworkers' Federation and the International Union of Food and Allied Workers attempted to match the influence of international capital by promoting multinational, industry-wide collective bargaining (Munck 2002: 145), in addition to lobbying intergovernmental regulatory bodies to establish corporate codes of conduct (Fairbrother & Hammer 2005; Hyman 2005: 148). Union representatives also attempted to bargain with TNCs firm by firm through World Company Councils (WCCs), which emerged in the late 1960s through the mid-1970s to promote multinational collective bargaining. The early excitement of labour leaders over the potential of TNCs carving out a new path for labour transnationalism was echoed by academics, whose writings theorized unions' "natural" progression from their present national structures into fully international organizations capable of multinational bargaining as traditional industrial relations institutions grew inadequate (Levinson 1972; Piehl 1974; Windmuller 1967).

Yet clearly that vision of the 1970s never materialized. The attempt to replicate national corporatist institutions on an international scale proved difficult, if not wholly unfeasible, and enthusiasm waned (Gumbrell-McCormick 2000). Although one can cite numerous practical reasons for why the European-led ITSs and WCCs failed to fulfil their vision, the key factor inhibiting the extension of union practices from a national to an international level was the enduring capacity of governments to guarantee their citizens decent living standards through national welfare states. Workers saw no need to supplant the national institutional structures through which their own interests were adequately met (Wills 2002).

Logue (1980) formalizes this point. According to his "rational self-interest" framework, "the greater the degree of trade union control over its national environment, the less likely it is to undertake international activity to achieve its members' goals" (Logue 1980: 21–2). This implies that unions will only embrace transnationalism when national-level institutions no longer enable them to serve the economic interests of their members. Nonetheless, while Logue might have been correct in his observation that national protections inhibit internationalist aspirations, it would later become clear that even in the face of crumbling social pacts and the far-reaching reconfiguration of national institutional arrangements, unions do not easily, let alone automatically, adopt transnational strategies.

The neoliberal era

Various interlinked macroeconomic pressures ushered in an era of crisis, backlash and breakdown in the 1970s, causing economic globalization to once again retract and reverse before rebounding with a fury as capital fled overseas in an attempt to regain the profit levels of the previous era (see Chapter 11). The most prominent of these pressures were the 1973 and 1979 OPEC oil shocks, which forced firms to raise prices and forced unions, in turn, to demand wages that kept pace with the rising cost of living. As wages rose and production costs soared, so too did layoffs and, consequently, the cost of government-funded social protections. The result was wage-price spiral and the unprecedented simultaneous rise of unemployment and inflation amidst a stagnant economy called "stagflation".

One major way to restore the profits rates of the "Golden Age of Capitalism" was for firms to cut labour costs by relocating production abroad and lowering labour costs at home. Employers did this by offshoring production, increasing the pace and intensity of work, implementing labour-displacing technologies and otherwise making labour as flexible as possible. Although Japan and the continental countries of Western Europe managed to maintain their more coordinated styles of capitalism well into the 2000s (Hall & Soskice 2001), left-wing political parties generally failed to find solutions that went beyond Keynesianism and traditional social democracy. Hence, even the most labour-friendly social democracies began to buckle under the weight of increased international capital mobility, deindustrialization and the growth of the service sector economy, and ageing populations increasing the costs of social expenditures (Iversen & Wren 1998).

In the UK and the US, the political and ideological shift to neoliberalism was in full swing by the 1980s. Neoliberalism called for "liberating" corporate and business power from government intervention and the re-establishment of market freedoms through deregulation and privatization (Harvey 2005). Governments slashed corporate taxes, privatized public enterprises, cut public expenditures and deregulated trade and finance on a massive scale, all while promoting a culture of individualism and personal responsibility. Weakening unions was high on the agenda of both the Reagan and Thatcher administrations, which notoriously broke the air traffic controllers' union in the case of the former and the miners' union in the latter. Alongside these trends arose the professional union avoidance industry, anti-union consultants that coach companies to use mandatory one-on-one meetings, captive audience employee training and other legal and illegal forms of intimidation to prevent workers from forming or joining unions (Logan 2006).

Given this massive shift in economic ideology and practice, unions in the advanced industrialized countries became preoccupied with evermore mobile capital, TNCs' growing autonomy from the state and the disappearance of secure, stable and gainful employment (Piven & Cloward 2000). The exodus of manufacturing jobs that had begun in the early 1960s but ramped up in the 1980s and 1990s, coupled with employment growth in the low-wage service sector, meant not only declining employment in unions' traditional strongholds but also rising employment in jobs not readily conducive to unionization, especially with the growth of part-time and temporary positions. The restructuring of jobs into more flexible and precarious positions also rendered workers overall far more vulnerable than in the postwar years, as did the abandonment of Keynesian economics, the use of macroeconomic austerity measures, and large-scale welfare state reforms that to varying degrees eroded traditional social protections (Huber & Stephens 2001; Pierson 2001; Scharpf & Schmidt 2000; Swank 2002).

To be clear, since the start of the twenty-first century, scholars have applied dozens of correctives to the ubiquitous "race to the bottom" narrative. Chief among them is the recognition that such a simplified view of global economic processes obscures the manifest reality that these pressures have had quite different impacts across and even within different countries, as unions, employers and policy-makers have responded to challenges in ways that comport with existing institutional structures, even as those institutions are themselves undergoing change (Garrett 1998; Hollingsworth, Schmitter & Streeck 1994; Thelen 2001). That is not to say, however, that the altered political climate and macroeconomic pressures of the two decades prior to and following the turn of the century did not present organized labour with new challenges, however different those challenges looked across countries and over time. Particularly in the liberal market economies – countries such as Australia, the UK, New Zealand and the US and in which business strategies were premised on firms' low coordination, minimal regulation and cost-driven competition (Hall & Soskice 2001) – unions experienced drastic declines in membership, reduced political influence and even a crisis of legitimacy (Fairbrother & Yates 2003; Goldfield 1989; Peetz 1998; Puette 1992). While unions in the coordinated market economies – countries characterized by high firm coordination, patient capital, diversified quality production and corporatist institutions typical of continental and northern Europe – did not experience as dramatic a decline in density and influence; nonetheless, many struggled to negotiate with employers from a weakened position and eventually saw the breakdown of industrial relations systems and welfare state institutions into patterns of economic dualization or embedded flexibilization (Thelen 2014).

The economic and political climate of the post-Cold War years contributed to an easing of ideological tensions in the labour movement globally (Harrod & O'Brien

2002: 6; Moody 1997), allowing workers in both the developed and developing countries to refocus their attention on the ills of neoliberal capitalism. Yet the pressures unions had been experiencing since the 1970s and 1980s also deepened in the 1990s, exacerbating the crisis of organized labour and creating new difficulties for vulnerable workers and incipient unions in developing countries. Unions' responses to these economic and political pressures varied widely in content and form. Overall, however, they were overwhelmingly national or local in scope. That is, rather than seeking out new strategies of transnational cooperation, unions tended to pursue relatively insular strategies such as lobbying for protectionist legislation (Hurd, Milkman & Turner 2003: 16) or otherwise falling back on existing institutional arrangements that had afforded them protection in the past, even if it meant making an increasing number of compromises in the form of wage freezes, more flexible working conditions, a two-tiered labour market or reduced collective bargaining power. Logue's (1980) hypothesis that national-level power would discourage unions from seeking out international alternatives was, thus, only partially borne out in the sense that its inverse – that a lack of national power should encourage internationalism – was not immediately evident. Hence, a decline in national power might be a necessary but not sufficient condition for embracing labour transnationalism.

New union transnationalism in the late neoliberal era

The enduring ideological grip of neoliberalism, the growing global influence of TNCs and the worldwide spread of the internet in the late 1990s and 2000s fuelled the most rapid and intense period of economic globalization yet known. Nevertheless, with international trade and finance penetrating more markets than ever before, the general public in both the Global North and Global South grew increasingly aware of globalization's dramatic downsides. The decades on either side of the millennial divide only added to neoliberalism's track record of devastating job losses, declining economic equality, dehumanizing austerity measures and the destruction of social solidarities due to internationally contagious financial crises plummeting whole swathes of populations into poverty and precarity, most notably in 1998 and 2008. Partly pushed by these pressures, and partly pulled by the prospects of a new transnationalism facilitated by cheaper and easier international transportation and communication, some unions began experimenting with new strategies beyond the national scale (Brookes 2019).

While unions by no means abandoned their national and local-level strategies, those that began to employ transnational tactics in the early twenty-first century shared a perception that traditional channels of politics and industrial relations had become ineffective or even problematic. Moreover, even though TNCs had existed for decades, unions were finding it more and more difficult to negotiate with employers whose structures were rapidly changing. As corporations expanded, merged or were bought out by foreign companies, unions found it "harder to organize in their backyards ... [The Hotel Employees and Restaurant Employees union in the US], for example, once faced local hotel and restaurant owners but now face powerful international corporations" (*In These Times* 16 February 2004). The Service Employees International Union (SEIU), whose well-known "Justice for Janitors" campaign had largely succeeded in winning contracts and organizing rights throughout major US cities, likewise struggled to cope with rapid change. As the SEIU's Stephen Lerner (2007: 28) observed, "even as we made these gains, the industry continued to mutate under the pressure of globalisation". The SEIU's then-president expressed similar sentiments about the security industry: "All of a sudden we found ourselves needing to talk to more CEOs in Europe than in the US" (Andrew Stern in *Workforce Management* 20 January 2006).

What began as a trickle of international activity in the labour movement at the turn of the century – including a handful of standout campaigns led by the United Steelworkers at Ravenswood, dockers in Liverpool, the Teamsters at UPS and dozens of different unions in anti-sweatshop campaigns in Asia and Latin America (Banks & Russo 1998; Brookes & McCallum 2017; Herod 1995; Johns & Vural 2000; Masur 2000; Sadler & Fagan 2004) – soon turned into a torrent with well over 100 transnational labour alliances forming and launching corporate campaigns throughout the 2000s and 2010s (Brookes 2019). This wave of transnational labour campaigns spanned nearly every continent and major industry in the global economy.

Neoliberalism's persistence in the twenty-first century prompted a strong sense of urgency and optimism not only in individual labour unions but also those organizations already operating on the international scale. In 2002, the old ITSs reformulated and rebranded themselves into the GUFs, organized by industry to formally link together national union federations, individual unions and works councils from different countries at the international level. The GUFs grew significantly throughout the 2000s, gaining more national union affiliates and a larger overall membership than ever before (Ford & Gillan 2015; Müller, Platzer & Rub 2010: 3). The ICFTU also underwent transformation, taking on a new agenda to add a layer of social regulation to global governance structures (Ghigliani 2005: 361) and merging with the World Confederation of Labour to form the International Trade Union Confederation (ITUC) in 2006. Early on, sceptics were quick to argue that the mere existence of formal structures such as the GUFs and the ITUC did not automatically constitute meaningful transnational action, let alone any new-found capability to challenge capital (Hodkinson 2005). Moreover, at times such formal organizations have promoted workers' national interests rather than creating international communities of interest and trust. By the 2010s, however, the GUFs had proven themselves to be indispensable for helping national unions develop shared interests, regularize communication and obtain financial resources and logistical support for transnational campaigns (Brookes 2019: 29; Croucher & Cotton 2009: 8; Ford & Gillan 2015; Goodman 2004: 112; Lillie 2006: 125).

Still, despite their diverse forms and global spread, there is nothing automatic or mechanical about unions widening their circles of struggle to include transnational strategies, which are contingent upon varying macroeconomic pressures, specific workplace circumstances, historically entrenched national institutions and a mix of challenges and opportunities provided by an ever-evolving economic globalization. Uncertainties about the future of labour transnationalism arose again in the early 2020s following a series of interlocking crises and consciousness shifts with the potential to both discourage and encourage international solidarities. These included the increased salience of intersectional identities most prominently captured in the global Movement for Black Lives, the profound economic and social disruption of the Covid-19 global pandemic, an alarming global decline in freedom of association and expression, a world-wide spike in populism and tendencies towards economic nationalism, and growing global awareness of climate change, environmental degradation and biodiversity loss as serious existential threats (Schulse-Cleven 2021).

UNDER WHAT CONDITIONS CAN CIRCLES EXPAND?

The broadening of labour organizing and activism into ever-wider circles has never been a simple process. Unions' transnationalism has depended as much on macroeconomic events and geo-politics as it has on union members' and leaders' personal proclivities towards international solidarity or national self-interest. As evidenced above, some facets of globalization have undermined worker power and acted as "push" factors for more expansive organizing, including international capital mobility, corporate restructuring, and increasingly flexible labour practices. Other features of globalization – such as advances in information and communication technologies, multilevel institutional frameworks and regional trade agreements – have acted as "pull" factors that enticed labour into more transnational activity.

Nonetheless, unlike TNCs, which are by definition international actors, unions have operated primarily at the national and subnational scales (Lillie & Martínez-Lucio 2004). The promise of economic globalization was that better transportation and communications technologies would facilitate more effective cross-national coordination for all economic actors (Carter *et al.* 2003; Lee 1996; Munck 2002: 158). While the growth of transnational labour alliances in the early twenty-first century was certainly a product of such enhanced mobility and connectivity, cheap flights and internet access were not enough to get unions to expand their circles of struggle. The teleology implicit in the term "globalization", thus obscures the fact that the phenomenal internationalization of trade and finance in the neoliberal era did not automatically create comparable expansions in international migration or transnational collective bargaining.

Class consciousness and ideology played a significant role in convincing unions and workers to take advantage of globalization to facilitate more effective cross-national coordination and transnational labour organizing. Class consciousness refers to the awareness among workers of their shared interests, common struggles and the recognition of their position within the broader capitalist system, and in this era of globalization such consciousness developed in part as better communications technologies helped workers broadcast and witness labour exploitation and precarious working conditions prevalent in different parts of the world, thus fostering a sense of solidarity and an understanding that their struggles were not isolated but part of a broader global context. The internet, email, social media and video conferencing platforms helped class consciousness develop in the late twentieth and early twenty-first century, as these technologies provided unprecedented opportunities for communication, information sharing and organizing among workers globally. These technologies enabled faster and more accessible exchange of ideas, strategies and experiences, helping to strengthen bonds and foster solidarity among workers across borders. Moreover, unions and workers could observe and learn from successful examples of transnational labour organizing that emerged in this era. Instances of cross-national solidarity, such as joint campaigns, coordinated strikes and transnational alliances demonstrated the potential effectiveness of such strategies. These success stories provided inspiration and practical models for unions and workers to follow, encouraging them to take advantage of technological advancements and embrace cross-national coordination for their own organizing efforts.

At the same time, ideological frameworks such as global labour solidarity and the new internationalism provided the intellectual basis for unions and workers to embrace cross-national coordination. These ideologies emphasized the importance of workers standing together across borders to challenge the power imbalances inherent in the globalized capitalist system. Global labour solidarity argued that workers' interests transcended national boundaries and that collaboration among workers of different countries was crucial for achieving meaningful change. Internationalist ideologies highlighted the need to unite against common enemies, including multinational corporations and exploitative labour practices. In contrast to past iterations of such ideological frameworks, however, this updated set of ideas addressed more concretely what were then considered new challenges for workers, such as the fragmentation of production, offshore outsourcing, extreme capital mobility, the race to the bottom in wages and working conditions, and the exploitation of labour across supply chains.

Finally, globalization's pull factors must be understood not only as those features of globalization that have made it easier for unions to expand their circles but also in terms of the features that created new opportunities for more effective labour activism. The power resources approach has been especially useful for understanding how globalization has afforded unions new tools for transnational activism (Schmalz, Ludwig & Webster 2018; see also Chapter 3). Specifically, as TNCs increased their reliance on just-in-time manufacturing, extensive chains of subcontracting and complex logistics networks, so too did they increase their vulnerability to deliberate disruptions in the production and delivery of goods and services. Likewise, while more globally mobile capital meant more corporate freedom to shop around for favourable rules and regulations, it also meant TNCs becoming entangled in multiple layers of legal and regulatory institutions to which they are held accountable in several countries and on the international scale. Finally, as TNCs grew to depend more and more upon international networks of suppliers, investors and consumers, so too did they grow more exposed to influence from these outside stakeholders. Unions have, thus, found themselves able to exercise power – that is, compel an employer to do something it otherwise would not do – by exploiting vulnerabilities in global supply chains (structural power), invoking national and international rules and regulations (institutional power), and leveraging the influence of various outside stakeholders (coalitional power) (Brookes 2013). When workers successfully expand the scale of conflict beyond the realm of local and national-level institutions, they can bring an employer to the bargaining table more quickly, since that employer is forced to devote attention and resources to labour activism on the international scale.

Not all factors affecting the scope of workers' struggles operate on the macro-level, however. In fact, the widest circles of transnational activism fundamentally depend on the smaller circles of union-to-union relationships within them. Before engaging in something as broad and bold as a global campaign, individual unions must first coalesce around common goals and jointly plan and execute tactics in an effort to attain those goals. Such cooperation remains difficult for unions even when globalzsation offers an ideal mix of push and pull factors. In particular, union-to-union coordination remains difficult due to three core challenges: conflicting material interests, collective action problems and practical problems.

Conflicting material interests

In some ways, the globalization of production in the era of neoliberalism has made it more difficult for workers to coordinate with each other across national borders. Widespread deregulation and privatization, corporations' constant restructuring, and the actual or threatened relocation of worksites by highly mobile TNCs rendered workers increasingly vulnerable to dislocation. These competitive pressures encouraged some unions to embrace protectionism and other inward-looking strategies that exacerbated international divides (Johns 1998). Moreover, the perception that states lost much of their power to regulate capital and provide social safety nets in the advanced economies (Pierson 2001; Scharpf & Schmidt 2000; Thelen 2014), and the fact that states had rarely played such a strong protective role in most developing countries (Piven & Cloward 2000: 418), heightened workers' sense of vulnerability, which bred intra-class distrust in some parts of the world just as readily as it incentivized transnational cooperation (Silver 2003).

Nevertheless, such competitive pressures affected some sectors more severely than others. Because the threat of capital flight was particularly strong in the manufacturing industries, manufacturing workers were often hesitant to help those with whom they were competing for jobs. Unions in service industries – such as education, healthcare and property services – grew more open to labour transnationalism since many public and private service jobs cannot be offshored (Evans 2010: 358). Still, these broad generalizations obscure the fact that many manufacturing unions, such as the United Steelworkers, played active roles in transnational campaigns, while there were plenty of service sector unions that did not. Moreover, not all service sector jobs are immune to offshoring or otherwise free of competitive pressures. Unions' proclivities towards transnationalism are therefore only partly a function of sector and industry.

Whether unions cooperated with each other also depended on the national institutional frameworks in which they were embedded (Larsson 2012; Lillie & Martínez Lucio 2004). Cross-national differences in labour rights and employment relations institutions have created conflicting material interests. For instance, institutions of co-determination and social partnership forged during the "Golden Age of Capitalism" gave unions in some countries not only more power to negotiate with management but also incentives to maintain their positive relationships with employers, regardless of any negative impact on other workers abroad (Greven 2008: 6). Even in the absence of formal co-determination, a union that has a good relationship with an employer probably has rational incentives to prioritize the interests of the employer over those of workers located elsewhere (Young & Sierra Becerra 2014).

The "home union" embodies this difficult paradox. The home union represents employees of a TNC within that company's home country and, therefore, has the closest relationship with that employer. Close relations with management can bias home unions towards the interests of the employer, leading home-country workers to emphasize national-, firm- or even plant-level interests over transnational solidarity (Dufour & Hege 2010). For instance, a home union might share management's interests in avoiding bad publicity and selling products at competitive prices, in contrast to workers at lower levels of a value chain who seek to expose poor working conditions and low wages (Lindberg 2010: 212). Home unions, therefore, sometimes have incentives to keep within their own circles. That said, the opposite trend is also sometimes true, as a home union could have a strategic interest in helping to level up wages elsewhere within a TNC's operations in order to prevent being undercut through offshoring. By advocating for higher wages and improved working conditions in other branches or subsidiaries of the TNC, the home union can help mitigate the incentive for offshoring and, thus, safeguard domestic jobs and maintain a more stable employment landscape. Home unions might also recognize that their own ability to negotiate favourable collective bargaining agreements can be strengthened by promoting equitable wages across the company's global operations. Thus, advocating for higher wages elsewhere helps the home union maintain its bargaining power and ensure that workers' interests are protected in negotiations with the TNC.

Furthermore, while the home union may prioritize the welfare of its own members, its leaders and members can also recognize the importance of solidarity and shared interests among workers globally, either through a compassionate understanding of workers' struggles across the TNC's subsidiaries or suppliers, or through an interest in maintaining a good reputation and positive public perception. Concerns related to exploitative labour practices, low wages and poor working conditions can lead to negative publicity, consumer backlash and potential damage to the TNC's brand image. Hence, campaigning for workers to be treated fairly and paid well throughout a TNC's operations can help build a positive public perception of the TNC, benefiting the company's long-term sustainability and, thus, the workers that the home union represents.

Collective action problems

In addition to such possible conflicting material interests, potentially allied unions might also face collective action problems. Like states, unions are subject to collective action problems insofar as their goals – such as an industry-wide floor wage or better safety regulations across a corporation's global operations – constitute public goods (Olson 1965). In other words, because the benefits of successful transnationalism often cannot be excluded from others and are not diminished when utilized by unions that did not participate in gaining them, it is rational for potentially allied unions to attempt to free-ride on each other's efforts, despite the fact that the potential gains from expanding the circle of struggle would be greater than those possible from acting alone.

The collective action problem is exacerbated by the fact that cooperating with one or more other unions tends to be more time-consuming, resource-draining and uncertain in outcome than a union acting on its own (Burgoon & Jacoby 2004). Union leaders and members have rational concerns that the effort required for effective cooperation is not worth their time, money and energy. This is especially true when the investment of resources is one-sided or when tactics seem geared towards benefiting only one of the unions involved. American unions in particular were criticized in the late 2000s for forming "ad hoc" and "peculiarly asymmetric" alliances in which they seemed "interested in obtaining support for their struggles from foreign unions, but have offered precious little support in return" (Hyde & Ressaissi 2009: 16). Such "last-minute, one-way solidarity" tended to be "merely tactical and short-lived" (Greven 2008: 6) and was not to the mutual benefit of partner unions.

Practical problems

Another major, albeit broad, category of barriers to effective union cooperation can be referred to as practical problems. Unions may have congruent material interests and even commit to contributing equal effort in mobilization action yet, nevertheless, encounter obstacles such as language barriers, lack of finances, limited access to technology or unstable political climates that put labour activists at physical or legal risk. Cross-national institutional differences can make it particularly hard to make circles of struggle more encompassing. Such differences range from relatively straightforward prohibitions on secondary strike action to the more idiosyncratic rules that preclude otherwise standard practices. In India, for example, legal restrictions against unions receiving funds from foreign donors were the legacy of a colonial-era court ruling interpreted to mean that Indian unions "are creatures of the state and that by accepting cross-border support they assert a challenge to state sovereignty" (Sukthankar & Kolben 2007: 66). National institutional differences can also be the source of simple misunderstandings.

Overcoming union-to-union conflicts

Conflicting material interests, collective action problems and practical problems all inhibit inter-union coordination. Of these three issues, those rooted in seemingly incompatible interests are arguably the most pernicious because they indicate a fundamental unwillingness on the part of potential union allies to engage in cooperative efforts, as opposed to a situation in which unions wish to cooperate but have rational incentives to free-ride (collective action problems) or a situation in which unions wish to cooperate, and are also willing to put in equal effort yet, nonetheless, encounter obstacles in the political or institutional settings in which they seek to act (practical problems). Essentially, if unions cannot find common ground materially, then the other two inter-union coordination problems are irrelevant.

Hence, the first step in expanding the circle of struggle beyond a single workplace or union is to foster shared interests among potential partners. Having clear material goals enables unions to focus on specific shared interests such as securing collective bargaining rights or improving safety standards at a particular company (Goodman 2004). Nevertheless, cooperation based solely on material goals may be too weak a basis for something as complex and drawn-out as a transnational campaign. Even closely matched material interests among unions are unlikely to completely overlap, and without some deeper basis for long-term commitment, workers might not be able to break free from collective action problems or address their practical problems. Beyond just focusing on immediate shared goals, then, unions need more sophisticated incentives towards active cooperation.

The strongest possible basis for inter-union coordination is having a history of genuine reciprocity between unions, which develops as unions establish a track record of taking action in support of each other over time. More than just an explicit recognition of long-term mutual interests, reciprocity is about unions internalizing norms of solidarity based – without any sense of superiority or status involved in relation to partners and allies – not just on shared interests but on their past experience of reciprocated support and mutual expectations of future assistance. Unions that have established a track record of taking direct, sometimes altruistic, actions in support of other unions time and time again tend to place immense value on international solidarity, which effectively deters free-riding even in campaigns that make great demands on unions' resources.

It is usually not possible, however, for unions seeking to cooperate with one another to first spend decades demonstrating their commitment to mutual support through piecemeal displays of solidarity. Still, strategic outlooks can solve this problem. When unions focus on long-term issues of mutual concern that extend beyond immediate material goals, they are more likely to put effort into cooperation even in the absence of an established history of solidarity. Although not a substitute for a record of genuine reciprocity, the development of a shared strategic outlook can serve as a strong basis for expanding the circle of struggle. Shared strategic outlooks develop through unions' conscious efforts to project others' perspectives beyond a single issue and towards big-picture, long-term issues of mutual concern. Nonetheless, unions often struggle to effectively develop and implement strategic actions due to various constraints (Hyman 2007). The diversity of interests among unions, historical legacies, power imbalances and practical barriers all contribute to the difficulties unions face in formulating and implementing effective strategies for any sort of collaboration, including transnational organizing and mobilization.

INDISPENSABILITY OF THE LOCAL SCALE

Establishing effective union-to-union cooperation – which, in turn, makes it possible to expand circles of labour organizing and action evermore widely – not only requires shared understandings between unions, but also requires sufficient solidarity and strength within each individual union involved. Evidence suggests that strong organization at the level of the individual workplace plays a vital role in helping workers achieve higher levels of organization and class consciousness. In particular, it is at the local scale that workers secure key resources, construct common identities and a sense of shared struggle, develop learning capabilities and foster supportive leaderships. These strengths then allow unions to overcome collective action problems and other intergroup conflicts when they attempt to cooperate on a larger scale. Notwithstanding the challenge of constructing a consciousness that collapses distance and difference between workers, this smaller circle of worker organization provides the stronger foundation for further expansion into more encompassing forms of organization, including the formation of transnational labour alliances and other modes of collaboration that include workers from around the world.

Most studies of transnationalism recognize the need for transnational actors to coordinate with each other across borders yet take for granted the internal cohesion of these actors within their own organizations. A transnational labour alliance is a collective actor, but so too is each union comprising that alliance. Each union in a transnational alliance, therefore, faces its own internal coordination challenges that must be resolved before that union can carry out actions in support of a campaign. As Anderson (2009: 961) argues, merely "forming transnational union relationships has little impact on power relations within firms, what matters is the ability of unions participating in such alliances to tap their locally- and nationally-scaled powers and project or diffuse them through the networks of TNCs". In other words, workers require not just a union or other formal organization but one with the capacity to consciously plan and collectively take action (Lévesque & Murray 2010: 336).

Union capacity consists of both resources – "fixed or path-dependent assets that an actor can normally access and mobilise" – and capabilities – "sets of aptitudes, competencies, abilities, social skills or know-how that can be developed, transmitted and learned" (Lévesque & Murray 2010: 335–6). These resources and capabilities afford workers the capacity to act collectively through their own organizations. Union capacity thus makes the difference between a union that merely exists and one that is capable of collective action.

The degree to which various resources and capabilities within unions matter for facilitating cooperation among unions varies from case to case. However, four aspects appear to be particularly important: basic material resources, common identity, learning capabilities and supportive union leadership. First, although fairly self-evident, the importance of basic material and human resources such as financial resources, offices or other meeting spaces, and dedicated research staff should not be underestimated (Lévesque & Murray 2010: 340; Piven 2008: 11). For unions that seek to engage employers through strategic campaigns, which "go far beyond traditional organizing and bargaining and develop creative and complex processes that pressure firms in a multitude of ways" (Juravich 2007: 17), research capacity is indispensable. Workers organized in a union with a well-coordinated division of labour, including dedicated research staff and campaign organizers, are able to map out corporations' production networks and key relationships to locate potential leverage points (Juravich 2007; Rainnie, Herod & McGrath-Champ 2011: 165). Such capacities also enable workers to consciously articulate strategies over time and across space by deciding which short- or long-term objectives to pursue and on what scales (Lévesque & Murray 2010: 343–4).

Second, common identity is equally important. As Silver (2003) notes, Marx and Engels were incorrect to suggest that the salience of age, gender, race and ethnicity would diminish as capitalism transformed all workers into a homogenous working class. "Indeed, precisely because the ongoing unmaking and remaking of working classes creates dislocations and competitive pressures on workers, there is also an endemic tendency for workers to draw nonclass borders and boundaries as a basis for claims for protection from the maelstrom" (Silver 2003: 22). Yet unions can counter these tendencies by promoting a unity of purpose and by emphasizing shared values, ideologies and stories that establish cohesive class identities within the union – that is, by fostering internal solidarity (Lévesque & Murray 2010: 336–40).

Third, learning – the ability to "reflect on ... past and current change in contexts and organisational practices and routines in order to anticipate and act upon change" (Lévesque & Murray 2010: 344) – is particularly relevant to TNCs, especially when unions are dealing with companies whose constant restructuring renders corporate structures "often unclear and shifting" (Lillie & Martínez Lucio 2004: 176). In such cases, unions must not only apply lessons from past experiences but also be able to learn quickly in the short term and rework strategies over time. Tendencies of workers to fall back on traditional repertoires, union leaders' resistance to change, and other forms of organizational inertia can prevent unions from adapting to changed circumstances and exercising power in novel ways (Johnston 1994: 37; Piven & Cloward 2000: 415). Indeed, effective action has often depended on unions' abilities to "unlearn" strategies that are no longer useful (Hyman 2010: 21).

Finally, leadership is another key factor within the innermost circle of struggle that is necessary to facilitate workers' collective action. Union leaders help members secure the material and human resources necessary for on-the-ground mobilization, develop proactive agendas, mediate internal conflicts, encourage democratic participation in union proceedings and convey shared values, ideologies and stories that establish cohesive class identities within the union (Brookes 2019: 33; Kelly 1998; Piven & Cloward 2000: 415; Tattersall 2010: 144). Nevertheless, as Hyman (2010: 23, 20, emphasis in original) argues: "Strategic capacity in trade unions is a product of *both* leadership *and* internal democracy", which exist in "a complex dialectic". Because collective action depends as much on membership-driven activity as it does on leadership guidance, unions need a balance between bottom-up and top-down approaches (Bieler & Lindberg 2010: 229; Heery 2005; Milkman 2006: 152–3). Although excessive decentralization inhibits collective action, so too does leadership that resists rank-and-file decision-making.

PUTTING IT ALL TOGETHER: CONCEPTUALIZING CONCENTRIC CIRCLES

The expansion of workers' organizing efforts over time, moving beyond individual workplaces to encompass larger circles of influence, can be diagrammatically conceived using concentric and non-concentric circles with directions of travel (centripetal and centrifugal). By mapping these concepts on to the dynamics of transnational labour organizing, we can more comprehensively understand the compatibilities and tensions that unions face in mobilizing and organizing on a global scale.

In this framework, concentric circles represent the expanding scope of worker organizing, moving from local to national, regional and global dimensions. Each circle represents a spatial dimension with its own centre of gravity, indicating the locus of power and influence within that particular sphere. The movement from smaller to larger circles signifies the increasing reach and interconnectedness of workers' organizing efforts. Centripetal forces within this model represent the factors that draw workers and unions together, fostering unity, collective action and shared interests. These forces can include common grievances, shared objectives and the recognition of the need for international solidarity in confronting globalized capital. Centrifugal forces, on the other hand, symbolize the factors that pull workers and unions apart, creating divisions and challenges to collective action. These forces may arise from diverse national interests, power imbalances, historical legacies and practical constraints that hinder the ability to coordinate and mobilize on a global scale. Non-concentric circles indicate that outward movement through centripetal forces to wider circles is limited, partial and stalled.

The interactions between concentric and non-concentric circles depict the flows of interest and action in reaching outwards. Unions and workers seek to extend their influence beyond individual workplaces and national boundaries, striving to organize across corporate structures, industries and global supply chains. This outward movement represents a centrifugal force as workers push against the limitations imposed by these boundaries. However, tensions and challenges emerge as workers encounter the complexities of organizing in non-concentric circles. Differences in language, culture, legal frameworks and labour relations systems pose significant obstacles to collective action and coordination. These tensions are represented by the non-concentric nature of the circles, symbolizing the divergent interests and challenges faced when expanding organizing efforts on a transnational scale.

In this framework, unions and workers navigate the dynamic interplay between centripetal and centrifugal forces, striving to build solidarity and overcome the obstacles inherent in non-concentric organizing. The compatibility and tension between these forces shape the possibilities and limitations of transnational labour organizing. Diagrammatically representing these dynamics can provide a visual representation of the complexities and interrelationships involved, highlighting the overlapping and diverging interests across different spatial dimensions. Doing so allows for a deeper understanding of the challenges faced by unions in mobilizing and organizing transnationally, as well as the potential strategies and pathways for achieving effective collective action on a global scale.

CONCLUSION

Throughout history, unions have gone back and forth in terms of their proclivities to expand outwardly. While the ebb and flow of labour transnationalism partially tracks with trends in the expansion and contraction of economic globalization and government intervention in national economies, labour organization on the international scale also depends on the robustness of unions, which sit within the innermost circle of class struggle. Seemingly paradoxically, effective global unionism requires strong local power (McCallum 2013). The smallest circle of worker organization, at the local level, plays a fundamental role in providing the necessary resources, building a common identity and class consciousness, developing learning capabilities and fostering supportive leadership. These elements are essential for workers to overcome collective action problems and inter-group conflicts, paving the way for broader forms of organization, mobilization and activism.

Transnational labour cooperation is more likely to succeed when each union involved has already established a solid base of power and organization at the local level. While external factors such as economic globalization and government policies can influence the expansion or contraction of labour transnationalism, it is ultimately the internal strength and cohesion of unions at the local level that underpin the success of organizing efforts on a global scale. The smallest circle of worker organization, thus provides the foundation for further expansion into more encompassing forms of organization, mobilization and activism.

REFERENCES

Anderson, J. 2009. "Labour's lines of flight: rethinking the vulnerabilities of transnational capital". *Geoforum* 40(6): 959–68.

Banks, A. & J. Russo 1998. "Development of international campaign-based network structures: a case study of the IBT and ITF world council of UPS unions". *Comparative Labour Law and Policy Journal* 20: 543.

Battista, A. 2002. "The unions and cold war foreign policy in the 1980s: the National Labour Committee, the AFL-CIO, and Central America". *Diplomatic History* 26(3): 419–51.

Bieler, A. & I. Lindberg 2010. *Global Restructuring, Labour, and the Challenges for Transnational Solidarity*. London: Taylor & Francis.

Brookes, M. 2013, "Varieties of power in transnational labour alliances an analysis of workers' structural, institutional, and coalitional power in the global economy". *Labour Studies Journal* 38(3): 181–200.

Brookes, M. 2019. *The New Politics of Transnational Labour: Why Some Alliances Succeed*. Ithaca. NY: Cornell University Press.

Brookes, M. & J. McCallum 2017. "The new global labour studies: a critical review". *Global Labour Journal* 8(3): 1–18.

Burgoon, B. & W. Jacoby 2004. "Patch-work solidarity: describing and explaining US and European labour internationalism". *Review of International Political Economy* 11(5): 849–79.

Carter, C. *et al.* 2003. "The polyphonic spree: the case of the Liverpool dockers". *Industrial Relations Journal* 34(4): 290–304.

Croucher, R. & E. Cotton 2009. *Global Unions, Global Business*. London: Middlesex University Press.

Dufour, C. & A. Hege 2010. "The legitimacy of collective actors and trade union renewal". *Transfer: European Review of Labour and Research* 16(3): 351–67.

Evans, P. 2010. "Is it labour's turn to globalize? Twenty-first century opportunities and strategic responses". *Global Labour Journal* 1(3): 352–79.

Fairbrother, P. & N. Hammer 2005. "Global unions: past efforts and future prospects". *Relations Industrielles/Industrial Relations* 60(3): 405–31.

Fairbrother, P. & C. Yates (eds) 2003. *Trade Unions in Renewal: A Comparative Study*. London: Routledge.

Ford, M. & M. Gillan 2015. "The global union federations in international industrial relations: a critical review". *Journal of Industrial Relations* 57(3). https://doi.org/10.1177/0022185615574271

Garrett, G. 1998. *Partisan Politics and the Global Economy*. New York: Cambridge University Press.

Garver, P. 1989. "Beyond the Cold War: new directions for labour internationalism". *Labour Research Review* 13(1): 61–71.

Ghigliani, P. 2005. "International trade unionism in a globalizing world: a case study of global economy". *Economic Geography* 71(4): 341–63.

Goldfield, M. 1989. *The Decline of Organized Labor in the United States*. Chicago, IL: University of Chicago Press.

Goodman, J. 2004. "Australia and beyond: targeting Rio Tinto". In *Labour and Globalisation: Results and Prospects*, edited by R. Munck, 105–27. Liverpool: Liverpool University Press.

Greven, T. 2008. *Competition or Cooperation? The Future of Relations between Unions in Europe and the United States*. Berlin: Friedrich Ebert Foundation.

Gumbrell-McCormick, R. 2000. "Facing new challenges: the International Confederation of Free Trade Unions 1972–1990s". In *The International Confederation of Free Trade Unions: A History of the Organization and its Precursors*, edited by A. Carew *et al.*, 341–517. Bern: Peter Lang.

Hall, P. & D. Soskice (eds) 2001. *Varieties of Capitalism: The Institutional Foundations of Comparative Advantage*. New York: Oxford University Press.

Harrod, J. & R. O'Brien 2002. *Global Unions? Theory and Strategies of Organized Labour in the Global Political Economy*. London: Routledge.

Harvey, D. 2005. *A Brief History of Neo-liberalism*. New York: Oxford University Press.

Heery, E. 2005. "Sources of change in trade unions". *Work, Employment and Society* 19(1): 91–106.

Herod, A. 1995. "The practice of international labour solidarity and the geography of the global economy". *Economic Geography* 71(4): 341–63.

Hodkinson, S. 2005. "Is there a new trade union internationalism? The international confederation of free trade unions' response to globalization, 1996–2002". *Labour, Capital and Society* 38(1/2): 36–65.

Hollingsworth, J., P. Schmitter & W. Streeck 1994. *Governing Capitalist Economies: Performance and Control of Economic Sectors*. New York: Oxford University Press.

Huber, E. & J. Stephens 2001. *Development and Crisis of the Welfare State: Parties and Policies in Global Markets*. Chicago, IL: University of Chicago Press.

Hurd, R., R. Milkman & L. Turner 2003. "Reviving the American labour movement: institutions and mobilization". *European Journal of Industrial Relations* 9(1): 99–118.

Hyde, A. & M. Ressaissi 2009. "Unions without borders: recent developments in the theory, practice and law of transnational unionism". *Canadian Labour and Employment Law Journal* 14(3): 47–104.

Hyman, R. 2005. "Shifting dynamics in international trade unionism: agitation, organisation, bureaucracy, diplomacy". *Labour History* 46(2): 137–54.

Hyman, R. 2007. "How can trade unions act strategically?" *Transfer: European Review of Labour and Research* 13(2): 193–210. https://doi.org/10.1177/102425890701300204

Hyman, R. 2010. "Trade unions, global competition and options for solidarity". In *Global Restructuring, Labour and the Challenges for Transnational Solidarity*, edited by A. Bieler & I. Lindberg, 16–30. London: Taylor & Francis.

Iversen, T. & A. Wren 1998. "Equality, employment, and budgetary restraint: the trilemma of the service economy." *World Politics* 50(4): 507–46. https://doi.org/10.2307/25054055

Johns, R. 1998. "Bridging the gap between class and space: U.S. worker solidarity with Guatemala". *Economic Geography* 74(3): 252–71.

Johns, R. & L. Vural 2000. "Class, geography, and the consumerist turn: UNITE and the Stop Sweatshops campaign". *Environment and Planning* 32(7): 1193–213.

Johnston, P. 1994. *Success While Others Fail: Social Movement Unionism and the Public Workplace*. Ithaca, NY: Cornell University Press.

Juravich, T. 2007. "Beating global capital: a framework and method for union strategic corporate research and campaigns". In *Global Unions: Challenging Transnational Capital Through Cross-Border Campaigns*, edited by K. Bronfenbrenner, 16–39. Ithaca, NY: Cornell University Press.

Kelly, J. 1998. *Rethinking Industrial Relations: Mobilisation, Collectivism and Long Waves*. London: Routledge.

Larsson, B. 2012. "Obstacles to transnational trade union cooperation in Europe: results from a European survey". *Industrial Relations Journal* 43(2): 152–70.

Lee, E. 1996. *The Labour Movement and the Internet*. London: Pluto.

Lerner, S. 2007. "Global unions: a solution to labour's worldwide decline". *New Labour Forum* 16(1): 23–37.

Lévesque, C. & G. Murray 2010. "Understanding union power: resources and capabilities for renewing union capacity". *Transfer: European Review of Labour and Research* 16(3): 333–50.

Levinson, C. 1972. *International Trade Unionism*. London: Allen & Unwin.

Lichtenstein, N. 2003. *State of the Union: A Century of American Labor*. Princeton, N.J.: Princeton University Press.

Lillie, N. 2006. *A Global Union for Global Workers: Collective Bargaining and Regulatory Politics in Maritime Shipping*. New York: Routledge.

Lillie, N. & M. Martínez Lucio 2004. "International trade union revitalization: the role of national union approaches". In *Varieties of Unionism: Struggles for Union Revitalization in a Globalizing Economy*, edited by C. Frege & J. Kelly, 159–74. Oxford: Oxford University Press.

Lindberg, I. 2010. "Varieties of solidarity: an analysis of cases of work action across borders". In *Global Restructuring, Labour and the Challenges for Transnational Solidarity*, edited by A. Bieler & I. Lindberg, 206–19. London: Taylor & Francis.

Logue, J. 1980. *Toward a Theory of Trade Union Internationalism.* Kent, OH: Kent Popular Press.

Logan, J. 2006. "The union avoidance industry in the United States". *British Journal of Industrial Relations* 44(4): 651–75.

Masur, J. 2000. "Labour's new internationalism". *Foreign Affairs* 79(1): 79–93.

McCallum, J. 2013. *Global Unions, Local Power: The New Spirit of Transnational Labour Organizing.* Ithaca, NY: Cornell University Press.

Milkman, R. 2006. *L.A. Story: Immigrant Workers and the Future of the U.S. Labor Movement.* New York: Russell Sage Foundation.

Moody, K. 1997. "Towards an international social-movement unionism". *New Left Review* 225: 52–72.

Müller, T., H.-W. Platzer & S. Rub 2010. *Global Union Federations and the Challenges of Globalisation.* Berlin: Friedrich Ebert Foundation.

Munck, R. 2002. *Globalisation and Labour: The New Great Transformation.* London: Zed.

Olson, M. 1965. *The Logic of Collective Action: Public Goods and the Theory of Groups.* Cambridge, MA: Harvard University Press.

Peetz, D. 1998. *Unions in a Contrary World: The Future of the Australian Trade Union Movement.* Melbourne: Cambridge University Press.

Piehl, E. 1974. *Multinationale Konserne und internationale Gewerkschaftsbewegung.* Frankfurt: EVA.

Pierson, P. 2001. *The New Politics of the Welfare State.* Oxford: Oxford University Press.

Piven, F. 2008. "Can power from below change the world?" *American Sociological Review* 73(1): 1–14.

Piven, F. & R. Cloward (2000). "Power repertoires and globalization". *Politics & Society* 28(3): 413–30.

Polanyi, K. 2001 [1944]. *The Great Transformation: The Political and Economic Origins of Our Time.* Second edn. Boston, MA: Beacon Press.

Puette, W. 1992. *Through Jaundiced Eyes: How the Media View Organized Labor.* Ithaca, NY: Cornell University Press.

Rainnie, A., A. Herod & S. McGrath-Champ 2011. "Review and positions: global production networks and labour". *Competition & Change* 15(2): 155–69.

Rodrik, D. 2012. *The Globalization Paradox: Democracy and the Future of the World Economy.* Reprint edn. New York: Norton.

Sadler, D. & B. Fagan 2004. "Australian trade unions and the politics of scale: reconstructing the spatiality of industrial relations". *Economic Geography* 80(1): 23–44.

Scharpf, F. & V. Schmidt 2000. *Welfare and Work in the Open Economy.* Oxford: Oxford University Press.

Schmalz, S., C. Ludwig & E. Webster 2018. "The power resources approach: developments and challenges". *Global Labour Journal* 9(2). https://mulpress.mcmaster.ca/globallabour/article/view/3569/3157

Schulse-Cleven, T. & T. Vachon (eds) 2021. *Revaluing Workers: Toward a Democratic and Sustainable Future.* Champaign, IL: Labour & Employment Research Association.

Silver, B. 2003. *Forces of Labor: Workers' Movements and Globalization Since 1870.* New York: Cambridge University Press.

Sukthankar, A. & K. Kolben 2007. "Indian labour legislation and cross-border solidarity in historical context". In *Global Unions: Challenging Transnational Capital Through Cross-Border Campaigns*, edited by K. Bronfenbrenner, 57–77. Ithaca, NY: Cornell University Press.

Swank, D. 2002. *Global Capital, Political Institutions, and Policy Change in Developed Welfare States.* Cambridge: Cambridge University Press.

Tattersall, A. 2010. *Power in Coalition: Strategies for Strong Unions and Social Change.* Ithaca, NY: ILR Press.

Thelen, K. 2001. "Varieties in labour politics in developed democracies". In *Varieties of Capitalism: The Institutional Foundations of Comparative Advantage*, edited by P. Hall & D. Soskice, 71–103. New York: Oxford University Press.

Thelen, K. 2014. *Varieties of Liberalization and the New Politics of Social Solidarity.* Cambridge: Cambridge University Press.

Thompson, D. & R. Larson 1978. *Where Were You, Brother? An Account of Trade Union Imperialism.* London: War On Want.

Waterman, P. 2001. *Globalization, Social Movements and the New Internationalisms.* London: Continuum.

Wills, J. 2002. "Bargaining for the space to organize in the global economy: a review of the Accor–IUF trade union rights agreement". *Review of International Political Economy* 9(4): 675–700.

Windmuller, J. 1967. "International trade union organizations: structure, functions, limitations". In *International Labour*, edited by S. Barkin *et al.*, 81–105. New York: Harper & Row.

Young, K. & D. Sierra Becerra 2014. "How 'partnership' weakens solidarity: Colombian GM workers and the limits of UAW internationalism". *WorkingUSA* 17(2): 239–60.

CHAPTER 20

UNIONS AND POLITICS: WHY UNIONS ARE NOT JUST THE ECONOMIC WING OF THE LABOUR MOVEMENT

Jörg Nowak and Roland Erne

ABSTRACT

The "apolitical" economic laws of supply and demand do not really work in the labour market. Instead, employment relations are first and foremost shaped by power relations between capital, labour and the state. Unions can, therefore, hardly afford to abandon the political terrain. This chapter first explains why politics plays such a central role in employment relations in capitalist societies, and then outlines the merits and limitations of the various political action repertoires of unions (private interest government, lobbying, protest action, corporatist political exchanges, alliances with sister parties, direct democratic citizens' initiatives and referendums).

Keywords: Politics; political exchange; challenges for unions

INTRODUCTION

Throughout their history, unions have played a key role in the making of working classes and political mobilizations across the world. Without the support of the associational networks provided by the union movement across towns and workplaces, left-wing labour parties would not have been able to flourish, nor would the left–right divide have become a central feature of democratic politics (Bartolini 2000; Thompson 1963). More than a century ago, the British socialist reformers, Sidney and Beatrice Webb (1897), made a clear distinction between unions and labour parties in their seminal book, *Industrial Democracy*. Whereas unions represented the economic wing of the labour movement and defended workers' economic interests through collective bargaining in their trade, the Labour Party was tasked to represent the cause of labour in the political sphere. This conception of union-party relations had been particularly strong in Britain and also shaped unions' relationships to centre-left parties in many other parts of the industrialized world (Allern & Bale 2017).

Unions, nevertheless, also continue to play an important *autonomous* role in politics. First, unions give rank-and-file workers a forum where they can learn to articulate and pursue their interests collectively. This makes unions intrinsically political, even where they delegate parliamentary politics to centre-left sister parties or claim to be politically, confessionally and ideologically neutral. Thus, unions can be important "schools of democracy" (Sinyai 2006) even when they target employers rather than governments. Regardless of their particular political orientation, unions provide workers with a space that helps them to act collectively not only at workplace level but also in the political sphere.

Second, given the importance of power relations in employment relations and the imbalance in them, unions cannot afford to neglect politics, including party politics. This explains why political scientists still regard the close relationship between unions and their sister parties as a key feature of interest-group politics (Allern & Bale 2017; Erne 2023). In this chapter, however, we do not simply describe the ongoing changes in the relationships between unions, political parties, governments and the state (see also Chapters 9 and 11). Rather, we first address broader conceptual questions about the political nature of capitalist employment relations and unions, before discussing different repertoires of political union action in practice.

WHY CAN UNIONS HARDLY AFFORD TO ABANDON THE POLITICAL TERRAIN?

Having been elected to lead the transnational British and Irish union, Unite, its General Secretary, Sharon Graham, wrote in *The Guardian* (12 September 2021) that Unite should focus more on fighting "for jobs, pay and conditions" rather than "hoping for the election of a Labour government to solve our members' problems". Workers, indeed, cannot rely upon "supreme saviours" and must, therefore, fan the flames of the forge themselves and "strike while the iron is hot", as stated in the left-wing anthem, "The Internationale". It is, however, equally true that unions can hardly afford to abandon the political terrain in their fight for better wages and working conditions, as argued below.

The myth of the unpolitical laws of supply and demand

Neoliberal economics have assumed that a politicization of labour relations would serve only to create market disequilibrium or distortions of an ideal market situation that, although never realized, would somehow lurk behind the market forces as a benevolent invisible hand. It follows from this assumption that any political intervention in the laws of supply and demand would disturb the "natural laws" of the economics of labour markets. Historical research, however, has shown that labour markets were created with political means, such as the enclosures of communal lands and the harsh repression of vagabonds in England (Marx 1867a; Polanyi 2001 [1944]). Labour markets are also maintained with political means, for which the lack of access to land is crucial. Curtailing access to land was central to the creation of a working class in the US, thereby pressuring many small farmers to give up farming and move to cities to obtain work. The special conditions in the US meant that this process started on a mass scale only in the late nineteenth century and was more or less completed during the Great Depression when large masses of small farmers had to leave Oklahoma and other farming states to seek work in California, as reflected in the novel and film, *The Grapes of Wrath*, written by John Steinbeck and directed by John Ford, respectively.

The crucial difference between other product markets and labour markets is that producers in the former have some ability to change the product that they are offering. For example, although independent farmers can produce only what their climate and soil conditions allow, they can shift production from crops to dairy farming, if the latter is more profitable. As a result, the supply of crops on farmers' markets declines until the supply and demand curves achieve equilibrium. By contrast, workers cannot detach themselves from their own labour power, which they might be able to sell on the market for temporary use by an employer. Thus, workers can offer only one product, namely their labour power. Even if workers acquire higher qualifications over time, their ability to do that outside the workplace is limited, as they usually need to work to sustain their living.

This points to another crucial difference between labour markets and conventional product markets. Farmers, for example, have the ability to leave the tomato market when the price for tomatoes goes down, as they can produce another crop instead. By contrast, workers normally can hardly leave the labour market if wages go down, as they need to earn their living. Some workers may have other means of subsistence at their disposal, such as family support, but in general workers cannot afford to drop out of the labour market if the price offered for their labour power goes down. Instead, workers (or other members of their household) may take on additional work to make ends meet, or look for other kinds of work, inside or outside the same national social formation. This means, however, that declining wages will not lead to a lower supply of labour power, as assumed by the neoliberal, free-market model. As a result, the supply and demand curves for labour power may evolve in parallel, setting off a "race to the bottom" in wages and labour standards.

To prevent this from happening, states are required to take political and legislative action to intervene in labour markets. This can be done by governments directly, by setting national minimum wage, welfare and labour standards, which remove the risk of a "race to the bottom" by fiat. Or, it can be done indirectly, by granting workers the right to form unions so that they can defend their interests collectively. The latter has happened even in liberal, free-market-oriented economies, including the US. American and UK competition law forbids producers' associations to fix prices for their produce in any circumstances, but workers' attempts to set wages collectively are exempt from it. This suggests that even free-market-oriented policy-makers implicitly have had to recognize that labour markets do not work like any other product market.

In the labour market, the neoliberal assumptions about supply and demand, indeed, apply only in a very partial manner. If the demand for labour power exceeds its supply, as happened in some sectors of the British economy after Brexit and the Covid pandemic, employers may be compelled to pay higher nominal wages. Even so, businesses and governments may still be able to offset the effect of wage increases on business profitably. Workers' nominal wage gains can evaporate quickly as a consequence of higher inflation or the devaluation of the workers' national currency. Likewise, wage developments in Central and Eastern Europe (CEE) have not followed the laws of supply and demand. Although many CEE workers left their countries of origin after they joined the European Union (EU) in the 2000s, the resulting declining supply of workers in CEE labour markets did not necessarily lead to higher wages for those workers who stayed at home (Stan & Erne 2016). Hence, the second assumption of the demand and supply model – higher demand for a product leads to higher prices for that product – does not apply to labour markets in a universal fashion either.

Because labour markets do not work according to the assumptions of the supply and demand model, they require political intervention to work in the manner in which they are supposed to: to provide labour power for employers. Marx (1867a) shows how the state in England had to intervene to introduce limits to working time, as excessive working times led to high rates of mortality, putting the physical reproduction of workers in question, and, therefore, also the reproduction of capital, which depends on the availability of labour power. Other actions to intervene in this way are minimum wages, a public health and education system, and public transport systems.

Furthermore, when labour markets faced a lack of available labour power after the Second World War, various Western European governments sponsored guestworker recruitment programmes with governments from southwestern Europe, Tunisia, Turkey, Yugoslavia, Jamaica and elsewhere (Castles & Kosack 1973; Sala 2007). The agreements made between these governments, thus, alleviated the labour markets and increased the buying power in sending countries via remittances sent by migrants. In the receiving countries, the employment of migrants created an underclass in factories that could be controlled relatively easily because of migrant workers' societal outsider status and the lack of equal social and political rights. For example, non-EU workers' guestworker contracts bound them to a particular employer and, thus, such workers lacked EU free movement rights. Furthermore, the recruitment of workers from abroad allowed governments to keep their labour markets closed to women, maintaining the status of native male workers as breadwinners (Türkmen 2010). Far from being just an extraordinary measure to balance out the external shock to markets from the high number of Second World War casualties, the decision to install guestworker systems was a deliberate decision as there was a large female workforce available in the receiving countries.

As these empirical examples demonstrate, in the labour market there is no such law of supply and demand that would work in the way usually assumed by neoliberal textbook economics. The creation of all existing markets involves manifold political decisions and interventions, and many of them include physical violence. As Polanyi (2001 [1944]: 141) argues, the "laissez-faire economy was the product of deliberate state action". The distinctive nature of the labour market, in contrast to other product markets, means that the destructive "race to the bottom" in wages and working conditions can be restrained only by countervailing actions of unions and anti-laissez-faire interventions by governments. Thus, the relations between workers and employers do not reflect market equilibriums but rather power relations between workers, employers and governments. Labour relations are, thus, primarily political relations rather than market relations. Unions, thus, operate in both the political and economic sphere, which constitute the political economy. This means that we must study the role of the political in capitalism as well as organized labour's capacity to act collectively.

Labour relations are power relations

In the preceding section, we explained how the most basic economic process in the labour market – supply and demand for labour power – is instituted and maintained via political action and domination. This demonstrates how the economic and the political are deeply intertwined. Poulantzas (1978: 42) conceptualizes the separation of the economic and the political as separate spheres of action as an effect of the structure of capitalist relations of production. The exact nature of this separation is not static and homogeneous across different societies and across time. It takes different forms.

As the separation of the economic and the political is a basic feature of capitalist societies, the actions of workers and unions are also generally conceptualized as economic struggles and political struggles. Economic struggles take specific frameworks of the economy as given and contest mainly income distribution and labour conditions, and at times decisions about hiring and firing workers. As labour conditions regularly have a deep impact upon workers' bodies and health, we can see how the effects of economic struggles go beyond what we usually conceive of as "economic". Political struggles at governmental level are, therefore, equally important. They affect public labour policies and laws that govern workplaces directly, but they also affect income distribution indirectly, as they can be linked to the state's welfare policies. They are directed at general rules that apply to the population of a political system in general, and not just to an industrial sector or workplace. Hence, political struggles target political executives at local, national and supranational level, and not management, as in the case of economic struggles.

Finally, following Lenin's (1902 [1961]) concept of ideological struggle, Burawoy (1979: 177) adds a third type of struggle, which consists of contesting the overall societal order. He rightly emphasizes the ideological dimensions of these transformative struggles, but we refer to them as systemic rather than ideological struggles, as they are directed at the whole power elite, namely governments as well as economic elites (Erne & Nowak 2022).

It makes sense to distinguish economic, political and systemic struggles of labour movements analytically, because this helps to distinguish the different modes of collective action operating within and across different spheres of society. But we should not forget that all labour struggles are political in the broad sense of the term as a social activity based on the use of power (Schmitter & Blecher 2020). Hence, all labour struggles are political, as they relate to power relations between conflicting social interests, regardless of their articulation at workplace, governmental or systemic level.

POWER AND THE LABOUR MARKET

The issues of labour and power are located on several levels, one of which we have already dealt with, namely, labour market bargaining power. Usually, labour markets within one national formation are segmented. This segmentation results from several factors, but it is also an effect of closure, which cuts off certain groups of the working class from the ability to move into better paid and less physically demanding jobs. The lines of segmentation can be along lines of race, ethnicity, nationality, gender or others (Braverman 1974; Edwards 1979; Emmenegger *et al.* 2012; Thomas 1982). Thus, segmentation serves to reproduce certain labour forces for certain segments of the labour market. It comes with divisions between parts of the working class, as workers in the better-off segments are keen to keep their position and often engage in closure mechanisms along lines of habitus and cultural capital such as clothing, behaviour and food habits (Bourdieu 1979).

Workers are usually in a favourable position when there is more demand than supply on the labour market, and it is often in such situations that workers are more prone to enter into conflict, as in Brazil in the early 2010s when a government programme of major infrastructural works led to a construction boom, a lack of construction workers and a series of large strikes (Nowak 2019). Because of this beneficial effect, workers in some economic sectors try to control the supply of labour, such as in the hiring halls of dockworkers in most of Europe's ports, thereby creating a better bargaining position for dockworkers. Employers continuously try to undermine this control of the labour supply with the employment of temporary dockworkers outside of the hiring halls (Engelhardt 2020).

POWER AND THE WORKPLACE: EFFORT BARGAINING AND STRIKES

Another important level involving power relations is the workplace itself, and here the effort bargain is central (Edwards 1986). Employers pay for work with a wage, usually for a certain number of hours, but often the intensity of the labour process is not clearly defined, and in a lot of jobs the effort required is hard to define. Thus, employers and managers will try to make workers work harder, whereas workers will try to find loopholes in order to work less intensely. Piece rates are a way for employers to control the effort bargain, but also often a way for workers to earn higher wages (Carswell & De Neve 2013). Another way in which employers can exert some influence on effort is by adding performance incentives to the fixed wage. Algorithmic control of waged work tries to quantify and measure workers' effort with tracking devices (Moore 2018). Much platform work is a new form of piece work, but without the financial incentives involved as rates are unilaterally controlled by an impersonal system.

The power within the labour process as embodied in the wage-effort bargain is accompanied by power over the labour process: workers can stop work and demand better conditions, but this usually requires considerable coordination, at least at the level of a company unit. The history of the labour movement is also the history of the strike becoming the main form of social protest since the mid-nineteenth century, reflecting the ascent of wage labour as a central institution of social life (Cronin 1979). Blocking production, or other forms of labour such as transport or waste disposal, can exert pressure on either employers or overall society and provides leverage for workers' demands. This, however, is not an automatic process in cases where employers or governments take a firm stance: the list of strikes without any positive results is notoriously quite long. A strike's success depends on many economic and political factors, such as particular social or political circumstances or long-term trends (Gall 2013; van der Velden *et al.* 2007).

Other forms in which labour relations and power relations are intertwined concern the action of workers at societal level and in organizations such as parties, unions or other associations, treated in the following sections of this chapter. But before we can assess different types of organized labour's actions in the governmental sphere in practice, we must distinguish two logics of collective action (Offe & Wiesenthal 1980), namely, those that are available for business interests and those that are available for labour interests.

TWO LOGICS OF COLLECTIVE ACTION

Offe and Wiesenthal (1980) distinguish two logics of collective action, given any interest group's need to act more or less collectively in order to affect public policy-making. This highlights that business interests have a structural advantage in capitalist societies, but not because of business groups' capacity to spend more money on political lobbying compared with other interest groups (Greenwood 2017; McMenamin 2013). Rather, businesses can leverage extraordinary political power in capitalist societies because the material basis of a given political system depends on capitalists' individual investment decisions.

Even individual investment decisions can influence economic growth in a territory, namely when they are taken by a company that has the power to affect people's livelihood or working lives in a given area. The territory's political representatives must, therefore, consider the views of capitalists regardless of whether or not they are organized collectively. In contrast to workers' strikes, investment strikes by firms do not require any collective organization. This gives capital interests a substantive advantage over labour. Whereas withdrawal of labour requires collective organization and the willingness of workers to act together, businesses can much more and far more easily influence policy-makers without having to coordinate their actions with their peers. This increases the political leverage of business groups considerably.

There are cases, however, in which businesses fail to translate the "pre-associational power asymmetry between businesses and labour into corresponding differentials in terms of associational capacities" (Traxler, Blaschke & Kittel 2001: 37), namely when firms compete with one another for government support, as in the case of competitive tenders for government contracts, government bailouts or privatization bids. At times, conservative politicians have pursued policies that went against the economic interests of most (albeit not all) business leaders, as it happened in the case of the UK's Brexit referendum. Overall, however, business interests retain a considerable structural advantage in capitalist societies, as they do not face the intricate collective action dilemmas faced by unions and other interest groups and social movements (Crouch 1982; Kelly 1998; Olson 2009 [1965]).

Whereas unions rely on their members' willingness to act collectively, business associations do not need to engage in collective action. Instead, they can confront policy-makers with individual investment strikes if governments fail to accommodate business interests. Sometimes, businesses can even turn their own imminent bankruptcy into an effective political weapon, as shown in 2008 when banks successfully lobbied governments around the world to bail them out (Erne 2012). Compared with non-economic social movements and interest groups, however, unions do possess additional levers that potentially give them more clout in politics.

If unions can overcome their collective action problems, they can – like business groups – affect governments not only directly through political lobbying or protests, such as non-economic interest groups and social movements. Given unions' double character as political and economic actors (Hyman 2001; Soll 1976), they can also affect governments indirectly, through economic actions that affect the material basis of the political system. This explains why unions have historically been able to deploy a greater variety of collective action repertoires than other interest associations and social movements. Unions' double character also explains why governments have at times tried to incorporate unions in corporatist governance arrangements in order to ensure a smooth functioning of public administration and the economy within a capitalist society (Crouch & Streeck 2006; Erne 2008; Molina & Rhodes 2002).

ACTION REPERTOIRES OF BUSINESSES, UNIONS AND NON-ECONOMIC SOCIAL INTERESTS

The two logics of collective action and unions' double character as economic and political actors discussed above provide us with key distinctions for an assessment of the different political action repertoires of interest groups. Given Offe and Wiesenthal's work (1980), we first distinguish different action repertoires of businesses, unions and non-economic interests based on their necessity to act collectively by asking: Does the political representation of an interest require collective action? Given unions' double character as both economic and political actors, we then distinguish interest groups based on their capacity to affect the material basis of a political system. Thus, we ask: Does the political representation of an interest rely on economic and political, or only political power resources? In other words, to what degree are interest groups capable of creating economic and political facts that governments cannot ignore? This leads us to schematizing the issue in a tabular form (see Table 20.1) as this will facilitate the subsequent discussion of unions' use of the repertoires one by one in the final section of this chapter (see also Chapter 3).

As the role of interest groups in policy-making depends also on the balance of power between the various groups, we cannot assess the role played by unions in labour politics in isolation. In the following section, we therefore assess unions' action repertoires in practice in comparison with those of other interest groups, as outlined in Table 20.1.

Table 20.1 Political action repertoires of interest groups

		Necessity of collective action	
		Low	*High*
Location of power resources	*Economic and political sphere*	**Business interests:** Private interest government Lobbying Protest action Corporatist political exchanges Reliance of sister parties Initiatives and referendums	**Labour interests:** Lobbying Protest action Corporatist political exchanges Reliance on sister parties Initiatives and referendums
	Political sphere		**Non-economic interests:** Lobbying Protest action Reliance on sister parties Initiatives and referendums

Source: Adapted from Erne (2023: fig 15.1).

UNIONS AND POLITICS IN PRACTICE

Unions' marginal role in private interest government

Table 20.1 shows that unions usually lack an important type of political action repertoire available to businesses, namely "private interest government" (Streeck & Schmitter 1985). As far back as the Middle Ages, states at times left the authority to make binding decisions to business interests, namely guilds that were able to police their markets autonomously (Crouch 1993; Erne 2023). Today, some businesses are again powerful enough to shape governance standards independently. Especially in the transnational sphere, we can detect a variety of forms of private governance (Gras & Nölke 2007). Given the lack of effective global governmental systems and the tendency of neoliberalism to deregulation, non-state actors – including business groups and transnational corporations (TNCs) themselves (Crouch 2011) – often establish common rules and standards; for example, to facilitate the functioning of global financial markets or cyberspace.

In response to labour, ecological and consumer protests against sweatshop conditions in global supply chains (Anner 2015), TNCs at times have agreed to codes of conduct (CoC) on corporate social behaviour. These CoC are usually based on core conventions of the tripartite International Labour Organization, which unites governments, employer associations and unions. Even so, CoC auditing and implementation are usually conducted by auditing firms employed by the TNCs' headquarters (Kuruvilla 2021), and are conducted in order to ensure that their supplier factories comply with them, but also to protect the TNC from any negative press coverage in the event of CoC violations. At times, TNCs have signed international framework agreements with global union federations (Papadakis 2011); for example, when TNCs were facing labour and consumer protests, as happened after the 2013 Rana Plaza disaster in Bangladesh that led to the death of 1,129 garment workers and injured a further 2,500 (Reinecke & Donaghey 2015).

Although unions have at times been able to join private interest government arrangements (Streeck & Schmitter 1985), they have been able to do so only when businesses felt obliged to co-opt them. This leads to the next sections, in which we discuss various action repertoires that unions may use to gain political power.

Unions and lobbying

For the pluralist theorist of *Democracy in America*, Alexis de Tocqueville (2006 [1835]), interest associations represent the lifeblood of democratic political life. The state should, therefore, guarantee its citizens' right of association to give all social interests equal access to the policy-making process. As Schattschneider (1960: 34–5) observes, however, "the flaw in the pluralist heaven is that the heavenly chorus sings with a strong upper-class accent". To gain political clout, workers and unions, therefore, cannot just rely on lobbying activities, in contrast to employers that can rely on the power of their arguments obliquely backed up by the power threats discussed above (Offe & Wiesenthal 1980). Instead, to be heard, unions – and progressive movements in general – must be able to shift the balance of power in society through collective action (Bieler, Jordan & Morton 2019). Therefore, next we discuss labour's role in contentious politics (Tilly & Tarrow 2015).

Unions as contentious social movements

Workers' organizations began to form in the seventeenth and the eighteenth centuries, often under the name of "combinations" in England and Australia, for example (Irving 2017; Quinlan 2017; see also Chapter 8). These combinations were quickly banned, but repression failed to dismantle them. Informal organizations among artisans and customary rules exercised a high level of control over the workplace (Dobson 1980; Rule 1981). The more formalized unions that swept the European continent from the mid-nineteenth century on organized primarily male workers, largely excluding women and children in their first phase of formation from about 1850 on Marx (1867b), emphasized that unions are a school of struggle but complained about their sectional and narrow politics of interest representation. Although main union bodies became institutionalized in many European countries from the early twentieth century on, strikes often continued to follow a violent trajectory. This was the case especially in Germany at the end of the First World War. Despite being led by a rather moderate union federation, metal workers staged large strikes in 1918 after seeing their leaders being conscripted into the army (Luban 2008), and in 1920, 100,000 armed workers staged a month-long uprising during the Ruhr crisis (Lucas 1973–78).

In the US, violent conflicts around strikes were also frequent. Organizations such as the International Workers of the World, also known as the Wobblies, refused to recognize employer organizations. Instead, the Wobblies wanted to organize the whole working class so that it could liberate itself from the profit-making "wage system" altogether (Dubofsky 2000 [1969]: 87). Although the Wobblies were able to gain a great following in a short time span, they were not able to consolidate the organization over the course of the 1920s and, therefore, gave way to other forms of labour organization. It was only in the 1950s that unions were recognized by most employers and that collective bargaining became the norm (Ross 1954), despite the adoption of the National Labor Relations Act 1935 (also called Wagner Act) by President Franklin D. Roosevelt's New Deal administration. The act established the National Labor Relations Board and gave most workers in the private sector the right to join unions and to bargain collectively with their employers (Katz & Colvin 2021).

The post-Second World War class compromise led to the rise of welfare capitalism in many Western capitalist core countries. This transformed many unions into "intermediary organizations" that "pursue a policy of intermediation between the opposing interests of labour and capital" (Müller-Jentsch 1985: 3). Accordingly, unions participated as social partners with employers and government representatives in the management of the economy and the welfare state (see section on neo-corporatism below and Chapter 9), while still mobilizing their members from time to time to demonstrate their power.

After the election of the Thatcher Conservative government in 1979 and Ronald Reagan as US Republican president in 1980, and the growing integration of national economies across borders, however, more and more business and political leaders abandoned the mid-twentieth century class compromise between capital and labour and promoted neoliberal policies instead (Harvey 2005; see also Chapter 11). This led to a crisis of social democratic unionism in Western Europe and the Western world more generally (Upchurch & Taylor 2009; see also Chapter 9), as the turn to neoliberalism unsettled unions' established paths of collective bargaining and political interest intermediation. To get back on the offensive, some unions pursued a return to "social movement unionism" in North America, Europe and even across borders (Erne & Nowak 2022; Gumbrell-McCormick & Hyman 2013: ch 3; Moody 1997; Turner & Hurd 2001).

In other types of quickly industrializing societies, unions remained contentious social movements, as in, for example, Japan until the 1950s. Countries such as South Korea, Brazil and South Africa saw large mobilizations in which unions joined with other social movement organizations during the 1970s and 1980s, with insurgent episodes such as the occupation of the Gwangju industrial area by Korean workers in 1980 (Park 2007) or the 1984 general strike in South Africa (Webster 1987). Since the 1990s, unions in all three countries have seen a similar trajectory of integration, although with much fewer welfarism elements than in core countries. South Korean unions remain under legal pressure, with the current leader of the largest union federation KCTU under arrest since September 2021, the 13th KCTU president in a row to be arrested (Seol 2021). The strikes in gold and platinum mines in South Africa in 2021 also saw widespread repression and violence, including a massacre by police of 34 mineworkers in Marikana in 2012, and several casualties among strike breakers who were killed by striking workers. However, although violent incidents like these continue to be a characteristic of union and worker activities in many emerging countries, such as India and Indonesia, their contentious character is increasingly of a particularistic nature given the lack of larger political currents or organizations aligning with workers' protests.

Unlike in earlier periods when contentious mobilizations aimed to fight for an alternative model of society that questioned prevailing capitalist social relations (whether that was called socialism, social democracy or the New Deal), these references are rather vague today. More recent insurgent popular protests in countries such as Algeria, Sudan, Colombia, Iran and Chile since 2018, however, still saw wide participation and important roles for unions and worker protests. To be sustainable, however, social movement unionism requires more than a shift to a mobilizing model based on "deep organizing" rather than ad hoc advocacy and mobilization campaigns (McAlevey 2016). It is equally important to have credible visons of another world that is not dominated by capitalist interests and that is worth fighting for. But precisely because transformative social movements must provide not only an alternative political outlook, but also an alternative economic outlook (Cox & Nilsen 2014; Della Porta 2015), labour movements can potentially play a significant role in them, given their position in both the economic and the political sphere (see Table 20.1). It is, thus, hardly surprising that some governments and business leaders offered labour leaders "political exchanges" (Pizzorno 1978) that gave them access to the management of the economic and the political system in order to stabilize them.

Unions as neo-corporatist, intermediary organizations

Witnessing the profound crisis of capitalism during and after the Great Depression of the 1930s and the rising socialist and communist labour movements, fascist, populist and post-revolutionary governments in Romania, Italy, Argentina, Brazil and Mexico developed a corporatist model of employment relations. This model rejected both the liberal market model opposing individual buyers and sellers and the Marxist view of antagonist social classes. Instead, corporatist regimes saw the national economy as an integrated body with separate organs (namely management and labour) that must work in unison, usually under the leadership of a head (of government or state), to guarantee their nation's economic success (Manoilesco 1934; Pitigliani 1933).

During the economic crisis of the early 1970s, Christian and social democratic national policy-makers in Western Europe again tried to co-opt organized labour and capital into the management of the economy and the welfare state, albeit without suspending the democratic rights of unions and employers' associations. To distinguish this new form of corporatism from its authoritarian predecessors, Schmitter (1974) terms it "neo-corporatism". The establishment of unions as social partners in tripartite forums led them in specific cases to become actors in neo-corporatist systems, so that they were perceived as part of the political establishment. Although this case has been seen as a characteristic mainly of core countries in the capitalist system, namely in Northern and Western Europe, corporatist functions of unions can also be observed in countries of the former Soviet bloc and of the Global South.

In Western Europe, the high tide of neo-corporatism was during the late 1970s, but vestiges of corporatism remained salient in many countries, often under the title of social partnership. Given the increasing concentration of capital, the increased necessity for economic state intervention in advanced capitalism, and the balance of class forces, it was considered that neo-corporatist systems of interest intermediation, compared with pluralist systems, would deliver superior economic results. Whereas in pluralist systems the most powerful interest would always succeed, neo-corporatist systems arguably aimed at balancing the power of competing (economic) interest groups – namely capital and labour – in order to ensure that the best arguments would succeed (Erne 2023).

Esser (1982) analysed the specific form of union integration into the West German social formation. He introduced the concept of unions as "intermediary organizations" (see also Müller-Jentsch 1985), with both an integrative and an oppositional role. This characterization reflects the selective representation of a part of the working class, namely unionized workers in manufacturing industries, effectively excluding the poorer third of the working class. This led to political tensions within the working class and within unions as organizations, thereby effectively shifting class conflicts on to a different terrain, creating competition between factions of the working class and, therefore, effectively masking the contradictions between the interests of workers and capital at large (Esser, Fach & Simonis 1980: 40). Some of these contradictions came to the foreground in Germany in the late 1970s. Esser, Fach and Simonis (1980: 42) located the causes of these developments in the West German export-led economic model, which required an integration of those industries' core workers at the expense of other sections of the working class. Esser and Fach (1981) stress that the establishment of neo-corporatist coalitions led to a higher capacity of the state at policy level, strengthening the executive and administrative apparatuses of the state at the expense of parliament. These analyses are in line with Poulantzas's (1978) comments on the greater relevance of the state apparatus in core capitalist countries in the 1970s. In the Global South, the corporatist inclusion of unions is often patchier, might occur only in the context of specific periods of government and affects only some unions.

Following the rise of neoliberalism and the increasing integration of national economies across borders, Streeck and Schmitter (1991) expected a decline of neo-corporatism. Despite the making of the European single market and monetary union, however, unions continued to be co-opted into social pacts and partnership arrangements (Grote & Schmitter 1999). To remain at the bargaining table, unions felt obliged to make ever more concessions; for example, wage moderation to boost firms' and national economies' competitiveness (Erne 2008). Yet after the 2008 financial crisis, many neo-corporatist arrangements collapsed (Erne 2023, tab. 15.3; Meardi 2018). Instead, European political and business elites imposed wage cuts, labour market and public sector reforms unilaterally, namely in the EU's periphery (Erne *et al.* 2024; Jordan, Maccarrone & Erne 2021; Stan & Erne 2023). Whereas the German government continued to co-opt unions into a new form of "crisis corporatism" (Ebbinghaus & Weishaupt 2021), in other countries workers and unions felt compelled to pursue their interests by other means, by a return to protest and party politics (Kriesi *et al.* 2020).

Unions and party politics

Since its beginning, the labour movement has fought for universal voting rights and democracy (Foot 2005; Rueschemeyer, Stephens Huber & Stephens 1992). Engels (1922 [1895]) describes the extension of the franchise as one of the strongest weapons of the labour movement for the emancipation of humankind. Engels' enthusiasm no longer features prominently in current accounts of the state of liberal democracy (Crouch 2004; Mair 2013). Nevertheless, although the share of working-class voters in the electorate has been declining in the Western world (Crouch 2015; Oesch & Rennwald 2018), centre-left parties can hardly afford to lose them. Although the union vote alone could never secure electoral victory, workers' and trade unionists' frustrations with social democratic governments can ensure their electoral defeat, as former German chancellor Gerhard Schröder found after his SPD–Green coalition government implemented the liberalizing Harts IV labour market reforms in the 2000s (Erne 2008: 32) or as Hungarian, Greek and French socialists or Irish Labour Party officials learned after they implemented harsh austerity cuts.

Without doubt, the rise of neoliberalism and the subsequent turn of many social democrats to more business-friendly economic policies strained the relations between unions and their sister parties in many countries. Overall, however, their relationship remains surprisingly strong, namely when parties, indirectly or directly, depend upon union resources (Allern & Bale 2017). Union members' voting rights in the UK's Labour Party facilitated Jeremy Corbyn's rise to party leadership, and the widespread informal presence of union leaders in the Swiss, Austrian or Scandinavian social democratic parties provides unions with political clout (Allern & Bale 2017; Erne & Schief 2017; Gumbrell-McCormick & Hyman 2013; Seymour 2017). In Ireland, Greece and Spain, however, workers increasingly question their unions' ties to their traditional sister parties and instead have supported new parties of the left, such as Sinn Féin, Syriza and Podemos.

As much as it is wrong to focus solely on industrial action without taking politics into account, it is equally wrong to believe that labour-friendly governments will solve the problems of union members. English dockworkers' struggles that relied exclusively on industrial action failed, as did those of Greek dockworkers who wrongly hoped that the election of the radical left Syriza to government would protect them. Portuguese dockers, however, managed to successfully resist attempts to deregulate their workplaces by combining political and industrial action at local, national and international level (Fox-Hodess 2017). Likewise, after a long battle, Ryanair pilots obtained union recognition in 2017 as a result of a transnational campaign that combined coordinated strike action with an intelligent use of political power resources, namely an EU health and safety law that limits pilots' flight times (Golden & Erne 2022). But how can unions influence government policy at national or even EU level when leading centre-left parties believe that they must ignore key union demands in order to be seen as "responsible" leaders (Mair 2013)?

Unions as an actor in initiatives and referendums

The more the traditional alliances between unions and their sister parties have deteriorated, the more unions have been forced to consider alternative strategies to make their voices heard in politics. An additional channel that allows unions to influence policy-making consists of referendums and other direct democratic initiatives. Particularly in Germany, Italy, Ireland, Slovenia and Switzerland, unions have been quite successful in using direct democratic instruments as a lever to advance their interests, regardless of whether direct democratic consultations were initiated by the government or by citizens' initiatives from below (Erne & Blaser 2018).

In Switzerland, for example, unions made their "Yes" vote in past EU referendums conditional upon the adoption of "flanking measures", which consisted of a package of national laws and procedures that considerably strengthened Swiss unions' capacity to enforce the "equal pay for equal work" principle for mobile EU workers in Switzerland. As a result, Switzerland's association with the EU's internal market and the corresponding introduction of free movement of EU workers in Switzerland significantly strengthened collective bargaining coverage and the effective labour rights of unions and migrant workers in Switzerland (Afonso 2016; Erne & Imboden 2015). Likewise, the Irish Congress of Trade Unions threatened to reject the EU's Lisbon Treaty if the Irish government dared to follow the Conservative government in the UK in opting out of the EU Charter of Fundamental Rights (Béthoux, Erne & Golden 2018). In both cases, unions were able to use EU referendums as a lever to compel neoliberal, pro-EU business groups and parties to accept social measures that went against their agenda, given their dependence on the working-class vote in EU referendums.

In some countries, unions were also able to initiate referendums, often in coalition with other social movements. In 2011, a union-social movement coalition successfully rejected the privatization of local water services in a national referendum in Italy (Bieler 2021). Likewise, German public sector unions won several referendums at municipal and regional level against the privatization of public services, and Slovenian unions defeated a rise in the pension age by using citizens' direct democratic participation rights (Erne & Blaser 2018). The introduction of the citizens' right to initiate referendums was also a key demand of the *gilets jaunes* movement in France in order to make its political leaders more responsive to the demands of the popular social classes (Bedock *et al.* 2020).

It would, nevertheless, be wrong to portray the use of direct democratic rights as a silver bullet, as shown by the unequal fate of two European Citizens' Initiatives (ECI) launched by sectoral European union federations at EU level (Szabó, Golden & Erne 2022). In 2013, the European Federation of Public Sector Trade Unions' "Right2Water" ECI reached far more than the required one million signatures, leading EU leaders to exempt water services from a new commodifying EU directive on the awarding of public *concession* contracts. In 2016, however, the European Transport Workers Federation's Fair Transport ECI failed. This shows that the success or failure of direct democratic initiatives depends, like collective action in general, on a variety of actor-centred and structural factors. These include, for example, unions' capacity to frame their campaigns in a way that mobilizes their own rank and file but also inspires larger sections of society, or the particular economic and political EU integration pressures that unions face at a particular time (Erne & Nowak 2022; Szabó, Golden & Erne 2022).

CONCLUSION

The different areas, arenas and aspects of unions' activities demonstrate why unions are political actors, despite an often self-imposed restriction on representing, first and foremost, their members' economic interests. Unions operate on three planes – economic, political and systemic (including ideological) (see also Chapter 7 on terrains). The inherently political nature of labour markets reinforces this political role in various shades as co-opted, integrated or oppositional union movements. Political splits between different sections of the union movement are legion and often affect the political action repertoires that unions pursue in interest politics. Unlike employers that can withhold investment based on corporations' individual decisions, unions are forced to foster cooperation between workers and eventually between different unions and between unions and other social and political actors to successfully pursue their interests (see Table 20.1).

Because of their double character as economic and political actors, nevertheless, unions occupy an important political position in society – depending on the number of their members and their ability to influence other social and political actors. The more unions can make alliances with other forces in society, the more their influence increases. Although their influence on centre-left sister parties has been declining in some countries, unions still play a central role in party–interest group relations in many parts of the world (Allern & Bale 2017). Furthermore, unions have started to use new political tools, such as direct democratic referendums and initiatives, and seek alliances with new partners in civil society, namely social movements, non-government organizations, charity and welfare organizations, or people from migrant, homeless or peasant organizations.

In summation, employment relations are intrinsically political and part of political economy. Unions can, therefore, hardly afford to act like an economic interest group that pursues only its vested economic interests. If they want to be effective, they must also reach out to society at large and act as a "sword of justice", as Flanders (1970) argued a long time ago.[1]

REFERENCES

Afonso, A. 2016. "Freer labour markets, more rules? How transnational labour mobility can strengthen collective bargaining". *Comparative Social Research* 32(1): 159–82.

Allern, E. & T. Bale (eds) 2017. *Left-of-Centre Parties and Unions in the Twenty-First Century*. Oxford: Oxford University Press.

Anner, M. 2015. "Labor control regimes and worker resistance in global supply chains". *Labor History* 56(3): 292–307.

 We acknowledge funding from the European Research Council under the EU's Horizon 2020 research and innovation programme (grant agreement No. 725240, www.erc-europeanunions.eu). We also thank Catherine O'Dea for making the chapter more readable.

Bartolini, S. 2000. *The Political Mobilization of the European Left, 1860–1980: The Class Cleavage*. Cambridge: Cambridge University Press.

Bedock, C. *et al.* 2020. "Une représentation sous contrôle: visions du système politique et réformes institutionnelles dans le mouvement des gilets jaunes". *Participations* 28(3): 221–46.

Béthoux, E., R. Erne & D. Golden 2018. "A primordial attachment to the nation? French and Irish workers and unions in past EU referendum debates". *British Journal of Industrial Relations* 56(3): 656–78.

Bieler, A. 2021. *Fighting for Water: Resisting Privatization in Europe*. London: S Books.

Bieler, A., J. Jordan & A. Morton 2019. "EU aggregate demand as a way out of crisis? Engaging the post-Keynesian critique". *Journal of Common Market Studies* 57(4): 805–22.

Bourdieu, P. 1979. *La Distinction: Critique sociale du jugement*. Paris: Minuit.

Braverman, H. 1974. *Labour and Monopoly Capital: The Degradation of Work in the Twentieth Century*. New York: Monthly Review Press.

Burawoy, M. 1979. *Manufacturing Consent: Changes in the Labor Process Under Monopoly Capitalism*. Chicago, IL: University of Chicago Press.

Carswell, G. & G. De Neve 2013. "Labouring for global markets: conceptualising labour agency in global production networks". *Geoforum* 44(1): 62–70.

Castles, S. & G. Kosack 1973. *Immigrant Workers and Class Structure in Western Europe*. London: Oxford University Press.

Cox, L. & A. Nilsen 2014. *We Make Our Own History: Marxism and Social Movements in the Twilight of Neoliberalism*. London: Pluto.

Cronin, J. 1979. *Industrial Conflict in Modern Britain*. London: Croom Helm.

Crouch, C. 1982. *Trade Unions: The Logic of Collective Action*. Glasgow: Fontana.

Crouch, C. 1993. *Industrial Relations and European State Traditions*. Oxford: Clarendon Press.

Crouch, C. 2004. *Post-Democracy*. Cambridge: Polity.

Crouch, C. 2011. *The Strange Non-Death of Neo-Liberalism*. Cambridge: Polity.

Crouch, C. 2015. *Governing Social Risks in Post-crisis Europe*. Cheltenham: Edward Elgar.

Crouch, C. & W. Streeck (eds) 2006. *The Diversity of Democracy: Corporatism, Social Order and Political Conflict*. Cheltenham: Edward Elgar.

Della Porta, D. 2015. *Social Movements in Times of Austerity: Bringing Capitalism Back into Protest Analysis*. Cambridge: Polity.

Dobson, C. 1980. *Masters and Journeymen: A Prehistory of Industrial Relations, 1717–1800*. London: Croom Helm.

Dubofsky, M. 2000 [1969]. *We Shall Be All: A History of the Industrial Workers of the World*. Abridged edn, edited by J. McCartin. Urbana, IL: University of Illinois Press.

Ebbinghaus, B. & J. Weishaupt (eds) 2021. *The Role of Social Partners in Managing Europe's Great Recession: Crisis Corporatism or Corporatism in Crisis?* London: Routledge.

Edwards, P. 1986. *Conflict at Work: A Materialist Analysis of Workplace Relations*. Oxford: Blackwell.

Edwards, R. 1979. *Contested Terrain: The Transformation of the Workplace in the Twentieth Century*. New York: Basic Books.

Emmenegger, P. *et al.* (eds) 2012. *The Age of Dualization: The Changing Face of Inequality in Deindustrializing Societies*. Oxford: Oxford University Press.

Engelhardt, A. 2020. "Logistische knotenpunkte – der schlüssel sur macht? Transnationale arbeitskämpfe im europäischen hafensektor". In *Transnationalisierung der Arbeit und der Arbeitsbesiehungen*, edited by H.-W. Platzer, M. Klemm & U. Dengel, 179–214. Baden-Baden: Nomos.

Engels, F. 1922 [1895]. *Introduction to Marx's Class Struggles in France*. Moscow: Foreign Languages Publishing House.

Erne, R. 2008. *European Unions: Labor's Quest for a Transnational Democracy*. Ithaca, NY: Cornell University Press.

Erne, R. 2012. "European unions after the global crisis". In *Economy and Society in Europe: A Relationship in Crisis*, edited by L. Burroni, M. Keune & G. Meardi, 124–39. Cheltenham: Edward Elgar.

Erne, R. 2023. "Interest groups". In *Comparative Politics*. Sixth edn, edited by D. Caramani, 277–93. Oxford: Oxford University Press.

Erne, R. & M. Blaser 2018. "Direct democracy and union action". *Transfer: European Review of Labour and Research* 24(2): 217–32.

Erne, R. & N. Imboden 2015. "Equal pay by gender and by nationality: a comparative analysis of Switzerland's unequal equal pay policy regimes across time". *Cambridge Journal of Economics* 39(2): 655–74.

Erne, R. & J. Nowak 2022. "Structural determinants of transnational solidarity: explaining the rise in socioeconomic protests across European borders since 1997". Working Paper No. 11, ERC Project "European Unions". Dublin: University College Dublin.

Erne, R. & S. Schief 2017. "Strong ties between independent organizations". In *Centre-Left Parties and Unions in the Twenty-First Century*, edited by E. Allern & T. Bale, 226–45. Oxford: Oxford University Press.

Erne, R. *et al.* 2024. *Politicising Commodification: European Governance and Labour Politics from the Financial Crisis to the Covid Emergency*. Cambridge: Cambridge University Press.

Esser, J. 1982. *Gewerkschaften in der Krise. Die Anpassung der deutschen Gewerkschaften an neue Weltmarktbedingungen*. Frankfurt: Suhrkamp.

Esser, J. & W. Fach 1981. "Korporatistische krisenregulierung im 'modell Deutschland'". In *Neokorporatismus*, edited by U. von Alemann, 158–79. Frankfurt: Campus Verlag.

Esser, J., W. Fach & G. Simonis 1980. "Grensprobleme des 'modells Deutschland'". *Prokla* 3(10): 40–63.

Flanders, A. 1970. *Management and Unions: The Theory and Reform of Industrial Relations*. London: Faber.

Foot, P. 2005. *The Vote: How It Was Won and How It Was Undermined*. London: Penguin.

Fox-Hodess, K. 2017. "Re-locating the local and national in the global: multi-scalar political alignment in transnational European dockworker union campaigns". *British Journal of Industrial Relations* 55(3): 626–47.

Gall, G. 2013. "Quiescence continued? Recent strike activity in nine Western European economies". *Economic and Industrial Democracy* 34(4): 667–91.

Golden, D. & R. Erne 2022. "Ryanair pilots: unlikely pioneers of transnational collective action". *European Journal of Industrial Relations* 28(4): 451–69.

Gras, J.-C. & A. Nölke (eds) 2007. *Transnational Private Governance and Its Limits*. London: Routledge.

Greenwood, J. 2017. *Interest Representation in The European Union*. Fourth edn. Cham: Springer.

Grote, J. & P. Schmitter 1999. "The renaissance of national corporatism: unintended side-effect of European economic and monetary union or calculated response to the absence of European social policy?". *Transfer: European Review of Labour and Research* 5(1/2): 34–63.

Gumbrell-McCormick, R. & R. Hyman 2013. *Unions in Western Europe: Hard times, Hard Choices*. Oxford: Oxford University Press.

Harvey, D. 2005. *A Brief History of Neoliberalism*. Oxford: Oxford University Press.

Hyman, R. 2001. *Understanding European Trade Unionism: Between Market, Class and Society*. London: Sage.

Irving, T. 2017. "History and the working class now: the collective impulse, tumult and democracy". *Journal of Working-Class Studies* 2(1): 105–15.

Jordan, J., V. Maccarrone & R. Erne 2021. "Towards a socialization of the EU's new economic governance regime? EU labour policy interventions in Germany, Italy, Ireland and Romania 2009–2019". *British Journal of Industrial Relations* 59(1): 191–213.

Katz, H. & A. Colvin 2021. "Employment relations in the United States". In *International and Comparative Employment Relations: Global Crises and Institutional Responses*. Seventh edn, edited by G. Bamber *et al.* London: Sage.

Kelly, J. 1998. *Rethinking Industrial Relations: Mobilisation, Collectivism and Long Waves*. London: Routledge.

Kriesi, H. *et al.* (eds) 2020. *Contention in Times of Crisis: Recession and Political Protest in Thirty European Countries*. Cambridge: Cambridge University Press.

Kuruvilla, S. 2021. *Private Regulation of Labor Standards in Global Supply Chains*. Ithaca, NY: Cornell University Press.

Lenin, V. [1902] 1961. "What is to be done?" In *Collected Works*, Vol. 5, 347–530. Moscow: Foreign Languages Publishing House.

Luban, O. 2008. "Die massenstreiks für frieden und demokratie im Ersten Weltkrieg". In *Streiken gegen den Krieg*, edited by C. Boebel & L. Wentsel, 11–26. Hamburg: VSA.

Lucas, E. 1973–78. *Märsrevolution 1920*, 3 Vols. Frankfurt: Verlag Roter Stern.

Mair, P. 2013. *Ruling the Void: The Hollowing of Western Democracy*. London: Verso.

Manoilesco, M. 1934. *Le siècle du corporatisme: Doctrine du corporatisme intégral et pur*. Paris: Felix Alcan.

Marx, K. 1867a. *Capital*, Vol. 1. Moscow: Progress Publishers.

Marx, K. 1867b. "Instructions for the delegates of the provisional general council. The different questions". In *Collected Works*, Vol. 20, 185–94. Moscow: Progress Publishers.

McAlevey, J. 2016. *No Shortcuts: Organizing for Power in the New Gilded Age*. Oxford: Oxford University Press.

McMenamin, I. 2013. *If Money Talks, What Does It Say? Corruption and Business Financing of Political Parties*. Oxford: Oxford University Press.

Meardi, G. 2018. "Economic integration and state responses: change in European industrial relations since Maastricht". *British Journal of Industrial Relations* 56(3): 631–55.

Molina, O. & M. Rhodes 2002. "Corporatism: the past, present and future of a concept". *Annual Review Political Science* 5: 305–31.

Moody, K. 1997. "Towards an international social-movement unionism". *New Left Review* 225: 52–72.

Moore, P. 2018. *The Quantified Self in Precarity: Work, Technology and What Counts*. London: Palgrave Macmillan.

Müller-Jentsch, W. 1985. "Unions as intermediary organizations". *Economic and Industrial Democracy* 6(1): 3–33.

Nowak, J. 2019. *Mass Strikes and Social Movements in India and Brazil*. London: Palgrave Macmillan.

Oesch, D. & L. Rennwald 2018. "Electoral competition in Europe's new tripolar political space: class voting for the left, centre-right and radical right". *European Journal of Political Research* 57(4): 783–807.

Offe, C. & H. Wiesenthal 1980. "Two logics of collective action: theoretical notes on social class and organizational form". *Political Power and Social Theory* 1(1): 67–115.

Olson, M. 2009 [1965]. *The Logic of Collective Action*. Cambridge, MA: Harvard University Press.

Papadakis, K. (ed.) 2011. *Shaping Global Industrial Relations: The Impact of International Framework Agreements*. Basingstoke: Palgrave Macmillan.

Park, M. 2007. "South Korean trade unionism at the crossroads: a critique of 'social-movement' unionism". *Critical Sociology* 33(2): 311–44.

Pitigliani, F. 1933. *The Italian Corporative State*. London: King & Son.

Pizzorno, A. 1978. "Political exchange and collective identity in industrial conflict". In *The Resurgence of Class Conflict in Western Europe since 1968*, edited by C. Crouch & A. Pizzorno, 277–98. London: Palgrave Macmillan.

Polanyi, K. 2001 [1944]. *The Great Transformation: The Political and Economic Origins of Our Time*. Boston, MA: Beacon Press.

Poulantzas, N. 1978. *State, Power, Socialism*. London: New Left Books.

Quinlan, M. 2017. *The Origins of Worker Mobilisation: Australia 1788–1850*. Abingdon: Routledge.

Reinecke, J. & J. Donaghey 2015. "After Rana Plaza: building coalitional power for labour rights between unions and consumption-based social movement organisations". *Organization* 22(5): 720–40.

Ross, A. 1954. "The natural history of the strike". In *Industrial Conflict*, edited by A. Kornhauser, R. Dubin & A. Ross, 26–36. New York: McGraw-Hill.

Rueschemeyer, D., E. Stephens Huber & J. Stephens 1992. *Capitalist Development and Democracy*. Cambridge: Polity.

Rule, J. 1981. *The Experiences of Labour in Eighteenth-Century England*. London: Croom Helm.

Sala, R. 2007. "Vom Fremdarbeiter zum Gastarbeiter: die Anwerbung italienischer Arbeitskräfte für die Deutsche Wirtschaft 1938–1973". *Vierteljahreshefte für Seitgeschichte* 1: 93–120.

Schattschneider, E. 1960. *The Semi-Sovereign People*. New York: Holt, Rinehart & Winston.

Schmitter, P. 1974. "Still the century of corporatism?". *Review of Politics* 36(1): 85–131.

Schmitter, P. & M. Blecher 2020. *Politics as a Science: A Prolegomenon*. London: Routledge.

Seol, K. 2021. "South Korea's top independent labor leader has been arrested". *Jacobin*, 9 August.

Seymour, R. 2017. *Corbyn: The Strange Rebirth of Radical Politics*. London: Verso.

Sinyai, C. 2006. *Schools of Democracy: A Political History of the American Labor Movement*. Ithaca, NY: Cornell University Press.

Soll, R. 1976. *Der Doppelcharakter der Gewerkschaften: Sur Aktualität der Marxschen Gewerkschaftstheorie*. Berlin: Suhrkamp.

Stan, S. & R. Erne 2016. "Is migration from Central and Eastern Europe an opportunity for unions to demand higher wages?". *European Journal of Industrial Relations* 22(2): 167–83.

Stan, S. & R. Erne 2023. "Pursuing an overarching commodification script through country-specific interventions? The EU's New Economic Governance prescriptions in healthcare (2009–2019)". *Socio-Economic Review*. DOI: 10.1093/ser/mwad053

Streeck, W. & P. Schmitter (eds) 1985. *Private Interest Government: Beyond Market and State*. London: Sage.

Streeck, W. & P. Schmitter 1991. "From national corporatism to transnational pluralism: organized interests in the single European market". *Politics & Society* 19(2): 133–64.

Szabó, I., D. Golden & R. Erne 2022. "Why do some labour alliances succeed in politicizing Europe across borders? A comparison of the Right2Water and Fair Transport European citizens' initiatives". *Journal of Common Market Studies* 60(3): 634–52.

Thomas, R. 1982. "Citizenship and gender in work organization: some considerations for theories of the labor process". In *Marxist Inquiries*, edited by M. Burawoy & T. Skocpol, 86–112. Chicago, IL: University of Chicago Press.

Thompson, E. 1991 [1963]. *The Making of the English Working Class*. London: Penguin.

Tilly, C. & S. Tarrow 2015. *Contentious Politics*. Second edn. Oxford: Oxford University Press.

Tocqueville, A. de. 2000 [1935]. *Democracy in America*. New York: Harper & Row.

Traxler, F., S. Blaschke & B. Kittel 2001. *National Labour Relations in Internationalized Markets: A Comparative Study of Institutions, Change, and Performance*. Oxford: Oxford University Press.

Türkmen, C. 2010. "Rethinking class-making. Sur historischen dynamik von klassensusammensetsung, gastarbeitsmigration und politik". In *Klassen im Postfordismus*, edited by H. Thien, 202–33. Münster: Westfälisches Dampfboot.

Turner, L. & R. Hurd 2001. "Building social movement unionism". In *Rekindling the Movement: Labor's Quest for Relevance in the Twenty-First Century*, edited by L. Turner, H. Katz & R. Hurd, 9–26. Ithaca, NY. Cornell University Press.

Upchurch, M. & G. Taylor 2009. *The Crisis of Social Democratic Trade Unionism in Western Europe*. London: Routledge.

van der Velden, S. *et al.* 2007. *Strikes Around the World, 1968–2005*. Amsterdam: Aksant.

Webb, S. & B. Webb 1897. *Industrial Democracy*. London: Longmans, Green.

Webster, E. 1987. "The rise of social-movement unionism: the two faces of the Black union movement in South Africa". In *State, Resistance and Change in South Africa*, edited by P. Frankel, N. Pines & M. Swilling, 174–96. London: Croom Helm.

CHAPTER 21

CONSTANTLY OUTPACED AND OUTGUNNED? LABOUR UNIONISM IN THE PLATFORM ECONOMY

Horen Voskeritsian

ABSTRACT

The rise of the platform economy has been hailed as a new phase of capitalist development, based upon the principles of freedom, flexibility, collaboration, community, egalitarianism and sharing. However, the realities of work in this hyper-flexible environment generate important concerns about the quality of work and its impact upon platform workers, posing important questions regarding their regulation and the role unions can have in this environment. In employment terms, platform work remains largely an uncharted and unregulated territory, which is further accentuated by the resistance of platforms to be regarded as something other than a neutral space that facilitates transactions between two independent economic agents. This chapter examines the challenges unions face to respond to the changing organizational terrain of capitalism.

Keywords: Platform capitalism; platform workers; cyber-work; labour organizing

INTRODUCTION

> "The future is already here – it's just not very evenly distributed."
> William Gibson, National Public Radio interview, 1999

Platform-mediated work may be a recent phenomenon, but it certainly is not a novel one. Using intermediaries to find a job is a practice as old as capitalism. Yet platforms are different from traditional, high-street employment agencies, in that their role as an intermediary party is much more liquid (Bauman 2000) than the latter; contrary to an employment agency, where roles, identities and expectations are clearly defined where the agency receives a commission to match employers and employees in the labour market, within a platform environment these roles and identities become somewhat blurred. This is not only due to the ambivalent way "work" is conceptualized by the participants in the platform, where in most cases platform workers are defined as "self-employed", but primarily because of the social dynamics that characterize the platform environment and that resemble those of a market rather than of an employment agency.

In that sense, the seller and the purchaser of labour power appear to enter the platform as independent and free agents, who treat the platform as a "neutral" digital ground to conclude their transactions (see also Chapter 7). Freedom and independence, in this case, resemble the Hayekian vision of the free market and, in most cases, this is indeed so: platform-mediated work is subjected to minimum – if any – regulation by the state, which is usually confined to the tax department. This feeling of agency – especially from the part of the worker – is crucial, as it distorts the reality of the platform's role in the digital marketplace and, as a consequence, the reality of the worker's own role and identity.

Far from providing a space where labour supply meets labour demand (see Chapter 20), for example, platforms actively interfere in the transactions between sellers and buyers, influencing the latter's behaviour and choice, and the former's probability of securing a job (or a task, to use the platform terminology). In that sense, platforms cannot be considered as neutral in the allocation of resources and in the resulting equilibria. On the contrary, they influence decisions to advance their own aims. In employment terms, platform work remains largely an uncharted and unregulated territory, which is not the case, for example, with agency work. This situation is further accentuated by the stringent resistance of platforms to be regarded as something other than or beyond a neutral space that facilitates transactions between two independent economic agents.

Consequently, various questions emerge regarding the nature and role of platforms in the labour market. Is platform-mediated work real or bogus self-employment? Who is the employer and who is the employee in this case? Are the platforms simple marketplaces, or neutral intermediaries, or should they be regarded as employers themselves? These are important questions that the judiciary and the labour unions, as well as some governments, have tried to navigate over in the past decade. However, defining the exact nature and role of the platform has proven to be a difficult political and legal exercise, the results of which largely depend on the specific institutional environment in which platform-mediated work takes place. For labour unions, the answer to these questions largely determines the potentiality for organizing and regulating this highly unregulated terrain. Unions, however, are also faced with an additional obstacle, in the form of an ideological cloak that may prevent platform workers from realizing their socio-economic position in the platform economy.

Despite the unfixed and liquid environment, instances of resistance and of successful challenging of the platforms' power are observable worldwide. Many of these actions are individual, as when platform workers discover ways to counteract the algorithmic control of platforms and to beat the algorithm. In other cases, spontaneous collective action has led to the birth of social movement-type unionism, whereas in other cases official labour unions have managed to represent platform workers. After many years of operating in the shadow of cyberspace, platform workers slowly but surely and decisively have begun to organize and demand better pay and working conditions, and to challenge their characterization as self-employed, independent contractors.

This chapter will examine the challenge of labour unions responding to the changing organizational terrain of capitalism. It will develop the earlier chapters on terrain (specifically chapters 7 and 12) to look more contemporaneously at the constant churn in organizational forms under neoliberal capitalism with the move to atomization and fragmentation (as opposed to individualization or collectivization) of workforces, necessitating that unions discover or rediscover the means by which they can renew, reset and re-establish inter-worker and intra-workforce relations as the basis for providing the processes of (re)building collective power.

The following section discusses the advancement of platform capitalism and the nature of platform-mediated work, and theorizes how it can be set apart from more traditional modes of employment. Despite its still peripheral role in the production of a country's gross domestic product, platforms and platform work continue to expand in new domains, especially after the Covid-19 pandemic and the rise of the "digital worker" and the "digital nomad". As such, appreciating and understanding its dynamics is crucial for labour unions and the state alike. The next section focuses on the ways organized labour can reclaim the new employment terrain by building on various examples of mobilization, organization, and regulation from across the globe, suggesting that change is not only possible but, most importantly, feasible, despite the seemingly powerful position of platforms in contemporary capitalism.

PLATFORM CAPITALISM AND PLATFORM-MEDIATED WORK

The rise of the platform economy has been hailed as a new phase of capitalist development, based upon the principles of freedom, flexibility, collaboration, community, egalitarianism and sharing (Howcroft & Bergvall-Kåreborn 2019). Its depiction comprises the core elements of neoliberal capitalism and of a cooperative movement, giving capitalism a more social and, perhaps, humane face. The description of this part of the economy as the "collaborative" or the "sharing economy" (Riso 2019), implies an element of collective ownership of these endeavours, a socialization and democratization of the design and outcomes of this economic activity, in contra-distinction to the faceless reality of non-platform capitalism.

"Freedom" and "ownership" are being promoted as the cornerstones of the users' activity on platforms; from social media, such as Facebook, Instagram or X, to marketplace platforms, such as eBay, users are in control of their content and of its distribution to a global audience. However, this apparent democratization of data conceals an element of control exercised over the users by the platforms – a control that may not be immediately discernible, since it usually occurs behind the screens by the algorithms that run and organize data on the platforms' mainframe. From bots, which decide what kind of information each user will be receiving based on their searches and profile characteristics, to algorithms that classify and categorize workers based on their (subjective, most of the times) performance, the ideal of information democratization rests on feet of clay.

Platforms, and platform-mediated work, are not something uniquely novel, however (Berg & De Stefano 2018), but build upon already-existing principles, which were developed and materialized in the late 1990s during the age of information capitalism and of the network society (Castells 2000). The rise of the internet and of global capitalism accentuated the emergence of cyber-marketplaces at first, and of cyber-labour markets later. This evolution was assisted by the foundations laid down by the erosion of traditional labour market institutions during the 1980s and 1990s, and the rise and ideological justification of various forms of precarious work that took place at around the same time. The advance of "flexible specialization" (Hyman 1988) and of dual labour markets created the necessary base for the successful accommodation of cyber-work in our collective (un)consciousness. Together with the digitalization of work, and the rise of what Huws (2003) called the "cybertariat", the transition from a physical labour market to a cyber one came almost naturally to a generation of workers who grew up perceiving flexibility and precariousness as normal states of a labour market.

According to the OECD (2020: 248), the term "online platform" "has been used to describe a range of services available on the Internet. These include marketplaces, search engines, social media, creative content outlets, app stores, communications services, payment systems, services comprising the "collaborative" or "gig" economy and much more." The literature classifies platforms in various ways. For instance, based upon the traded commodity (hence, capital platforms versus labour platforms), or the different kind of products traded online (for example, services versus material products), whether they are for-profit or non-profit, or if they offer virtual and global services (such as Amazon's MTurk) versus more localized and physical services (such as Uber) (Drahokoupil & Fabo 2016; for an extensive review see Riso 2019).

Eurofound (2018a: 19) defines platforms as "digital networks that coordinate transactions in an algorithmic way". The digital networking characteristic of platforms allows them to have a global reach and to operate transnationally without the limitations that "physical" capital face. At the same time, users interact in a borderless way as well, thus further enhancing the ideal of freedom and flexibility the platform economy is being identified with. But, as the Eurofound definition purports, platforms also perform a coordinating function in the realm of cyber-transactions. Coordination can be understood in various ways. A market, for example, "coordinates" activities based on the principles of supply and demand: temporary equilibria are being reached depending on the circumstances that characterize supply and demand at each specific time-period and/or space-location. Coordination may also imply a collaborative or mediating approach, as when an agent facilitates a transaction between agents that would not have met otherwise. Coordination, however, also entails an element of control, in the sense that agents' behaviours or actions are being orchestrated by an overseeing authority.

Although platforms may be perceived and promoted as marketplaces, they are more than that. Contrary to a market, platforms actively interfere in the allocation and distribution of resources, and in influencing decision-making. Even when they act as intermediaries, bringing together sellers and purchasers of goods and services, they do not do so passively or in a neutral manner. On the contrary, control is a core characteristic of most platforms, from the way information is being depicted, to the creation of preferences and the manipulation of consumer behaviour through algorithmic management (Vandaele 2018).

Interestingly, despite their immediacy and apparent penetration in the everyday fabric of society, measuring and identifying the extent of the platform economy is a complicated exercise. Most countries do not keep detailed statistics about the economic activity generated in platforms, and international institutions such as the Organisation for Economic Co-operation and Development (OECD) base their evaluations of the size of this part of the economy on estimations. The situation becomes more complicated when trying to estimate the size and extent of platform-mediated work. As Piasna (2020: 6) argues: "The paradox in measuring the platform economy is that, although its operations generate a wealth of data, with all transactions being digitally recorded, one of the biggest unknowns is still the scale of platform work." Apart from the fact that national statistical services do not explicitly measure platform work, an additional complication arises from the lack of a commonly agreed definition of what constitutes platform work.

The OECD (2019: 5) defines platform workers as those who "use an app or a website to match with customers in order to provide a service (rather than goods) in return for money". In a similar vein, Eurofound (2018b) defines platform work as "a form of employment that uses an online platform to enable organizations or individuals to access other organizations or individuals to solve problems or to provide services in exchange for payment". Thus, according to Eurofound (2018b), the main characteristics of platform work are the following: (1) paid work is organized through an online platform; (2) three parties are involved (the online platform, the client and the worker); (3) the aim is to carry out specific tasks or solve specific problems; (4) the work is contracted out; (5) jobs are broken down into tasks; and (6) services are provided on demand.

The above definitions are quite broad and capture a variety of activities that take place on platforms: routine and non-routine tasks; physical versus cognitive tasks as well as physical versus digital delivery (OECD 2019). The defining characteristic of platform-mediated work is the existence of some type of contract between one party and another to conduct paid work. This definition, however, does not clearly differentiate between self-employed and those working under a bogus self-employment regime. Hence, although in some cases some work has the characteristics of a professional activity that can count as manifestly self-employment, in many cases the limits between what constitutes self-employment or not are quite blurred.

Platform work can take many different forms and can be conducted by very different types of workers. Some platform work, such as micro-tasking (OECD 2019), is very trivial and monotonous, replicating a Taylorist approach to the organization of the labour process, albeit in cyberspace. Workers who offer services such as data entry or translation fall into this category. Others engage in crowd-work, even though this term can be broken down into four distinct types of work (Howcroft & Bergvall-Kåreborn 2019), namely online task crowd-work; asset-based services; playbour – combining play and labour – crowd-work; and profession-based freelance crowd-work. Finally, others may engage in gig work or on-demand work via apps – the most common example being delivery riders and drivers (Johnston 2020). Within these categories one can find both high-skilled as well as low-skilled workers (Drahokoupil & Fabo 2016), those working in platforms to supplement existing income and others who depend entirely on platform work for their livelihoods. For some, their platform profile acts as a cyber-firm, with considerable time and money being spent to create a digital profile and a reputation that will attract future clients, whereas others may navigate cyberspace in a relative anonymity, trying to attract gigs or tasks on a day-by-day basis.

Despite its apparent prevalence, OECD (2020: 255) estimates that "platform-mediated employment is still a small share of overall employment, typically about 0.5 per cent to 3 per cent", with different countries experiencing larger or smaller participation of their workforce in the platform economy. In the Netherlands, for example, it is estimated that 18 per cent of the labour force have attempted at least once to find work via a platform, whereas in Sweden 12 per cent of workers have participated in platform-mediated work (Howcroft & Bergvall-Kåreborn 2019). In the UK, platform-mediated work has doubled since 2016, with similar trends observed in other EU countries as well (Huws, Spencer & Coates 2019).

Undeniably, platform-mediated work is a very dynamic segment of the labour market with high growth potential in the coming years (OECD 2020). Some are attracted to platform work to supplement their income (Vallas & Schor 2020), but also because the structural characteristics of the platforms (especially their global reach and the low operating costs for users) provide opportunities of employment to people who find entrance to more traditional labour markets difficult. As Huws, Spencer and Coates (2019) show, adverse labour market conditions and poverty contribute to the participation in platform work, especially in countries where the welfare state does not provide adequate protection from labour market rigidities. Yet others are attracted to this kind of work by the prospects of flexibility, economic independence and income growth potential. Although this prospect may materialize for some, for most platform workers the reality is very different. Platform work is associated with low income, an increase in non-standard forms of employment (OECD 2020) and low job quality (Wood *et al.* 2019); whereas in many cases, especially in low-skilled tasks, the promise of flexibility and freedom is quickly replaced by an electronic panopticon-type of control that severely constrains autonomy (Woodcock 2020). In this highly dynamic, precarious and unregulated environment, the questions of exploitation, resistance, representation and regulation inevitably emerge.

REGULATING CYBER-WORK: ORGANIZATION AND MOBILIZATION OF PLATFORM WORKERS

Despite its rapid growth potential and increasing attractiveness to a proportion of the labour force, the platform economy remains largely an unregulated terrain, especially with regards to labour rights (De Stefano 2016). A global market, characterized by a liquid nature and high labour turnover, personifies the core characteristics of international, networked, capital, operating globally without always requiring a spatial fix or national investments, and generating profits with little or no investment in human capital. Combining the use of highly sophisticated artificial intelligence (AI) software and of adversarial management, platforms create an environment where organizing and mobilizing workers becomes an increasingly challenging and complicated affair.

In recent years, however, various attempts to regulate this market, both at the sphere of production, as well as at the sphere of the market and the sphere of politics, have been made (Wright 2015). For example, gig workers have organized themselves in social networks to challenge the control of the platform over their work and their identity; labour unions have become involved in the organization and mobilization of workers, by signing recognition agreements and engaging into collective bargaining;[1] and the state – either through the judiciary or through government policies – has attempted to challenge and regulate the modus operandi of the platform economy.[2] Yet the problems associated with working in these environments are still persistent, namely the spatial diffusion of the workforce; the highly individualized and atomized nature of jobs and tasks; the blurred limits of who is an employer and who, if any, is an employee; and the casualization of work, with all the socio-psychological and economic consequences that this type of work brings about.

The social realities of work in the platform economy have been well documented by now. Platform work is precarious and casualized, characterized by high levels of stress and monotonous activities (especially in the case of low-skilled crowd-work), with workers doing unpaid labour to get paid work (Wood *et al.* 2019), and eventually receiving low hourly wages – most of the time below the level of the national minimum wage (Doellgast 2018; Kahancová, Meszmann & Sedláková 2020). Platform work is often characterized by an intensity, with important health and safety implications: workers may not be taking breaks or resting properly in order to fulfil tasks on time, or to comply to performance indicators set by the platforms (Eurofound 2018a), further contributing to their mental and physical fatigue. Moreover, platform workers experience social and psychological isolation, which enhances their feelings of vulnerability and replaceability (Johnston 2020), and places them in a position of constantly having to "reinvent" and "reposition" themselves in the market, in what Arvidsson (2019) describes as the passage from "industrial" to "industrious" capitalism.

These are not, by any means, unique characteristics of the platform labour market, nor something that organized labour has not encountered before. Precarious forms of employment, be it atypical employment, or temporary agency work, or zero-hours contracts, have dominated the organization of the labour market in advanced capitalist economies for the past quarter of a century. In certain industries – such as call centres or

1. For an overview of unions activities with respect to platforms in Europe, see https://digitalplatformobservatory.org/
2. Although the discussions about the regulation of AI and platform work in the US are still at an embryonic stage, in the EU two proposals about the regulation of AI (Proposal for a Regulation of the European Parliament and the Council laying down harmonized rules on Artificial Intelligence; Artificial Intelligence Act, COM/2021/206 final) and of platform-mediated work (Proposal for a Directive on improving working conditions in platform work; COM/2021/762 final) are in the final stages of consultation with the EU social partners. Additionally, and independently of these EU acquis – the collection of common rights and obligations that constitute the body of EU law, which are incorporated into the legal systems of EU member states – various member states (e.g. Greece and Spain), have laid down rules regarding the functioning of platform work.

(non-platform) delivery workers – workers experience very similar conditions to those of platform workers (Moore & Newsome 2018; Shire *et al.* 2009). Meanwhile, in the Global South, precarity and insecurity constitute the work reality for the vast majority of the population (Kumar 2020). But an important difference between existing modes of production and platform-mediated work, however, lies not so much in the organization of work, but rather in the way platforms impose and perpetuate a hegemonic ideology to their workforce. To achieve this, platforms build on three elements: the organization of the labour process, which is characterized by an extreme atomization and fragmentation of labour and the spatial liquidity of surplus value generation; the control of the labour process, through the use of algorithmic management and the refusal to characterize work in the platform environment as employment; and, finally, the advancement of a modernist, techno-deterministic, ideology that aims to provide meaning and justification to the structure of platform work.

One of the core visible characteristics of platform work, be it crowd-work or on-demand app work, is its highly individualized, fragmented and atomized nature (Lehdonvirta 2018). The work offered in, and by, the platforms is taken up by individuals who carry it out in a completely atomized way – be that riding a bicycle to deliver a pizza or translating documents for a client. In many cases, the job comprises a variety of tasks, which may not be delivered by the same individual contractor but may be allocated to several platform users. In that sense, platforms survive and grow on a paradox: although they help organize labour collectively, as capital does, they do not necessarily create a community of workers. That way, platforms can recreate the benefits of the collective organization of labour, namely the production and appropriation of surplus value, and to eliminate the generation of the social forces that define a community, namely the socialization and the formation of a shared identity (Tassinari & Maccarrone 2020). Platforms, therefore, push what Joyce (2020) theorizes as the "subsumption of labour and the cash nexus" to new extremes: not only by imposing a certain mode of production upon the workers but, perhaps more crucially, by transforming, through a hegemonic process, the perceptions and realizations platform workers hold for themselves as workers, and for the platforms as media of employment.

In a sense, platform work is alienation par excellence: the worker does not only participate in a Taylorist-like production process, fulfilling a set of diverse, dispersed and highly individualized tasks under conditions of strict control, but is doing so willingly and, in many cases, agreeably. Contrary to traditional industrial production, however, where the workers know their role and status in the labour process, and where the production and appropriation of the final product is observable, in the case of platform-mediated work, the workers usually have no knowledge of their role in the production of the final product – they are only aware of the very specific task they are assigned to perform. To them, the product they produce and the process of production are tautological and equally fragmented. At the same time, the appropriator of their surplus value is not one, but two entities: the "client", who is the final recipient of the service or product produced, and the platform, which charges a premium for its "services" to the worker. The limits of who the employer is and how the workers self-identify themselves become blurred from the very instant one engages in platform work.

The atomized nature of platform-mediated work is closely linked to a second characteristic, namely the lack of a spatial fix and of a physical workplace (Johnston 2020). Platform workers operate from within their individualized work "shell", be it their house, a café, their bicycle or their car, usually with little communication with others. Depending on the type of platform work and, in many cases, the platform environment used, the extent and level of social interaction differs. Riders, for example, who engage in a physical activity that requests their presence in the wider society, can socialize both with fellow riders and staff at the shops they collect their deliveries from, or in urban hubs (Tassinari & Maccarrone 2020). Even workers who may not need to leave their room while working may have the opportunity to communicate with a wider worker community, either via the platform they use to offer their services or via social media. The latter interactions, however, are quite fragmented and usually restricted to an exchange of know-how or troubleshooting for the job at hand.

The crucial point for labour unions in their attempts to organize and mobilize platform workers concerns the quality of these social interactions. Are those interactions simply exchanges of cordiality, bestowed with a generic social meaning, or can they be perceived as breeding grounds of worker or class consciousness and of the realization of one's position in the labour market? Are these (virtual or physical) social groups communities with shared beliefs, concerns and worries about their conditions, or are they ephemeral constructs with a narrowly defined purpose (such as troubleshooting advice for example)? And, if the latter, can they be transformed into a community of workers? As we know from mobilization theory (see Kelly 1998), the emergence of a community and of a shared and clear purpose directed towards action lies at the core of any successful mobilization and organizing activity – if workers do not perceive themselves as a class in themselves with shared interests, bringing them together becomes increasingly difficult.

A third, and perhaps very crucial, characteristic of platform-mediated work is that platforms and, in many cases, the workers themselves, do not regard this type of work as dependent employment. For the platforms, the people offering services through their infrastructures are independent contractors, or "entrepreneurial partners" and, as such, not liable to protection from labour law. This classification is also shared by many of the workers themselves, who also perceive themselves as entrepreneurs, partially attracted to this mode of employment by the promises of independence and autonomy (Purcell & Brook 2020). This is a similar motif that we also observe in other, non-platform, modes of dependent self-employment, as in the case of couriers (Moore & Newsome 2018). Although there are instances where the "self-employed" label makes sense, in many cases this is not so – and certainly it does not become so because a platform stipulates it in the content agreement with the workers (Fredman & Du Toit 2019). Interestingly, the persistence of platforms to promote themselves as "intermediaries" between a seller and a buyer is challenged by the practices the platforms use to regulate the flow of information within their domain. To put it differently, although one may claim that platforms are cyber-markets, their practices distort the very notion of the free market they aim to promote and perpetuate. Nowhere is this more evident than in the case of algorithmic management.

A central function of algorithmic management is the algorithm's control over work allocation (Wood *et al.* 2019). Depending on the platform, the process changes, but at its core remains the fact that an AI algorithm, using certain variables, allocates and manages work, or suggests workers to a client. The moment this happens, the platform ceases being an impartial free market and becomes at best a marketing tool, at worst a controlling employer. Moreover, the algorithm's decisions aim primarily to serve the business interests of the platform: work is allocated to the more productive riders, exactly because they generate better profits for the platform; contractors with excellent reviews are promoted at the top of a search because they generate more income for the platform as well. Contrary to a market, therefore, where information is collected and collated by the two transacting parties, the platform has its own logic and serves its own interests, generating an artificial equilibrium through its control and manipulation of information to both ends of its users. Markets, be what they may, certainly are not functionalist: equilibria are reached because of behaviours taking place within the market (i.e. by the market actors) and not by the market itself.

Like a magician, therefore, the platform's management tries to misdirect our attention from the reality of social relations that are developed within its realm. The misdirection uses ideas, such as objectivity and transparency, to persuade users that the decisions about the allocation of tasks and the scheduling of the work are bias-free. Yet this is far from the case, since the promised autonomy is severely curtailed by the implementation of a strict algorithmic management (Lehdonvirta 2018; Veen, Barratt & Goods 2020; Wood *et al.* 2019), creating instead a condition of "fictitious freedom" (Shibata 2020).

At the same time, the mantra of objectivity is further accentuated by the management's attempt to shed its physical presence and to be perceived as the very algorithm that makes decisions on the everyday running of the cyber shop floor. By taking the human agent out of the equation and becoming a faceless entity, management creates the illusion of impartiality and passes the responsibility for future work tasks not to the algorithm per se, which in this case is just an intermediary, but to the worker: to continue earning, workers must conform to the performance goals set by the platform. In other words, workers enter a mental state of self-management to avoid being punished by the algorithm. Being nowhere to be seen, but apparently omnipresent, platform management creates a Kafkaesque work environment, where the worker willingly succumbs to the power of the platform.

The hegemonic position of platforms over the workers, and the perpetuation of their current modus operandi, rests on an ideological superstructure that is strategically exploited by the platforms' media gurus; namely, the marriage of neoliberal discourse and of technological determinism to promote the ideas of progress, growth and personal achievement. Ideology operates on two levels in this case. First, the combined discourse of "flexibility", "autonomy" and "independence", and the promise of financial reward, attracts workers to the platform, helps to justify and legitimize their status as "independent contractors", and promotes the individualization and atomization of work. Combined in some cases, such as Uber's, with an overt paternalism, or the adoption of a "playful" discourse and of gamification techniques that aim to revert attention from the reality of work (Tassinari & Maccarrone 2020), platforms further reinforce their dominance over the relations of production.

Second, as a socio-economic project, platform work and the gig economy are attached to the vehicle of technological determinism and to its consequent ideas: modernity, the progressive nature of technology and its dominating inevitability. Although technology has undoubtedly improved the quality of human life, the idea that it leads to progress and, consequently, to economic growth and prosperity is not necessarily self-evident. Contrary to the linear futurology advocated by technological determinists, the course of capitalism has shown that the way technology enters the production process is usually the result of a negotiated compromise with organized labour (Moore, Akhtar & Upchurch 2018). Moreover, the emergence of new technologies does not necessarily mean that they will be incorporated in the production process. As Noble (1995: 102) argues: "to the extent that people do strive primarily for the highest return on their investment, there is no guarantee that they will invest in new means of production if other, more profitable routes are available" – meaning, in this case, cheap labour.

Finally, the effects of technology on labour may not always be progressive – deskilling, labour substitution, lowering of wages or technological unemployment may be some of the consequences of technological advancements. Notwithstanding the above, equating technology with progress (and, subsequently, dismissing any critique of technology as Luddism), creates a justificatory context for the redefinition of labour by platforms: working in platforms is promoted as the latest technological trend, contributing to a techno-positive identity like that of owning the latest technological gadget.

Despite its hegemonic prevalence, however, the ideological superstructure promoted by platforms does not remain unchallenged. Workers across platforms realize that the promise of progress is, in fact, unequally distributed. Feelings of discontent emerge, but they do not always materialize in mobilization, since, to become mobilized, a worker needs to be able to attribute directly their grievance to a specific agent and then, through collective organization and leadership, to move towards challenging the source of the aforementioned grievance (Kelly 1998). Thus, the first hurdle to be overcome is to identify the agent: workers may still be prone to blame themselves, or the "platform", for their situation, something that is counterproductive, as self-blame and abstract constructs do not lead to action other than, perhaps, exiting the situation.

Indeed, platform-mediated work is characterized by high turnover rates, and many platform workers have low commitment to their job. Others, in their inability to blame a specific agent, may resort to rationalization: they justify their situation and support their decision to engage in this work because they may find it difficult to escape it, or psychologically hard to negotiate its disadvantages (Muszyski *et al.* 2021). Yet others challenge the platform's hegemony, either in the form of individual resistance or, especially in the case of on-demand work, through collective organization (Bessa *et al.* 2022). Across Europe, for example, labour unions have made several steps towards regulation, either by signing recognition and collective bargaining agreements or by pushing for changes at the regulatory level. In other cases, workers and their unions appeal to the judiciary to challenge their employment status.

Successfully organizing and mobilizing platform workers is not without its challenges, however. Adversarial management aside, unions need to tackle a variety of factors that may inhibit the workers' motivation to organize collectively. One such factor is the lack of exposure to labour unions. Although workers may not be negatively inclined towards unionization, they may find it difficult to organize due to the lack of union activity in their ranks (Vandaele, Piasna & Drahokoupil 2019). In other cases, workers may be reluctant to organize collectively either because they do not perceive themselves as "workers" or because they are not sufficiently committed to the job to invest time and effort in collective action. The lack of a physical workspace, which helps in the formation of a community and the development of social relations, is yet another hurdle that unions face, as is the feeling of powerlessness that platform workers may experience. Their social isolation, the structuring of the labour process and the panopticon effect of algorithmic management contributes to the acceptance of the status quo and the belief that change may be futile.

However, platform workers have a disruptive capacity exactly because of the way their labour process is organized (Vandaele 2018). Although the platform may appear as omnipotent, the workers can log out unexpectedly, or they may refuse to take orders when they come in – two strategies that delivery workers have utilized extensively across Europe (Tassinari & Maccarrone 2020). As Kimeldorf (2013) shows in the case of the US, the disruptive capacity of workers does not only lead to successful outcomes but can also facilitate union growth. Thus, although platform workers may be easily replaceable and may work in relatively isolated conditions, when they organize collectively they can potentially take advantage of two core characteristics of the platform's business model: its high dependency upon customer satisfaction and reviews, and the fact that it proves very difficult to replace in real-time workers that may decide not to collaborate with the algorithmic organization of production.

Successfully constructing and exercising power rests upon the development of a common identity and a common of sense of interests and purpose among platform workers, which in turn implies the emergence of solidarity towards fellow workers. Although the fragmented organization of platform-mediated work makes the development of solidarity difficult, it can, nevertheless, emerge in a hyper-individualistic and neo-Taylorist labour process (Tassinari & Maccarrone 2020). Developing a discourse of solidarity among the rank and file, but also across wider society and with other workers, is crucial for the success of organizing activities among precarious workers (Carver & Doellgast 2021). The successful mobilization of delivery riders in the UK to improve their working conditions (Cant & Woodcock 2020; Tassinari & Maccarrone 2020), or of the Italian food riders to be reclassified as employees (Chesta, Zamponi & Caciagli 2019; Marrone & Finotto 2019) are cases in point. The success of these motions rested on the ability of the organizers to nurture social relations by utilizing existing social and community networks, and to communicate with workers via the various "free" spaces (such as waiting areas outside restaurants or social media) to overcome the inherent atomization and individualization of the tasks. Moreover, the experience of mobilization per se was fundamental in shaping consciousness and the riders' social identification as "workers" and not "self-employed" contractors. The discourse that was used in various communications and in the social media played a crucial rule in the reframing of their identity.

Grassroots unions of a syndicalist ideological orientation were central in these processes (Borghi *et al.* 2021). Sharing similar characteristics with the more spontaneous social movement-type actions that usually spring out of this body of workers, and with extensive experience in representing and organizing precarious workers (cf. Kretsos 2011), they managed to put together the necessary resources to mobilize resistance. Grassroots unions appealed better to a substantial body of workers since they represented an alternative to the more "established" unions, which many workers in the periphery saw as at best indifferent and at worst antagonistic towards their interests.

Nevertheless, mainstream labour unions have also realized the importance of engaging with platform workers, both to represent them from a social justice perspective and to ensure a better regulation of the domestic market to counteract social dumping and negative wage effects. Across Europe and the US, labour unions represent workers in court actions and push for regulatory changes at the state level. In Italy and the UK, the judicial decision about Deliveroo and Uber (Fredman & Du Toit 2019), created an expectation about how delivery workers should be treated. In Spain, the High Court of Aragon ruled in March 2021 that Deliveroo riders are to be classified as workers, whereas in Sweden, Unionen, which organizes white-collar workers within the private sector, has negotiated collective agreements with various platforms.[3] This is an emerging trend in Europe, with the EU Commission also committed to taking steps towards the regulation of this market (Pochet 2021). Despite these developments, however, the struggle is very far from over. The fine line of who is an employee and who is an employer is still diligently negotiated at court level, not only in the UK with its common law tradition, but also in civil law judiciaries such as in France, where the courts have rejected claims of Uber or Deliveroo contractors to be reclassified as employees (Borghi *et al.* 2021).

An unexpected ally that platform workers have in various European countries are the consumers. For instance, the mobilization of the Italian food riders extensively engaged the customers of the platforms via social media, to support the riders' initiative and put an indirect pressure on the platforms' management. However, although the consumer is a force to be reckoned with, it is not necessarily one that can always be trusted. The consumers' diverse socio-demographic characteristics and the lack of a class identity make them prone to opportunistic behaviours. Nowhere was this more evident than in the case of California's Uber drivers and, the infamous by now, Proposition 22 (De Stefano 2021). Whereas the Californian courts had ruled in favour of the classification of Uber drivers as dependent employees, an aggressive marketing campaign from Uber, combined with the threat of leaving California, asked for a referendum – known as Proposition 22 – to exempt Uber drivers from minimum wage legislation. The outcome of the referendum was in Uber's favour, allowing the company to introduce a company-devised policy with regards to the remuneration of Uber drivers, further expanding its paternalistic reach to its workers.

The strategies labour unions use to represent platform workers vary depending on a country's institutional context and the kind of power unions have. In contexts of low institutional power, labour unions may follow a conflict-based path of representation, whereas in contexts with strong institutional power, unions pursue a social partnership approach (Carver & Doellgast 2021). In the first case, unions engage in combative action, alliance-building with wider communities, developing and promoting solidarity among workers, and representing minority groups, such as migrant workers. In the latter, unions utilize existing institutions to their benefit, engaging in sectoral collective

3. See https://digitalplatformobservatory.org/ and https://english.elpais.com/economy_and_business/2021-05-12/spain-approves-landmark-law-recognizing-food-delivery-riders-as-employees.html

bargaining to constrain the employers' social dumping strategies or by appealing to the state to regulate the market. Many unions also engage into "soft representation" strategies (Doherty & Franca 2020), such as providing various services to members.

The distinction between conflict-based and social partnership approaches to representation is not always clearly defined. The kind of strategy that will be used depends on the power dynamics that characterize the spheres of production, market and politics at certain historical moments. A lesson drawn from the representation of peripheral workers is that they cannot always rely on mainstream structures to represent them adequately, even in contexts where labour enjoys sufficient institutional power (Doellgast & Greer 2007). To make things more complicated, unions attempting to represent platform workers must navigate a much more liquid terrain, where the traditional roles of an employment relationship are not clearly defined. To raise consciousness and to adequately mobilize resources, unions use every available tool in their toolbox. Hence, we usually observe a coexistence of both these approaches across institutional contexts, reflecting the different kinds of power and strategic games initiated at each sphere of representation (Voskeritsian & Kornelakis 2019).

CONCLUSION

"A middleman's business is to make himself a necessary evil", William Gibson pointed out in the Neuromancer almost three decades ago; and when the middleman's business is being challenged, one can expect a stout reaction. Platform companies will be on the offensive in the coming years, as platform workers become increasingly aware of their socio-economic status and become organized in unions, and national governments and intra-state institutions proceed to begin to regulate this liquid terrain. This offensive will be played out in the courts, in the corridors of power and in the public domain – the example of Uber in California is such a case in point. Building on the ideological superstructure of flexibility, freedom, prosperity and the inevitability of progress, implementing aggressive neo-Taylorist management practices and utilizing the platform's immobile mobility, platform companies will attempt to continue to blur the limits of agency and responsibility. In many respects, platform-mediated work bares uncanny similarities to any other kind of precarious employment – from part-time and zero-hours contract workers, to the bogus self-employed blue-collar and white-collar workers. Contrary to these cases, however, platforms operate in a – seemingly – virtual world, where social relationships between workers may be totally inexistent, and where the hegemonic ideology of the platform distorts the ways individuals perceive their roles and positions within the labour process.

To that extent, organized labour is faced with additional challenges in its attempt to represent platform workers. Contrary to a mobile and global cyber-capital, labour unions' power remains entrenched in their national boundaries. International coordination is, therefore, required if social dumping is to be challenged. Although the regulation of crowd-work may seem a distant reality, regulating on-demand app work is more readily achievable, especially within the EU, where cross-border workers' participation mechanisms exist. The establishment of a European Works Council representing riders or drivers working in platforms with a cross-national presence could be a possible next step forward, for example. To reach this stage, however, the roles of the two parties need to be reimagined and redetermined across several national contexts.

Mobilizing and organizing the shop floor – be it through social networks, grassroot or mainstream unions – is an important step in the right direction. As is the case with mobilization, the rank and file must believe that change is not only required, but that it is, indeed, feasible. Although the space where this realization occurs is usually the shop floor, engaging the platforms only at the sphere of production or the sphere of the market is not enough. Unions need to continue to campaign for institutional change at the national and international levels for this terrain to be properly regulated.

REFERENCES

Arvidsson, A. 2019. *Changemakers: The Industrious Future of the Digital Economy*. Cambridge: Cambridge University Press.

Bauman, Z. 2000. *Liquid Modernity*. Cambridge: Polity.

Berg, J. & V. De Stefano 2018. "Employment and regulation for clickworkers". In *Work in the Digital Age*, edited by M. Neufeind, J. O'Reilly & F. Ranft, 175–84. London: Rowman & Littlefield.

Bessa, I. *et al.* 2022. "A global analysis of worker protest in digital labour platforms". ILO Working Paper 70. Geneva: International Labour Organization.

Borghi, P. *et al.* 2021. "Mind the gap between discourses and practices: platform workers' representation in France and Italy". *European Journal of Industrial Relations* 27(4): 425–43.

Cant, C. & J. Woodcock 2020. "Fast food shutdown: from disorganisation to action in the service sector". *Capital & Class* 44(4): 513–21.

Carver, L. & V. Doellgast 2021. "Dualism or solidarity? Conditions for union success in regulating precarious work". *European Journal of Industrial Relations* 27(4): 367–85.

Castells, M. 2000. *The Rise of the Network Society*. Second edn. London: Blackwell.

Chesta, R., L. Zamponi & C. Caciagli 2019. "Labour activism and social movement unionism in the gig economy: food delivery workers struggles in Italy". *Partecipazione e Conflitto* 12(3): 819–44.

De Stefano, V. 2016. "The rise of the just-in-time workforce: on-demand work, crowd-work, and labor protection in the gig-economy". *Comparative Labor Law & Policy Journal* 37(3): 471–504.

De Stefano, V. 2021. "Regulation is not an a la carte menu: insights from the Uber judgment". UK Labour Law. https://uklabourlawblog.com/2021/03/02/regulation-is-not-an-a-la-carte-menu-insights-from-the-uber-judgment-by-valerio-de-stefano/

Doellgast, V. 2018. "Rebalancing worker power in the networked economy". In *Work in the Digital Age*, edited by M. Neufeind, J. O'Reilly & F. Ranft, 199–208. London: Rowman & Littlefield.

Doellgast, V. & I. Greer 2007. "Vertical disintegration and the disorganization of German industrial relations". *British Journal of Industrial Relations* 45(1): 55–76.

Doherty, M. & V. Franca 2020. "The (non)response of trade unions to the 'gig' challenge". *Italian Labour Law E-Journal* 13(1): 125–40.

Drahokoupil, J. & B. Fabo 2016. "The platform economy and the disruption of the employment relationship". Policy Brief 5. Brussels: ETUI.

Eurofound 2018a. *Automation, Digitisation and Platforms: Implications for Work and Employment*. Luxembourg: Publications Office of the European Union.

Eurofound 2018b. *Definition of Platform Work*. Luxembourg: Publications Office of the European Union.

Fredman, S. & D. Du Toit 2019. "One small step towards decent work: Uber v Aslam in the court of appeal". *Industrial Law Journal* 48(2): 260–77.

Howcroft, D. & B. Bergvall-Kåreborn 2019. "A typology of crowd-work platforms". *Work, Employment and Society* 33(1): 21–38.

Huws, U. 2003. *The Making of a Cybertariat: Virtual Work in a Real World*. New York: Monthly Review Press.

Huws, U., N. Spencer & M. Coates 2019. *The Platformisation of Work in Europe: Highlights from Research in Thirteen European Countries*. Brussels: Foundation for European Progressive Studies.

Hyman, R. 1988. "Flexible specialization: miracle or myth?". In *New Technology and Industrial Relations*, edited by R. Hyman & W. Streeck, 48–60. Oxford: Blackwell.

Johnston, H. 2020. "Labour geographies of the platform economy: understanding collective organizing strategies in the context of digitally mediated work". *International Labour Review* 159(1): 25–45.

Joyce, S. 2020. "Rediscovering the cash nexus, again: subsumption and the labour–capital relation in platform work". *Capital & Class* 44(4): 541–52.

Kahancová, M., T. Meszmann & M. Sedláková 2020. "Precarization via digitalization? Work arrangements in the on-demand platform economy in Hungary and Slovakia". *Frontiers in Sociology* 5(3).

Kelly, J. 1998. *Rethinking Industrial Relations: Mobilisation, Collectivism and Long Waves*. London: Routledge.

Kimeldorf, H. 2013. "Worker replacement costs and unionization origins of the U.S. labor movement". *American Sociological Review* 78(6): 1033–62.

Kretsos, L. 2011. "Union responses to the rise of precarious youth employment in Greece". *Industrial Relations Journal* 42(5): 453–72.

Kumar, A. 2020. *Monopsony Capitalism*. Cambridge: Cambridge University Press.

Lehdonvirta, V. 2018. "Flexibility in the gig economy: managing time on three online piecework platforms". *New Technology, Work & Employment* 33(1): 13–29.

Marrone, M. & V. Finotto 2019. "Challenging Goliath: informal unionism and digital platforms in the food delivery sector. The case of riders union Bologna". *Partecipazione e Conflitto* 12(3): 691–716.

Moore, P., P. Akhtar & M. Upchurch 2018. "Digitalisation of work and resistance". In *Humans and Machines at Work: Monitoring, Surveillance and Automation in Contemporary Capitalism, Dynamics of Virtual Work*, edited by P. Moore, M. Upchurch & X. Whittaker, 17–44. Cham: Springer.

Moore, S. & K. Newsome 2018. "Paying for free delivery: dependent self-employment as a measure of precarity in parcel delivery". *Work, Employment and Society* 32(3): 475–92.

Muszyski, K. *et al.* 2021. "Coping with precarity during COVID-19: a study of platform work in Poland". *International Labour Review* 161(3): 463–85.

Noble, D. 1995. *Progress Without People: New Technology, Unemployment and the Message of Resistance*. Toronto: Between the Lines.

OECD 2019. *Measuring platform mediated workers*. Paris: OECD Publishing.

OECD 2020. *OECD Digital Economy Outlook 2020*. Paris: OECD Publishing.

Piasna, A. 2020. "Counting gigs: how can we measure the scale of online platform work?". Brussels: ETUI.

Pochet, P. 2021. "A battle won, but not yet the war". ETUI News. Brussels: ETUI.

Purcell, C. & P. Brook 2020. "At least I'm my own boss! Explaining consent, coercion and resistance in platform work". *Work, Employment and Society* 36(3): 391–406

Riso, S. 2019. *Mapping the Contours of the Platform Economy*. Dublin: Eurofound.

Shibata, S. 2020. "Gig work and the discourse of autonomy: fictitious freedom in Japan's digital economy". *New Political Economy* 25(4): 535–51.

Shire, K. *et al.* 2009. "Collective bargaining and temporary contracts in call centre employment in Austria, Germany and Spain". *European Journal of Industrial Relations* 15(4): 437–56.

Tassinari, A. & V. Maccarrone 2020. "Riders on the storm: workplace solidarity among gig economy couriers in Italy and the UK". *Work, Employment and Society* 34(1): 35–54.

Vallas, S. & J. Schor 2020. "What do platforms do? Understanding the gig economy". *Annual Review of Sociology* 46(1): 273–94.

Vandaele, K. 2018. "Will trade unions survive in the platform economy?" Working Paper. Brussels: ETUI.

Vandaele, K., A. Piasna & J. Drahokoupil 2019. *Unwilful Ignorance: Attitudes to Trade Unions Among Deliveroo Riders in Belgium*. Brussels: ETUI.

Veen, A., T. Barratt & C. Goods 2020. "Platform-capital's 'app-etite' for control: a labour process analysis of food-delivery work in Australia". *Work, Employment and Society* 34(3): 388–406.

Voskeritsian, H. & A. Kornelakis 2019. "Power, institutional change, and the transformation of Greek employment relations". In *Greek Employment Relations in Crisis: Problems, Challenges and Prospects*, edited by H. Voskeritsian, P. Kapotas & C. Niforou, 14–30. London: Routledge.

Wood, A. *et al.* 2019. "Good gig, bad gig: autonomy and algorithmic control in the global gig economy". *Work, Employment and Society* 33(1): 56–75.

Woodcock, J. 2020. "The algorithmic panopticon at Deliveroo: measurement, precarity, and the illusion of control". *Ephemera: Theory & Politics in Organization* 20(3): 67–95.

Wright, E. 2015. *Understanding Class*. London: Verso.

CHAPTER 22

WHEN MAY THE INTERESTS OF LABOUR AND CAPITAL ALIGN?

Mark Bray and Johanna Macneil

ABSTRACT

This chapter argues that an alignment of interests between capital and labour, demonstrated in voluntary cooperative relationships between enterprise managers and unions, is possible in liberal market economies, although relatively rare. The central question is: under what circumstances do these relationships develop and bear fruit for all participants? The chapter begins by selectively reviewing the literature and identifying challenges to measuring "success" in alignment of interests. Next, three related sets of factors that may support (or thwart) development of cooperation are identified: context in which the parties operate; agency of the parties; and historical path of relationships. Experiences of participants in successful examples of cooperation, drawn from primary data and published accounts, are discussed. These reveal how manifestations of the three factors have come together in particular times and places to support cooperation. They also reveal the significant challenges to sustaining alignment of interests in individual enterprises.

Keywords: Partnership; union interests; challenges for unions

INTRODUCTION

The point of departure for this chapter is the recognition that an alignment in the interests of labour and capital at an enterprise level, leading to the establishment of cooperative relationships and collaborative action by managers and unions, is possible, even though it is rare and often short-lived. Moreover, where and when these unusual examples of labour-management cooperation occur, they have benefited the enterprise, workers and unions through outcomes like financial stability of the enterprise, greater job security, improved job satisfaction and enhanced union influence and membership. We explore this by identifying some examples of alignment from various "liberal market economies" and show how they are largely exceptional, depending on how "success" and/or "alignment of interests" are defined and measured. So, the key question is: under what circumstances does this alignment occur? Our answer to this explanatory question is not a simple one and focuses upon three sets of factors, which must come together at a particular time and in a particular place. First, the context in which the parties operate; second, the agency of the parties (i.e. the particular representatives of labour and capital involved); and third, the historical path by which relations within the enterprise evolve.

In exploring these three sets of factors and the way they come together, we draw on a well-established English-language research literature and primary data from over 140 semi-structured interviews conducted in Australia between 2015 and 2021. Interviewees were union representatives (full-time officials and part-time workplace representatives), managers and tribunal members, all of whom were personally involved in the transition to cooperative relations in at least one of eight case study enterprises (see, e.g. Bray & Macneil 2021a, 2021b, 2021c; Bray, Macneil & Stewart 2017). The value of this approach is that it allows us to develop a broader account but also to illustrate our argument with the views of the parties from the "coalface".

ISLANDS OF "SUCCESS" IN SEAS OF HOSTILITY?

There is a long tradition of case study research in English-speaking countries (such as the UK, US, Ireland, Australia and New Zealand) describing and seeking to explain examples of alignment between labour and capital at enterprise level. These examples, which have been referred to as "islands of success in a hostile context" (Eaton, Cutcher-Gershenfeld & Rubinstein 2016), provide the evidence upon which our initial assertion is based.

The US probably provides the most conspicuous evidence of companies and unions working cooperatively to produce what were called "mutual gains". Kochan and Osterman's (1994) study cites numerous well-known instances from the 1970s and 1980s, including Xerox, Motorola, NUMMI, Saturn, Hewlett-Packard and Federal Express. More recently, the most important and sustained example of a mutual gains enterprise – namely, Kaiser Permanente Health Services and its coalition of unions – has received close attention from researchers (Kochan *et al.* 2008; Kochan *et al.* 2009). The long-term evolution of cooperation between unions and managers at the Ford automobile company has also been documented (Cutcher-Gershenfeld, Brooks & Mulloy 2015). In all of these American case studies, researchers have found that union–management cooperation has produced genuine mutual gains, although the nature and distribution of those gains has varied significantly. And yet, consistent with Eaton, Cutcher-Gershenfeld & Rubinstein's (2016) conclusion, these cases remained isolated in a larger context of adversarialism.

In the UK, examples of collaboration between labour and capital at the enterprise level have come from two literatures. The first involves case studies from an important state agency promoting collaboration, namely the Advisory Conciliation and Arbitration Service (ACAS), which has commissioned case studies of "workplace projects" conducted in the voluntary, public and private sectors (Acas 2021; Cooper 2011). Success in these enterprises in developing greater cooperation has consistently produced medium-term improvements as well as positive longer-term outcomes being evident (Broughton, Pearmain & Cox 2010).

The second UK literature focuses on the development of union–management "partnerships", especially during the period of the Blair-led Labour government between 1997 and 2007 (Johnstone 2016; Stuart & Martínez Lucio 2005). Many of the individual enterprises demonstrated the mutual gains that can accrue from partnership relationships (e.g. Johnstone 2015), although there was controversy among UK commentators over the outcomes of these cases (Dobbins & Dundon 2015; Jenkins 2008; Kelly 2004). Formal partnership agreements were estimated to have numbered 248 by 2007, covering just less than 10 per cent of the workforce (Bacon & Samuel 2009), although whether these corresponded in practice with genuine union–management cooperation varied. Moreover, they seemed to rapidly fall out of fashion under Conservative-led governments that were in office after 2010.

Ireland saw many examples of union–management cooperation during the years between 1987 and 2009, when its employment relations were governed by "a regime of national social partnership" (Roche & Teague 2014: 781). The cases included unions and employers at the Irish airport authority, Aer Rianta (Roche & Geary 2006) and an alumina refinery, Aughinish Alumina (Dobbins & Dundon 2016). A distinguishing feature of the Irish experience, however, was that the adoption of the cooperative model was limited in the private and concentrated in the public sector. There was also a paradox: "while widely credited with being effective, it, [the cooperative model] appeared unappealing to employers and unions in general" (Roche & Teague 2014: 782). By 2016 it was reported that "workplace partnership appears very much a minority practice and exists alongside the much more prevalent tendency for managerial-led and dominated employee involvement and HRM practices" (Dobbins & Dundon 2016: 110).

In Australia, most early examples of labour-capital collaboration have come from government-sponsored schemes. During the 1990s, for example, the "Best Practice Program" funded by the federal government saw union–management cooperation in many enterprises, especially in manufacturing, that produced outcomes advantageous to both enterprises and workers (Macneil, Haworth & Rasmussen 2011; Rimmer *et al.* 1996). State-government schemes in Victoria and Queensland in the first decade of the new millennium had similar successes (Schneider Australia 2006; Thompson & Booth 2008). By 2013, however, the limited incidence and accomplishments of management-union cooperation led some commentators to suggest that "the impetus for partnerships appears to have reached a 'dead end' in policy terms" (Townsend *et al.* 2013: 254). More recently, a novel form of third-party intervention by industrial tribunals – not unlike interventions in the UK by ACAS – saw enterprise managers working together cooperatively with union representatives to achieve a range of benefits for the enterprise, workers and unions (Bray, Macneil & Stewart 2017).

We have researched eight enterprises across a range of sectors and over the period between 2014 and 2021 that have established and, in some cases, sustained cooperation over significant periods of time with good (although not always perfect) outcomes for management, workers and unions. Again, however, the number of enterprises seeking assistance from the Fair Work Commission was limited, let alone the (much smaller) number whose cooperative efforts proved successful.

New Zealand offers a slightly different story, although the overall conclusion is familiar. As Delaney and Haworth (2016a) observe, union–management partnership largely failed in the face of a "bleak environment". This failure, however, came after government support during the 1980s and 1990s, especially in the form of funding until 2011 for pro-cooperation third-party agents (Macneil, Haworth & Rasmussen 2011). These institutional mechanisms produced positive achievements among public sector organizations (such as Air New Zealand and Auckland City Council) and some private sector companies (such as a major dairy company, Fonterra; see Delaney & Haworth 2016b). By the end of the 2010s, however, both major political parties showed "little interest" in cooperation. Moreover, support was "rare" among private employers and "examples of sustained and successful workplace reform involving workplace partnership are not common" (Delaney & Haworth 2016b: 305).

DEFINING AND MEASURING "SUCCESS"

Beyond the obvious conclusion that union–management cooperation is rare across all these countries, our highly selective review traverses contested terrain. In particular, there has been considerable disagreement between researchers over what constitutes "success" in such cases, in at least three ways.

First, there is dispute over who gets what and whether the benefits are equally distributed. Johnstone (2016: 89–90) provides a good summary of this debate in the UK, where he argues that critics of partnership have tended to focus on the distribution of "hard" outcomes (jobs, wages, hours worked, etc.), while supporters have tended to focus on "softer" outcomes (quality of relationships and the perceived fairness of organizational processes and procedures). More broadly, critics (especially those of a radical frame of reference) argue that management inevitably gets more out of cooperation than workers, or some groups of workers benefit more than others, while supporters consider that benefits are more evenly distributed.

Second, measuring the benefits of cooperation is complicated by the wide range of potential issues and interests involved. How, for example, can the benefits of productivity improvement or cost reductions for the employer be compared with saving workers' jobs, improved job satisfaction, changes in the intensity of work or wage increases? Similarly difficult to reconcile are the diversity of interests among different groups of managers (such as line managers versus specialists or site managers versus corporate managers) and groups of workers (potentially segmented by occupational group, age, employment tenure and gender). And what about whether the costs and benefits might accrue to individuals or unions as organizations? All this makes assessments of the "balance of advantage" flowing to labour versus capital much more complex than considered by many analyses.

Third, there is the time dimension. How long does cooperation need to last for it to be called a "success"? We consider it important for cooperation to go beyond short-term transactions to involve cooperative and ongoing relationships. But sometimes union-management cooperation is considered a "failure" if it is not permanent or it does not last for many years. Previously, we have argued that "cooperation doesn't have to be perfect or last forever" (see Bray, Macneil & Stewart 2017: 250–53). In these ways, the longevity of a cooperative relationship is more difficult to define than many suggest and tends to be specific to the circumstances rather than meeting a universal test.

THREE FACTORS NECESSARY FOR ALIGNING THE INTERESTS OF LABOUR AND CAPITAL

Irrespective of how "success" is measured, the cases cited above – and others – represent examples of the interests of labour and capital aligning, even if temporarily. How is this exceptionalism to be explained? We argue there are three types of factors that explain the emergence and sustainability of union–management cooperation. The first is context. These are factors beyond the control of the managers and unions directly involved in cooperation, but highly relevant because the parties are deeply affected by them and must respond to them. Contextual factors most commonly constrain the alignment of interests, especially in "liberal-market" capitalist nations, making cooperation more difficult. But sometimes they act in the opposite direction by assisting cooperation. The second is agency, which focuses on the perceptions of the parties – especially their interpretations of the context in which they operate and how it affects their interests – and the actions they take. These two factors, namely context and agency, however, cannot be considered independently, despite their separate exposition below. Explanation requires analysis of both and especially how they interact, with one influencing the other. This "dialectical" analysis of context and agency has a long tradition in the social sciences generally (Berger & Luckman 1967; Giddens 1995) and, as Hyman (1989: 76) argues, in industrial relations specifically:

> There is ... a complex two-way process in which our goals, ideas and beliefs influence and are influenced by the social structure. To do justice to its complexity, [researchers] must be attuned to this dynamic interaction between structure and consciousness. A static or one-way analysis necessarily distorts social reality, and is therefore an inadequate basis for understanding industrial behaviour or predicting its development.

We have, however, added a third element of explanation: path dependency. This recognizes the complex and contingent nature of the analysis in which process is important. Relations between capital and labour – within nations as well as within enterprises – are dynamic, evolving over time in sometimes unpredictable ways. In particular, cooperation between unions and management – and therefore them acting on an alignment of interests – never happens immediately. It takes time for trust to build and cooperative relationships to develop. Understanding these historical processes helps to understand the development of union–management cooperation.

CONTEXTUAL FACTORS

There are at least three types of contextual factors that affect the alignment of interests between capital and labour: the nature of capitalism itself, what we have called elsewhere "the institutional tide of adversarialism", and other external influences such as the operation of markets within which enterprises operate.

The nature of capitalism

Some commentators, especially those from the critical or radical schools of thought, argue that the very nature of capitalism makes the alignment of labour and capital's interests difficult, and even impossible (e.g. Bacon & Blyton 2006; Danford *et al.* 2008; Dobbins & Dundon 2015; Jenkins 2008; Kelly 2004). Danford and Richardson (2016: 66), for example, conclude: "The underlying argument that frames our critique of partnership is that, in the final analysis, the conflict of class will inevitably frustrate and, in time, invalidate attempts to create sustainable partnerships between employers and labour." There is much to applaud in these accounts. There is no doubt that various features of capitalism represent significant contextual factors that make union–management cooperation difficult to attain and sustain and render its outcomes problematic (especially for workers and unions). Among other things, the pressures of competition in capitalist markets and the imperative towards management control of the labour process are seen to ensure that the interests of capital and labour do not align (Thompson 2003).

Such accounts, however, are sometimes based on unstated assumptions, sometimes on universal logic (i.e. explicitly identified features of capitalism as it operates everywhere) and sometimes on empirical findings of specific instances of union-management cooperation. They suffer from corresponding weaknesses. First, where they rest on unspoken assumptions, they are difficult to rebut: one set of assumptions can only be countered by an alternate set of assumptions, leading to sterile debate. Second, where their analysis of capitalism is more explicit, it tends to treat capitalism as if it were the same everywhere. As the "varieties of capitalism" literature, among others, has demonstrated, this is not necessarily the case. Apart from anything else, institutional formations differ between forms of capitalism as well as nation states and they impact differently on the alignment of interests and perceptions of alignment. Third, empirical assessments of union–management cooperation in specific circumstances are often questionable because they encounter some of the definitional problems discussed above; for example, how can the outcomes of cooperation for various stakeholders be measured and compared, and how long must cooperation continue for it to be assessed as a "success"? Finally, both logical and empirical arguments of this ilk tend to overgeneralize: in particular, they cannot easily explain the instances of successful cooperation, at national or enterprise level, even if they are rare or temporary.

Institutions

Another type of contextual factor that helps to explain the alignment (or otherwise) of interests between labour and capital focuses on institutions. In particular, a wide variety of institutional arrangements are considered by many commentators to impede, if not prevent, an alignment of interests and the development of union–management cooperation. The most common example is the law, especially employment relations law. Kochan and Osterman (1994), for example, lament the unsupportive nature of US labour law and explore potential remedies – themes consistent with the "hostile context" highlighted by Eaton, Cutcher-Gershenfeld and Rubinstein (2016). Roche and Teague (2014) offer a similar assessment of the poor "macro-support" (which is mostly a lack of hard law) for "micro-efforts" at cooperation in Ireland. Mitchell and O'Donnell (2008) explore the extent to which cooperation was embedded in Australian and UK labour law, finding the former to be largely unsupportive. We argue elsewhere (Bray, Stewart & Macneil 2018) that the Fair Work Act 2009 in Australia largely failed to promote cooperation, focusing more on encouraging "adversarial pluralism" (in the form of collective bargaining) rather than "collaborative pluralism".

The effect of specific institutions, however, is deepened because – again, as the varieties of capitalism literature demonstrated, among others – they tend to be "complementary" (Hall & Soskice 2001: 17–18). Put in abstract terms, this concept expresses the idea that "certain institutional forms, when jointly present, reinforce each other and contribute to improving the functioning, coherence or stability of specific institutional configurations, varieties or models of capitalism" (Amable 2016: 79). Applied to the employment relationship, this implies that adversarialism between capital and labour in the labour market and labour process will be reinforced by adversarialism in other spheres – such as skill formation or the financial system – thereby making cooperation more difficult.

We refer elsewhere (see Bray, Macneil & Stewart 2017: 30) to these overlapping and reinforcing contextual factors as producing an "institutional tide" that drags unions and managers away from cooperation and towards adversarial relationships:

> The pressures created by these institutions make it difficult for individuals and organisations that seek to create and sustain more cooperative relationships at work. They must "swim against the tide" of adversarialism. Swimmers can make gains, especially if they are motivated, skilled, strong and well directed. They may even reach their destination. However, the tide can easily drag back towards adversarialism those who falter, whose motivation weakens, whose skills become diluted or whose strength is drained by encounters with new or unexpected challenges.

The impact of the institutional tide towards adversarialism can be seen most vividly in two circumstances where "outsiders" to a cooperative relationship work against it because they are unfamiliar with or unsympathetic towards what is going on "inside". Frequently, for example, parties within an organization who have adopted cooperative strategies (often at some risk to the individuals' reputations and careers) experience considerable pressure from their superiors or colleagues outside the organization to revert back to more traditional adversarial attitudes and behaviours. This pressure bears on managers, for example, from company boards or, where the organization is part of a larger enterprise, from more senior managers, all of whom are unfamiliar with or actively hostile towards unions. A survey of board members of Australian companies, for example, found that over 70 per cent considered their company to be in partnership with its employees, but this was mostly conceived in unitarist rather than pluralist terms (Jones & Marshall 2008: 212).

Similarly, union officials and shop stewards often find themselves at odds with union committees of management or more senior officials, whose experience has been only of the adversarial approach and who fail to appreciate the benefits of cooperation. Australian union officials who were key insiders to the longstanding cooperation at Opal Fibre Packaging, for example, observed how they received:

> ... little active support from their colleagues elsewhere in the union ... This was evident in two ways. First, there are very few open conversations within the union about collaboration. Instead, the Opal officials were treated with suspicion. In particular, they felt that "the talk behind our backs is that we're doing deals and just bullshit because they don't understand it ...". Second, when other union officials occasionally experimented with collaborative methods, they often failed for reasons that could have been avoided if more open discussions had occurred amongst officials about the collaborative experience at Opal. In both cases, these external forces by no means prevented cooperation at Opal, but they did make the sustainability of cooperation more difficult. Continued success required commitment and persistence on behalf of its supporters. (Bray & Macneil 2021c: 20)

The other example is the challenge to the sustainability of cooperation when key individuals leave. As observed in many studies (e.g. Kochan *et al.* 2008: 41; Rimmer *et al.* 1996; Roche & Geary 2006: 240–41), individuals who replace these long-term participants in cooperative projects frequently do not understand the cooperative history of the organization and fall back on adversarial views acquired outside the organization that make continued cooperation difficult. As a union official commented on new site managers being appointed at Opal Fibre Packaging:

> [It] falls down at a site level with the changing of the guard ... because they don't know how to do it [i.e. collaboration] ... You get a lot of new managers that are coming in ... going, "Why the hell are we doing this? At [a competitor], we just do ABC and we get it done." And it's true. [They] have gotten away with doing things in an unscrupulous way and have effectively weakened the union movement within the organisation and to the detriment of ... workers. So, I see why managers come across and say that ... "Well, why are we doing this? It's ridiculous. Why do we need their permission? Why do we need to have the conversation? It's management prerogative." So, it's [i.e. sustaining cooperation] unlearning behaviours that are still going on today.
> (Bray & Macneil 2021c: 16)

So, again, these institutional arguments contribute to a strong case that the alignment of interests – and cooperative relationships – is difficult. But it is not impossible. Indeed, these arguments help identify some of the social pressures that must be overcome if cooperation is to be successful.

Successful cooperation, thereby, reveals how institutional arguments over-generalize in at least three ways. First, not all institutions work against cooperation: some are explicitly designed to encourage cooperation and they can even succeed in producing union–management cooperation in practice. Heery and Noon (2008) argue that some institutions – quintessentially in Germany – served as "beneficial constraints". This argument was summarized in Heery & Noon (2008) as:

> ... institutions that constrain employer behaviour, and which may initially be resisted as a consequence, but which generate beneficial consequences for business in the longer term. Streeck's initial example was the system of codetermination in Germany, which inhibits employer freedom to restructure businesses and requires them to share information and consult with worker representatives. The effect of this restriction, Streeck argues, has been to encourage the development of a long-term cooperative relationship between German employers and their employees, which in turn has provided the basis for competition on the basis of quality enhancement and high value-added and allowed German manufacturing to thrive in world markets.

In countries such as the UK and Australia, some institutional arrangements are designed to also be "beneficial constraints" by promoting cooperation between capital and labour, although they tend to rely on "soft regulation" rather than "hard law" (e.g. Macneil, Haworth & Rasmussen 2011). Good examples would be policy packages, such as the partnership policies of the Blair government in the UK (Johnstone 2016) and the productivity-promoting policies of Australian and New Zealand governments (Macneil, Haworth & Rasmussen 2011). There is also the role of third parties, such as ACAS in the UK and the Fair Work Commission in Australia, which are obliged by law to work with employers and unions to encourage cooperation (see Bray, Macneil & Stewart 2017, ch. 5). Whether these institutional arrangements meet their objectives is another matter, but ignoring them unnecessarily simplifies the analysis of context in these countries and makes more difficult explanations of successful cooperative ventures. At the very least, the interplay between competing, even contradictory, contextual influences needs to be accommodated.

A second weakness of institutional explanations is that they can be rigid, in that they neglect that institutions can change over time. There is, for example, no doubt that periods of social democratic government in countries such as the UK, Australia and New Zealand have seen laws and policies more favourable to labour and more sympathetic to cooperation between capital and labour (Macneil, Haworth & Rasmussen 2011). The period of The Accord in Australia provides a good example (see Bray, Macneil & Stewart 2017: 33–5). Institutional supports for cooperation can also operate within specific regions within nations, as occurred in the Hunter region of Australia (Bray, Macneil & Stewart 2017, ch. 6) or in Scotland and Wales – but not England (Gall 2021) – with these devolved governments' promotion of "Fair Work". During these periods and in these places, labour-capital cooperation is more likely to succeed. Failing to recognize such explanatory factors – even if government policies or laws prove temporary or region-specific – reduces the capacity to understand examples of cooperation where they occur.

Third, institutional arguments run the risk of partial analysis because they fail to take into account agency – the perceptions and actions of the parties. As the next section will show, agency must combine with context to produce proper explanation.

Specific environmental factors

A final set of contextual factors affecting the alignment of interests between capital and labour is more prosaic. Rather than being systemic, these factors are specific to particular circumstances of single enterprises or to multiple enterprises in the same industry. Particularly important in many accounts of cooperation are external threats, which can create an imperative for internal change within the enterprise; these include crises in the product market, political demands (especially for public sector enterprises) and changes in ownership of the enterprise (e.g. Bray, Macneil & Stewart 2017; Cutcher-Gershenfeld, Brooks & Mulloy 2015: 7–8; Kochan *et al.* 2008; Purcell 1981). They may even be driven by public health events, such as Covid-19. Kessler and Purcell (1994: 3–4) emphasize the crisis nature of these external pressures:

> [It] was only in the extreme circumstances of organisational crisis or trauma that management and unions, previously locked in low trust relations, were likely to be thrown together to deal with issues on a more integrative basis, often as the only means of ensuring survival.

Despite the potency of these more specific contextual imperatives, however, they give little clue as to why the organizational responses by employers (and unions) court cooperation. Similar pressures at other organizations – indeed, arguably most other enterprises – produced adversarial responses. Why? In summary, contextual factors provide powerful – but incomplete – explanations for the lack of union–management cooperation. They are not, and cannot be, determinative of cooperation because they do not explain the character or the impact of the organizational response. In some situations, the parties see them as actually serving to align their respective interests, but in others they intensify conflict between the parties.

AGENCY OF THE PARTIES AT ENTERPRISE LEVEL

The agency of the parties represents another set of factors helping to explain the circumstances under which the interest of capital and labour may align. This encompasses several elements, including the perceptions of the parties, values or frames of reference, the actions the parties take, the way the parties interact together and the role of third parties.

Perceptions

To understand the (exceptional) examples of labour-management cooperation, it is important to recognize the parties' subjective perceptions of their interests and how those perceptions lead to action. Focusing solely on objective interests imposed on the situation by researchers will never effectively explain these situations because individuals, groups and organizations act on the basis of their own perceptions, not the interpretations of external observers and commentators. This is also not just a matter of good social science, but especially important in understanding instances of union-management cooperation because it is necessarily "voluntary"; the parties must want to engage in cooperation.

By way of example, key concepts in many accounts of successful union–management cooperation are "legitimacy" and "trust" – not only must managers and unions see each other as sufficiently legitimate and trustworthy to embark on a cooperative response to external pressures (Kochan *et al.* 2008: 41), but these main parties must also see third parties (such as tribunals or consultants) as legitimate and trustworthy before they accept their overtures and work with them (Bray, Macneil & Stewart 2017: 78–81, 222–5). Subjective perceptions are inherent in legitimacy and trust. A good example of subjective responses that are frequently neglected in studies of union–management cooperation is the relentless negative feelings and even exhaustion generated by adversarialism – feelings that often leave individuals and groups receptive to the prospect of greater workplace cooperation and almost joyous when cooperation is delivered. An Australian union official responsible for Aruma, a large Australian disability service provider, for example, observed:

> There's a limit to how negative you can be in a relentless way. People want to actually enjoy their work and be happy at work. From a union perspective too, it's exhausting for us just to have a relentlessly defensive, negative attack message. It wears people down. The company, they wear themselves down as well and just feel constantly on guard and everything's done in secret and everything's imposed on people and there's always a fallout cost to that transaction (Bray & Macneil 2021a: 13)

A common response to questions from researchers about the outcomes of union-management cooperation in Australia was a perception that they produced better work relationships and greater job satisfaction among managers and union representatives. At Sydney Water, for example, a manager observed: "It's a much better work environment when managers and staff feel that they can discuss things ... it's a much more pleasant, satisfying job" (Bray, Macneil & Stewart 2017: 188). A senior union official was more expansive: "The real success at Sydney Water is how the culture of work has changed away from a culture where people hated working there and they hated their managers ... [Now], people actually like going to work" (Bray, Macneil & Stewart 2017: 188–9).

Values

The agency exercised by the parties is affected by their values. Since Fox's (1969, 1973) work on management ideologies, others have further interpreted and developed the concept of "frames of reference" to explain approaches to the employment relationship (Budd & Bhave 2008; Heery 2016; Johnstone & Wilkinson 2018). Our previous work on the "many meanings of cooperation" in the employment relationship (Bray, Budd & Macneil 2020), for example, provides a useful typology to explore values and ideologies that are most likely to reject, accept or promote labour-capital cooperation. There are instances, for example, when the values of managers lead them to accept and even encourage a role of unions. The CEO of Aruma revealed fulsomely pluralist values when he acknowledged an advantage of the partnership with unions in his organization:

> ... there is huge value for us [i.e. management] in hearing the voice of our workforce through the Australian Services Union and its delegate system, as a really important complement to hearing the voice of the workforce through the more traditional employer things, like surveys and workshops ... People tend to tell their union delegates more than they might tell their boss, for whatever reason, and so we see it as really valuable and constructive feedback. (Bray & Macneil 2021a: 14)

At the same time, however, values are far from a determining factor. Indeed, union representatives and managers holding values that seem to reject the notion of alignment have often participated in cooperation, especially in the short term. Managers holding unitarist values and radical union representatives, for example, may set aside their values and focus pragmatically upon cooperation as a compromise approach that delivers outcomes for their constituents in specific circumstances. The union secretary central to cooperation at Sydney Water, for example, who previously adopted highly adversarial strategies towards management, was subjected to significant criticism from some union members for supporting cooperation. But, as one union delegate observed, this was a pragmatic position driven by a concern to protect jobs:

> She stood up at one meeting and said to this one guy, "You know it's either you want a job or you don't want a job, so unless we take these changes on board and we try and make them work, you're not going to have a job. That's the bottom line." She copped a fair bit of personal flak, but she tells it as it is. (cited in Bray, Macneil & Stewart 2017: 244)

Another Australian union official, similarly well-versed in traditional adversarialism, expressed very similar pragmatic sentiments about how the union should respond to the situation at Orora Fibre Packaging:

> So, you can either ... sit at the back of the bus and watch them [i.e. management] drive it off the cliff. Or you can get up the front and start driving the bus with them. And we might actually turn the corner instead of going off the cliff. (cited in Bray, Macneil & Stewart 2017: 244)

The longer-term sustainability of labour-capital cooperation, however, is most likely where union representatives and managers hold complementary values that are supportive of cooperation (Budd, Pohler & Huang 2022).

Actions of the parties

The successful alignment of interests must be to some degree manufactured. Actions aimed at achieving this goal comprise diverse initiatives that range from the personal leadership exercised by key individuals to the introduction and embedding of institutional mechanisms, such as works councils, collective bargaining and consultation procedures. Both, however, must recognize the roles of capital and labour. Purcell (1981: 55) recognized this need for "reciprocity" long ago: "[I]n the cooperative pattern each party seeks to preserve the other, giving support in the other's intra-organisational bargaining, engaging in actions and behaviour which help form attitudes in a positive and supportive manner in the other's organisation as well as one's own." In other words, genuinely pluralist collaboration between labour and capital requires "jointness": joint leadership and the adoption of institutional mechanisms that foster joint problem-solving and decision-making (see, e.g. Bray, Macneil & Stewart 2017; Purcell 1981; Roche & Geary 2006). The attainment of aligned interests, but especially the sustainability of these arrangements, is most likely when these agency factors go beyond the role of individual leaders and the embedded mechanisms meld to form an integrated and robust system of joint governance.

The issues being addressed

The preparedness of the parties at enterprise/workplace level to perceive an alignment of interests and act cooperatively is affected by the issues at stake, especially if those issues are considered to be of "common interest". As discussed above, extreme crises like those that threaten the survival of an enterprise or a specific work site are most likely to break down historic adversarialism and promote cooperation. Certainly, cooperation between the union and management at Orora Fibre Packaging was generated by the threat of enterprise closure and was subsequently credited with avoiding this outcome (Bray & Macneil 2021c).

Edwards, Bélanger and Wright (2006) developed a distinction between "control" and "developmental" concerns. They argue that the former control issues (such as the wage-effort bargain, working conditions and the limits of managerial power) were more likely to be subject to contestation, while the latter developmental issues (such as efficiency improvements, skill development, process improvement and organizational innovation) had more potential for cooperation between workplace parties. Research on training and organizational learning often supports these conclusions. Stuart and Wallis (2007), for example, found that cooperation on "learning" was common among European countries. These authors, however, also argue that the alignment of interests on training is more complex: successful cooperation depends partially on the broader approach taken towards restructuring and on support from institutional arrangements.

A similar argument was found with respect to another issue on which cooperation might be expected to be flourish; namely, workplace health and safety. In a study of Australian coal mines, Walters *et al.* (2016) found that cooperative strategies by unions towards this issue foundered if managers were unsympathetic and sought to exclude unions from decision-making of these issues.

Relationships

The parties' perceptions and their actions are affected by the relationships between them: if, for example, there is a long history of adversarial relationships between the parties, then they are unlikely to trust each other sufficiently to enter into cooperative arrangements. An important role of third parties, therefore, is to break down historical antagonisms, recognize common interests and create opportunities for cooperation. A senior human resources manager at an Australian disability service provider, for example, reflected on the value of early discussions facilitated by a third party:

> All those conversations and those workshops we had with [the third party] were the key moments for me ... It was the "aha" moments for managers, and trying to reset the culture and the mindsets of some of those managers and representatives of the union. ... It meant that we could be more open and honest, upfront, in discussions with the union, whereas in the past it was keep your cards way close to your chest and only expose them when the proverbial hit the fan. (Bray & Macneil 2021a: 8)

Interestingly, a history of adversarial relationships is more likely when unions within an enterprise are strong. In these situations, especially when the parties are confronted with external pressures that make "business as usual" impossible, "taking on" the unions in adversarial action is seen by managers as unlikely to succeed. Instead, they turn to cooperation. In this way, Roche and Teague (2014) saw Irish managers turning to cooperation when "the alternative strategy of marginalizing unions was simply not viable because they were well organised and difficult to displace". Similarly, Kochan *et al.* (2008: 38) talk about "inter-dependence" being an essential condition for cooperation:

> ... neither party has the option of escaping from the relationship rather than dealing with its pressures. In practice, this means that the union must be sufficiently powerful to make it costly or impossible for the employer to achieve its objectives by abandoning or terminating the union–management relationship or working around the union in some other way.

In contrast, and somewhat counter-intuitively, cooperation is less likely when unions are weak. Unions without strong workplace organization, for example, are generally unable to deliver their side of a bargain to employers. They are, therefore, less attractive as a partner to cooperation even if weakness can be seen to compel a willingness to cooperate. Managers are also more able to assert their own – often unitarist – solutions to problems rather than being forced by the strength of the union – to pursue cooperative arrangements.

Third parties

A much-neglected aspect of union–management cooperation is the agency of third parties in bringing the two sides together, in recognizing the need for change, identifying common interests, overcoming historical hostilities and then facilitating ongoing relationships. Our recent study in Australia explored in great detail the actions of industrial tribunals in performing this mediating role (see Bray, Macneil & Stewart 2017), but third parties are mentioned in many earlier studies without receiving much direct attention. Kochan *et al.* (2008: 57), for example, acknowledge the role of private consultants in facilitating cooperation at Kaiser Permanente; Kessler and Purcell (1994) explore the impact of ACAS acting as a third party; Roche and Geary (2006: 228–9) mention the role of third-party facilitators in supporting cooperation at the Irish Airport's Authority (Aer Rianta) and the impact of their withdrawal on declining cooperation; and, more broadly, government-sponsored programmes have supported the provision of third parties who been important to the promotion of cooperation across many countries (see, e.g. Macneil, Haworth & Rasmussen 2011).

PATH DEPENDENCY

The final set of factors affecting the alignment of interests between labour and capital involves the progress of time and the sequential unfolding of events in the enterprise. The process by which contextual and agency factors interact – the "dialectic", some might say – sometimes unfolds in a predictable pattern, but it is also sometimes unpredictable and even the result of chance. This reflects the precariousness of cooperation, the complexity of matters being juggled by the parties and the contingent nature of success. Any element can fail at any time, making vital an understanding of the historical process.

The contingent nature of the many factors that must come together to produce an alignment of interests and effective cooperation means that the solution is partly a matter of bringing together the "right people" into the "right place" at the "right time". But this rarely happens immediately. Also, the "trust" between the parties that is necessary for cooperation to succeed takes time to build, especially in a context of widespread adversarialism. Getting the parties to begin "swimming" against the tide of adversarialism is one part of the challenge, as one manager at an Australian energy generator said:

> Where you've got that sort of [adversarial] industrial history that's been established, it's very hard ... for the first party to extend the hand ... It takes somebody to break the nexus with the past, and say: "Okay, well I'm going to stick my neck out here and we're going to do this one a little bit differently and just see how it goes." (Bray, Macneil & Stewart 2017: 159)

Once started, the progress of cooperation requires the parties to maintain motivation, effectively frame the issues on which they collaborate, build collaborative skills that were never part of their adversarial history, share information, develop and implement joint decision-making structures and define the roles of individuals and collective parties within them (Bray, Macneil & Stewart 2017). All this takes time and each step can easily fail.

The importance of path dependency has been highlighted in different ways in other studies. A well-used framework for understanding the process of cooperation building is the change model of "unfreezing-change-refreezing". It traces the process from dissatisfaction with existing employment relationships and recognizing the need for change to negotiating the change and then embedding the changes in ways that ensure the ongoing improvements continue (e.g. Cutcher-Gershenfeld, Brooks & Mulloy 2015, ch. 1; Kochan *et al.* 2008: 38).

Kochan *et al.* (2008: 58–61) also recognized the importance of path dependency when they highlighted the reliance of sustainable cooperation at Kaiser Permanente upon the achievement of "concrete results" and the impact of "pivotal events". With respect to the former, managers will not enter a cooperative arrangement if they do not perceive it as likely to produce outcomes that are valuable for the organization, while union leaders will not support such arrangements if they do not hope or expect them to provide rewards for their members and the union as an organization. According to Kochan *et al.* (2008: 39–40), these concrete results must go beyond improvements in "interpersonal and interorganizational relationships" to be "substantive gains" that are "clear, tangible, and of high priority to each of the key interests involved".

The need for concrete results can also be seen in the impact of early "wins" that support nascent cooperative initiatives and spawn further experimentation (see Bray, Macneil & Stewart 2017). Concrete results, however, continue to be important to sustaining cooperation: gains for both labour and capital must be reproduced over time in order to reinforce the alignment of interests. If these rewards are not forthcoming, then cooperation is unlikely to last. Moreover, the continuing delivery of rewards is far from guaranteed, and failure at any stage can lead to the dissolution of cooperation. Tracing the often uncertain process of results and responses to those results is vital to understanding the (relatively unusual) success of some cooperative projects.

The notion of "pivotal events" also demonstrates the role of path dependence. Drawing on earlier work by Cutcher-Gershenfeld, Kochan *et al.* (2008: 41) see pivotal events as "crisis situations or situations where a problem emerges that is not addressed successfully will threaten the continuity of the partnership", although if successfully resolved, then "the experience [of these events] tends to strengthen the commitment of the parties to the partnership and often expands its scope". Examples of such events include "a change in business conditions that threatens a layoff, the turnover of key partnership champions, a decision to outsource work, or the emergence of conflict in another part of the labour management relationship that then holds the partnership effort hostage" (Kochan *et al.* 2008: 41). Again, tracing the progress of pivotal events – and these events are features of all partnerships, even if not always discussed in these terms (see, e.g. Cutcher-Gershenfeld, Brooks & Mulloy 2015; Roche & Geary 2006: 17–18) – helps to understand the nature of the partnership.

As vital as they are, however, these "bigger" pivotal moments underestimate the "smaller", everyday moments that constantly arise if union–management cooperation is to be established and sustained. A lower-level manager at Orora Fibre Packaging captured this well when he explained that the new cooperative way of working was often difficult but ultimately better than the old adversarialism:

> It's not like [the senior leaders] had a conversation and the world changed. It's not like that at all. It's not even close to like that. Every day is difficult, every single day is difficult and it's difficult for a thousand reasons. But so long as everybody is working together through those difficulties and [we're] discussing issues and trying to do what we can – that is a significant change. (Bray & Macneil 2015: 7)

CONCLUSIONS

The alignment of the interests of labour and capital is undoubtedly rare in capitalist societies, especially in liberal market economies such as those that have been the subject of this chapter and in the enterprises operating within these economies. And yet alignment still happens. Moreover, this alignment and the cooperative relationships that emerge when alignment occurs are always fragile and often temporary. And yet alignment still happens. This chapter has focused on why – specifically, the circumstances under which – union–management cooperation occurs. Bringing together three sets of factors explaining the alignment of interests – namely, context, agency and historical process – we are able to offer an analysis that sees cooperation as complex and highly contingent but, ultimately, achievable.

One powerful force – what we call the "institutional tide" – that must be overcome for cooperation to be achieved is contextual. The nature of capitalism itself along with the (usually complementary) institutional structures around it push the parties away from cooperation and towards adversarialism. The institutional tide is relentless. And yet it is not omnipresent. There are qualifications and exceptions, even if they are specific to moments of time or place. These alternative contextual factors can be important. They can help to motivate the parties within the enterprise to seek cooperation and provide opportunities to succeed in the cooperative project. But they do not – and cannot – determine whether cooperation will be attempted or whether it will succeed. The key is the agency of the parties and whether they choose to "swim against the institutional tide".

Whether they embark on such a venture and whether it will be successful depends on many factors: how they perceive their circumstances, including the relationships between the parties; their values; the nature and effectiveness of the actions they take; and the support their receive from third parties. Moreover, the contingent nature of these many factors involved – both contextual and agency – and how they come together unfolds in ways that are invariably unique to the situation of each venture. This requires attention to path dependency – the historical process by which managers and unions in each enterprise pursue cooperation – is central to an understanding of how and why "success" is achieved.

REFERENCES

ACAS 2021. Official website. www.acas.org.uk/

Amable, B. 2016. "Institutional complementarities in the dynamic comparative analysis of capitalism". *Journal of Institutional Economics* 12(1): 79–103.

Bacon, N. & P. Blyton 2006. "Union cooperation in a context of job insecurity: negotiated outcomes from teamworking". *British Journal of Industrial Relations* 44(2): 215–38.

Bacon, N. & P. Samuel 2009. "Partnership agreement adoption and survival in the British private and public sectors". *Work, Employment and Society* 23(2): 231–48.

Berger, P. & T. Luckman 1967. *The Social Construction of Reality*. New York: Anchor.

Bray, M., J. Budd & J. Macneil 2020. "The many meanings of cooperation in the employment relationship and their implications". *British Journal of Industrial Relations* 58(1): 114–41.

Bray, M. & J. Macneil 2015. *Facilitating Productive Workplace Cooperation: A Case Study of Sydney Water and the ASU Water Division*. Melbourne: Fair Work Commission.

Bray, M. & J. Macneil 2021a. *The Development and Benefits of a Collaborative Model: A Case Study of Aruma and the ASU*. Melbourne: Fair Work Commission.

Bray, M. & J. Macneil 2021b. *Sustaining a Collaborative Model: A Case Study of Aruma and the ASU*. Melbourne: Fair Work Commission.

Bray, M. & J. Macneil 2021c. *Sustaining a Collaborative Model: A Case Study of Opal Fibre Packaging and the AMWU Print Division*. Melbourne: Fair Work Commission.

Bray, M., J. Macneil & A. Stewart 2017. *Cooperation at Work: How Tribunals Can Help to Transform Workplaces*. Sydney: Federation Press.

Bray, M., Stewart, A. & Macneil, J. 2018. "Bargaining, cooperation and 'new approaches' under the Fair Work Act". In *Collective Bargaining Under the Fair Work Act: Evaluating the Australian Experiment in Enterprise Bargaining*, edited by S. McCrystal, B. Creighton & A. Forsyth. Sydney: Federation Press.

Broughton, A., D. Pearmain & A. Cox 2010. *An Integrated Evaluation of ACAS Workplace Projects*. London: ACAS.

Budd, J. & D. Bhave 2008. "Values, ideologies, and frames of reference in industrial relations". In *The Sage Handbook of Industrial Relations*, edited by P. Blyton *et al.*, 92–112. Los Angeles, CA: Sage.

Budd, J., D. Pohler & W. Huang 2022. "Making sense of mismatched frames of reference: a dynamic cognitive theory of instability in HR practices". *Industrial Relations* 61(3): 268–89.

Cooper, J. 2011. *Looking Back to Move Forward: Assessing the Impacts of ACAS Advisory Projects*. London: ACAS.

Cutcher-Gershenfeld, J., D. Brooks & M. Mulloy 2015. *Inside the Ford-UAW Transformation: Pivotal Events in Valuing Work and Delivering Results*. Cambridge, MA: MIT Press.

Danford A. *et al.* 2008. "Partnership, high performance work systems and quality of working life". *New Technology, Work and Employment* 23(3): 151–66.

Danford, A. & M. Richardson 2016. "Why partnership cannot work and why militant alternatives can: historical and contemporary evidence". In *Developing Positive Employment Relations*, edited by S. Johnstone & A. Wilkinson, 49–74. London: Palgrave Macmillan.

Delaney, H. & N. Haworth 2016a. "Battling in a bleak environment: the New Zealand context for partnership". In *Developing Positive Employment Relations*, edited by S. Johnstone & A. Wilkinson, 181–206. London: Palgrave Macmillan.

Delaney, H. & N. Haworth 2016b. "Partnership in practice: improving productivity in Fonterra's Whareroa site". In *Developing Positive Employment Relations*, edited by S. Johnstone & A. Wilkinson, 305–26. London: Palgrave Macmillan.

Dobbins, T. & T. Dundon 2015. "The chimera of sustainable labour-management partnership". *British Journal of Management* 28(3): 519–33.

Dobbins, T. & T. Dundon 2016. "Workplace partnership in Ireland: irreconcilable tensions between an 'Irish third way' of voluntary mutuality and neoliberalism". In *Developing Positive Employment Relations*, edited by S. Johnstone & A. Wilkinson, 101–24. London: Palgrave Macmillan.

Eaton, A., J. Cutcher-Gershenfeld & S. Rubinstein 2016. "Labour-management partnerships in the USA: islands of success in a hostile context". In *Developing Positive Employment Relations*, edited by S. Johnstone & A. Wilkinson, 125–54. London: Palgrave Macmillan.

Edwards. P., J. Bélanger & M. Wright 2006. "The bases of compromise in the workplace: a theoretical framework". *British Journal of Industrial Relations* 44(1): 125–45.

Fox, A. 1969. "Management's frame of reference". In *Collective Bargaining: Selected Readings*, edited by A. Flanders, 390–409. Harmondsworth: Penguin.

Fox, A. 1973. "Industrial relations: a social critique of pluralist ideology". In *Man and Organization: The Search for Explanation and Social Relevance*, edited by J. Child, 185–233. London: Allen & Unwin.

Gall, G. 2021. "'Fair work' in Scotland – a critical assessment". Jimmy Reid Foundation, Glasgow.

Giddens, A. 1995. *A Contemporary Critique of Historical Materialism*. Stanford, CA: Stanford University Press.

Hall, P. & D. Soskice 2001. "An introduction to varieties of capitalism". In *Varieties of Capitalism: The Institutional Foundations of Comparative Advantage*, edited by P. Hall & D. Soskice, 1–68. Oxford: Oxford University Press.

Heery, E. 2016. *Framing Work: Unitary, Pluralist and Critical Perspectives in the 21st Century*. Oxford: Oxford University Press.

Heery, E. & M. Noon 2008. *A Dictionary of Human Resource Management*. Third edn. Oxford: Oxford University Press. www.oxfordreference.com/view/10.1093/oi/authority.20110803095458551

Hyman, R. 1989. *Strikes*. Basingstoke: Macmillan.

Jenkins, J. 2008. "Pressurised partnership: a case of perishable compromise in contested terrain". *New Technology, Work and Employment* 23(3): 167–80.

Johnstone, S. 2015. "The case for partnership". In *Finding a Voice at Work? New Perspectives on Employment Relations*, edited by S. Johnstone & P. Ackers, 153–74. Oxford: Oxford University Press.

Johnstone, S. 2016. "Participation and partnership in the UK: progress and prospects". In *Developing Positive Employment Relations*, edited by S. Johnstone & A. Wilkinson, 77–100. London: Palgrave Macmillan.

Johnstone, S. & A. Wilkinson 2018. "The potential of labour–management partnership: a longitudinal case analysis". *British Journal of Management* 29(3): 554–70.

Jones, M. & S. Marshall 2008. "What directors think about partnerships between companies and their employees?". In *Varieties of Capitalism, Corporate Governance and Employees*, edited by S. Marshall, R. Mitchell & I. Ramsay, 188–220. Melbourne: Melbourne University Press.

Kelly, J. 2004. "Social partnership agreements in Britain". *Industrial Relations* 43(1): 267–92.

Kessler, I. & J. Purcell 1994. "Joint problem solving and the role of third parties: an evaluation of ACAS advisory work". *Human Resource Management Journal* 42(1): 1–21.

Kochan, T. & P. Osterman 1994. *The Mutual Gains Enterprise: Forging a Winning Partnership Among Labor, Management and Government*. Boston, MA: Harvard Business School Press.

Kochan, T. *et al.* 2008. "The potential and precariousness of partnership: the case of the Kaiser Permanente labor management partnership". *Industrial Relations* 47(1): 36–65.

Kochan, T. *et al.* 2009. *Healing Together: The Labor-Management Partnership at Kaiser Permanente*. Cambridge, MA: MIT Press.

Macneil, J., N. Haworth & E. Rasmussen 2011. "Addressing the productivity challenge? Government-sponsored partnership programs in Australia and New Zealand". *International Journal of Human Resource Management* 22(18): 3813–29.

Mitchell, R. & A. O'Donnell 2008. "What is labour law doing about 'Partnership at Work'? British and Australian developments compared". In *Varieties of Capitalism, Corporate Governance and Employees*, edited by S. Marshall, R. Mitchell & I. Ramsay, 95–129. Melbourne: Melbourne University Press.

Purcell, J. 1981. *Good Industrial Relations: Theory and Practice*. London: Macmillan.

Rimmer, M. *et al.* 1996. *Reinventing Competitiveness: Achieving Best Practice in Australia*. Melbourne: Pitman.

Roche, W. & J. Geary 2006. *Partnership at Work: The Quest for Radical Organizational Change*. London: Routledge.

Roche, W. & P. Teague 2014. "Successful but unappealing: fifteen years of workplace partnership in Ireland". *International Journal of Human Resource Management* 25(6): 781–94.

Schneider Australia Consulting 2006. *Evaluation of Partners at Work and Better Work and Family Balance Programs*. Melbourne: Industrial Relations Victoria.

Stuart, M. & M. Martínez Lucio (eds) 2005. *Partnership and Modernisation in Employment Relations*. London: Routledge.

Stuart, M. & E. Wallis 2007. "Partnership approaches to learning: a seven-country study". *European Journal of Industrial Relations* 13(3): 301–21.

Thompson, C. & A. Booth 2008. *Smart Workplaces: Pilot Projects Investigating the Scope for Co-Operative Workplace Relationships in Pursuit of Productive Outcomes*. Brisbane: Department of Employment and Industrial Relations, Queensland.

Thompson, P. 2003. "Disconnected capitalism: or why employers can't keep their side of the bargain". *Work, Employment and Society* 17(2): 359–78.

Townsend, K. *et al.* 2013. "Has Australia's road to workplace partnership reached a dead end?". *International Journal of Comparative Labour Law and Industrial Relations* 29(2): 239–56.

Walters, D. *et al.* 2016. "Cooperation or resistance? Representing workers' health and safety in a hazardous industry". *Industrial Relations Journal* 47(4): 379–95.

CONCLUSION

Gregor Gall

There is a relatively well-known saying on the left in the Anglophone world: "The cause of labour is the hope of the world." "Labour" here is not any Labour (political) party but, rather, the largest existing social group of human beings, namely workers whose labour is held to create the wealth of the world but that never receives anything approximating to its "fair" share of that wealth. If labour can achieve economic equality and social justice for itself, then it would achieve economic equality and social justice for the vast majority of human beings.

In 1894, English artist and book illustrator and sometime collaborator with William Morris, Walter Crane, created the illustration of the "The Workers' Maypole" for the socialist newspaper, *Justice*, from which the dictum is derived. Among the demands adorning the maypole are "socialization", "solidarity" and "humanity". These are expressed in the aforementioned saying along with another one on the maypole being "the hope of labour is the welfare of all". Only a few years earlier in 1875, Karl Marx (1950:17–18) in the *Critique of the Gotha Programme* had boldly stated the notion that workers create the wealth of the world and that exploitation as theft of the fruits of labour creates a separate social class: "Since labour is the source of all wealth, no one in society can appropriate wealth except as the product of labour. Therefore, if he himself does not work, he lives by the labour of others ..."

If both assertions are true, and there is much evidence and inclination to support them, some of which can be found in the preceding pages, then it is politically and practically correct that most of the analysis and assessment made in this handbook has concerned itself with not just diagnosis but also prognosis. This is because, in general and notwithstanding certain places in time and space including Russia in October 1917, not only has labour as an organized force through labour unions (and any associated political parties) been historically weaker than capital but in the new millennium labour is weaker than it has been for many decades.

Understanding the weakness is, of course, the first necessary but in itself insufficient step to be taking in seeking to right the wrong. That weakness is derived not just from what capital – and the compliant state – do and do not do to further their interests but also from how organized labour reacts and responds. The terrain on which labour acts, while quintessentially the same at the subterranean level, is constantly subject to change at the surface level. And not only has there always been an uphill (external) battle by organized labour to respond adequately and effectively to capital and the state, but there has also been an extensive internal battle within organized labour to arrive at the point by which an adequate and effective response can be made. In this situation, and rather than trying to draw together the many and varied threads that run through the preceding chapters, the remainder of this short conclusion will focus upon the primary and contextual organizational aspects of labour unionism.

Thus, labour unions are far more complex organizations than their comparators, namely units of capital. This brings significant challenges for labour unions. By comparison, labour unions are voluntary and democratic collective organizations that are under-resourced and, even taking into account their bureaucratic tendencies, without "command and control" structures – underpinned by contracts in law – by which their "generals" can order their "foot soldiers" into battle. To be active members, union members must undertake a "double shift" – be in work, working, and then additionally be active in their own non-work time – whereas units of capital employ organizers called managers solely to manage and do so on their behalf. And although unions employ paid officers, and facility time does exist, as a foundation stone of labour unionism, this "double shift" is far more critical than the "great doubling" (Freeman 2007).

So too is that this organizing to create and exercise self-agency takes place upon another foundation stone that labour unions did not themselves create. Not only are labour unions "secondary" and "intermediary" organizations (Müller-Jentsch 1985; Offe & Wiesenthal 1980), they are also heavily coloured by being reactive organizations where not just power but the power of initiative ordinarily lies with units of capital. In an age of management's heightened psychological resilience and resistance, unions often find that the "power to" disrupt the operations of the bargaining partner – as a result of associational and structural power resources – does not necessarily generate "power over" the bargaining partner to gain the union's bargaining demands. The case of the RMT rail union in the UK during its prolonged strikes from 2022 onwards is a clear example of this (see Gall 2024).

Labour unions, it can be argued, are also quintessentially social democratic (and not socialist) organizations by ideology and worldview, while units of capital are patently not. Labour unions wish to challenge the status quo of capitalism by seeking to intervene in its processes and outcomes of the market-mechanism economy without necessarily seeking the abolition of capitalism itself. They are clearly, then, not inherently revolutionary because they do not seek to abolish wage labour but merely to bargain over its terms. Whether this trajectory can change in a period of revolutionary upheaval remains a point of much continued contention.

Notwithstanding their wish to have state support (economically, industrially, politically), units of capital seek to maintain the status quo of the "free market" by resisting labour unionism, industrially and politically. This suggests that for labour unionism in the short to medium term to reflower, the social democratic state must also reflower. Ergo, that means neoliberalism as an ideological theory and systemic practice must be consigned to the "dustbin of history". Until it does there will always be very definite limits to what labour unionism can achieve in terms of the procedural and substantive aspects of defending and advancing its members' interests across space and time, whether conceived in terms of vested self-interest or the wider sword of social justice. And all this is said in full recognition that the social democracy of any variety or in any place in the postwar period should not be viewed as the height of what is possible for progressive human endeavour.

The chapters in this book provide what might be termed a "deep dive" into the issues based upon concepts and theory, extant literature and empirical research. It is to be hoped that, as intimated in the Introduction, this *Handbook of Labour Unions* is able to reach its intended audiences and that those audiences are among the active agents in helping to realize the attainment of the politics and practice of labour unionism.

REFERENCES

Freeman, R. 2007. "The great doubling: the challenge of the new global labor market". In *Ending Poverty in America: How to Restore the American Dream*, edited by J. Edwards, M. Crain & A. Kalleberg, 55–65. New York: The New Press.

Gall, G. 2024. *Mick Lynch: The Making of a Working-Class Hero*. Manchester: Manchester University Press.

Marx, K. 1950. "Critique of the Gotha Programme". In K. Marx & F. Engels. *Selected Works in Two Volumes (Volume II)*, 13–45. London: Lawrence and Wishart.

Müller-Jentsch, W. 1985. "Trade unions as intermediary organizations". *Economic and Industrial Democracy* 6(1): 3–33.

Offe, C. & H. Wiesenthal 1980. "Two logics of collective action: theoretical notes on social class and organizational form". *Political Power and Social Theory* 1(1): 67–115.

INDEX